T0142158

Lecture Notes of the Institute for Computer Sciences, Social Informatics and Telecommunications Engineering 337

More information about this series at http://www.springer.com/series/8197

Xiaofei Wang · Victor C. M. Leung ·
Keqiu Li · Haijun Zhang ·
Xiping Hu · Qiang Liu (Eds.)

6GN for Future Wireless Networks

Third EAI International Conference, 6GN 2020
Tianjin, China, August 15–16, 2020
Proceedings

Editors
Xiaofei Wang
College of Intelligence and Computing
Tianjin University
Tianjin, Tianjin, China

Keqiu Li
School of Computer Science
and Technology
Tianjin University
Tianjin, Tianjin, China

Xiping Hu
Shenzhen University
Shenzhen, China

Victor C. M. Leung
Department of Electrical and Computer
Engineering
The University of British Columbia
Vancouver, BC, Canada

Haijun Zhang
University of Science and Technology
Beijing, China

Qiang Liu
National University of Defense Technology
Changsha, China

ISSN 1867-8211 ISSN 1867-822X (electronic)
Lecture Notes of the Institute for Computer Sciences, Social Informatics
and Telecommunications Engineering
ISBN 978-3-030-63940-2 ISBN 978-3-030-63941-9 (eBook)
https://doi.org/10.1007/978-3-030-63941-9

This Springer imprint is published by the registered company Springer Nature Switzerland AG
The registered company address is: Gewerbestrasse 11, 6330 Cham, Switzerland

Preface

We are delighted to introduce the proceedings of the Third European Alliance for Innovation (EAI) International Conference on B5G/6G for Future Wireless Networks (6GN 2020). This conference has gathered research contributions that address the major opportunities and challenges in the latest cellular network and 5G/B5G/6G technologies, including novel air interfaces, cellular spectrum, networking architectures and techniques, cellular-driven modern network systems, network measurement and experiment, application and services, etc., with emphasis on both new analytical outcomes and novel application scenarios.

The technical program of 6GN 2020 consisted of 54 full papers. The main conference tracks were: Track 1 – Network Scheduling and Optimization; Track 2 – Intelligent Resource Management; Track 3 – Future Wireless Communications; Track 4 – Intelligent Applications; Track 5 – Wireless System and Platform; Track 6 – Network Performance Evaluation; and Track 7 – Cyber Security and Privacy. Aside from the high-quality technical paper presentations, the technical program also featured four keynote speeches and three technical workshops. The four keynote speakers were Prof. Mohsen Guizani from Qatar University, Qatar; Prof. Dusit Niyato from Nanyang Technological University, Singapore; Prof. Sherman Shen from the University of Waterloo, Canada; and Prof. Tarik Taleb from Aalto University, Finland. The three workshops organized were the Workshop on Intelligent Computing for Future Wireless Network and Its Applications (ICFWNIA); the Workshop on Enabling Technologies for Private 5G/6G (ETP5/6G); and the First International Workshop on 5G/B5G/6G for Underground Coal Mine Safety and Communication (UCMSC 2020). The ICFWNIA workshop aimed to bring together researchers, developers, and technical experts to share their works and experiences of intelligent system, deep learning, supervised and unsupervised learning, and emerging applications for future wireless networks. The ETP5/6G workshop aimed to bring together researchers, developers, and technical experts to share their works and experiences of enabling technologies for private 5G/6G including network architecture, service/business models, and data analytics/AI. The UCMSC aimed to bring together researchers, developers, and technical experts to share their works and experiences of 5G/B5G/6G for Underground Coal Mine Communication.

Coordination with the steering chair, Imrich Chlamtac, was essential for the success of the conference. We sincerely appreciate his constant support and guidance. It was also a great pleasure to work with such an excellent organizing Committee team and we thank them for their hard work in organizing and supporting the conference. In particular, the Technical Program Commitee (TPC), led by our TPC co-chairs, Prof. Xiaofei Wang, Xu Chen, and Gongliang Liu, who completed the peer-review process of technical papers and made a high-quality technical program. We are also grateful to the conference manager, Angelika Klobusicka, for her support and all the authors who submitted their papers to the 6GN 2020 conference and workshops.

We strongly believe that the 6GN conference provides a good forum for all researcher, developers, and practitioners to discuss all science and technology aspects that are relevant to 5G/B5G/6G. We also expect that future 6GN conferences will be as successful and stimulating as indicated by the contributions presented in this volume.

October 2020

Xiaofei Wang
Victor C. M. Leung
Keqiu Li
Haijun Zhang
Xiping Hu
Qiang Liu

Organization

Steering Committee

Imrich Chlamtac — University of Trento, Italy

Organizing Committee

General Chair

Xiaofei Wang — Tianjin University, China

General Co-chairs

Victor C. M. Leung — The University of British Columbia, Canada
Keqiu Li — Tianjin University, China
Haijun Zhang — University of Science and Technology Beijing, China
Xiping Hu — Shenzhen Institutes of Advanced Technology, Chinese Academy of Sciences, China

TPC Chair and Co-chairs

Xiaofei Wang — Tianjin University, China
Xu Chen — Sun Yat-sen University, China
Gongliang Liu — Harbin Institute of Technology, China

Sponsorship and Exhibit Chair

Chao Qiu — Tianjin University, China

Local Chairs

Chao Qiu — Tianjin University, China
Chenyang Wang — Tianjin University, China

Workshops Chair

Zhengguo Sheng — University of Sussex, UK

Publicity and Social Media Chair

Xiuhua Li — Chongqing University, China

Publications Chair

Qiang Liu — National University of Defense Technology, China

Web Chair

Wei Cai	The Chinese University of Hong Kong, China

Technical Program Committee

Alagan Anpalagan	Ryerson University, Canada
Cai, Wei	The Chinese University of Hong Kong, China
Chen, Wei	China University of Mining and Technology, China
Chen, Xu	Sun Yat-sen University, China
Chi, Yuanfang	Alibaba Group, China
Ekram Hossian	University of Manitoba, Canada
Fan, Hao	Tianjin University, China
Fang, Weidong	Fujian University of Technology, China
Han, Yiwen	Tianjin University, China
Hu, Xiping	Shenzhen Institutes of Advanced Technology, Chinese Academy of Sciences, China
Jiang, Chunxiao	Tsinghua University, China
Li, Ruibin	Tianjin University, China
Li, Xi	Beijing University of Posts and Telecommunications, China
Li, Xiuhua	Chongqing University, China
Liu, Gongliang	Harbin Institute of Technology, China
Liu, Qiang	National University of Defense Technology, China
Liu, Zhicheng	Tianjin University, China
Mehdi Bennis	University of Oulu, Finland
Miao Pan	University of Houston, USA
Mingyi Hong	Iowa State University, USA
Ning Lei	Shenzhen Technology University, China
Niyato, Dusit	Nanyang Technological University, Singapore
Qiu, Chao	Tianjin University, China
Ren, Jianji	Henan Polytechnic University, China
Ren, Xiaoxu	Tianjin University, China
Shen, Shihao	Tianjin University, China
Sheng, Zhengguo	University of Sussex, UK
Shuai Zheng	Henan Polytechnic University, China
Sun, Weifeng	Dalian University of Technology, China
Taleb, Tarik	Aalto University, Finland
Tian, Daxin	Beihang University, China
Tingting Hou	Henan Polytechnic University, China
Wang, Chenyang	Tianjin University, China
Wang, Rui	Tianjin University, China
Wang, Xiaofei	Tianjin University, China
Wang, Zehua	The University of British Columbia, Canada
Weisi Guo	University of Warwick, UK
Xiaoli Chu	The University of Sheffield, UK

Xiong, Zehui	Nanyang Technological University, Singapore
Yanru Zhang	University of Houston, USA
Yao, Haipeng	Beijing University of Posts and Telecommunications, China
Zhang, Haijun	University of Science and Technology Beijing, China
Zhang, He	Tianjin University, China
Zhang, Heli	Beijing University of Posts and Telecommunications, China
Zhang, Heng	Tianjin University, China
Zhang, Hengda	Tianjin University, China
Zhang, Xiaohong	Henan Polytechnic University, China

Contents

Network Scheduling and Optimization

Intelligent Resource Management

Future Wireless Communications

Intelligent Applications

Wireless System and Platform

Network Performance Evaluation

Cyber Security and Privacy

The 1st International Workshop on 5G/B5G/6G for Underground Coal Mine Safety and Communication (UCMSC)

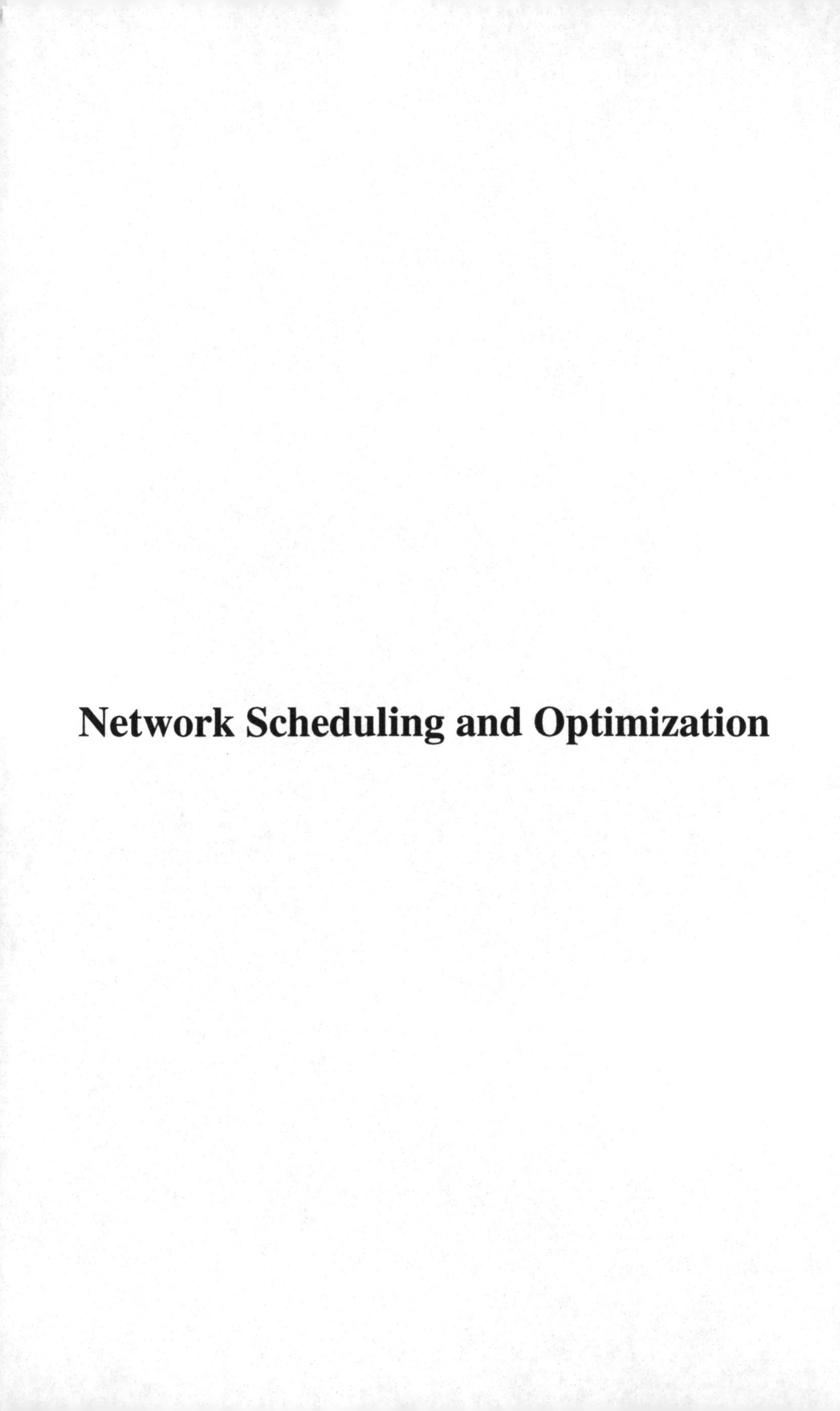

Network Scheduling and Optimization

Collaborative Mobile Edge Caching Strategy Based on Deep Reinforcement Learning

Jianji Ren, Tingting Hou, and Shuai Zheng(✉)

HeNan Polytechnic University, Jiaozuo, Henan, China
zhengshuai@home.hpu.edu.cn

Abstract. Recently, with the advent of the 5th generation mobile networks (5G) era, the emergence of mobile edge devices has accelerated. Nevertheless, the generation of massive edge data brought by massive edge devices challenges the connectivity and cache computing capabilities of the internet of things (IoT) devices. Therefore, mobile edge caching, as the key to realize efficient prefetc.h and cache of edge data and improve the performance of data access and storage, has attracted more and more experts and scholars' attention. However, the complexity and heterogeneity of the devices in the edge cache scenario make it unable to meet the low latency requirements of 5G. In order to make the mobile edge caching more intelligent, based on the widely deployed macro base stations (ξBSs) and micro base stations (μBSs) in 5G scenarios, the ξBS cooperation space and μBS cooperation space is conceived in this paper. Besides, deep reinforcement learning (DRL) algorithms with perception and decision-making capabilities are also used to implement collaborative edge caching. DRL agents perform original and high-dimensional observation training on high-dimensional edge cache scenes, which can effectively solve the dimensionality problem. Then, we jointly deployed federated learning (FL) locally to train DRL agents, which not only solved the problem of resource imbalance, but also realized the localization of training data. In addition, we formulate the energy consumption problem in the collaborative cache as an optimization problem. The simulation results show that the solution greatly reduces the cost of caching and improves the user's online experience.

Keywords: Mobile edge caching · Deep reinforcement learning · Federated learning

1 Introduction

With the arrival of 5G and artificial intelligence (AI), we will enter the intelligent age of the internet of everything (IoE). AI makes it convenient for people to interact with everything, while 5G makes the IoT a reality. The three application scenarios of 5G (eMBB, uRLLC and mMTC) have high requirements on bandwidth, delay and connectivity of devices in IoT, which requires 5G network to be

X. Wang et al. (Eds.): 6GN 2020, LNICST 337, pp. 3–15, 2020.
https://doi.org/10.1007/978-3-030-63941-9_1

more decentralized and intelligent. The traditional centralized storage method cannot meet the low latency requirements required in the 5G scenario. Therefore, it is necessary to deploy small or portable data centers on the edge of the network and conduct intelligent local processing of terminal requests with the help of AI technology to meet the ultra-low delay requirements of uRLLC and mMTC. However, considering the high expenses of deploying a data center on the edge of the network, the mobile edge caching is a promising solution.

As an extended concept of mobile edge computing [1], mobile edge caching not only caches resources in edge nodes (ENs) closer to users, but also provides real-time and reliable data storage and access for edge computing. It is regarded as a data storage method to improve network efficiency and alleviate the high demand for radio resources in the future network. Considering the localization mode of edge caching, which is faster to access than cloud storage, and it can be used offline even when the internet connection is interrupted, so local applications that rely on edge storage are more resilient to service interruptions. However, a particular concern in edge caching is that when the local node cannot satisfy the user's request, the user's internet experience will be reduced. The above problems can be solved through the cooperation between ENs. Through the cooperation among ENs, the geographically distributed ENs can cooperate together to form a more intelligent distributed storage network, so as to address the problem of resource imbalance and ensure quality of service (QoS) at the same time.

To make the cooperation between ENs more intelligent, not only the perception ability but also the decision-making ability is needed. As an AI method closer to the way of human thinking, DRL combines the perception ability of deep learning (DL) with the decision-making ability of RL. With complementary advantages, it can directly learn control strategies from high-dimensional original data. Therefore, in this paper, we conceived the ξBS cooperation space and μBS cooperation space based on the commonly deployed ξBSs and μBSs in 5G scenario and made the perception and decision in the cooperation of edge nodes with the help of DRL. When the user equipment (UE) sent a cache request, based on the perception of the DRL agent placed at the node, The DRL agent learns the optimal strategy, and then designs an optimal resource allocation plan between the cache requester and the cache provider. However, since the training of the DRL agent requires a large amount of multi-dimensional data, which may involve cross-enterprise data transmission and cause excessive pressure on the network in the process of training data uploading, distributed training DRL agent is a promising choice. FL, as an implementable path and "data island" solution for machine learning under privacy protection, allows the construction of collection models from data distributed across data owners without getting through the data to meet the requirements of joint modeling. Therefore, in the process of training the DRL agent, this paper deploys FL local training the DRL agent, which can not only share the model and solve the problem of resource imbalance but also ensure the localization of training data and the security of data.

The main contributions of this paper are as follows:

- Based on the generally dense deployment of ξBSs and μBSs in 5G scenarios, the ξBS cooperation space and μBS cooperation space is proposed, and devices in the collaboration space are used for collaborative transmission and caching.
- Due to heterogeneous resources and the complexity of a large number of devices in mobile edge nodes, mobile edge caching involves complex scheduling problems. Therefore, when solving the problem of cooperative transmission of content, we use the perception and decision-making ability of AI algorithm, namely DRL algorithm, to place DRL agent in each node to decide the best cooperative mode.
- To ensure the security and privacy of the data, when training the DRL agent of each node, the FL uses local data to perform training and parameter update of the DRL model without prior knowledge of the global data, to realize the sharing of the model.
- Simulation results show that the DRL and FL based collaborative edge caching strategy has better performance than the centralized edge caching strategy.

The rest of this paper is organized as follows. In the next section, we review the related work. In Sect. 3, we present the system model of collaborative edge caching and formulate the energy minimization problem. In Sect. 4 we introduce the background and fundamentals of DRL and FL, then we describe the process of training DRL agents with FL. Simulation results are discussed in Sect. 5. Finally, Sect. 6 concludes the paper and look forward to the future research direction.

2 Related Work

The increase of massive intelligent devices brings a huge burden on the internet, which poses a great challenge to the current internet infrastructure. This also causes users to suffer severe network congestion and delays frequently, so the emergence of cloud caching [2] has alleviated this situation. As a centralized storage model of long-distance data transmission, cloud caching requires data to cross the geographical location limitation, and has a significant delay in data transmission and the possibility of network fluctuation, which is difficult to meet the real-time requirements of edge applications. Fortunately, edge caching [3] distributes data across adjacent edge storage devices or data centers, dramatically reducing the physical distance between data generation, data calculation, and data storage, thereby overcoming the high latency, network dependency and other issues caused by long-distance data transmission in cloud storage, providing high-speed, low-latency data access for edge computing.

Edge caching is not a new topic and has generally been extensively studied by many communities. With aim to reduce the delay of data transmission by IoT devices, [4] focused on the pre-caching of video to reduce the amount of repeated data, a collaborative joint caching and processing strategy for on-demand video streaming in mobile edge computing network was proposed to effectively select the cache video and its version to solve the problem of code switching between different video. The author described the collaborative joint caching and processing problem as an integer linear programming that minimizes backhaul network costs when limited by cache storage and processing power. [5] proposed a cooperative cache allocation and computational offloading scheme to cope with the delay and backhaul pressure caused by a large number of data exchanges in the 5G application scenario, and the cache and computing resources on the MEC server are allocated according to their needs and payments. For service requesters, mobile network developers allocate resources according to weighted proportions to maximize resource utilization. Based on the size and popularity of cached content, the authors in [6] introduced an objective function to assess the popularity of content to ensure global cache hit ratios, and use 0–1 knapsack dynamic programming to maximize wireless The local cache hit ratio of the access points (APs), thereby reducing the content acquisition delay in the wireless network and improving the network throughput.

Inspired by the success of deep reinforcement learning in solving complex control problems the author in [7] proposed a DRL-based content cache strategy, to improve the long-term cache hit ratio of base-station stored content without knowing the popularity of content in advance. Considering the multi-level structure of the network, [8] proposed a network cache setting that abstracts the server into a parent node that is connected to multiple leaf nodes, using leaf nodes closer to the characteristics of the end-user, through locally stored files. Or get the file from the parent node to process the request. Besides, the author proposes an efficient caching strategy using DRL, and proves that the strategy can learn and adapt to the dynamic evolution of file requests and the caching strategy of leaf nodes.

Also, as a data storage method that provides real-time and reliable data storage and access for edge computing, mobile edge caching has been studied by many scholars. The scholars in [9] have studied the edge buffer and computational offload in the mobile edge system and proposed the "In-Edge AI" architecture. With the cooperation between the device and the edge nodes, the learning parameters are exchanged to better train and infer the model for dynamic system-level optimization and application-level enhancement while reducing unnecessary system communication load. Inspired by artificial intelligence algorithm, the author in [10] considered the calculation unloading problem in edge computing. The DRL was used to indicate the decision-making of IoT devices and conducted distributed training on DRL agents. In [11], aiming at the complexity of unloading drive caused by dynamic topological changes caused by vehicle mobility in the internet of vehicles, considering the dynamic road condition information and with the help of mobile edge cache, the author proposed a predictive mode transmission scheme for uploading task files, which reduced the cost of task transmission and improved the efficiency.

The difference between this paper and related work is as follows: I) The cooperative space of ξBS and μBS is conceived, which can realize the cooperative caching of mobile edge scene. II) The problem of energy consumption in a collaborative cache is formulated as an optimization problem. III) DRL and FL are combined to make decisions in the process of collaboration based on DRL perception. To ensure the privacy and security of data, federated learning and localization are adopted to train DRL agents, to share models.

3 System Architecture

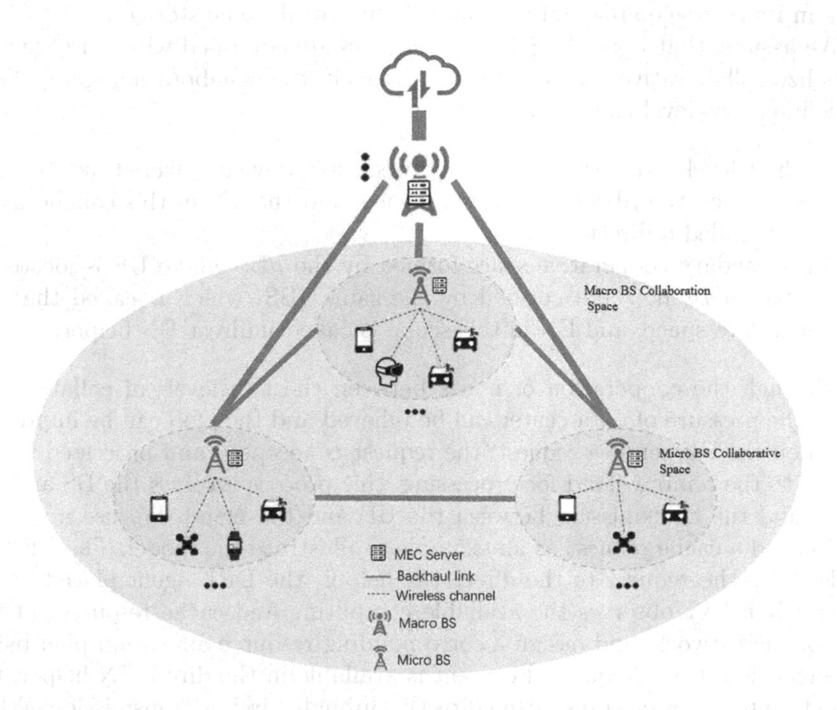

Fig. 1. Collaborative mobile edge caching architecture

3.1 Scenario Analysis

In 5G scenarios, the following three factors need to be considered in order to improve the network capacity: bandwidth, spectrum efficiency, and the number of scenarios covered by the network. As is known to all, the frequency band in 5G scenario is higher than that in 4G, ranging from 3.5 GHz to 30 GHz, or even higher. However, the problem existing at the same time is that the higher the frequency band, the worse the penetration capability and the smaller the coverage area. That also means that 5G network needs a huge number of μBSs, so μBSs have become the key to solving network coverage and capacity in the future. Therefore, in this paper, we focus on the cooperation of user nodes among a large number of commonly deployed μBSs to improve the QoS.

As shown in Fig. 1, the architecture of collaborative mobile edge cache includes several ξBSs and several μBSs within the coverage of a specific ξBS, which constitute a cellular network. There are three types of network connection: I) the UE creates communication and data transmission with EN through the wireless channel assigned by the μBS within the coverage of the μBS. II) The μBS and the ξBS are connected through a backhaul link. III) the backhaul link connects the ξBS to the internet through the gateway. Each ξBS covers the users in its coverage area and is equipped with local cache storage.

We assume that both the ξBS and the μBS are equipped with MEC servers to realize collaborative caching of user devices in the collaboration space. Then we define a two-level collaboration space:

- The first-level collaboration space between IoT devices covered by the same μBS is called the μBS collaboration space, and the EN in this collaboration space is called a direct EN helper.
- The secondary cooperative space formed by the μBS where UE is located is counted in L_b hop and covered by the same ξBS, which is called the ξBS cooperative space, and EN in this scope is called indirect EN helper.

Through the cooperation of nodes between the two levels of collaborative space, the pressure of data center can be relieved and the QoS can be improved.

When the UE sends a request, the request is accepted and processed by BS or sent to the remote cloud for processing, this process involves the BS and the cloud, and the transmission between the UE and the cloud. We use a specific UE to send a cache request as an example to illustrate the model. The UE first sends the cache request to the direct EN helper, the DRL agent placed in the direct EN helper observes the available computing and cache resources at the edge of the network, and design a corresponding resource allocation plan based on perception. If the requested content is available in the direct EN helper, the cached content can be transmitted directly through wireless transmission. When the method cannot meet the requirements of the UE, the ξBS covering the UE needs to be used, searching for requested content at other indirect EN helper, and if the content exists in a indirect EN buffer, then send the requested content to the content requester. If the above steps do not meet the user's needs, the agent will send content requests to the cloud data center.

We give two examples from different areas to illustrate the scenarios, which are strong proof of the advantages of collaborative caching in real scenarios: i) Video stream collaboration cache scenario: video stream caching often occurs in real life. With the help of direct EN helper and indirect EN helper, it can reduce the download of duplicate data, alleviating the backhaul of the backbone network and the pressure of the cloud. ii) Unmanned vehicle cooperation scenario: 5G has high transmission speed and millisecond delay characteristics, with the help of ξBS collaboration space and μBS collaboration space through the vehicle networking will realize data sharing between cars, can take the initiative to avoid obstacles, cooperative path planning, overspeed monitoring, etc., greatly improve the response data and scene recognition accuracy.

3.2 Dynamic System Architecture

Table 1. Notations used in experiment

Notation	Definition
$\mathcal{U}$	The users equipment set
$\mathcal{W}$	A set of wireless channels in the same μBS
$\mathcal{N}$	A EN set in the same μBS coverage as UE
$\mathcal{M}$	A set of μBSs within the coverage of the same ξBS
L	The maximum radio resources
p	The transmit power of each UE in each assigned subchannel
q_f	User request for content f
S_i	Cooperative transmission mode indicator
d_i	The amount of data requested for transmission
v_i	The total number of CPUs required to transfer the requested content
r_i^n	The maximum uplink rate can be achieved by the UE in the subchannel
g_i	The channel gain of UE$_i$ in each subchannel
t_i^n	The shortest transmission time of data in the μBS cooperative space
e_i^n	The energy consumption of data transmission in the μBS cooperative space
e_c	The energy consumed during resource transmission between μBS using indirect EN helper
p_i^a	The transmitted power of the requested target content in the indirect EN helper
p_i^b	The transmitting power of the μBS where ue resides
r_i^a	The channel transmission rate corresponding to the indirect wireless EN helper collaboration
r_i^b	The channel transmission rate of the μBS to the UE where the UE is located
e_i	The energy consumed to assist device$_i$ in content transmission

In this paper, the system model of 5G IoT with ENs is adopted for analysis, as shown in Fig. 1. The popularity of content cached at mobile nodes is described by the Zipf distribution. We assume that the IoT devices in Fig. 1 all have computing and caching resources and communication capabilities, and that the wired communication capability between the μBSs within the coverage of the same ξ station is very strong. Each BS has a mobile edge caching server, which we consider to be a small data center with computing and storage capabilities that can aggregate DRL agent model parameters. For the clarity of the following discussion, the key notations are summarized in Table 1.

The DRL agent placed at EN determines whether the content requested by UE is stored at EN by sensing user preferences, social relationships, and so on. In our simulation, content requests from UE are generated by the Poisson distribution and expressed by q_f, where f represents the requested content. We represent the set of user equipments as $\mathcal{U} = \{1, 2, ..., U\}$, the EN set within the coverage of the same μBS as UE is $\mathcal{N} = \{1, 2, ..., N\}$, and the wireless channel in the same μBS is $\mathcal{W} = \{1, 2, ..., W\}$. $\forall U \in \mathcal{U}$ can establish communication with $\forall N \in \mathcal{N}$ and collaborate with the wireless bandwidth of WHz allocated randomly.

The transmission mode indicator $S_i = \{0, 1\}$, $S_i = 0$ means that the request sent by UE$_i$ is executed in the direct EN helper mode within the μBS collaboration space, $S_i = 1$ means the cooperative transmission in the indirect EN helper mode within the ξBS collaboration space. We represent the requested content transfer of UE$_i$ as (d_i,v_i), where d_i represents the data amount of the requested content transfer, and v_i represents the total CPUs required for the content transfer.

Considering that the channel gain fading in a small range is average and only the fading in a large range can affect it, we assume that the channel gain of different subchannels in the μBS collaboration space is the same for UE and can be different for different UE. Therefore, the power allocated to each subchannel is equal. Then, the uplink rate of the UE in each subchannel is:

$$r_i^n = W \cdot log_2[(1 + p \cdot g_i/(W \cdot K)] \tag{1}$$

Where p is the transmit power of each UE in each assigned subchannel, g_i is the channel gain of UE$_i$ in each subchannel. W is the subchannel bandwidth and K is the noise power spectrum.

The shortest transmission time of the data in the μBS cooperation space is:

$$t_i^n = d_i/r_i^n \tag{2}$$

Therefore the corresponding transmission energy consumption is:

$$e_i^n = p_i^n/t_i^n \tag{3}$$

p_i^n represents the transmission power of the UE$_i$ transmission content distribution.

When the resource transmission between μBSs is performed by means of the indirect EN helper, the corresponding energy consumed is:

$$e_c = e_i^a + e_i^b \tag{4}$$

That is

$$e_c = P_i^a \cdot d_i/r_i^a + P_i^b \cdot d_i/r_i^b \tag{5}$$

Where P_i^a is the transmitted power of the requested target from indirect EN helper, and r_i^a is the corresponding channel transmission rate in this cooperative mode. Similarly, P_i^b is the transmitting power of the μBS where UE resides, and r_i^b is the channel transmission rate from the content provider to UE. We ignored the data transfer time and the energy transfer between the BSs, considering the short distance of the wire transfer time is negligible.

To minimize the transmission consumption, combined with the above formula, the energy consumed by the UE$_i$ to perform device transmission is formalized into the following form:

L is the maximum radio resource, the optimization is:

$$
\begin{aligned}
& e_i = \min_{S_i \theta} \sum_{i=1}^{U} S_i (P_i^a \cdot \tfrac{d_i}{r_i^a} + P_i^b \cdot \tfrac{d_i}{r_i^b}) + (1 - S_i)(P_i^a \cdot \tfrac{d_i}{r_i^a}) \\
& s.t. \\
& S_i \in \{0, 1\} \\
& \sum_{i=1}^{U} \theta_i \leq L \\
& 0 \leq \theta_i \leq S_i L
\end{aligned}
\tag{6}
$$

4 Deep Reinforcement Learning and Federated Learning

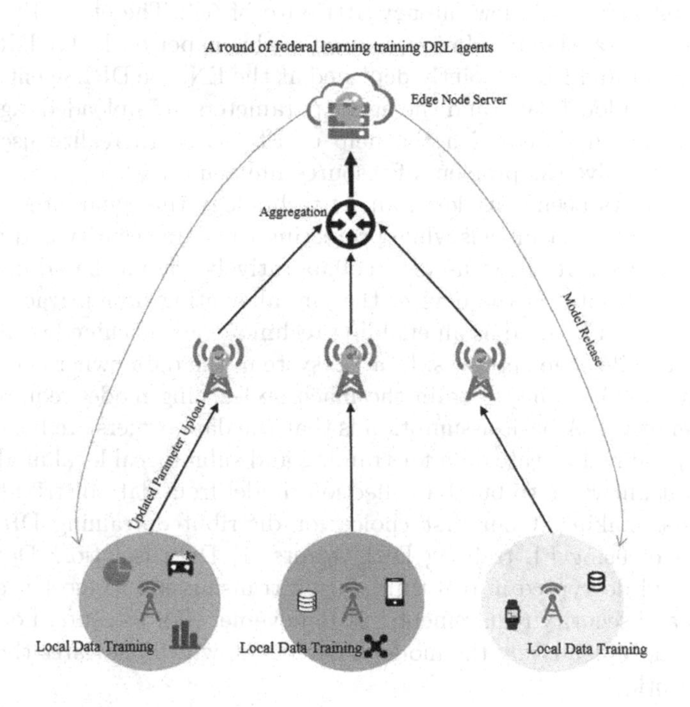

Fig. 2. DRL agent training process based on FL

Due to the complexity of edge caching and EN assistance, with the help of AI to obtain efficient resource scheduling strategies in complex environments with heterogeneous resources and a large number of devices, we introduce the Double Deep Q Network (DDQN) algorithm to solve the search problem of the requested content. As an end-to-end perception and control algorithm, the learning process of DRL is as follows : i) at each moment, agents interact with the environment to obtain a high-dimensional observation, and use DL method to perceive the observation, so as to obtain an abstract and concrete representation of state features; ii) evaluate the value function of each action based on the expected report, and map the current state into the corresponding action through strategies; iii) react to the environment and actions and get the next observation. Through the above steps, the optimal strategy is finally obtained. It is effective for DRL technology to find the optimal strategy for a dynamic edge system, but it also needs a lot of computing resources. Therefore, we should consider the issue of DRL agent deployment (Fig. 2).

In a traditional DRL training process, a central server located at a BS can access the entire training data set within its coverage. Therefore, the server can divide the μBS covered by it into a subset that follows a similar distribution. These subsets are then sent to the participating nodes for distributed training. However, this centralized training method may face privacy and security risks, and mass data uploading process will increase the pressure on the network, which is contrary to the low latency attribute of 5G. Therefore, FL, which is considered a "data island" solution, is used in this paper to deploy DRL agents. The DRL agent and FL are jointly deployed at the EN, the DRL agent is trained locally based on local data, and the agent parameters are uploaded, aggregated, downloaded and updated with the help of FL, so as to realize user privacy protection and solve the problem of resource imbalance.

FL has always been considered an AI technology that guarantees that participants have the best models while protecting the data security and privacy of all parties involved. It allows users to collaboratively train a shared model while saving personal data on the device, thereby alleviating user privacy concerns. Therefore, FL can be used as an enabling technique for machine learning model training in mobile edge networks. In a FL system, the data owner acts as an FL participant, collaborating to train the machine learning model required by the aggregation server. A basic assumption is that the data owners are honest, which means they use real private data for training and submit real local models to the FL server. It allows us to build a collection model from data distributed across data owners, making it our first choice for distributed training DRL agents. Advantages of using FL training DRL agents: 1) Data isolation: Data can be encrypted and decrypted in real time during transmission to meet user privacy protection and security requirements. 2) Relieve network pressure: Locally train the model and upload only the model parameters, which alleviates the pressure on the network.

5 Simulation Test and Results

5.1 Experimental Setting

In the experiment, Zipf distribution was used as the content popularity. With the help of the fitting value of real data captured in [2], we set the Zip parameter $\partial = 1.58$. Requests from UE were expressed by using Poisson distribution. Once the EN receives a request from UE, the DRL agents located in the indirect and direct EN helpers will determine the location of the requested content. We set the total bandwidth between UE and EN as W= 4.8 Hz, and divide it into

Algorithm 1. DRL agent training and aggregation based on FL.

1: **Initialization:**
2: With respect to the global DRL agent in the EN:
3: Initialize the DRL agent with random weight λ_0;
4: Initialize the gross training times T_0 of all devices;
5: With respect to each user equipment $U \in \mathcal{U}$:
6: Initialize the experience replay memory $\mathcal{M}_0^U$;
7: Initialize the local DRL model λ_0^U;
8: Download θ_0 from the EN and let $\lambda_0^U = \lambda_0$;
9: **Iteration:**
10: **For** each round $t = 1$ **to** T **do**;
11: $E_t \leftarrow \{$ random set of m available user equipments$\}$;
12: **For** each user equipment $U \in E_t$ in parallel do;
13: Fetch λ_t from the EN as let $\lambda_t^U = \lambda_t$;
14: Sense and update $\mathcal{M}_t^U$;
15: Train the DRL agent locally with λ_t^U on $\mathcal{M}^U$;
16: Upload the trained λ_{t+1}^U to the EN;
17: Notify the EN the times T_t^U of local training;
18: **End For**
19: With respect to the EN:
20: Receive all model updates;
21: Refresh the statistical $T_t = \sum_{U \in E T_t^U}$;
22: Perform model parameter aggregation as:
23: $\lambda_{t+1} \leftarrow \sum_{U \in E} (T_t^U / T_t) \cdot \lambda_{t+1}^U$;
24: **End For**

subchannels of equal bandwidth according to the number of ENs in the μBS cooperative space where the UE resides. We investigated a collaborative cache in an IoT scenario with 7 ξBSs, 16 μBSs, and the amount of UE was randomly generated and distributed in μBSs to evaluate the capabilities of our proposed approach.

For the training of DDQN agents placed in EN, BS and cloud, we set relevant parameters as follows: exploration probability 0.001, replay memory capacity 5000, learning rate 0.005, discount factor 0.9, full connection of two layers, 200 neurons in each layer activated by tanh function, replacement of target Q network every 250 times, minimum batch of 200. To evaluate our proposed collaboration strategy, we also implemented in the experiment the strategy that UE's request is satisfied in a centralized way, that is, all sensor data collected by IoT devices are uploaded to a central server for centralized DRL training.

Algorithm 1 shows the process of training DRL agent using FL.

5.2 Simulation Results Analysis

In this section, we present simulation results to evaluate the performance of the collaborative caching strategy presented in this paper. First, we evaluate the total energy consumption of the UE collaborative cache in the collaborative

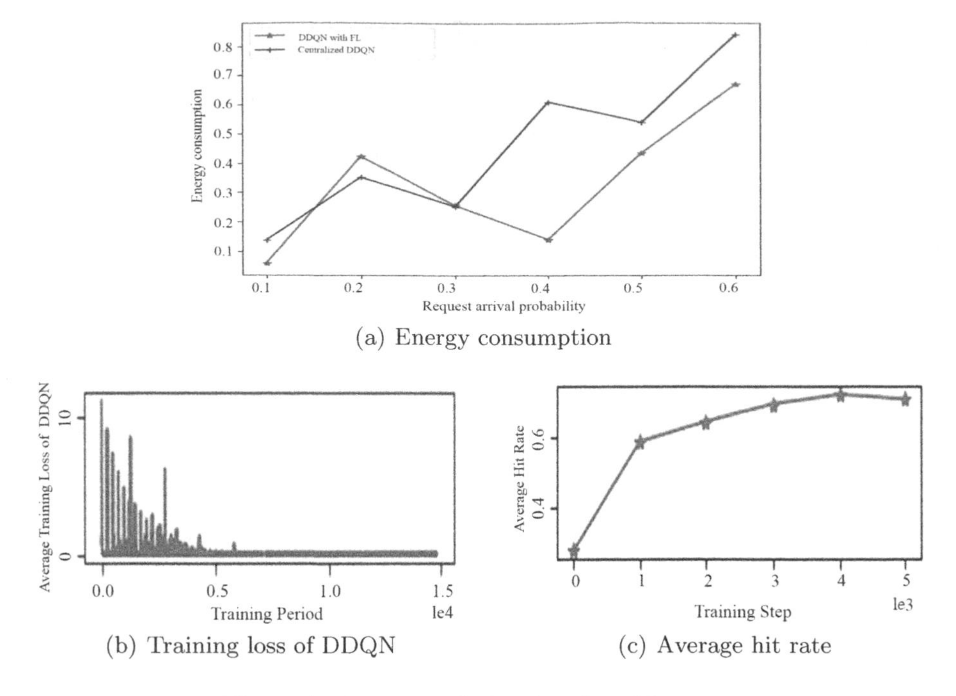

Fig. 3. Performance of DRL based on FL training

scenario proposed in this paper, and compare it with the energy consumption of the traditional transmission method. The result is shown in Fig. 3(a). With the increase of the probability of the arrival of the cache task, the energy consumption of the FL-based DDQN cooperative cache scheme proposed in this paper is lower than that of the traditional centralized training DRL agent method, which indicates that the collaborative cache strategy proposed in this paper has better energy-saving performance. Then we evaluated the change of DDQN training loss with training duration. It can be seen from Fig. 3(b) that as the training time increases, the training loss gradually decreases. Finally, we assume that the task request probability is the same, select three UEs, and evaluate the cache hit rate according to the changes in the number of training. As shown in Fig. 3(c), we can see that the hit rate is gradually increasing, which shows the effectiveness of the proposed scheme and can guarantee QoS.

6 Conclusion and Future Work

In this paper, we consider the collaborative edge caching problem for multiple users in 5G scenarios. Firstly, we propose the ξBS cooperation space and μBS cooperation space and carry out collaborative edge caching of UE by EN cooperation in the cooperative space. Then, we deploy the DRL agent at the EN, and use the DRL's perception and decision-making ability to make collaborative decision-making through the agent's perception. Afterward, to ensure data security and ease network stress, FL is used to localize DRL agents. Furthermore, we formulated the energy consumption problem in the collaborative cache as an optimization problem. The final simulation results verify that our strategy can reduce energy consumption in the cache and QoS. However, the current model proposed in this paper has not considered device to device (D2D) cooperative caching. In future work, we will consider D2D collaborative edge caching to improve user experience and save spectrum resources. Besides, we will further improve the structure and experiment proposed in this paper.

References

1. ETSI. Mobile Edge Computing - Introductory Technical White Paper, September 2014
2. Dash, D., Kantere, V., Ailamaki, A.: An economic model for self-tuned cloud caching. In: 2009 IEEE 25th International Conference on Data Engineering. IEEE (2009)
3. Lobo, A.R., Nadgir, D.K., Kuncolienkar, S.T.S.: Intelligent edge caching. U.S. Patent Application 13/548,584[P]. 2014-1-16
4. Tran, T.X., Pandey, P., Hajisami. A., et al.: Collaborative multi-bitrate video caching and processing in mobile-edge computing networks. In: 2017 13th Annual Conference on Wireless On-Demand Network Systems and Services (WONS), pp. 165–172. IEEE (2017)
5. Ndikumana, A., Ullah, S., LeAnh, T., et al.: Collaborative cache allocation and computation offloading in mobile edge computing. In: 2017 19th Asia-Pacific Network Operations and Management Symposium (APNOMS), pp. 366–369. IEEE (2017)
6. Ren, J., Hou, T., Wang, H., et al.: Increasing network throughput based on dynamic caching policy at wireless access points. Wirel. Netw. **26**, 1577–1585 (2020). https://doi.org/10.1007/s11276-019-02125-0
7. Zhong, C., Gursoy, M.C., Velipasalar, S.: A deep reinforcement learning-based framework for content caching. In: 2018 52nd Annual Conference on Information Sciences and Systems (CISS), pp. 1–6. IEEE (2018)
8. Sadeghi, A., Wang, G., Giannakis, G.B.: Deep reinforcement learning for adaptive caching in hierarchical content delivery networks. IEEE Trans. Cogn. Commun. Netw. **5**(4), 1024–1033 (2019). https://doi.org/10.1109/TCCN.2019.2936193
9. Wang, X., Han, Y., Wang, C., et al.: In-edge AI: intelligentizing mobile edge computing, caching and communication by federated learning[J]. IEEE Netw. **33**(5), 156–165 (2019)
10. Ren, J., Wang, H., Hou, T., et al.: Federated Learning-Based Computation Offloading Optimization in edge computing-supported Internet of Things. IEEE Access **7**, 69194–69201 (2019)
11. Zhang, K., Mao, Y., Leng, S., et al.: Mobile-edge computing for vehicular networks: a promising network paradigm with predictive off-loading. IEEE Veh. Technol. Mag. **12**(2), 36–44 (2017)

A Joint Scheduling Scheme for Relay-Involved D2D Communications with Imperfect CSI in Cellular Systems

Zhixiang Hu[1], Zhiliang Qin[2], Ruofei Ma[1,2](✉), and Gongliang Liu[1]

[1] Harbin Institute of Technology, Weihai 264209, Shandong, China
160200505@stu.hit.edu.cn, {maruofei,liugl}@hit.edu.cn
[2] Beiyang Electric Group Co. Ltd., Weihai, Shandong, China
qinzhiliang@beiyang.com

Abstract. The introduction of device-to-device (D2D) communication brings many benefits to cellular systems, especially when relay-assisted (RA) D2D modes are included. This paper focuses on the joint scheduling problem when some channel state information (CSI) are not known, involving relay selection, probabilistic access control, power coordination, mode selection, and resource allocation. Since the fading components of the channel appear in the model as random variables in imperfect channels, we make some modifications to access control and channel allocation based on previous works. By using the existing algorithm and corresponding mathematical optimization theories, the transformed integer programming (IP) problem is decomposed into two stages to be solved separately. At the same time, in order to improve the solution efficiency, we transform the original NP-hard problem into a linear programming (LP) problem that can be easily solved. Simulation results validate the performance of the joint scheduling scheme and influence of imperfect CSI from the perspective of cell capacity.

Keywords: Device-to-device (D2D) communication · Joint scheduling · Imperfect CSI · Relay

1 Introduction

In order to cope with the massive connections of devices and a huge amount of data transmissions in the fifth generation (5G) cellular system, device-to-device (D2D) communication is proposed to achieve better performances in terms of spectrum efficiency and system capacity, which opens new horizons of device-centric communication. New research directions such as (vehicle-to-vehicle) V2V

Ruofei Ma—This work was supported supported partially by National Natural Science Foundation of China (Grant No. 61801144, 61971156), Shandong Provincial Natural Science Foundation, China (Grant No. ZR2019QF003, ZR2019MF035), and the Fundamental Research Funds for the Central Universities, China (Grant No. HIT.NSRIF.2019081).

X. Wang et al. (Eds.): 6GN 2020, LNICST 337, pp. 16–30, 2020.
https://doi.org/10.1007/978-3-030-63941-9_2

communication, D2D communications on mmWave spectrum band, and D2D network based on social trust, are also constantly proposed in the evolution of 5G D2D network. For two proximity D2D-capable user equipments (DUEs) who are relatively close to each other in the cell, their data can be transmitted directly bypassing the base station (BS) under unified scheduling. The introduced D2D pairs can switch communication modes flexibly and make efficient use of the limited wireless resources, so as to realize the significant improvement of data transmission performance of cellular system under controllable interference.

Most of the existing works on D2D communications included only dedicated and reused D2D modes [1] without fully exploiting the potential benefits of involving relay-aided (RA) D2D modes. A RA D2D mode demonstrates that two DUEs perform data transmissions with assistance of an idle relay-capable UE (RUE) to adapt to long distances and poor channel conditions. In this paper, we only consider the more implementable single-way RA D2D, as interference control and timing management for other RA D2D schemes are too complex [2]. Overall, the research on RA D2D are still in the stage of end-to-end transmission performance analysis, and few of them investigate the joint scheduling issue involving relay selection, power coordination, resource allocation, and mode selection. In [1], scheduling decisions obtained by such a heuristic scheme could not be strictly optimal, because all issues were not jointly modelled. The contribution of [3] was that the issue of mode selection involving RA and direct D2D modes was modeled as an integer programming (IP) problem which was then transformed into a linear programming (LP) problem for effective solution. The deficiency was that relay selection and power coordination were not considered. The branch and bound (BB) method mentioned in [4] can also be used to solve the mode selection problem in D2D communications, but the process was relatively complex and the solution efficiency was not high.

In the process of solving mode selection or power coordination, existing papers usually assumed that BS can acquire channel state information (CSI) of all relevant links. However, in actual situations, only those direct links between UEs and the BS correspond to easy CSI reporting, whereas the CSI acquisition for other links between UEs will correspond to severe signaling overhead. If the number of cellular UEs (CUEs) and D2D pairs are large enough, the severe signaling overhead can not be ignored, which means the assumption of a perfect CSI may be no longer reasonable. In order to better model the actual communications, we partially introduce imperfect CSI situation [5] to replace the traditional perfect CSI situation.

In conclusion, it is more reasonable to consider the assumption of imperfect CSI in designs of efficient joint scheduling schemes involving RA D2D modes, which is more suitable for real communication scenarios. Based on these considerations, this paper designs a more applicable scheduling scheme for a fully loaded single cell jointly considering relay selection, access control, power coordination, and mode selection. On the basis of [6], aiming to increase the cell-wise throughput, we form the joint scheduling issue into a mathematical optimization problem. First, the optimal relay node is selected from the candidate RUE set, which can maximize end-to-end transmission rate of the relay path. Access control then allows DUEs to meet the constraints on power and signal to interference plus noise ratio (SINR) before power coordination. Finally, in order to obtain the final result of channel allocation and mode selection, we relax the original IP problem to a LP problem and solve it further by simplex method or interior point method. Certainly, for some imperfect channels conditions, it is necessary to make corresponding modifications in access control and transmission rate calculation.

The rest of this paper is outlined as follows. Section 2 describes the system model and problem formation for the joint scheduling issue. The mathematical way to solve the formed problem is presented in Sect. 3. Section 4 depicts the simulation results, followed by conclusions in Sect. 5.

2 System Model and Problem Formation

In this section, we first introduce the fully loaded single-cell system model in which imperfect CSI situation is assumed, i.e., the BS does not know exact CSI for some channels, and then formulate the joint scheduling problem into a mathematical optimization model.

2.1 System Model

Considering a fully loaded single-cell with M CUEs, $C_1, C_2, \cdots, C_M$. There are K pairs of DUEs that can perform communications only by reusing the CUEs' uplink channels. In each D2D pairs, let $S_1, S_2, \cdots, S_K$ denote the source DUEs (SDUEs) and $D_1, D_2, \cdots, D_K$ denote the destination DUEs (DDUEs), respectively. In the full loaded cell, all channels have been occupied by the CUEs, and each channel has the same bandwidth, which can only be reused by one new D2D pair. Apart from directly multiplexing the uplink channel of CUEs in direct D2D mode, two distant DUEs can also establish connections with assistance of a selected RUE. But the source-to-relay and relay-to-destination transmissions must performed in two time segments and the two hops use the same channel. $R_{k,1}, R_{k,2}, \cdots, R_{k,R}$ are R uniformly distributed candidate relays for the k-th D2D pair and we denote the optimal RUE for the k-th D2D pair as R_k. On this basis, the types of UEs, available modes, and corresponding data and interference links in the cell are depicted in Fig. 1.

In this scenario, DUEs can measure the link qualities between D2D pairs during the device discovery process. The CSI is fed back to BS when a D2D connection is requested. Meanwhile, BS can obtain the CSI of the links that directly connects itself and UEs through channel estimation method, so channel gains $g_{C_m,B}$, $g_{S_k,B}$, g_{S_k,R_k}, g_{R_k,D_k}, and g_{S_k,D_k} can be assumed to be known. The subscripts in these symbols represents corresponding link's source and destination node, respectively.

However, for the channel gains of the interference links, i.e., g_{C_m,D_k} and g_{C_m,R_k}, acquiring their exact values in each scheduling subframe will lead to very

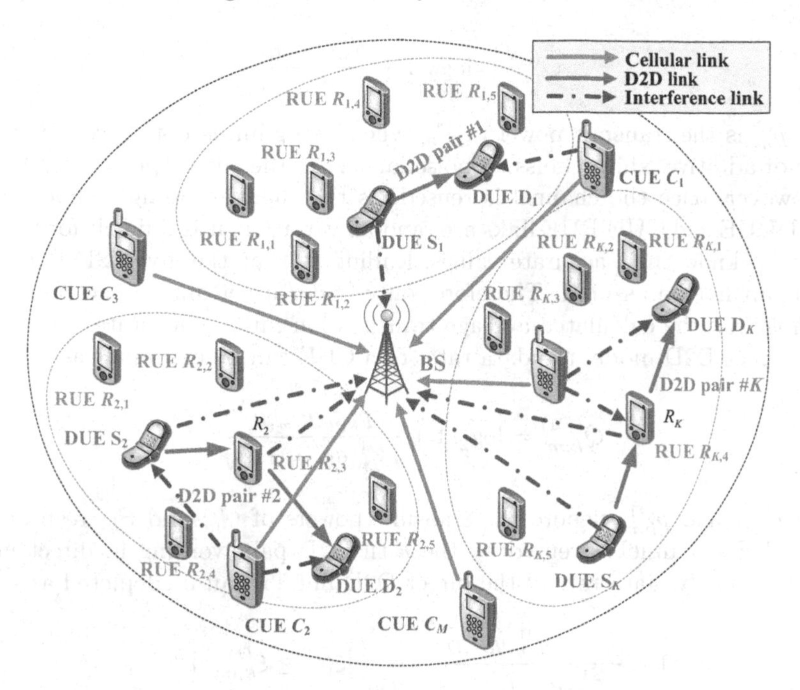

Fig. 1. Fully loaded single-cell scenario involving direct and relay-assisted D2D communications.

severe signaling overhead. Hence, to make the model closer to actual situations, it can be assumed that BS only grasps partial channel information of these interference links according to the distance based path loss, instead of the perfect CSI containing instantaneous channel fading. Therefore, such channels are called imperfect channels. The instantaneous channel gain can be expressed as:

$$g_{i,j} = K \cdot \beta_{i,j} \cdot L_{i,j}^{-\alpha} \tag{1}$$

where K is a constant for path loss, $L_{i,j}$ is the distance between transmitter i and receiver j, and α is the path loss exponent. $\beta_{i,j}$ denotes fading component of the channel, which is a stationary ergodic random variable. The part of the formula based on path loss is $\bar{g}_{i,j} = K \cdot L_{i,j}^{-\alpha}$.

In a Rayleigh fading channel, $\beta_{i,j}$ obeys exponential distribution whose mean is λ, and its cumulative probability function and probability density function are respectively represented as $F(\beta_{i,j})$ and $f(\beta_{i,j})$. $F(\beta_{i,j})$ can be expressed as follows.

$$F(\beta_{\mathrm{i,j}}) = 1 - e^{-\lambda\beta_{i,j}},\ \beta_{i,j} \geq 0 \tag{2}$$

2.2 Problem Formation

For CUE C_m whose uplink channel is not reused, since its radio resource is orthogonal to other UEs in the cell, its transmission rate can be expressed as follows:

$$Q_m^{(C)} = \log_2(1 + \frac{p_m^C g_{C_m,B}}{\delta_N^2}) \quad (3)$$

where p_m^C is the transmit power of C_m when its uplink is not reused, δ_N^2 is the power of additive white Gaussian noise, and B in the subscript denotes BS.

However, once the channel is reused, as the channel fading components in CUE-DDUE and CUE-RUE links are random variables, it is difficult for the base station to know their accurate values, leading to that the given SINR requirements can not be assured. Therefore, only under a certain outage probability threshold can the calculated average transmission rate be accurate.

In direct D2D mode, the data rates of a CUE can be expressed as

$$Q_{k,m}^{(C1)} = \log_2(1 + \frac{p_{k,m}^{C1} g_{C_m,B}}{p_{k,m}^{D1} g_{S_k,B} + \delta_N^2}) \quad (4)$$

where $p_{k,m}^{C1}$ and $p_{k,m}^{D1}$ denote the transmit powers of C_m and S_k, respectively, when CUE's channel is reused by the k-th D2D pair working in direct mode. Correspondingly, data rate of the direct D2D link [5] can be depicted as

$$\begin{aligned} EQ_{k,m}^{(D1)} &= \{\log_2(1 + \frac{p_{k,m}^{D1} g_{S_k,D_k}}{p_{k,m}^{C1} \bar{g}_{C_m,D_k} \beta_{k,m} + \delta_N^2}) \,|\xi_k^{D1} \geq \xi_{k,\min}^{D1}\} \\ &= \{\log_2(1 + \frac{p_{k,m}^{D1} g_{S_k,D_k}}{p_{k,m}^{C1} \bar{g}_{C_m,D_k} \beta_{k,m} + \delta_N^2}) \left| \frac{p_{k,m}^{D1} g_{S_k,D_k}}{p_{k,m}^{C1} \bar{g}_{C_m,D_k} \beta_{k,m} + \delta_N^2} \geq \xi_{k,\min}^{D1}\right\} \\ &= \{\log_2(1 + \frac{p_{k,m}^{D1} g_{S_k,D_k}}{p_{k,m}^{C1} \bar{g}_{C_m,D_k} \beta_{k,m} + \delta_N^2}) \,|\beta_{k,m} \leq \eta\} \\ &= \int_0^{\eta} [\log_2(1 + \frac{p_{k,m}^{D1} g_{S_k,D_k}}{p_{k,m}^{C1} \bar{g}_{C_m,D_k} \beta_{k,m} + \delta_N^2})] \frac{f(\beta_{k,m})}{F(\eta)} d\beta_{k,m} \end{aligned} \quad (5)$$

where, $\eta = \frac{p_{k,m}^{D1} g_{S_k,D_k} - \delta_N^2 \xi_{k,min}^{D1}}{p_{k,m}^{C1} \bar{g}_{C_m,D_k} \xi_{k,min}^{D1}}$ is the cutoff value of $\beta_{k,m}$.

In RA D2D mode, the transmission rate of a CUE is as

$$Q_{k,m}^{(C2)} = \frac{1}{2}\left[\log_2(1 + \frac{p_{k,m}^{C2_1} g_{C_m,B}}{p_{k,m}^{R2_1} g_{S_k,B} + \delta_N^2}) + \log_2(1 + \frac{p_{k,m}^{C2_2} g_{C_m,B}}{p_{k,m}^{R2_2} g_{R_k,B} + \delta_N^2})\right] \quad (6)$$

where $p_{k,m}^{C2_1}$ and $p_{k,m}^{C2_2}$ are the transmit powers of C_m when its channel is reused by the source-to-relay and relay-to-destination links of the k-th D2D pair, respectively. $p_{k,m}^{R2_1}$ and $p_{k,m}^{R2_2}$ denote transmit powers of the SDUE S_k and the selected RUE R_k of the k-th D2D pair in RA D2D mode, respectively, while reusing a CUE's channel. In such a case, date rate of the RA D2D path within a scheduling time slot can be estimated as

$$EQ_{k,m}^{(SR)} = \{\log_2(1 + \frac{p_{k,m}^{R2_1} g_{S_k,R_k}}{p_{k,m}^{C2_1} \bar{g}_{C_m,R_k} \beta_{k,m} + \delta_N^2}) \,|\xi_k^{SR2} \leq \xi_{k,\min}^{SR2}\} \quad (7)$$

$$EQ_{k,m}^{(RD)} = \{\log_2(1 + \frac{p_{k,m}^{R2_2} g_{R_k,D_k}}{p_{k,m}^{C2_2} \bar{g}_{C_m,D_k} \beta_{k,m} + \delta_N^2}) \,|\, \xi_k^{RD2} \leq \xi_{k,\min}^{RD2}\} \tag{8}$$

$$EQ_{k,m}^{(D2)} = \frac{1}{2} \min\left\{EQ_{k,m}^{(SR)}, EQ_{k,m}^{(RD)}\right\} \tag{9}$$

Based on the expressions of data rates, we can formulate an optimization problem for joint scheduling, whose objective is to maximize the cell-wise throughput while ensuring the minimum SINR requirements of all links. Use $\mathbf{X} = \{\mathbf{X}^{(1)}, \mathbf{X}^{(2)}\}$ to represent the mode selection and channel allocation matrix. $\mathbf{X}^{(1)}$ and $\mathbf{X}^{(2)}$ are $K \times M$ indicator matrices of channel allocation in direct and RA D2D modes respectively. $x_{k,m}^{(1)}$ ($x_{k,m}^{(2)}$) is an element of $\mathbf{X}^{(1)}$ ($\mathbf{X}^{(2)}$), whose value is 0 or 1. If $x_{k,m}^{(1)} = 1$ ($x_{k,m}^{(2)} = 1$), it indicates that the k-th D2D pair works in direct (RA) D2D mode and reuses the channel of C_m; otherwise it indicates that the D2D pair is not allowed to resue the channel of CUE. Use $\mathbf{P}$ to denote the power matrix.

Therefore, the joint scheduling problem at the BS side can be formulated as

$$\begin{aligned}(\mathbf{P}^*, \mathbf{X}^*) = \arg\max_{\mathbf{P},\mathbf{X}} \sum_{k=1}^{K} \sum_{m=1}^{M} [(Q_{k,m}^{(C1)} + EQ_{k,m}^{(D1)}) x_{k,m}^{(1)} + (Q_{k,m}^{(C2)} + EQ_{k,m}^{(R2)}) x_{k,m}^{(2)}] \\ + \sum_{m=1}^{M} (1 - \sum_{k=1}^{K} x_{k,m}^{(1)} - \sum_{k=1}^{K} x_{k,m}^{(2)}) Q_m^{(C)}\end{aligned} \tag{10}$$

subject to:

$$x_{k,m}^{(1)}, x_{k,m}^{(2)} \in \{0, 1\} \quad \forall k, m \tag{11}$$

$$\sum_{k=1}^{K} (x_{k,m}^{(1)} + x_{k,m}^{(2)}) \leq 1 \quad \forall m \tag{12}$$

$$\sum_{m=1}^{M} (x_{k,m}^{(1)} + x_{k,m}^{(2)}) \leq 1 \quad \forall k \tag{13}$$

$$x_{k,m}^{(1)} p_{k,m}^{D1} + x_{k,m}^{(2)} \cdot \max\left\{p_{k,m}^{R2_1} p_{k,m}^{R2_2}\right\} \leq P_{\max}^{D} \; \forall k, m \tag{14}$$

$$\begin{aligned}(1 - \sum_{k=1}^{K} x_{k,m}^{(1)} - \sum_{k=1}^{K} x_{k,m}^{(2)}) p_m^C + \sum_{k=1}^{K} x_{k,m}^{(1)} p_{k,m}^{C_1} \\ + \sum_{k=1}^{K} x_{k,m}^{(2)} \cdot \max\left\{p_{k,m}^{C2_1}, p_{k,m}^{C2_2}\right\} \leq P_{\max}^{C} \; \forall m\end{aligned} \tag{15}$$

$$\begin{aligned}& x_{k,m}^{(1)} \cdot \min\left\{\frac{p_{k,m}^{D1} g_{S_k,D_k}}{p_{k,m}^{C1} g_{C_m,D_k} + \delta_N^2}, \frac{p_{k,m}^{C1} g_{C_m,B}}{p_{k,m}^{D1} g_{S_k,B} + \delta_N^2}\right\} + x_{k,m}^{(2)} \cdot \\ & \min\left\{\frac{p_{k,m}^{C2_1} g_{C_m,B}}{p_{k,m}^{R2_1} g_{S_k,B} + \delta_N^2}, \frac{p_{k,m}^{C2_2} g_{C_m,B}}{p_{k,m}^{R2_2} g_{R_k,B} + \delta_N^2}, \frac{p_{k,m}^{R2_1} g_{S_k,R_k}}{p_{k,m}^{C2_1} g_{C_m,R_k} + \delta_N^2}, \frac{p_{k,m}^{R2_2} g_{R_k,D_k}}{p_{k,m}^{C2_2} g_{C_m,D_k} + \delta_N^2}\right\} \\ & \geq \xi_{\min} \; \forall k, m\end{aligned} \tag{16}$$

where $P_{\max}^C$ and $P_{\max}^D$ denote the maximum transmit powers of CUEs and DUEs/RUEs, respectively. $\xi_{\min}$ is the minimum SINR/SNR requirement for each link.

3 Joint Scheduling Algorithm

In this section, we further decompose the formulated optimization problem into two solvable sub-problems and derive the optimal scheduling results. We will ensure the SINR of all involved links while maximizing the overall throughput of the system through relay selection, access control, and power coordination in the first stage and joint mode selection and channel allocation in the second stage.

3.1 Relay Selection

With a goal of maximizing the end-to-end data rate of the RA D2D mode, the optimal RUE is selected from the candidate set. Since the end-to-end transmission of RA D2D mode consists of two hops, i.e., the source-to-relay (first hop) and relay-to-destination (second hop) transmissions, objective function for optimal relay selection for the k-th D2D pair can be formulated as

$$f(R_k) = \max_{R_{k,r}} \left\{ Q^{(D2)} \right\} = \max_{R_{k,r}} \left\{ \frac{1}{2} \min \left\{ Q_{k,m}^{(SR)}, Q_{k,m}^{(RD)} \right\} \right\} \tag{17}$$

where $Q_{k,m}^{(SR)} = \log_2(1 + \frac{p_{k,m}^S g_{S_k,R_{k,r}}}{\delta_N^2})$, $Q_{k,m}^{(RD)} = \log_2(1 + \frac{p_{k,m}^R g_{R_{k,r},D_k}}{\delta_N^2})$. The result obtained from the formulation is considered as the optimal RUE for the k-th D2D pair, and the all selected RUEs are marked as $R_1, R_2, R_3, \cdots, R_R$.

3.2 Probabilistic Access Control

For DUEs who want to access the full-loaded cell, they must reuse the uplink channels occupied by CUEs. In order to meet quality of service (QoS) requirements, the following constraints [5] must be met, i.e.,

$$\begin{cases} \frac{p_{k,m}^C g_{C_m,B}}{p_{k,m}^D g_{S_k,B} + \delta_N^2} \geq \xi_{m,\min}^C \\ \Pr \left\{ \frac{p_{k,m}^D g_{S_k,D_k}}{p_{k,m}^C g_{C_m,D_k} + \delta_N^2} < \xi_{k,\min}^D \right\} \leq \psi \\ p_{k,m}^C \leq P_{\max}^C, p_{k,m}^D \leq P_{\max}^D \end{cases} \tag{18}$$

where $\Pr\{\cdot\}$ denotes probability, $\xi_{k,\min}^D$ and $\xi_{m,\min}^C$ denote the minimum SINR required for the k-th D2D pair and C_m, respectively, and ψ denotes the maximum acceptable outage probability for D2D users.

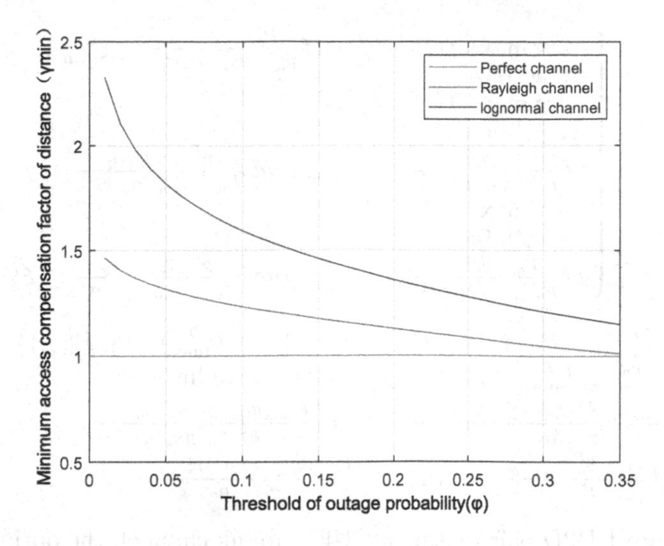

Fig. 2. The minimum access distance compensation factors under under different fading channels

According to [5], the shortest access distance corrected according to the imperfect channel condition is

$$\hat{L}_{C_m,D_k}^{\min}=\begin{cases}\gamma_{\min}\left[\frac{K\xi_{\min}^{C}\xi_{\min}^{D}P_{\max}^{C}g_{S_k,B}}{(P_{\max}^{C}g_{C_m,B}-\xi_{\min}^{C}\delta_N^2)g_{S_k,D_k}-\xi_{\min}^{C}\xi_{\min}^{D}\delta_N^2g_{S_k,B}}\right]^{\frac{1}{\alpha}}\\ if\ \frac{P_{\max}^{C}g_{C_m,B}}{P_{\max}^{D}g_{S_k,B}+\delta_N^2}<\xi_{\min}^{C}\\ \gamma_{\min}\left[\frac{K\xi_{\min}^{C}\xi_{\min}^{D}(\delta_N^2+P_{\max}^{D}g_{S_k,B})}{g_{C_m,B}(P_{\max}^{D}g_{S_k,D_k}-\xi_{\min}^{D}\delta_N^2)}\right]^{\frac{1}{\alpha}}\\ if\ \frac{P_{\max}^{C}g_{C_m,B}}{P_{\max}^{D}g_{S_k,B}+\delta_N^2}\geq\xi_{\min}^{C}\end{cases}\tag{19}$$

Only when $L_{C_m,D_k} \geq \hat{L}_{C_m,D_k}^{\min}$ is satisfied, can DUEs reuse the channel of CUEs to access the cellular network.

The minimum access distance compensation factor $\gamma_{\min}$ is dependent on the probability distribution of channel fading components $\beta_{k,m}$. Figure 2 shows the minimum access compensation factors corresponding to different thresholds of outage probability under different fading channels. In Rayleigh fading channel, it can be expressed as $\gamma_{\text{rayleigh}} = \left[\frac{1}{\lambda}\ln\left(\frac{1}{\psi}\right)\right]^{\frac{1}{\alpha}}$.

3.3 Power Coordination

For the DUEs who are allowed to access, power coordinations are performed based on the algorithm presented in [7]. According to the minimum SINR requirements for different communication links, we can derive the optimal transmit powers of different UEs in different situations. The derivation for the optimal transmit powers can be finally expressed as follows.

$$
(P_m^{C^*}, P_k^{D^*}) = \begin{cases} \underset{(p_m^C, p_k^D) \in \mathbf{\Omega_1}}{\arg\max} \ f(P_m^C, P_k^D), if \frac{P_{\max}^C g_{C_m,B}}{p_{\max}^D g_{S_k,B} + \sigma_N^2} \leq \xi_{m,\min}^C \\ \underset{(p_m^C, p_k^D) \in \mathbf{\Omega_2}}{\arg\max} \ f(P_m^C, P_k^D), \\ if \frac{P_{\max}^C g_{C_m,B}}{p_{\max}^D g_{S_k,B} + \sigma_N^2} > \xi_{m,\min}^C \ and \ \frac{p_{\max}^D g_{S_k,D_k}}{P_{\max}^C g_{C_m,D_k} + \sigma_N^2} < \xi_{m,\min}^D \\ \underset{(p_m^C, p_k^D) \in \mathbf{\Omega_3}}{\arg\max} \ f(P_m^C, P_k^D), \\ if \frac{P_{\max}^C g_{C_m,B}}{p_{\max}^D g_{S_k,B} + \sigma_N^2} > \xi_{m,\min}^C \ and \ \frac{p_{\max}^D g_{S_k,D_k}}{P_{\max}^C g_{C_m,D_k} + \sigma_N^2} \geq \xi_{m,\min}^D \end{cases} \tag{20}
$$

where $\mathbf{\Omega_1} = \{(P_{\max}^C, P_1), (P_{\max}^C, P_2)\}$, $\mathbf{\Omega_2} = \{(P_3, P_{\max}^D), (P_4, P_{\max}^D)\}$, and $\mathbf{\Omega_3} = \{(P_{\max}^C, P_1), (P_{\max}^C, P_{\max}^D), (P_4, P_{\max}^D)\}$, in which we have
$P_1 = \frac{(P_{\max}^C \bar{g}_{C_m,D_k} F^{-1}(1-\psi) + \sigma_N^2)\xi_{k,\min}^D}{g_{S_k,D_k}}$, $P_2 = \frac{P_{\max}^C g_{C_m,B} - \xi_{m,\min}^C \sigma_N^2}{\xi_{m,\min}^C g_{S_k,B}}$,
$P_3 = \frac{P_{\max}^D g_{S_k,D_k} - \xi_{m,\min}^D \sigma_N^2}{\xi_{m,\min}^D \bar{g}_{C_m,D_k} F^{-1}(1-\psi)}$, and $P_4 = \frac{(P_{\max}^D g_{S_k,B} + \sigma_N^2)\xi_{m,\min}^C}{g_{C_m,B}}$

For the direct D2D pair reusing a CUE's uplink channel, the optimal powers for the source DUE and the corresponding CUE can be obtained directly according to the above method. For RA D2D pair reuse a CUE's uplink channel, since both the source-to-relay and relay-to-destination links reuse the same channel, the optimal powers for the source DUE, the RUE, and the corresponding CUE can also be calculated in each time slot using the above method.

3.4 Mode Selection and Resource Allocation

Based on the results of the relay selection, access control, and power coordination in the first stage, we can further simplify the original optimization problem by removing the constraints on transmit powers and minimum SINR requirements. Accordingly, the simplified optimization problem covers only the issues on resource allocation and communication mode selection, which can be expressed as follows.

$$
\begin{aligned} (\mathbf{X}^*) = \arg\max_{\mathbf{X}} \Bigg\{ & \sum_{k=1}^{K} \sum_{m=1}^{M} (Q_{k,m}^{(C1)} + Q_{k,m}^{(D1)} - Q_m^{(C)}) x_{k,m}^{(1)} \\ & + \sum_{k=1}^{K} \sum_{m=1}^{M} (Q_{k,m}^{(C2)} + Q_{k,m}^{(R2)} - Q_m^{(C)}) x_{k,m}^{(2)} \Bigg\} \end{aligned} \tag{21}
$$

subject to:

$$
\begin{aligned} & x_{k,m}^{(1)}, x_{k,m}^{(2)} \in \{0,1\}, \quad \forall k, m; \\ & \sum_{m=1}^{M} (x_{k,m}^{(1)} + x_{k,m}^{(2)}) \leq 1, \quad \forall k; \\ & \sum_{k=1}^{K} (x_{k,m}^{(1)} + x_{k,m}^{(2)}) \leq 1, \quad \forall m. \end{aligned} \tag{22}
$$

It can be easily identified that the above optimization problem is a $0-1$ integer programming problem and we can rewrite it into a general form as

$$\begin{aligned} &\max \mathbf{C}^T\mathbf{X} \\ &\text{subject to}: \mathbf{AX} \leq \mathbf{B}, \\ &\qquad\qquad \mathbf{X} \in \{0,1\}^{2MK}, \end{aligned} \tag{23}$$

where $\mathbf{C}^T$ is a row vector with $2MK$ elements corresponding to the coefficients of $x_{k,m}^{(1)}$ and $x_{k,m}^{(2)}$ in (21). $(\cdot)^T$ denotes the transpose operation. $\mathbf{X}$ is a column vector with $2MK$ elements corresponding to the values of $x_{k,m}^{(1)}$ and $x_{k,m}^{(2)}$. $\mathbf{A}$ is the constraint matrix with a size of $(K+M)\times 2MK$, which corresponds to the left sides of the two inequalities of (22). $\mathbf{B}$ is a right-hand-side vector with a length of $(K+M)$, and all its elements are 1.

It is known that such an IP problem is NP-hard owing to its large feasible domain and branch-and-bound (BB) method is generally used to search the optimal solution. As the BB method can be considered as an improved exhaustive solution, it corresponds to extremely high computational complexity. In contrast, if we can transform the above IP problem into a LP problem mathematically, the optimal solution might be able to be easily obtained, as there have been some very efficient ways to solve LP problems. For the IP problem depicted by (23), its feasible domain's characteristic is determined by the constraint matrix $\mathbf{A}$, and it can be proved that if $\mathbf{A}$ is a totally unimodular matrix (TUM), the above IP problem can be completely transformed into a LP problem whose optimal solution is definitely the optimal solution for the initial IP problem. The definition of TUM and corresponding judgement conditions are demonstrated as follows.

Definition 1: If every square sub-matrix of matrix $\mathbf{Z}_{I\times J}$ is with a determinant of -1, 0, *or* 1, $\mathbf{Z}$ *can be called totally unimodular matrix.*

Whether a matrix $\mathbf{Z}_{I\times J}$, with each element z_{ij} equal to -1, 0, or 1, is a TUM or not can be decided according to the following conditions [8]: (a) Each column of $\mathbf{Z}$ has no more than two non-zero elements; (b) There exists a division $(\mathbf{Z}_{upper}, \mathbf{Z}_{lower})$ on the rows of $\mathbf{Z}$ such that in each column, the sum of non-zero elements in the upper part is equal to that in the lower part, i.e., $\sum_{i\in\mathbf{Z}_{upper}} z_{ij} = \sum_{i\in\mathbf{Z}_{lower}} z_{ij}, \forall j$.

It can be easily proved that the constraint matrix $\mathbf{A}$ in (23) can satisfy the two judgement conditions of TUM, so the original IP problem can be further relaxed to a LP problem which is expressed as (24).

For such a LP problems, there are some mature and effective solving schemes, such as simplex method and interior point method. In this paper, the widely used simplex method is adopted to obtain the final solution. When the final indicator matrix for channel allocation and communication mode selection is fixed, combining with the optimal transmit powers derived in the first stage, we can say the whole optimization problem has been solved. The solving process corresponds definitely to the design of the joint scheduling scheme.

$$(\mathbf{X}^*) = \arg\max_{x} \left\{ \sum_{k=1}^{K} \sum_{m=1}^{M} (Q_{k,m}^{(C1)} + Q_{k,m}^{(D1)} - Q_{m}^{(C)}) x_{k,m}^{(1)} + \sum_{k=1}^{K} \sum_{m=1}^{M} (Q_{k,m}^{(C2)} + Q_{k,m}^{(R2)} - Q_{m}^{(C)}) x_{k,m}^{(2)} \right\} \tag{24}$$

subject to:

$$0 \leq x_{k,m}^{(1)} \leq 1 \quad 0 \leq x_{k,m}^{(2)} \leq 1 \quad \forall k, m$$

$$\sum_{m=1}^{M} (x_{k,m}^{(1)} + x_{k,m}^{(2)}) \leq 1 \quad \forall k$$

$$\sum_{k=1}^{K} (x_{k,m}^{(1)} + x_{k,m}^{(2)}) \leq 1 \quad \forall m$$

4 Performance Evaluation

In this section, the performance of the joint scheduling scheme are evaluated via simulations. We establish the simulation scenario according to fully loaded single-cell system model shown in Fig. 1. The influences of some major parameters, such as the number of D2D pairs, the maximum transmit power of UEs, and the distance between the source and destination DUEs, on the cell-wise throughput performance are simulated and analyzed. The major parameters used in the simulations are listed in Table 1.

Table 1. Major parameters used in performance evaluation.

Parameter	Value
Path loss exponent (α)	4
Path loss constant (K_0)	0.01
Cell radius	500 m
D2D cluster radius	Uniformly distributed in [10, 200] m
Number of active CUEs (M)	20
Number of D2D pairs (K)	10(if fixed)
Noise power spectral density	−174 dBm/Hz
Maximum transmition power of UEs	24 dBm (if fixed)
Minimum SNR/SINR for each link	10 dB
Outage probability threshold	0.01

The established simulation scenario is a single cell with a radius of 500 meters. BS is located in the center, and all CUEs and D2D pairs are randomly distributed in the cell. For each D2D pair, the destination DUE is randomly located on a circle centered on the source DUE, and a certain amount of RUEs are assumed for each D2D pair as candidate relay UEs to ensure the possibility on performing RA D2D communication. The candidate RUEs for each D2D pair are randomly distributed in the circle area bounded by the possible location of the destination DUE. Note that for a set of predefined parameters, 1 000 tests are performed to get and average result. We mainly analyze the effect of bringing the imperfect channel condition and the advantage of further considering RA D2D mode in the joint scheduling.

Figure 3 shows the influence of UE's maximum transmit power on the total capacity of the cell under different scheduling scenarios, where "Perfect channel" means the BS can acquire precisely all its needed CSI, while "Imperfect channel" means that the precise CSI for the interference links between UEs are not known at the BS. "CUEs without D2D" means D2D communications are not included, "Direct D2D" means only direct D2D mode is considered, and "Direct and RA D2D" means both the direct and RA D2D modes are considered. It is seen that, with the increase of the maximum transmit power of UEs in the five cases, the cell-wise system capacity increases greatly, and further introduction of the RA D2D mode is helpful to improving the cell-wise system capacity. In case of the channel reuse between a CUE and a D2D pair, increasing all UEs' maximum transmit power means increasing the feasible region of the power optimization problem, which may leads to further improvements on the power coordination results. For CUE whose uplink channel is not reused by D2D links, its optimal transmit power is just CUE's maximum transmit power. Therefore, when the maximum transmission power increases, the system performance will definitely increase. At the same time, due to the imperfection of CSI, the performance of system capacity under imperfect channel conditions are slightly worse than that under perfect channel conditions.

Figure 4 illustrates the performance of cell-wise system capacity increase with varying distances between SDUEs and DDUEs. To describe the distance variation, all D2D pairs are set the same separation distance between the source and destination DUEs. It is seen from Fig. 4, when the source DUE is closer to the destination DUE, the increase of system capacity brought in by D2D communication is larger. With the supplement of RA D2D mode, the system capacity can be further improved. The more RUEs available for each D2D pair, the more obvious the advantage. However, as the source to destination distance for all D2D pairs increases, the system capacity increase brought in by the D2D communication decreases. Therefore, if the such a separation distance is further enlarged, the cell-wise capacity gain might reduce to zero. When the channel capacity generated by D2D communications can not compensate for the channel capacity loss caused by channel reuse, there is no capacity growth on the whole.

Figure 5 shows the increase in cell-wise system capacity versus the number of D2D pairs. It can be seen that increase of the number of D2D pairs leads to obvious increase of the cell-wise system capacity, and the system capacity with the introduction of relay D2D mode increases more. Such a advantage is more significant when increasing of the number of D2D pairs in the cell. In addition, comparing the performance on cell-wise system capacity increase

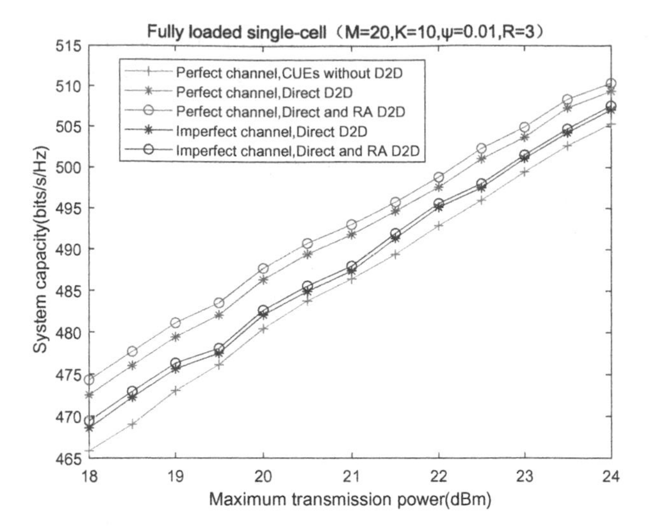

Fig. 3. System capacity versus maximum transmit power of UEs, where the distances between SDUEs and DDUEs are randomly selected from 10 to 200 m.

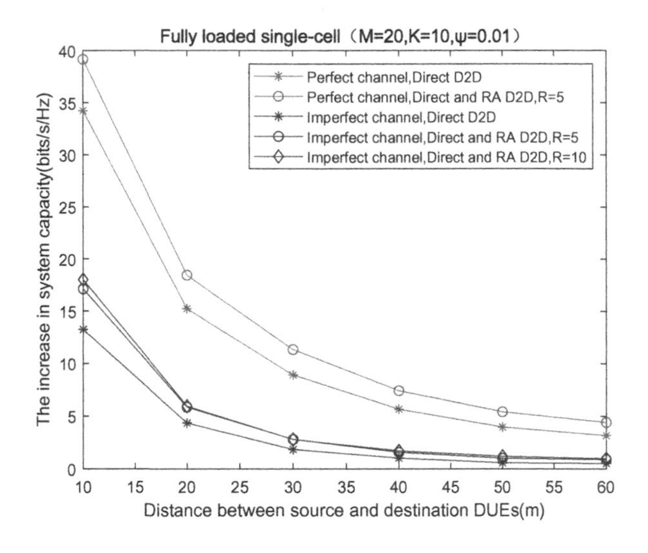

Fig. 4. Increase in system capacity versus distance between SDUEs and DDUEs, where $P^C_{\max} = P^D_{\max} = 24$ dBm.

under the imperfect CSI conditions with that under the perfect CSI conditions, we can observe that the gap becomes larger when the number of D2D pairs is increased. This is reasonable, as increase of the number of D2D pairs means more uncertainties on the interference links are involved, which might limit the improvement space for the cell-wise system capacity. But such a result clearly shows us the influence of channel uncertainties on the accuracy of the scheduling,

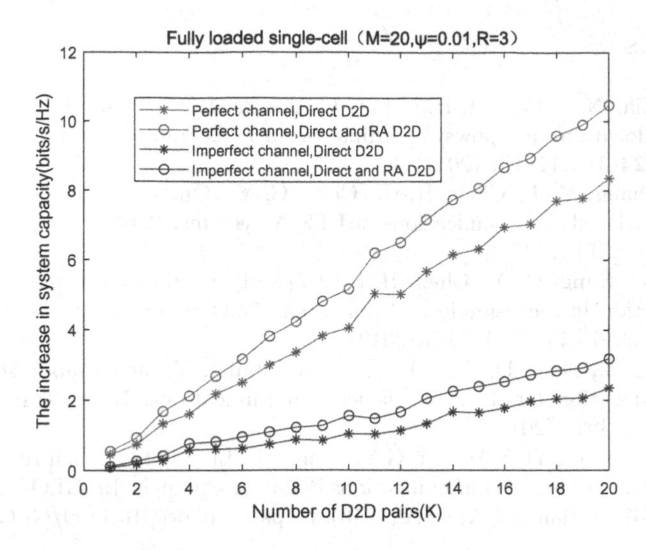

Fig. 5. Increase in system capacity versus number of D2D pairs, where distances between SDUEs and DDUEs are randomly selected from 10 to 200 m and $P_{\max}^{C} = 10\,P_{\max}^{D} = 24$ dBm.

and it prompts us to seek more effective solutions to overcome the influences of imperfect CSI on the accuracy of the jointly scheduling at the BS side.

5 Conclusions

This paper investigated the joint scheduling issue for cellular multi-mode D2D communications involving both direct and RA D2D modes under imperfect CSI condition. With an aim to maximize the cell-wise system capacity, joint scheduling on relay selection, access control, power coordination, mode selection, and resource allocation at the BS side in a fully loaded single-cell was formulated as an optimization problem which was then solved in two stages. The relay selection, access control, and power coordination issues were solved in the first stage, while the mode selection and channel allocation issues were solved in the second stage. By using probabilistic model to depict the fading component of uncertain interference channels, the condition of imperfect CSI was taken into account in the design of joint scheduling schemes. By transforming the IP problem into LP problem, the joint mode selection and resource allocation problem was solved with low complexity. Simulation results showed that including RA D2D mode into the scheduling can further increase the cell-wise system capacity, even under the imperfect CSI condition. Moreover, under both the perfect and imperfect CSI conditions, the cell-wise system capacity increased when the number of candidate RUEs for each D2D pair and the maximum transmit powers for involved UEs were increased. Meanwhile, the imperfect CSI condition did have influence on the accuracy of the joint scheduling and such an influence was more obvious while the number of D2D pairs was increased.

References

1. Ma, R., Xia, N., Chen, H.-H., Chiu, C.-Y., Yang, C.-S.: Mode selection, radio resource allocation, and power coordination in D2D communications. IEEE Wirel. Commun. **24**(3), 112–121 (2017)
2. Ma, R., Chang, Y.-J., Chen, H.-H., Chiu, C.-Y.: On relay selection schemes for relay-assisted D2D communications in LTE-A systems. IEEE Trans. Veh. Technol. **66**(9), 8303–8314 (2017)
3. Chen, C.-Y., Sung, C.-A., Chen, H.-H.: Capacity maximization based on optimal mode selection in multi-mode and multi-pair D2D communications. IEEE Trans. Veh. Technol. **68**(7), 6524–6534 (2019)
4. Yu, G., Xu, L., Feng, D., Yin, R., Li, G.Y., Jiang, Y.: Joint mode selection and resource allocation for device-to-device communications. IEEE Trans. Commun. **62**(11), 3814–3824 (2014)
5. Feng, D.Q., Lu, L., Yi, Y.W., Li, G.Y., Feng, G., Li, S.Q.: Optimal resource allocation for device-to-device communications in fading channels. In: GLOBECOM 2013, pp. 3673–3678, Atlanta, GA, USA (2013). https://doi.org/10.1109/GLOCOM.2013.6831644
6. Chen, C.-Y., Sung, C.-A., Chen, H.-H.: Optimal mode selection algorithms in multiple pair device-to-device communications. IEEE Wirel. Commun. **25**(4), 82–87 (2018)
7. Feng, D., Lu, L., Yi, Y.-W., Li, G.Y., Feng, G., Li, S.: Device-to-device communications underlaying cellular networks. IEEE Trans. Commun. **61**(8), 3541–3551 (2013)
8. Wolsey, L.A.: Integer Programming, 1st edn. Wiley Interscience, New York (1998)

Hybrid Duplex Relay Selection in Cognitive Radio Network

Yaru Xue[1(✉)], Weina Hou[2], Zhaoyi Li[1], and Shuaishuai Xia[1]

[1] School of Communication and Information Engineering, Chongqing University of Posts and Telecommunications, Chongqing, China
yaru0330@foxmail.com, 1771690650@qq.com, 844841121@qq.com

[2] School of Optoelectronic Engineering, Chongqing University of Posts and Telecommunications, Chongqing, China
houwn@cqupt.edu.cn

Abstract. The existing cognitive single relay selection technology is lack of a full-duplex/half-duplex mode switching criteria, and only maximizes the system performance of secondary users in the single-duplex mode relay selection process, which result in a single relay mode and the overall performance of single primary networks and secondary networks is not optimal. To tackle these problems, the criteria for switching the hybrid duplex mode of the relay node is first derived. In the case of adopting the amplify-and-forward (AF) and decode-forward (DF) protocols, each candidate relay terminal selects the optimal duplex mode according to the signal-to-noise ratio (SNR) of the self-interference channel and the critical point of self-interference. Then propose a low interference hybrid duplex relay selection algorithm, the relay node uses the comprehensive performance loss of the primary network performance and the performance gain of the secondary network user as the indicator to optimize the relay selection, which maximizes the sacrifice value of the primary network. The simulation results show that the proposed algorithm can significantly improve the performance of the primary network system and the whole system. Furthermore, the performance of the secondary network is slightly reduced (or even improved).

Keywords: Cognitive network · Relay selection · Hybrid duplex mode · Low-interference

1 Introduction

With the development of mobile communication,the demand for spectrum resources has increased dramatically. Due to limited spectrum resources, the shortage of spectrum resources becomes serious [1–3].

This work is supported by Chongqing University of Posts and Telecommunications Ph.D. (A1029-17) and Chongqing Municipal Education Commission Project (KJQN201900645).

X. Wang et al. (Eds.): 6GN 2020, LNICST 337, pp. 31–42, 2020.
https://doi.org/10.1007/978-3-030-63941-9_3

Cognitive relay selection technology has been widely studied for future communication development [4]. It enables secondary users to effectively use licensed spectrum to complete their own communications while ensuring the quality of primary users communication. The full-duplex relay technology can simultaneously receive and transmit signals of the same frequency [5]. Therefore, cognitive relay technology combines the two technologies to improve spectrum utilization. There are many alternative candidate relays in cognitive relay networks. When different nodes are selected as relays, different performance will be brought to the system [6]. It is very meaningful to study how to effectively select the appropriate node as the communication relay.

In the Underlay mode cognitive relay network, optimizing the secondary network performance indicators as the target of selecting relay nodes. Most of the existing cognitive relay selection schemes follow a principle: when the interference caused by the secondary user to the primary user satisfies the interference temperature, the relay selection scheme can be used to increase the capacity of the system or reduce the probability of interruption. Traditional max-min relay selection (MMRS) relay selection method is applied to cognitive half-duplex relay networks to maximize secondary system capacity [7–9]. Based on the interrupt performance of the orthogonal frequency division multiplexing network, within each subcarrier, the relay network adopts a full-duplex AF relay protocol and the relay selection mode follows the MMRS criterion [10]. Research on optimal relay selection (ORS) mode in cognitive relay network based on imperfect channel information imperfection (CII), as the number of candidate relays increases, the system performance of the secondary network will increase significantly [11]. The cognitive full-duplex relay selection method is studied in detail to optimize the performance of the secondary user network [12–15]. However, in the Underlay mode cognitive network, even if the secondary users can meet the strict interference temperature limit, user networks also sacrifice their performance or power to tolerate additional interference from secondary user network access. In addition, the existing works does not involve the hybrid duplex mode switching problem of cognitive relay networks. This paper in Underlay mode cognitive relay network based on AF protocol and DF protocol respectively, proposing a low-interference hybrid duplex relay selection algorithm to maximize the primary user's sacrifice value for full/half duplex mode switching and optimal relay selection.

2 System Model

Relay selection model for cognitive relay network, including a primary network base station P_T, a primary user destination terminal P_R, a secondary network base station S, K secondary candidate relays $R(R_1,R_2...R_K)$, and a secondary destination terminal D. Considering that each relay equips with two antennas, one of which is used to receive signals and the other is used to transmitting signals, moreover, each antenna can communicate both in half-duplex mode and full-duplex mode. Figure 1 shows the system model considered in this paper.

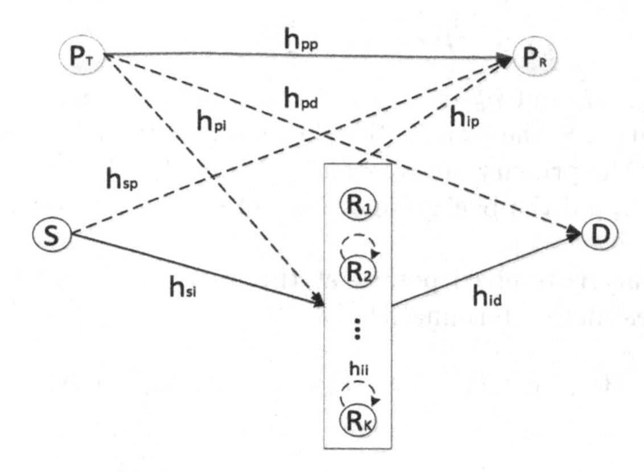

Fig. 1. Relay selection model in cognitive wireless networks.

In Underlay mode, the primary user allows the secondary user to access the licensed spectrum while satisfying the interference temperature I_{th} and treats the signals of the secondary user as background noise. It is supposed that in the cognitive relay network, S and D cannot communicate directly, because of the secondary user's transmit power is limited. The data interaction between S and D need to be completed with the assistance of the relay terminal, so the direct transmission path between S and D can be ignored. Assuming that the transmit powers of $\mathrm{P_p}$, $\mathrm{P_s}$, and $\mathrm{P_i}$ of $\mathrm{P_T}$, D, and i are constants. In the Fig. 1, the solid line indicates the transmission link of the useful signal and the dotted line indicates the interference link. In addition, we presume that the noise at each node obeys an additive white Gaussian noise with a mean of zero and a variance of N_0.

3 Performance Analysis of Secondary Network and Primary Network

In the Underlay mode cognitive relay network, the relay node can work in both half-duplex and full-duplex modes. The two modes will bring different gains to the capacity of the secondary network [16]. The capacity of the secondary network in the two modes will be analyzed separately.

3.1 Full Duplex Mode

In the Underlay mode cognitive relay network, when the relay node adopts the full-duplex relay mode, the relay node can simultaneously receive and retransmit signals at the same frequency. Therefore, the signal transmission will cause self-interference to the signal reception. The received signals at the secondary candidate relay node i and the secondary destination node D are

$$y_i = \sqrt{P_s}h_{si}x_s + \sqrt{P_i}h_{ii}x_i + \sqrt{P_p}h_{pi}x_p + n_i \tag{1}$$

$$y_d = \sqrt{P_i} h_{id} x_i + \sqrt{P_p} h_{pd} x_p + n_d \tag{2}$$

where x_s, x_i, x_p, n_i, and n_d denote the transmitting signal of the secondary network base station S, the retransmitted signal of secondary relay i, the transmission signal of the primary network base station $\mathrm{P_T}$, the background noise of secondary relay i, and the background noise of secondary destination terminal D, respectively.

The instantaneous received powers at the secondary candidate relay i and the secondary destination terminal D are

$$\varepsilon\{|y_i|^2\} = P_s|h_{si}|^2 + P_i|h_{ii}|^2 + P_p|h_{pi}|^2 + N_0 \tag{3}$$

$$\varepsilon\{|y_d|^2\} = P_i|h_{id}|^2 + P_p|h_{pd}|^2 + N_0 \tag{4}$$

When the secondary relay network adopts the AF protocol, the relay node directly amplifies the received signal, and the amplification factor β_1 is defined as

$$\beta_1 = \sqrt{\frac{P_i}{P_s|h_{si}|^2 + P_p|h_{pi}|^2 + P_i|h_{ii}|^2 + N_0}} \tag{5}$$

When the secondary relay network adopts the DF protocol, the relay node decodes the received signal first and then re-encodes for forwarding. Therefore, the resending signal of the relay node can be expressed as

$$x_i = \begin{cases} \beta_1 y_i, & \text{with AF} \\ \sqrt{\frac{P_i}{P_s}} x_s, & \text{with DF} \end{cases} \tag{6}$$

where h_{pi}, h_{ii}, and h_{si} are the channel coefficients of $\mathrm{P_T}$ to i, i to i and S to i respectively, when node i is selected as the relay. In the cognitive full-duplex relay network, the Signal to Interference Plus Noise Ratio (SINR) in the hops of i and D can be defined as

$$\gamma_i^{FD} = \frac{P_s^2|h_{si}|^2}{P_p^2|h_{pi}|^2 + P_i^2|h_{ii}|^2 + \mathrm{N}_0} \tag{7}$$

$$\gamma_{d-i}^{FD} = \frac{P_i^2|h_{id}|^2}{P_p^2|h_{pd}|^2 + \mathrm{N}_0} \tag{8}$$

The end-to-end SINR of secondary networks in cognitive half-duplex relay mode can be defined as

$$\gamma_{CFD-i} = \begin{cases} \frac{\gamma_i^{FD}\gamma_{d-i}^{FD}}{\gamma_i^{FD} + \gamma_{d-i}^{FD} + 1}, & \text{with AF} \\ \min(\gamma_i^{FD}, \gamma_{d-i}^{FD}), & \text{with } DF \end{cases} \tag{9}$$

The system rate is

$$R_{CFD-i} = W\log_2(1 + \gamma_{CFD-i}) \tag{10}$$

When the secondary user does not access the spectrum of the primary user, the SNR of the primary network communication can be defined as

$$\gamma_{pp} = P_p|h_{pp}|^2/N_0 \tag{11}$$

If the secondary user accesses the spectrum of the primary user in the full duplex relay mode, the secondary network base station S and the relay node i work simultaneously, which commonly causing interference to the primary user receiver PD is

$$I_{CFD-i} = P_s|h_{sp}|^2 + P_i|h_{ip}|^2 \tag{12}$$

The SNR of the primary network is

$$\gamma_{pp-i}^{FD} = \frac{P_p^2|h_{pp}|^2}{P_s^2|h_{sp}|^2 + P_i^2|h_{ip}|^2 + N_0} \tag{13}$$

The primary system rate is

$$R_{pp-i}^{FD} = W\log_2(1 + \gamma_{pp}^{FD}) \tag{14}$$

where h_{id}, h_{pd}, h_{pp}, h_{sp}, and h_{ip} are the channel coefficients of i to D, $\mathrm{P_T}$ to D, $\mathrm{P_T}$ to $\mathrm{P_R}$, S to $\mathrm{P_T}$ and i to $\mathrm{P_T}$, respectively. W is the system bandwidth.

3.2 Half Duplex Mode

When a relay node adopts the half-duplex relay mode, it operates in two orthogonal time slots or frequencies (this is assumed to be a time slot). The received signals at the secondary candidate relay node i and the secondary destination node D are

$$y_i = \sqrt{P_s}h_{si}x_s + \sqrt{P_p}h_{pi}x_{p1} + n_i \tag{15}$$

$$y_d = \sqrt{P_i}h_{id}x_i + \sqrt{P_p}h_{pd}x_{p2} + n_d \tag{16}$$

The retransmit signal of the relay node d is expressed as

$$x_i = \begin{cases} \beta_2 y_i, & \text{with AF} \\ \sqrt{\frac{P_i}{P_s}}x_s, & \text{with DF} \end{cases} \tag{17}$$

The amplification factor β_2 of the AF protocol is

$$\beta_2 = \sqrt{\frac{P_i}{P_s|h_{si}|^2 + P_p|h_{pi}|^2 + N_0}} \tag{18}$$

The instantaneous received signal power of the secondary candidate relay and the secondary destination terminal D are expressed as (19) and (4), respectively

$$\varepsilon\{|y_i|^2\} = P_s|h_{si}|^2 + P_p|h_{pi}|^2 + N_0 \tag{19}$$

The SINR of i and D hops can be expressed as

$$\gamma_i^{HD} = \frac{P_s^2|h_{si}|^2}{P_p^2|h_{pi}|^2+N_0} \tag{20}$$

$$\gamma_{d-i}^{HD} = \frac{P_i^2|h_{id}|^2}{P_p^2|h_{pd}|^2+N_0} \tag{21}$$

The end-to-end SINR of secondary networks in cognitive half-duplex relay mode is expressed as

$$\gamma_{CHD-i} = \begin{cases} \frac{\gamma_i^{HD}\gamma_d^{HD}}{\gamma_i^{HD}+\gamma_d^{HD}+1}, & \text{with AF} \\ \min(\gamma_i^{HD}, \gamma_d^{HD}), & \text{with DF} \end{cases} \tag{22}$$

The system rate is

$$R_{CHD-i} = W\log_2(1 + \overline{\gamma_{CHD}}) \tag{23}$$

When the secondary user accesses the spectrum of the primary user in the half-duplex relay mode, the interference caused to the primary is expressed as

$$I_{CHD-i} = \max(P_s|h_{sp}|^2, P_i|h_{ip}|^2) \tag{24}$$

The SNR of the primary network is

$$\gamma_{pp-i}^{HD} = \min(\frac{P_p\gamma_{pp}}{P_s\gamma_{sp}+1}, \frac{P_p\gamma_{pp}}{P_i\gamma_{ip}+1}) \tag{25}$$

The primary network rate is

$$R_{pp-i}^{HD} = W\log_2(1 + \gamma_{pp}^{HD}) \tag{26}$$

4 Mode Switching

The difference between the half-duplex relay mode and the full-duplex relay mode is reflected in the presence or absence of self-interference, so the SNR γ_{ii} of the self-interference channel is a key factor in mode switching. The necessary and sufficient condition for the full-duplex mode to be better than the half-duplex mode is that γ_{ii} is lower than the critical point γ_{ii}^{th} of self-interference. In this scenario, it is assumed that the secondary network base station S is farther away from the primary destination terminal than the secondary candidate relay, that is, i is more disturbing to the primary user than S. Therefore, the secondary network uses the half duplex to cause interference to the primary user is $I_{CHD-i} = P_i|h_{ip}|^2$.

Lemma 1: When the secondary relay network adopts the AF protocol, if (27) is established, the performance of the relay node in full-duplex mode is better than that in the half-duplex mode. Otherwise, the half-duplex mode is adopted, which the performance is better than performance in full-duplex mode.

$$\gamma_{ii} < \gamma_{ii-AF}^{th} = \frac{P_s\gamma_{si}(\gamma_{d-i} - \overline{\gamma_{CHD-i}} - P_s\gamma_{sp})}{P_i(\overline{\gamma_{CHD-i}} + \gamma_{sp})(\gamma_{d-i} + 1)} - \frac{P_p\gamma_{pi} + 1}{P_i} \tag{27}$$

Lemma 2: When the secondary network adopts the DF protocol, if the formula (28) is established, the performance of the relay node in the half-duplex mode is better than that in the full-duplex mode. If the Eq. (29) is established, then the performance of the relay node in full-duplex mode is better than that in half-duplex mode.

$$\gamma_{d-i}^{FD} < \overline{\gamma_{CHD-i}} + P_s\gamma_{sp} = (1 + \gamma_{CHD-i})^{1/2} - 1 + P_s\gamma_{sp} \tag{28}$$

$$\gamma_{ii} < \gamma_{ii-DF}^{th} = \frac{P_s\gamma_{si}}{P_i(\overline{\gamma_{CHD-i}} + P_s\gamma_{sp})} - \frac{P_p\gamma_{pi} + 1}{P_i} \tag{29}$$

When the secondary relay network adopts the AF protocol, each secondary candidate relay node performs mode selection according to Lemma 1. When the secondary network adopts the DF protocol, each secondary candidate relay node performs mode selection according to Lemma 2. Ensuring that the sacrifices made by the primary are more valuable optimizes the performance of the entire network, including the secondary user network and the primary network.

5 Low Interference Mode Switching Relay Selection Algorithm

A low-interference hybrid duplex mode switch relay selection (HDMSRS) algorithm is proposed. In this section, aiming to optimize the mode switching and relay selection of the whole network. The utility function can be defined as

$$C_i = \arg \max_{i=1\sim n} (\gamma_{ete}^i - \gamma_{s\to p}^i) \tag{30}$$

where γ_{ete}^i is the revenue function of the entire system

$$\gamma_{ete}^i = \begin{cases} \overline{\gamma_{CHD-i}} & \text{with HD} \\ \gamma_{CFD-i} & \text{with FD} \end{cases} \tag{31}$$

$\gamma_{s\to p}^i$ is the cost function of the entire system

$$\gamma_{s\to p}^i = \begin{cases} \max\left(\frac{P_s|h_{sp}|^2}{N_0}, \frac{P_i|h_{ip}|^2}{N_0}\right) & \text{with HD} \\ \frac{P_s|h_{sp}|^2 + P_i|h_{ip}|^2}{N_0} & \text{with FD} \end{cases} \tag{32}$$

In order to maximize the overall performance of both secondary and primary systems, a hybrid relay mode is proposed to switch the appropriate mode according to the instantaneous CSI. The utility function for each candidate relay is

$$C_{Hybrid} = \max\ (C_{i-HD}, C_{i-FD}) \tag{33}$$

where C_{i-HD} is the utility function of candidate relay nodes operating in half-duplex mode; C_{i-FD} is the utility function of candidate relay nodes operating in full-duplex mode.

Each candidate relay C_i selects a duplex mode according to its own utility to optimize overall performance.

The specific HDMSRS algorithm flow is described in Algorithm 1.

Algorithm 1. The HDMSRS algorithm flow is as follows

Step1 Initializing secondary candidate half-duplex and full-duplex relay terminal sets H and F in the secondary network base station S, then send relay request information by broadcast.

Step2 The secondary network base station S and each candidate relay terminal i that receives the relay request, estimate channel state information of itself to the primary destination terminal D first, determine whether the interference temperature limit can be met when i works in half-duplex mode and full-duplex mode respectively:

Step2.1 If only the work is satisfied in the half-duplex (full-duplex) mode, the half-duplex (full-duplex) relay mode is adopted directly;

Step2.2 If the work is not satisfied in the half duplex mode and the full duplex mode, the candidate relay terminal cannot communicate as a relay;

Step2.3 If the work is satisfied in both the half-duplex mode and the full-duplex mode, mode selection is performed by Lemma 1 or Lemma 2.

Step3 Calculate its own utility value according to equation (30) and feed it back to the F and H sets at S.

Step3.1 If i selects half-duplex mode, add this node to the set H.

Step3.2 If i selects full-duplex mode, add this node to the set F.

Step4 Select the maximum value according to equation (34) and equation (35) in the H and F sets at S.

$$C_i^{FD*} = \max\ (C_1^{FD}, C_2^{FD} ... C_n^{FD}) \tag{34}$$

$$C_i^{FD*} = \max\ (C_1^{FD}, C_2^{FD} ... C_n^{FD}) \tag{35}$$

Step5 Compare the utility values corresponding to the two candidate relay terminals selected in **Step4** according to the formula (36), and the larger one is selected as the relay node.

$$U_i^* = \max\ (C_i^{HD*}, C_i^{FD*}) \tag{36}$$

6 Simulation and Performance Analysis

The simulation results of the proposed mode switching and relay selection in cognitive hybrid duplex network will be discussed in this section. The system simulation is run on Matlab. Considering one primary network base station, one PN terminal, one secondary network base station, secondary network terminal

Table 1. Main simulation parameters

Parameter	Values
Transmission power of P_P	1 dB
Interference temperature I_{th}	3 dB
Road loss index η	3
System bandwidth W	1 MHz
Additive Gauss white noise variance N_0	1
Channel coefficient expression	$h_{m,n} = k * {d_{m,n}}^{-\eta}$

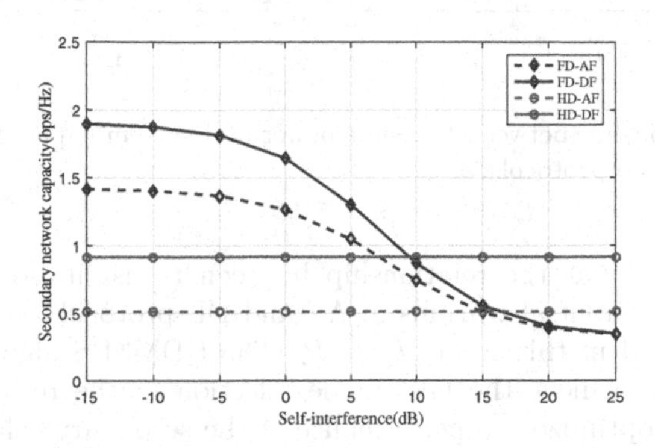

Fig. 2. The relationship between self-interference and secondary network system capacity under different duplex mode and relay protocol.

in the simulation scenario. There are many secondary candidate rely nodes uniformly distributed within the SN BS and SN DT. Assuming that the channel coefficients between each node can be expressed as $h_{m,n} = k * {d_{m,n}}^{-\eta}$, where k denotes the attenuation factor, which obeys a complex Gaussian distribution with a mean of zero and a variance of one, $d_{m,n}$ denotes the distance from node to node n, and η is the path loss index. The main simulation parameters are detailed in Table 1.

In Fig. 2 showing the relationship between the change in self-interference value and the capacity of the secondary network system when the secondary relay network adopts different duplex modes and relay protocols. The transmit power of the secondary network base station and the secondary relay are both $P_s = P_i = 1\,\text{dB}$. When the half-duplex relay mode is adopted, there is no self-interference at the relay node. The system capacity of the relay network is independent of the change of the self-interference value, and the relationship is expressed as a straight line. When the secondary relay network accesses the spectrum of the primary user in full-duplex mode, whether it adopts AF protocol or DF protocol, the system capacity of the secondary network decreases as the self-interference value increases. It can be seen from the figure that the necessary and sufficient condition for the performance of the full-duplex relay mode be superior to the half-duplex mode is that the self-interference value γ_{ii} is less than the two intersection points, i.e. the self-interference thresholds γ_{ii-DF}^{th} and γ_{ii-AF}^{th}.

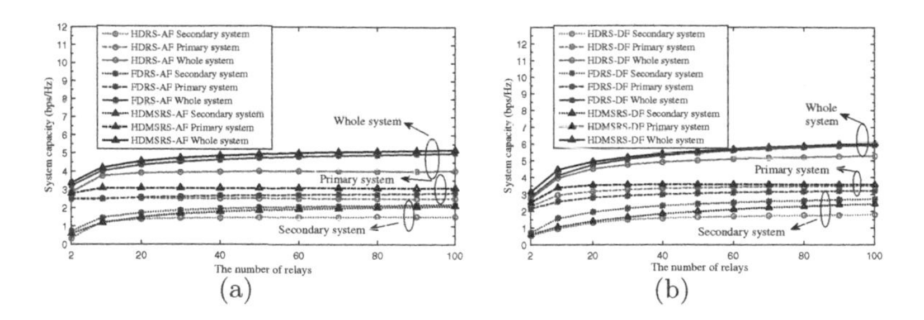

Fig. 3. The relationship between transmit power and system capacity under different algorithms under two protocols.

Figure 3 show that the relationship between transmit power and system capacity under different algorithms in AF and DF protocols. Setting the number of relays $n = 4$ in this scene, $P_s = P_i$. The HDMSRS algorithm proposed in this paper can achieve the best mode selection in the relay selection process. In order to optimize the performance of the secondary relay network, the relay will transmit in full-duplex mode with small P_i. As the P_i increases, the self-interference is large, and the relay will switch to half-duplex mode for transmission. In addition, the algorithm proposed in this paper can be significantly improved the primary system capacity while lightly reducing (or even enhancing) the capacity of the secondary system with the comparison of the HDRS and FDRS algorithms based on MMRS criteria. For example, when $P_s = 6\,\text{dB}$, the capacity of the secondary system, is reduced by 8.3% compared with the DRS algorithm, while increasing by 3.1% compared with the FDRS algorithm. For the capacity of the primary system, the proposed HDMSRS algorithm is improved by 38.9% compared with the HDRS algorithm, and is increased by 41.1% compared with the FDRS algorithm.

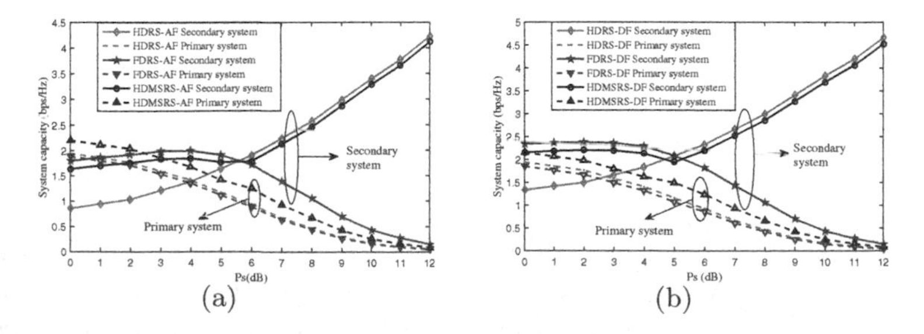

Fig. 4. The relationship between the capacity of system and the number of candidate relays under two protocols.

Figure 4 examine the relationship between the capacity of system and the number of candidate relays. Setting $P_s = P_i = 1\,\text{dB}$, what is more, in order to distinguish the capacity curve of the secondary user and the primary user, let $P_p = 2\,\text{dB}$. As can be seen from the above figures, the HDMSRS algorithm proposed in this paper can significantly improve the system capacity of the primary user and improve the whole system capacity under the condition of slightly reducing the capacity of the secondary system. For example, when the number of relay is 60, the capacity of the secondary system of the proposed HDMSRS algorithm is improved by 26.7% compared with the HDRS algorithm, while compared with the FDRS algorithm is reduced by 10.5%. For the capacity of the primary system, the proposed HDMSRS algorithm is improved by 26.5% compared with the HDRS algorithm, compared with the FDRS algorithm, the proposed algorithm is improved by 14.8%. For the capacity of the whole system, the HDMSRS algorithm can be increased by 31.6% and 13.1% compared with the HDRS algorithm and the FDRS algorithm, respectively.

7 Conclusion

In this paper, to solve the single duplex mode of the relay node, the criteria of full-duplex/half-duplex mode switching has been derived. Moreover, the proposed a low interference hybrid duplex relay selection algorithm which chooses optimal relay based on the combined performance of primary network performance loss and secondary network user performance gains, has improved the performance of the primary network system and the entire system. The simulation results have been performed to verify the superiority of the algorithm on performance improvement of primary network system and the whole system while the performance of secondary user network is slightly decreased (or even promoted).

References

1. Chen, X., Chen, H.H., Meng, W.: Cooperative communications for cognitive radio networks – from theory to applications. IEEE Commun. Surv. Tutor. **16**(3), 1180–1192 (2014)
2. Goldsmith, A., Jafar, S.A., Mari, I.: Breaking spectrum gridlock with cognitive radios: an information theoretic perspective. Proc. IEEE **97**(5), 894–914 (2009)
3. Jian, C., Lu, L., Liu, Y.: Energy efficient relay selection and power allocation for cooperative cognitive radio networks. Iet Commun. **9**(13), 1661–1668 (2015)
4. Liang, Y.C., Chen, K.C., Li, G.Y.: Cognitive radio networking and communications: an overview. IEEE Trans. Veh. Technol. **60**(7), 3386–3407 (2011)
5. Shim, Y., Wan, C., Park, H.: Beamforming design for full-duplex two-way amplify-and-forward MIMO relay. IEEE Trans. Wirel. Commun. **15**(10), 6705–6715 (2016)
6. Lee, J., Wang, H., Andrews, J.G.: Outage probability of cognitive relay networks with interference constraints. IEEE Trans. Wirel. Commun. **10**(2), 390–395 (2011)
7. Si, J.B., Li, Z., Huang, H.Y.: Capacity analysis of cognitive relay networks with the PU's interference. IEEE Commun. Lett. **16**(12), 2020–2023 (2012)

8. Zhao, R., Yuan, Y., Fan, L.: Secrecy performance analysis of cognitive decode-and-forward relay networks in Nakagami-m fading channels. IEEE Trans. Commun. **65**(2), 549–563 (2016)
9. Li, D.: Outage probability of cognitive radio networks with relay selection. IET Commun. **5**(18), 2730–2735 (2011)
10. Rajkumar, S., Thiruvengadam, J.S.: Outage analysis of OFDM based cognitive radio network with full duplex relay selection. IET Signal Process. **10**(8), 865–872 (2016)
11. Ho-Van, K.: Outage analysis of opportunistic relay selection in underlay cooperative cognitive networks under general operation conditions. IEEE Trans. Veh. Technol. **65**(10), 8145–8154 (2016)
12. Zhong, B., Zhang, Z., Chai, X.: Performance analysis for opportunistic full-duplex relay selection in underlay cognitive networks. IEEE Trans. Veh. Technol. **64**(10), 4905–4910 (2015)
13. Yin, C., Doan, T.X., Nguyen, N.P.: Outage probability of full-duplex cognitive relay networks with partial relay selection. In: 2017 International Conference on Recent Advances in Signal Processing, Telecommunications and Computing (SigTelCom), Da Nang, pp. 115118 (2017)
14. Zhong, B., Zhang, Z.: Opportunistic two-way full-duplex relay selection in underlay cognitive networks. IEEE Syst. J. **12**, 1–10 (2016)
15. Bang, J., Lee, J., Kim, S.: An efficient relay selection strategy for random cognitive relay networks. IEEE Trans. Wirel. Commun. **14**(3), 1555–1566 (2015)
16. Riihonen, T., Werner, S., Wichman, R.: Hybrid full-duplex/half-duplex relaying with transmit power adaptation. IEEE Trans. Wirel. Commun. **10**(9), 3074–3085 (2011)

High-Discrimination Multi-sensor Information Decision Algorithm Based on Distance Vector

Lingfei Zhang[1,2] and Bohang Chen[2(✉)]

[1] College of Physics and Electronic Information Engineering, Qinghai Nationalities University, Xining, China

[2] College of Physics and Electronic Information Engineering, Qinghai Normal University, Xining, China

cbh1207@163.com

Abstract. In the process of sensor target recognition, attitude estimation and information decision-making, most of the current sensor information decisions require probability conversion or weight calculation of sensor data. The calculation process is complex and requires a large amount of computation. In addition, the decision result is greatly affected by the probability value. This paper proposes a multi-sensor information decision algorithm with high-discrimination based on distance vectors. At the same time, the support function, dominance function and discrimination function for the algorithm are presented. The dominance function is obtained through the normalization processing of the support matrix, and then the dominance function after normalization is sorted. The maximum value is taken as the optimal solution. The discrimination function mainly provides the basis for the evaluation of the algorithm. The simulation results show that the discrimination degree of this method in sensor information decision-making reaches more than 0.5, and the decision-making effect is good. Compared with the classic D-S evidence theory, this algorithm can effectively avoid the phenomenon that D-S evidence theory contradicts with the actual situation when making a decision. It is less affected by a single sensor and the decision result is stable. Compared with the probabilistic transformation of the initial data of the sensor in the decision-making process, it has obvious advantages.

Keywords: Information decision · Distance vector · Support matrix · Dominance function · Discrimination function

1 Introduction

The rapid development of artificial intelligence, big data and 5G technologies has accelerated the process of "Internet of Everything" [1]. As an important medium for the machine to perceive the world [2], the sensor network directly affects the development and progress of society [3] because of its security, intelligence and advanced nature. An integrated sensor network (WSN) is usually composed of multiple sensor nodes, these sensor nodes seem to be independent. In fact, due to the correlation of the monitoring

X. Wang et al. (Eds.): 6GN 2020, LNICST 337, pp. 43–55, 2020.
https://doi.org/10.1007/978-3-030-63941-9_4

range, method and object, there is a potential internal connection between each data [4]. Using this internal connection, a relatively correct result is automatically determined by the algorithm, which makes a reference for people's scientific research [5].

Ref. [6] combined support vector machine (SVM) and improved D-S evidence theory to propose multi-classifier information fusion. Analysis shows that this method can effectively fuse classifier information from different SVM, it has strong robustness and high accuracy. Ref. [7] proposed a new divergence measure for the confidence function for the high-conflict problem of D-S evidence theory to measure the difference between the basic probability distributions in D-S evidence theory. Analysis shows that this method provides a promising solution for measuring the difference between belief functions in evidence theory. Aiming at the problem that the conflict weight of D-S evidence theory cannot be effectively reduced, Ref. [8] improved the D-S evidence theory and introduced a combination rule, which effectively reduced the conflict weight assigned by the basic probability allocation ordering. Ref. [9] proposed a trust model to improve D-S evidence theory. This model defines the evidence variance according to the Jousselme distance, modifies the evidence before the fusion, and then conducts the fusion through D-S evidence rules. The simulation results show that the model can accurately find the malicious nodes in the sensor network, which is conducive to improving the security and robustness of the network. Ref. [10] proposes a comprehensive information fusion method based on evidence theory and group decision-making. The weight of combination and disjunction rules are adjusted according to the consistency function of focus elements. Simulation result shows that this method can obtain reasonable and reliable decision results. Ref. [11] proposes an improved strategy of evidence theory based on confidence to solve the conflict problem of D-S evidence theory. The algorithm extracts weight from the preliminary predicted values of four neural networks, constructs BPAs function reasonably, and verifies the application effect of the system through examples. Ref. [12] proposed a similar Jaccard coefficient matrix partitioning processing method, which corrected the evidence source by calculating the weight of each sensor node, this method effectively solved the high conflict problem existing in D-S evidence theory, and reduced the decision-making risk. Ref. [13] proposed a multi-source information fusion fault diagnosis method based on D-S evidence theory. This method uses fuzzy subjection functions to construct the basic probability assignment of three evidence bodies. At the same time, in order to solve the conflict problem of D-S evidence theory, a D-S evidence theory based on the similar distance of new evidence is proposed. The analysis shows that the fault diagnosis results can be improved effectively. According to the conflict of D-S evidence theory. Ref. [14] proposes a testability evaluation method based on improved D-S evidence theory, establishes the density distribution function of equipment testability fault index detection rate, and constructs the quality function. At the same time, Lance and Williams distance are introduced to improve the information fusion of D-S evidence theory. The test results showed that this method had a good evaluation effect. Ref. [15] uses rough set theory and cloud parameter calculation, and combines D-S evidence theory to fuse multi-source information and identify faults. Simulation results show that this method can accurately identify faults; Ref. [16] makes use of the proximity of evidence in the process of sensor information decision-making, and keeps important information. At the same time, the conflict problem of D-S evidence theory is the weighting processing. Simulation analysis shows that this method can reduce the impact of conflict and improve the decision accuracy. The above methods have achieved good results in the process of sensor information decision-making, but the calculation process is relatively complex, most of the algorithms need to assign probability to the data collected by sensors, and seldom consider the characteristics and evaluation criteria of sensors.

In this paper, we make full use of the evaluation grade range of sensor data in the monitoring system, combine with the actual measured values, and calculate the support degree among the sensor data by using the distance between the actual values and the standard values, and finally get the system decision result. The main content and structure of the contents are organized and shown below. In Sect. 2 to 4, the fusion level, fusion process and decision model are reviewed. In Sect. 5 and 6, a more detailed description of proposed algorithms are given based on the distance vector. The performance analysis and discussion are described in Sect. 7 and concluded in the Sect. 8.

2 Multi-sensor Information Fusion Level

Multi-sensor information fusion includes research on transmission protocol and sensor network node information fusion, both of which are used to solve the redundancy problem of information acquisition in multi-sensor system, so as to improve the accuracy of information acquisition and reduce the amount of data sent by nodes. Figure 1 is a functional model of multi-sensor information fusion [17], whose fusion process can be divided into multiple levels. Firstly, the source data collected by the sensor is preprocessed. Secondly, the fusion results are obtained by further processing according to the application field. The technologies involved in fusion include data association, target tracking and identification, situation estimation, impact assessment and process assessment, among which the higher-level fusion technologies all involve information decision-making.

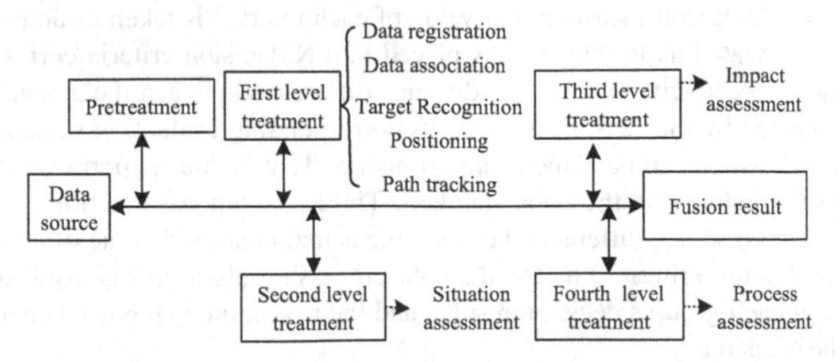

Fig. 1. Multi-sensor information fusion function model

3 Multi-sensor Information Fusion Process

Figure 2 shows the basic fusion process of multi sensor information [18], and the fusion process of data, features and decision-making is given in Fig. 2. First, the data collected by the sensor is preprocessed, including signal rectification, filtering and amplification. The first level of fusion directly sends the preprocessed data to the fusion center to obtain the fusion result. In the second level of fusion, the preprocessed data are extracted for features and sent to the fusion center to get the fusion results. This level of fusion is usually applied to data association and target recognition. The third level of fusion is to add the preliminary decision based on the second level of fusion, and then send the decision results to the fusion center to get the output results. This level of information fusion is often used in the field of information decision-making, which belongs to a higher level of information fusion, and the difficulty is increased to a certain extent compared with the first two levels of fusion.

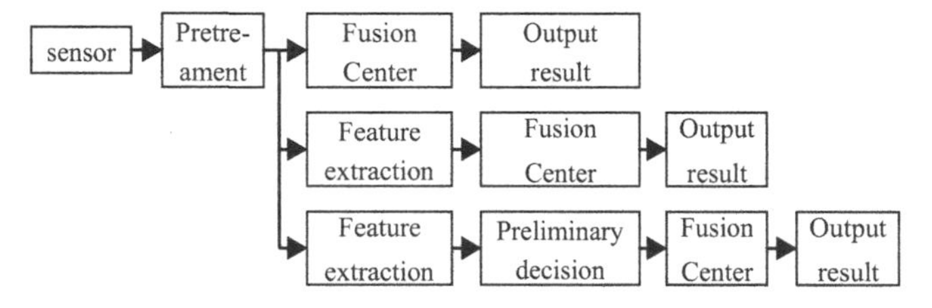

Fig. 2. Basic fusion process

4 Multi-sensor Information Decision Model

Figure 3 is a multi-sensor information decision model. There are n decision criteria and m sensor nodes. First, the measurable range of each sensor is divided into several intervals, and the central measurement value of each interval is taken as an evaluation standard (abbreviated as ICV) which is placed into N decision criteria corresponding to the sensor respectively. Second, the distance difference between the actual value of the data collected by the sensor node and its corresponding evaluation standard needs to be obtained, which can be named support matrix. That is, the support of each actual measured value relative to the node standard. Third, the support function is obtained by summing the distance difference between the actual measured value of each sensor and its n evaluation criteria. Finally, the data processing algorithm is used to get the superiority of each group's decision results, and the maximum value is taken as output result of the decision.

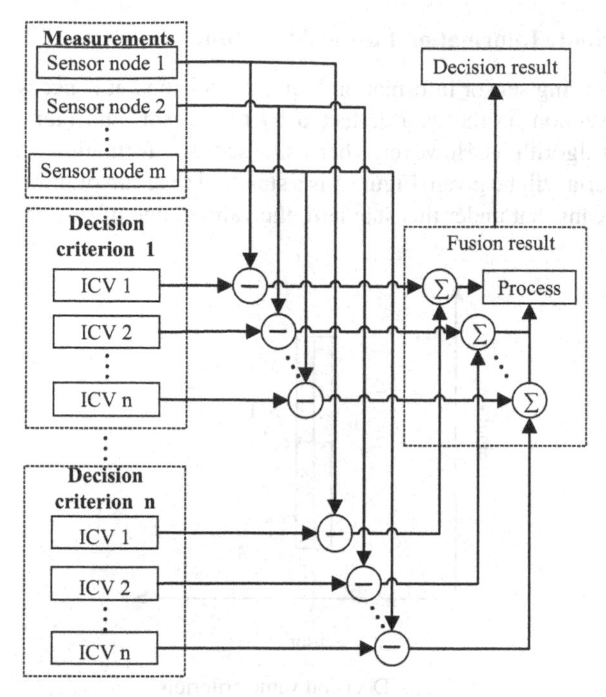

Fig. 3. Multi-sensor information decision model

5 Algorithm Implementation Principle

5.1 Single Attribute Information Fusion Algorithm

Figure 4 is a schematic diagram of single attribute sensor information decision. Assuming that the same kind of sensors are used for measurement in a monitoring system, there are four evaluation levels in the system, the value range of the first evaluation level is $a_1 \pm \Delta t$, and the value ranges of the other three evaluation levels are shown in Fig. 4. Let b be the actual measured value of the sensor, calculate the absolute value of the distance between b and the center value of each value range, and take the evaluation level range where the minimum absolute value is located as the final decision result. As shown in Fig. 4, is decision result 2 (abbreviated as DR2).

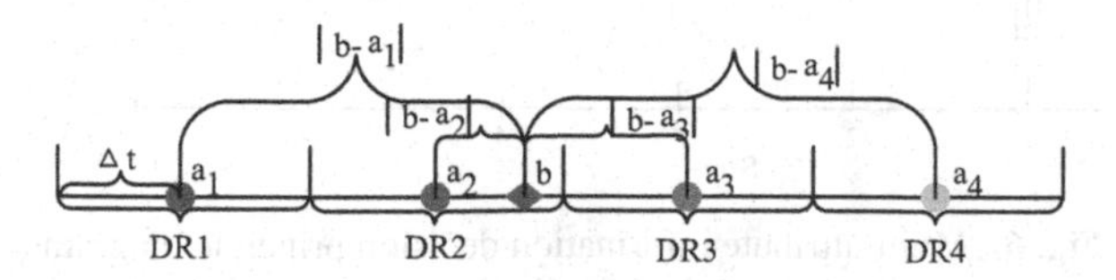

Fig. 4. Single attribute information decision schematic diagram

5.2 Multi-attribute Information Fusion Algorithm

Generally, when using sensor information to make decisions, it is necessary to perform probability conversion on the data collected by the sensor, and then make decisions through relevant algorithms. However, when using sensor information to make decisions, the decision criteria will be given. Figure 5 is a standard diagram of information decision value, and a_1^1 means that under this standard, the value range of sensor 2 is $a_1^1 + \Delta t$.

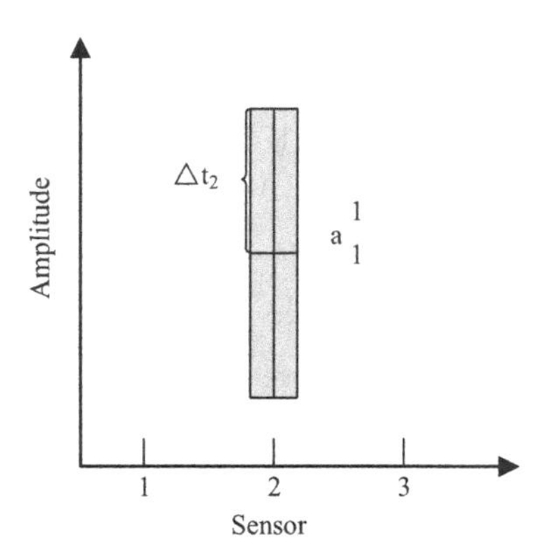

Fig. 5. Decision value criterion

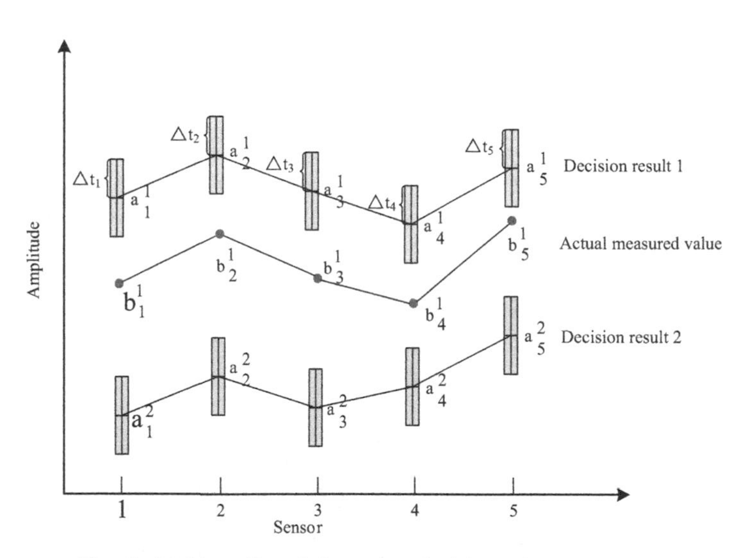

Fig. 6. Multi-attribute information decision principle diagram

The principle of multi-attribute information decision-making is shown in Fig. 6. Where $a_1^1, a_2^1, \cdots a_5^1$ are the central value of the value range for each sensor corresponding to DR1. Similarly, and $a_1^2, a_2^2, \cdots, a_5^2$ are the central value of the value range for each sensor corresponding to DR2, and $b_1^1, b_2^1, \cdots, b_5^1$ are the actual measured values of the first group. Next, the single attribute decision making method is extended horizontally to multi-attribute information decision making, and the distance vector modulus of the actual measured value and the central value of the standard decision range is vertically calculated, and the sum of the distance vector modulus of each decision result is calculated horizontally. The smaller the value is, the closer the calculation result is to the decision result.

6 Multi-sensor Decision Algorithm Based on Distance Vector

The specific process of the multi-sensor decision algorithm based on distance vector is described as below.

Step 1. If there are n evaluation levels and m sensors in the evaluation system, the value range of the $i-th$ sensor corresponding to each level j is $a_i^j \pm \Delta t_i$, where a_i^j is the center value, then the level j can be vector as $G_{\mathrm{j}} = \left(a_1^j, a_2^j, \cdots, a_m^j\right)$.

Step 2. Suppose the maximum value of each sensor is $max(a_i)$ and the minimum value $min(a_i)$, the actual measured value is a vector as $B = (b_1, b_2, \cdots, b_m)$, 且 $min(a_i) \leq b_i \leq max(a_i)$.

Step 3. Let N_j be the support degree of j to the measured values of each sensor, $\Phi_{ji} = \left(A_j, A_i\right) = b_i - a_i^j$ represents the distance from the measured value b_i to the evaluation a_i^j, where $i \leq m, j \leq n$.

Then the support matrix can be expressed as follow,

$$N\left(A_i, A_j\right) = \begin{pmatrix} \Phi(A_1, A_1) & \Phi(A_1, A_2) & \cdots & \Phi(A_1, A_m) \\ \Phi(A_2, A_1) & \Phi(A_2, A_2) & \cdots & \Phi(A_2, A_m) \\ \vdots & \vdots & \cdots & \vdots \\ \Phi(A_n, A_1) & \Phi(A_n, A_2) & \cdots & \Phi(A_n, A_m) \end{pmatrix} \tag{1}$$

The support function is obtained from the above Eq. (1) as follow,

$$\delta(A_i) = \sum\nolimits_{i=1}^{m} \Phi\left(A_j, A_m\right) \tag{2}$$

Step 4. Normalization processing,

$$T_j' = \frac{\delta\left(A_j\right)}{\delta} \tag{3}$$

where $\delta = \sum_{j=1}^{n} \delta\left(A_j\right),\ T' \in [0, 1]$.
then the result of the dominance function is,

$$T_i = 1 - T_j' \tag{4}$$

Step 5. According to the calculation scheme of Definition 1–4, the normalized function $s(A_i)$ is:

$$S_j = \frac{T_j - \min\limits_{j \leq n}\left\{T_j\right\}}{\max\limits_{j \leq n}\left\{T_j\right\} - \min\limits_{j \leq n}\left\{T_j\right\}} \tag{5}$$

Sort the calculation results in S_j, and take $max\{S_j\}$ as the optimal solution, that is, $S_j = 1$.

Step 6. Let φ be the discrimination function, then,

$$\varphi = max\{\varphi_j\} = \max_{j \le n}\{S_j\} - S_j \tag{6}$$

Step 7. When m = 1, $\delta(A_j) = \Phi(A_j, A_1)$, then, $T_j = 1 - \frac{\delta(A_j)}{\sum_{j=1}^{n} \delta(A_j)}$, the optimal solution scheme can be solved by Eq. (5).

7 Simulation and Analysis

Example 1. Assume that the temperature of a certain water area is to be graded, and the evaluation criteria are shown in Table 1. It is tested with a temperature sensor, and the temperature of current water is set to 15 °C.

Table 1. Temperature evaluation standard

Level	First level	Second level	Third level	Fourth level
Temperature range	30–35 °C	20–30 °C	10–20 °C	−5–10 °C

The calculation results of the normalized function in Table 2 can be obtained from Table 1, where DDV represents the distance difference between the measured value and the center value.

Table 2. The result is calculated by the normalized function

Level	First level	Second level	Third level	Fourth level
Temperature range	30–35 °C	20–30 °C	10–20 °C	−5–10 °C
Central value	32.5 °C	25 °C	15 °C	2.5 °C
DDV	17.5 °C	10 °C	0 °C	12.5 °C
T_j	0.5625	0.75	1	0.6875
S_j	0	0.4286	1	0.2857

Then sort the calculation results of S_j, and take $max\{S_j\}$ as the optimal solution, that is, the water area has three levels. The calculation results show that the water area is of third level. According to Eq. (6), the discrimination degree is 0.5714.

Example 2. It is assumed that a comprehensive evaluation of the pollution of a certain water area is required. The evaluation criteria are shown in Table 3. The environmental

Table 3. Evaluation criteria for water pollution

Level	Temperature	pH	Turbidity	Conductivity
First level	30–35 °C	9–10	35–40%	0.8–1
Second level	20–30 °C	8–9	25–35%	0.5–0.8
Third level	10–20 °C	6–8	10–25%	0.3–0.5
Fourth level	−5–10 °C	4–6	0–10%	0.1–0.3

parameters of water quality that collected by sensor nodes including temperature, pH, turbidity, conductivity, etc. and the actual measurement results are 12 °C, 7.5, 20%, and 0.4 respectively.

The following vector expressions are obtained from Table 3,

$$G_1 = (32.5, 9.5, 37.5, 0.9)$$

$$G_2 = (25, 8.5, 30, 0.65)$$

$$G_3 = (15, 7, 17.5, 0.4)$$

$$G_4 = (2.5, 5, 5, 0.2)$$

Then get the support matrix as,

$$N(A_i, A_j) = \begin{pmatrix} 20.5 & 2 & 17.5 & 0.5 \\ 13 & 1 & 10 & 0.25 \\ 3 & 0.5 & 2.5 & 0.2 \end{pmatrix}$$

According to Eqs. (1), (2) and (3), the following results can be obtained,

$$T_1 = 0.5865,\ T_2 = 0.7507,\ T_3 = 0.9388,\ T_4 = 0.7223$$

and according to Eqs. (4), the following results can be obtained,

$$S_1 = 0,\ S_2 = 0.4661,\ S_3 = 1,\ S_4 = 0.3555$$

First, sort the calculation results of S_j, and then take $max\{S_j\}$ as the optimal solution. The results show that the water area meets the third-level standard and are consistent with common sense.

Example 3. Suppose that a recognition framework is composed of A, B and C, and the evaluation criteria of the corresponding evidence are shown in Table 4. If the actual measured values of the sensors are 60, 38, 1.5 and 4.8 respectively, the basic probability distribution based on the measured values are shown in Table 5. In addition, the fusion results comparison of the proposed algorithm and D-S evidence theory as shown in Table 6.

Table 4. Evaluation criteria under the identification framework in Example 3

	A_1	A_2	A_3	A_4
A	90–100	0–15	0.5–0.8	3–5
B	75–90	15–30	0.8–1.3	1–3
C	50–75	30–50	1.3–1.8	−0.5–1

Table 5. Evidence data under the identification framework in Example 3

	$m(A_1)$	$m(A_2)$	$m(A_3)$	$m(A_4)$
A	0.1	0.1	0.1	0.9
B	0.3	0.32	0.45	0.1
C	0.6	0.58	0.45	0

Table 6. Comparison of fusion results of different methods

Method	m_1, m_2	m_1, m_2, m_3	m_1, m_2, m_3, m_4
D-S	$m(H_1) = 0.022$	$m(H_1) = 0$	$m(H_1) = 0.1724$
	$m(H_2) = 0.2115$	$m(H_2) = 0.4515$	$m(H_2) = 0.8276$
	$m(H_3) = 0.7665$	$m(H_3) = 0.7665$	$m(H_3) = 0$
	$m(\theta) = 0$	$m(\theta) = 0$	$m(\theta) = 0$
Proposed method	$m(H_1) = 0$	$m(H_1) = 0$	$m(H_1) = 0$
	$m(H_2) = 0.4508$	$m(H_2) = 0.4515$	$m(H_2) = 0.2204$
	$m(H_3) = 1$	$m(H_3) = 0.7665$	$m(H_3) = 1$
	$m(\theta) = 0$	$m(\theta) = 0$	$m(\theta) = 0$

It can be seen from the data in Table 6 that when the number of evidences is less than or equal to 3, each evidence shows that the recognition result is the most likely to be C, which is the same as the result of the method in this paper, and the recognition result is C. When the number of evidence is 4, the judgment result of D-S evidence theory is greatly affected by the basic probability distribution, which is contrary to the facts and has no reference value. However, the identification result of the method for proposed method is still C, which is consistent with common sense. The relationship between the recognition rate of C and the number of evidences in the two fusion methods are shown in Fig. 7. If the recognition rate of C is used to analyze the index, the recognition rate of this method is 1 when the number of evidence is 2, 3, and 4, while the recognition rate of D-S evidence rule is significantly lower when the number of evidence is 2, 3. And when the number of evidence is 4, the histogram of D-S evidence rule disappears, that is, the recognition rate is 0. The reason for analysis is that when the evidence is A_4, the basic probability assignment $m_4(A_4)$ appears a big conflict.

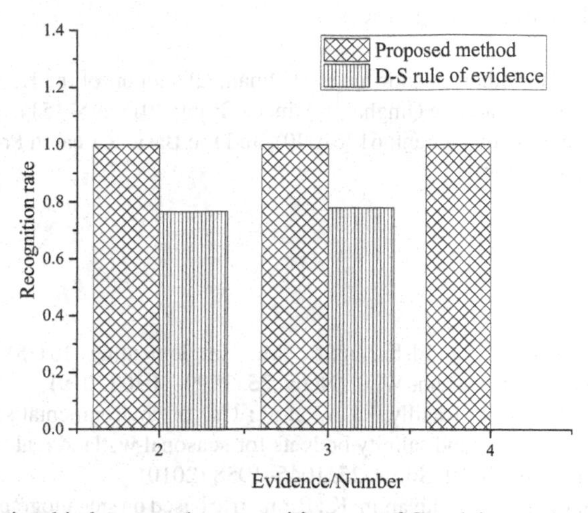

Fig. 7. Relationship between the recognition rate of C and the number of evidences

Furthermore, the degree of discrimination between the two methods is calculated separately, as shown in Table 7. The discrimination of the two methods is above 0.5, but when the number of evidence is 4, the discrimination of the method in this paper reaches 0.7796. However, the D-S evidence theory loses comparability due to the inconsistency between decision results and facts.

Table 7. Discrimination Comparison of different methods

Method	m_1, m_2	m_1 , m_2, m_3	m_1 , m_2, m_3, m_4
D-S	0.5550	0.5648	0.6552
Proposed	0.5592	0.5685	0.7796

8 Conclusion

This study proposes a multi-sensor information decision algorithm with high-discrimination based on distance vector, and the principle of the algorithm is analyzed from the perspective of single attribute and multi-attribute. The support matrix, dominance function and discrimination function of the algorithm can be used to obtain the optimal solution of the decision result and the evaluation basis of the algorithm performance. The analysis results show that the discrimination degree of the proposed method can reach more than 0.5, it can effectively avoid the conflict problem in the decision-making process compared with the traditional D-S evidence theory, and has obvious advantages over the application scenarios that need to assign the initial probability of the sensor in the decision-making process.

Acknowledgments. The authors acknowledge the financial support of the Key Projects of R&D and Achievement Transformation in Qinghai Province (Grant: 2018-NN-151), the National Natural Science Foundation of China (Grant: 61761040), and the Basic Research Program of Qinghai Province (Grant: 2020-ZJ-709).

References

1. Onasanya, A., Lakkia, S., Elshakankiri, M.: Implementing IoT/SWN based smart Saskatchewan healthcare system. Wirel. Netw. **25**, 3999–4020 (2019)
2. Quinn, N.W.T., Ortega, R., Rahilly, P.J.A., et al.: Use of environmental sensors and sensor networks to develop water and salinity budgets for seasonal wetland real-time water quality management. Environ. Model. Softw. **25**, 1045–1058 (2010)
3. Karthika, P., Ganesh Babu, R., Jayaram, K.: Biometric based on steganography image security in wireless sensor networks. Procedia Comput. Sci. **167**, 1291–1299 (2020)
4. Rodríguez, S., De Paz, J.F., Villarrubia, G., et al.: Multi-agent information fusion system to manage data from a WSN in a residential home. Inf. Fusion **23**(5), 43–57 (2015)
5. Zhou, X., Peng, T.: Application of multi-sensor fuzzy information fusion algorithm in industrial safety monitoring system. Saf. Sci. **122**, 1–5 (2020)
6. Pan, Y., Zhang, L., Wu, X., et al.: Multi-classifier information fusion in risk analysis. Inf. Fusion **60**(8), 121–136 (2020)
7. Xiao, F.: A new divergence measure for belief functions in D-S evidence theory for multisensor data fusion. Inf. Sci. **514**(4), 462–483 (2020)
8. Wang, J., Qiao, K., Zhang, Z.: An improvement for combination rule in evidence theory. Future Gener. Comput. Syst. **91**(2), 1–9 (2019)
9. Yang, K., Liu, S., Shen, J.: Trust model based on D-S evidence theory in wireless sensor networks. In: China Conference on Wireless Sensor Networks, Advances in Wireless Sensor Networks, pp. 293–301 (2014)
10. Leung, Y., Ji, N.-N., Ma, J.-H.: An integrated information fusion approach based on the theory of evidence and group decision-making. Inf. Fusion **14**, 410–422 (2013)
11. Si, L., Wang, Z., Tan, C., et al.: A novel approach for coal seam terrain prediction through information fusion of improved D-S evidence theory and neural network. Measurement **54**, 140–151 (2014)
12. Zhao, G., Chen, A., Guangxi, L., et al.: Data fusion algorithm based on fuzzy sets and D-S theory of evidence. Tsinghua Sci. Technol. **25**(1), 12–19 (2018)

13. Chuanqi, L., Wang, S., Wang, X.: A multi-source information fusion fault diagnosis for aviation hydraulic pump based on the new evidence similarity distance. Aerosp. Sci. Technol. **71**(12), 392–401 (2017)
14. Si, L., Wang, Z., Tan, C., et al.: An approach to testability evaluation based on improved D-S evidence theory. In: ACM International Conference Proceeding Series, pp. 155–159 (2019)
15. Mi, J., Wang, X., Cheng, Y., et al.: Multi-source uncertain information fusion method for fault diagnosis based on evidence theory. In: Prognostics and System Health Management Conference, Qingdao, China (2020)
16. Gao, X., et al.: Collaborative fault diagnosis decision fusion algorithm based on improved DS evidence theory. In: Wang, Y., Martinsen, K., Yu, T., Wang, K. (eds.) IWAMA 2019. LNEE, vol. 634, pp. 379–387. Springer, Singapore (2020). https://doi.org/10.1007/978-981-15-2341-0_47
17. Han, C., Zhu, H., Duan, Z., et al.: Multi-Source Information Fusion, 2nd edn. Tsinghua University Press, Beijing (2010)
18. Ma, L., Xu, C., He, Z.: System detection technology based on multi-sensor information fusion. In: Third International Conference on Measuring Technology and Mechatronics Automation, Shanghai, China, pp. 625–628 (2011)

MPR Selection Based on Link Transmission Quality and Motion Similarity in OLSR for FANETs

Ziheng Li, Shuo Shi(✉), and Xuemai Gu

Harbin Institute of Technology, Harbin, China
liziheng@stu.hit.edu.cn, {crcss,guxuemai}@hit.edu.cn

Abstract. With the development and popularization of 5G networks and unmanned aerial vehicle (UAV) applications, UAV networks have received more and more attention. Routing protocols have always been a key technology in mobile ad hoc net-works (MANETs), especially in Flying Ad-hoc Networks (FANETs). The rapid changes in network topology brought about by the high-speed mobility of UAV nodes make the network performance more susceptible, which poses a greater challenge to FANETs routing technology. This paper mainly focuses on the optimization problem of OLSR protocol in FANETs under planar topology, considering the link transmission quality and link stability between nodes, uses a weighted index to replace the node connection degree as the MPR set selection criterion, an MPR set selection algorithm based on motion similarity and link transmission quality is proposed. Then in the NS2 emulator, a comparative analysis of packet delivery ratio, average end-to-end delay, and routing overheads are performed to verify the performance improvement of the proposed algorithm.

Keywords: FANETs · OLSR · Multi point relay · Mobility

1 Instruction

FANETs is an extended application of MANETs in the field of drones, usually consisting of a ground control station (GCS) and multiple UAV nodes. FANETs can be seen as a special kind of MANETs, has the commonalities of traditional MANETs, such as no center, multi hop, self-organization, and self-healing, but it also has its own application form, network structure and network characteristics. UAV nodes are aerial nodes, and their movement is not easily affected by terrain factors. Compared with MANETs, UAV nodes have faster speed and higher freedom of movement. The distance between UAV nodes is mostly far, and the density of UAV nodes in a certain airspace is relatively low. These problems have caused the link stability and link quality between nodes to be more easily affected, and the network topology changes more frequently.

As we all know, the fast and accurate data transmission mechanism between UAV nodes is an important part of FANETs. The quality of the communication protocol directly determines the network transmission performance of FANETs, which in turn

X. Wang et al. (Eds.): 6GN 2020, LNICST 337, pp. 56–65, 2020.
https://doi.org/10.1007/978-3-030-63941-9_5

affects the application effect of FANETs. However, due to the characteristics of fast movement of UAV nodes, low node density and limited energy, the routing protocols used in traditional MANETs and VANETs are usually not able to directly adapt to the highly changing scenarios of topology and routing information in FANETs. In turn, it affects the communication performance of the entire network. Therefore, a core research point in FANETs is the study of routing protocols. How to overcome the adverse effects of FANETs' high mobility, frequent topology changes, and overall energy constraints, and design an excellent one suitable routing protocols for FANETs is especially critical.

2 Related Works

2.1 OLSR Protocol

The optimized link state routing protocol, OLSR protocol [1], is a proactive routing protocol with a flat topology. Compared with the LSR protocol, it mainly has the following two characteristics: Using multi-point relay technology (MPR), only those nodes selected as MPR in the network can forward TC messages, which reduces the number of forwarding control packets and overhead. OLSR can distinguish the old and new routing messages through the sequence number ANSN field in the TC messages, and use this to maintain and update the topology of the entire self-organizing network.

Many researchers have conducted detailed research and improvement on the OLSR protocol. K. Singh [2] provides ideas for the evaluation method of FANETs routing protocol and optimizes for the OLSR protocol under different mobile models. Pardeep Kumar [3] proposed an AOLSR routing protocol that optimizes the MPR selection criterion based on the location of the destination node of the data to be sent. Yi Zheng [4] proposed an OLSR protocol based on mobility and load awareness, that is, the ML-OLSR protocol, and introduced a mobile sensing algorithm and a load sensing algorithm. De-gan Zhang [5] proposed an artificial neural network OLSR protocol QG-OLSR based on quantum genetics to optimize the selection of MPR sets.

2.2 Paparazzi Mobility Model (PPRZM)

This paper investigates the existing FANETs mobility models and finds that there are two main types: one is the modification and adjustment of the traditional MANETs, making it suitable for FANETs, such as Random Walk (RW) model [6], Random Way Point (RWP) model [7], Gauss-Markov (GM) model [8], etc. RWP model is used in most UAV ad hoc network simulation scenarios.

Although the above mobile models can largely reflect the mobility of drones, FANETs are mostly application-oriented, the mobility of flight nodes is often more diverse, and traditional models often cannot accurately reproduce their behavior. Therefore, the second type is specially designed for FANETs, with its unique constraints, such as Semi-random Circular Movement (SRCM) model [9], Paparazzi Mobility (PPRZM) model [10].

Paparazzi mobility model is designed based on the Paparazzi UAS. Similar to the RWP model, PPRZM is also a path-based random mobility model, but PPRZM has more node movement behaviors that are closer to FANETs. The distribution uniformity of UAV nodes, regional communication frequency, number of link connections and other parameters are more in line with the actual situation.

3 Optimization Algorithm of MPR Set

The MPR set selection method of the traditional OLSR protocol is based on the node degree. However, due to the high mobility of UAV nodes, the stability of the links between the nodes decreases and the nodes with high connectivity may quickly leave the communication range of other nodes, and the larger coverage of the FANETs also makes the link quality different. As a result, it is inappropriate to adopt a single node connection degree as the selection criterion of the MPR set.

3.1 The Central Idea of the Optimization Algorithm

In FANETs, the relative movement between each node can usually represent the link survival time and stability between the nodes, and it is easy to be obtained and calculated. According to the link evaluation index in article [11], the link transmission quality between nodes can be characterized by the Expected Transmission Count (*ETX*). Therefore, this paper will introduce the node motion similarity and the transmission quality of nodes links as the criterion for measuring the link performance between nodes and selecting MPR sets.

This paper assumes that each node can know its real-time position and speed information, in order to calculate the motion similarity between nodes. Motion similarity refers to how similar a UAV node is to another neighboring UAV node in moving speed. Generally speaking, the greater the degree of similarity in the movement behavior between nodes, these two nodes are considered to be better able to maintain the connection state. Because this paper mainly focuses on planar topology, the vertical velocity of the UAV is not considered.

Now suppose that there are two UAV nodes i and j, with speed $\mathbf{v_i}$ and $\mathbf{v_j}$ respectively, then referring to the mobility measure of Euclidean distance, the calculation of the motion similarity of these two nodes as shown:

$$\theta_{ij} = 1 - \frac{|\mathbf{v_i} - \mathbf{v_j}|}{|\mathbf{v_i}| + |\mathbf{v_j}|} \tag{1}$$

The link transmission quality between nodes is calculated by *ETX*. The smaller the value of *ETX*, the higher the link quality of network communication. *ETX* is obtained by calculating the values of link quality (*LQ*) and neighbor link quality (*NLQ*) according to the proportion of two nodes receiving HELLO messages sent by its neighbors within a certain period. Then the calculation method of *ETX* of a node is as follows:

$$ETX = \frac{1}{LQ \times NLQ} \tag{2}$$

However, after analysis, it is found that *LQ* and *NLQ* cannot be directly calculated, and furthermore, *ETX* cannot be calculated. Therefore, this paper defines Mutual Link

Quality (*MLQ*) as an intermediate quantity to calculate the *ETX* of two nodes. For local nodes, the *ETX* calculation method is as follows:

$$MLQ\_L = \frac{i \text{ received the number of HELLO sent by } j}{\text{Number of HELLO } i \text{ sent to } j}$$
$$MLQ = \frac{j \text{ received the number of HELLO sent by } i}{\text{Number of HELLO } j \text{ sent to } i} \tag{3}$$

$$ETX = \frac{1}{MLQ\_L \times MLQ} \tag{4}$$

Through further analysis, it is found that the range of node motion similarity is [0, 1], and the larger the value, the better the link stability, while the value range of *ETX* is $[1, +\infty]$, but the smaller value means that the link transmission quality is higher. This leads to the inconvenience of normalizing the two indexes and the weighted average, which affects the calculation of the optimal link performance index. Therefore, this paper defines a new link transmission quality index *R_ETX*, with value range [0, 1].

$$R\_ETX = MLQ\_L \times MLQ \tag{5}$$

Therefore, the two indicators of link stability between nodes and link transmission quality between nodes are comprehensively considered, and since MPR nodes are used as relay nodes to cover 2-hop neighbors, their transitivity and relay amplification must be considered. In this paper, the comprehensive link evaluation index *L* is used to replace the node degree in the standard MPR selection algorithm as the decision basis to optimize the MPR set selection method. The calculation formula of *L* is as follows:

$$L(y_i) = \alpha[\theta_{Ay_i}\, average(\theta_{y_i s_j})] + \beta[R\_ETX_{Ay_i}\, average(R\_ETX_{y_i s_j})] \tag{6}$$

Among them, *A* is the node that performs the MPR set calculation, y_i is the 1-hop neighbor node of node *A*, and s_j is the strictly symmetric 2-hop node reachable by y_i in node 2-hop neighbor set; average() represents the arithmetic average of the link indicators between node y_i and all strictly symmetric 2-hop neighbor nodes s_j reachable through this node; α and β are weighting coefficients, and satisfying $\alpha + \beta = 1$, which can be adjusted for different network emphasis directions.

3.2 Algorithm Implementation Process

In order to obtain the numerical values described in the formula and add them to the OLSR protocol, it is necessary to modify the frame format of the HELLO and the format of the local link set and neighbor set to add some new parameters.

Because this paper assumes that all UAV nodes are on the same horizontal plane, the node only needs to know the horizontal speed of its neighbor nodes to calculate the motion similarity between the nodes. The node also needs to know the *MLQ* of its neighbor node to calculate the *R_ETX* between the two nodes, thereby updating the node's local link set and neighbor set. In addition, the node needs to know the motion similarity and *R_ETX* between a certain neighbor node and the 2-hop neighbor node

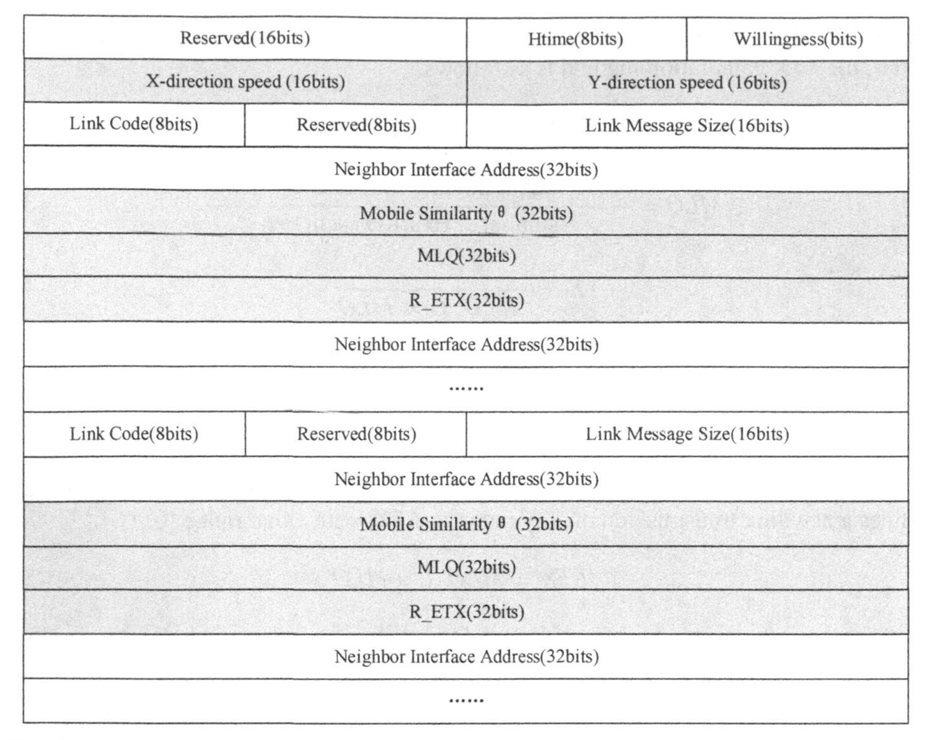

Fig. 1. Modified HELLO message format, which is added parameters such as node speed, movement similarity and *R_ETX*.

reachable via the HELLO to establish a 2-hop neighbor set. In summary, the frame format modification of the HELLO is shown in Fig. 1. Similarly, these parameters will also be added to the node local link set, neighbor set, and 2-hop neighbor set.

Based on the optimized packet format of the HELLO message and the table format of the neighbor information base, the improvement of the MPR set selection algorithm is implemented. Now suppose that the MPR set calculation is node *A*, and its MPR set is described by *M*; the set of 1-hop neighbor nodes of node *A* is *N*, and the set of 1-hop neighbor nodes is *y*; the set of 2-hop neighbor nodes s is *N2*, which not includes the 2-hop neighbor nodes that are reachable only through the nodes in set *N* whose forwarding intention is WILL_NEVER, the node *A* itself that performs MPR set calculation, and all 1-hop symmetric neighbor nodes of node *A*.

The process of optimizing the MPR set selection algorithm is as follows:

(1) First, add all nodes in node *A*'s 1-hop neighbor set *N* whose forwarding intention is WILL_ALWAYS, that is, the value of N_willingness is 7 to MPR set *M*;
(2) If there is a node in the set *N2* that is not covered by at least one node in the set M, then calculate the integrated link index $L(y_i)$ of each node y_i in the set *N*.
(3) Compare the link index $L(y_i)$ of each node in set *N*, and add the node with the largest $L(y_i)$ value to set *M*. If the value of $L(y_i)$ is the same, the node with the

largest N_willingness is selected to join the set *M*. Then, the nodes in the *N2* set that are covered by the nodes in the existing *M* set are removed.

(4) If the *N2* set is not empty, go back to step (2); if the *N2* set is empty, then set *M* is the final MPR set of node *A*.

4 Simulation and Performance Analysis

4.1 Simulation Setup and Metrics

In this paper we analyze OLSR, OLSR with L routing protocol using NS2 (Network Simulator 2) version 2.35 and the performance analysis is done using AWK script. In order to accurately simulate the movement scenes of FANETs, after analyzing the actual situation, the mobility model of UAV nodes uses PPRZM. For the simulated wireless propagation model, the two-path propagation model, the Rice fading model, can be used to more realistically simulate the actual fading of FANETs communication, corresponding to the Two Ray Ground model in NS2. The parameters of the simulation scenario are shown in the following Table 1.

Table 1. Simulation parameters

Parameters	Values
Simulation tools	NS2 (version 2.35)
Simulation duration	200 s
Mobility model	PPRZM
Propagation model	Two Ray Ground
MAC protocol	IEEE 802.11
Simulation area	3000 m × 3000 m
Number of nodes	50
Speed	0–30 m/s
Transmission range	400 m
Data type	CBR
Number of connections	10
Data packet size	512 Bytes

This paper focuses on three performance parameters: Average End-to-end Delay, Packet Delivery Ratio (PDR), and Routing Overheads. PDR is the ratio of the number of successfully received packets to the total number of packets sent during network data transmission. Average End-to-end Delay is the average time taken by the entire message to travel from source to destination. Routing Overheads is the ratio of the number of routing control messages sent through the network over the data messages received. However, since the main research object of this paper is OLSR protocol, the size of data packet in OLSR is uncertain. In addition, the packet format of Hello message is modified, which further leads to the change of routing control packet size. Therefore, in order to accurately measure the cost of routing control, this paper uses the ratio of the number of packets received by the destination node to the number of bits contained in the routing control packet as the routing overheads measurement standard, that is, the number of routing control packets required for successful transmission of one bit data packet.

4.2 Performance Analysis

According to the NS2 simulation parameters described above, the simulation scene is built, and the experimental results are obtained as follows. Figure 2 shows a comparison of PDR. Compared with the original OLSR protocol, OLSR after MPR set selection optimization has a certain degree of improvement in PDR. With the increase of the maximum speed of the nodes, the change of the network topology is further accelerated, the overall PDR shows a downward trend. However, the OLSR that uses node link stability and link transmission quality instead of node degree as the new MPR set selection criterion can better adapt to the rapid topology change scenario. Its performance is better and more stable at high node speed. The main reason is that the movement of the nodes is considered when selecting the MPR node, so that the relative mobility of the link is smaller and the link survival time is longer.

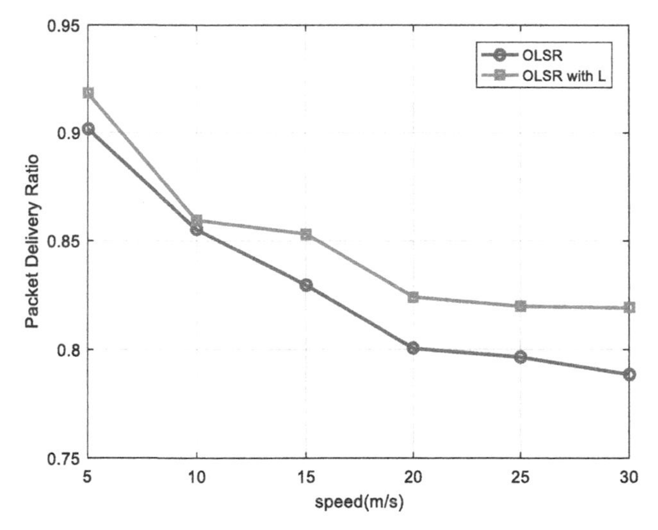

Fig. 2. PDR simulation results under the change of node movement speed

Figure 3 shows the average end-to-end delay simulation of the two routing protocols. From this figure, it can be seen that the average end-to-end delay of the optimized OLSR protocol is less than the original OLSR protocol. When the node speed is low, the algorithm proposed in this paper refers to the link transmission quality when selecting MPR nodes, so although the average delay performance of the two is not much different, it still has improvement. When the node speed is high, the average end-to-end delay performance improvement is more obvious. The consideration is that the selection algorithm of MPR set in this paper is more stable, which making the communication link more stable, so as to reduce the delay.

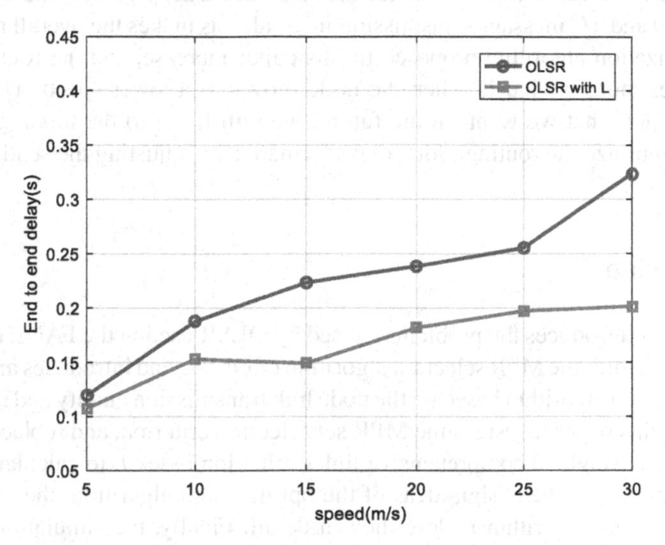

Fig. 3. Average end-to-end delay simulation results under the change of node movement speed

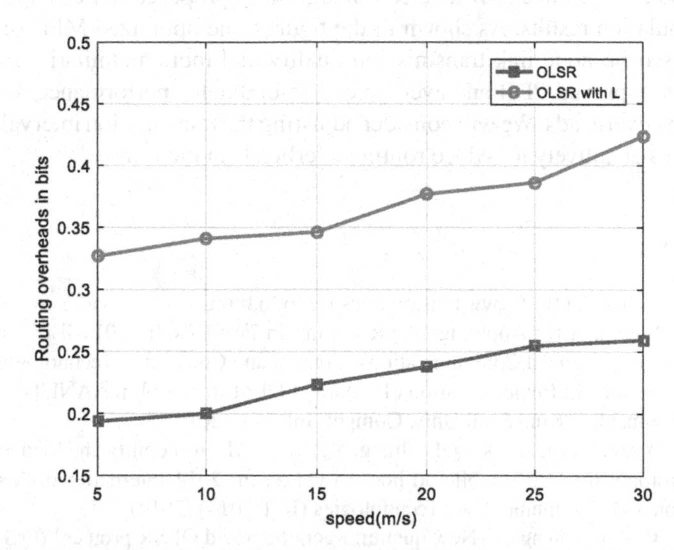

Fig. 4. Routing overheads simulation results under the change of node movement speed

As shown in Fig. 4, it shows the performance comparison of the two routing protocols in terms of routing control overhead. The MPR selection optimization algorithm based on motion similarity and link transmission quality can effectively reduce the frequency of MPR selector set member changes and reduce the frequency of TC message transmission during route maintenance, so that the cost of route control has been reduced to a certain extent. However, due to the modification of the HELLO packet format in this paper, the size of the HELLO has more than doubled the original OLSR protocol, and OLSR uses a fixed HELLO and TC message transmission interval, this makes the overall routing cost of the optimization algorithm proposed in this paper increase, and the routing control cost increases more obviously when the node moves at a lower speed. This problem is obviously not what we want. In the future, we will listen to the topology changes, and further optimize the routing protocol performance by adjusting the sending interval adaptively.

5 Conclusion

This paper first introduces the problems caused by OLSR used in the FANETs scenario, and secondly studies the MPR selection algorithm of OLSR, and introduces an optimized MPR set selection algorithm based on the node link transmission quality and motion similarity. Then, this paper redesigns the MPR set selection criterion, and replaces the node degree with the weighted comprehensive link evaluation index L to calculate the MPR set. After introducing the design ideas of the optimization algorithm, the implementation process of the algorithm is described in detail. Finally, the simulation compares the three performances of the average end-to-end delay, PDR, and routing overheads of the OLSR and the optimized MPR selection algorithm proposed in this paper, and analyzes the simulation results. As shown in the figures, the optimized MPR set selection algorithm based on node link transmission quality and motion similarity proposed in this paper has improved PDR and average end-to-end delay performance, but it brings greater routing overheads. We will consider adjusting the transmission interval of routing control packets adaptively to reduce routing overheads in the future.

References

1. OLSR [OL] (2003). http://www.faqs.org/rfcs/rfc3626.htmL
2. Singh, K., Verma, A.K.: Applying OLSR routing in FANETs. In: 2014 IEEE International Conference on Advanced Communications, Control and Computing Technologies (2014)
3. Kumar, P., Verma, S.: Implementation of modified OLSR protocol in AANETs for UDP and TCP environment. J. King Saud Univ. Comput. Inf. Sci. (2019)
4. Zheng, Y., Wang, Y., Li, Z., Dong, L., Jiang, Y., Zhang, H.: A mobility and load aware OLSR routing protocol for UAV mobile ad-hoc networks. In: 2014 International Conference on Information and Communications Technologies (ICT 2014) (2014)
5. Zhang, D., Cui, Y., Zhang, T.: New quantum-genetic based OLSR protocol (QG-OLSR) for mobile ad hoc network. Appl. Soft Comput. J. **80**, 285–296 (2019)
6. Manimegalai, T., Jayakumar, C.: A conceptual study on mobility models in MANET. Int. J. Eng. Res. Technol. (IJERT) **2**(11), 3593–3598 (2013)
7. Bettstetter, C., Hartenstein, H., Pérez-Costa, X.: Stochastic properties of the random waypoint mobility model. Wirel. Netw. **10**(5), 555–567 (2004)

8. Liang, B., Haas, Z.J.: Predictive distance-based mobility management for PCS networks. In: Proceedings of the Eighteenth Annual Joint Conference of the IEEE Computer and Communications Societies, INFOCOM 1999, vol. 3, pp. 1377–1384. IEEE (1999)
9. Wang, W., Guan, X., Wang, B., Wang, Y.: A novel mobility model based on semi-randomcircular movement in mobile ad hoc networks. Inf. Sci. **180**(3), 399–413 (2010)
10. Bouachir, O., Abrassart, A., Garcia, F., Larrieu, N.: A mobility model for UAV ad hoc network. In: 2014 International Conference on Unmanned Aircraft Systems (ICUAS), pp. 383–388. IEEE (2014)
11. Mohapatra, S., Tripathy, T.: MM-OLSR: Multi metric based optimized link state routing protocol for wireless ad-hoc network. In: Signal Processing, Communication. Power and Embedded System (SCOPES), pp. 153–158. IEEE (2016)

Intelligent Resource Management

Deep Reinforcement Learning-Based Joint Task Offloading and Radio Resource Allocation for Platoon-Assisted Vehicular Edge Computing

Yi Chen[1(✉)], Xinyu Hu[2], Haoye Chai[2], Ke Zhang[2], Fan Wu[2], and Lisha Gu[1]

[1] Research Institute of Highway Ministry of Transport, Mail Address No. 8 Xitucheng Road Haidian District, Beijing, China
{yi.chen,ls.gu}@rioh.cn

[2] School of Information and Communication Engineering, University of Electronic Science and Technology of China, Chengdu, China
huxinyu97@163.com, haoyechai@163.com, {zhangke,wufan}@uestc.edu.cn

Abstract. Platoons, formed by smart vehicles driving in the same patterns, bring potential benefits to road traffic efficiency while providing a promising paradigm to execute computation tasks with onboard computing resources. However, constrained resources of individual vehicles (IV), limited wireless coverage of vehicular communication nodes as well as high mobility of running platoons pose critical challenges on task scheduling and resource management. To address these challenges, we propose a platoon-based vehicular edge computing mechanism, which exploits computation capabilities of both platoons and edge computing enabled Roadside Units (RSUs), and jointly optimizes task offloading target selection and resource allocation. Taking aim at minimize delay cost and energy consumption of the platoon-based task execution, we leverage deep deterministic policy gradient (DDPG) to design a learning algorithm, which efficiently determines target computation servers and obtains optimized resource scheduling strategies. Numerical results demonstrate that our algorithm significantly reduces delay and energy costs in comparing its performance to that of benchmark schemes.

Keywords: Vehicular edge computing · Task offloading · Resource allocation · Platoon

1 Introduction

The Internet of vehicles (IoV) is a vital application of the Internet of things (IoT) in the automotive industry and is regarded as the information foundation for the next generation of intelligent transportation system with great potential [1]. To

Supported by organization x.

X. Wang et al. (Eds.): 6GN 2020, LNICST 337, pp. 69–85, 2020.
https://doi.org/10.1007/978-3-030-63941-9_6

reduce the traffic accident rate, as well as improve traffic efficiency and traveling convenience, the amount of intelligent transportation applications such as automatic driving, intelligent auxiliary driving for vehicles have been increasing consistently. However, these emerging applications not only require high computational complexity but also have strict delay sensitivity [2]. The current limited computing capability and storage capacity of the on-board equipment may not fully meet the requirements. On the other hand, the vehicular edge computing (VEC) [3] which combines edge computation and vehicle networks is widely considered as a promising approach to handle the computation-intensive tasks. In this case, the computation-intensive tasks of the vehicles can be offloaded to edge severs in proximity to them instead of the remote cloud servers, and thus the processing delay of tasks will be reduced significantly.

Compared with mobile edge computing (MEC), VEC has more challenge due to the high mobility and distributed nature of vehicles [4]. The author of [5] introduced a software-defined vehicular edge computing (SD-VEC) architecture where a controller guides both the vehicles task offloading strategy and the edge cloud resource allocation strategy. They devised a mobility-aware greedy algorithm (MGA) that determines the amount of edge cloud resources allocated to each vehicle. [6] studied the task offloading problem from a matching perspective based on three vehicular mobility models to simulate the movement of vehicles. To minimize the network delay, they proposed a pricing-based one-to-one matching algorithm and pricing-based one-to-many matching algorithms for the task offloading. However, all the mentioned works ignore the fact that compared with the smart mobile phones or tablets, vehicles have much powerful computing capacity, and the aggregate computing capacity will grows with the number of vehicles. Therefore, if the vehicles can be utilized to provide task offloading services, the computing performance of the vehicular networks will be subsequently improve.

Considering how to utilize the computing ability of vehicles, [7] introduced the federated offloading of vehicle-to-infrastructure (V2I) and vehicle-to-vehicle (V2V) communication in MEC-enabled vehicular networks. They aimed at minimizing the total latency for the moving vehicles. In [8], a computation offloading method named V2X-COM is proposed, which employs vehicle-to-everything (V2X) technology for data transmission in edge computing. Non-dominated sorting genetic algorithm III (NSGA-III) is adopted to generate balanced offloading strategies. The research above address to utilize the excess computing capacity of a single vehicle. As a matter of fact, facing the huge computing requirements for IoV, the computing capacity of a single vehicle is very limited. We have to develop an approach to aggregate the computing capacity of multiple vehicles to fulfill task offloading requirements. So how to centralize the use of multiple vehicles is still a subject to be studied.

With the rapid development of artificial intelligence, more and more researches apply reinforcement learning to solve the problem of offloading. [9] provided two novel approaches termed as distributed deep deterministic policy gradient (DDDPG) and sharing deep deterministic policy gradient (SDDPG) based on DDPG algorithm, solving the multi-agent learning and non-cooperative power allocation problem in D2D-based V2V communications. The author of [10] formulated the offloading decision as a resource scheduling problem with single or multiple objective function and constraints, proposing a knowledge driven (KD) service offloading framework by exploring the deep reinforcement learning (DRL) model to find the long-term optimal service offloading policy. However, to our best knowledge, there is few research to discuss how to apply the reinforcement learning algorithms to the computing resource management for platoon.

Inspired by the above observations, we exploit a platoon-based VEC to offload part of computing load to platoons. In this case, an IV with requirements can select a proper offloading server, i.e., a RSU with MEC server or a platoon, according to its computation capability and its continuous wireless connection duration with the IV. In addition, in order to develop the computation capability of a platoon, the leader of the platoon is responsible for unified management of the radio resource owned by this platoon and the computing resource distributed across each vehicle within the platoon. To this end, the tasks offloaded to the platoon will be further divided into series of subtasks with different computational complexity and assigned to the platoon members associated with different proportions of the radio resource by the leader of the platoon.

In this paper, we jointly optimize the task offloading selection, inner-platoon radio resource and computation resource allocation to minimize latency and energy consumption in the platoon-based VEC. In summary, the main contributions of this paper include:

1) We devise a platoon-based VEC system model for the tasks offloaded from multiple IVs to reduce system task-offloading delay and the energy consumption. Many practical conditions and constraints for the vehicular networks are considered in our proposed model, especially including the duration of communication connection between the vehicle and a platoon or an RSU, the radio resource limitation of each platoon, and the difference of computing capability of each platoon member.
2) We formulate an optimization problem for joint task offloading target selection, inner-platoon radio resource and computing resource allocation with the objective of minimizing the overall system average offloading cost, which is defined as the weighted sum of average task-offloading delay and energy consumption.
3) Since the formulated problem is a well-known NP-hard problem (i.e., maximum cardinality bin packing problem [11]), and we have no priori-knowledge of arrival distribution of computation tasks, channel state and vehicles locations in practical. We propose a DDPG-based joint task offloading decision algorithm by integrating deep neural networks and reinforcement learning approaches.

The remainder of this paper is organized as follows. Section 2 gives the system model. In Sect. 3, we introduce a DDPG-based approach for task offloading and resource allocation. Simulation results and discussions are given in Sect. 4. Finally, we conclude this paper and propose some future works in Sect. 5.

2 System Model

In this section, we first describe the multi-lane scenario and present the system model for platoon-based VEC, including network model, communication model and computation model. Then the problem of jointly optimizing the offloading and resource allocation decision is formulated in details.

2.1 Network Model

As illustrated in Fig. 1, we consider a vehicular network consisting of m RSUs and n platoons and v IVs that do not belong to any platoon. A platoon consists of a group of vehicles travelling in the same lane and maintaining constant relative velocity. Vehicles belonging to the same platoon referred to as platoon members and each platoon has a platoon leader, which is responsible for the resource allocation within the platoon. The RSU provides computation resources (e.g. CPU cycles per second) for vehicles within its radio coverage. Let $\mathcal{S}erver = \left\{ \underbrace{S_1, S_2, ..., S_m}_{RSU}, \underbrace{S_{m+1}, S_{m+2}, ..., S_{m+n}}_{Platoon} \right\}$ denote the set of servers. The coordinates of of server i at time slot t is (a_i^t, b_i^t), a_i^t is the horizontal coordinates while b_i^t is the vertical coordinate. As for platoon, (a_i^t, b_i^t) is the coordinates of leader. The computational capability of each server is f_i^t. The platoon size and platoon length for platoon i is N_i and L_i, respectively. The j-th vehicle in platoon i is denoted by V_i^j and the leader is V_i^0. The computational capability of V_i^j at time slot t is $f_{i,j}^t$.

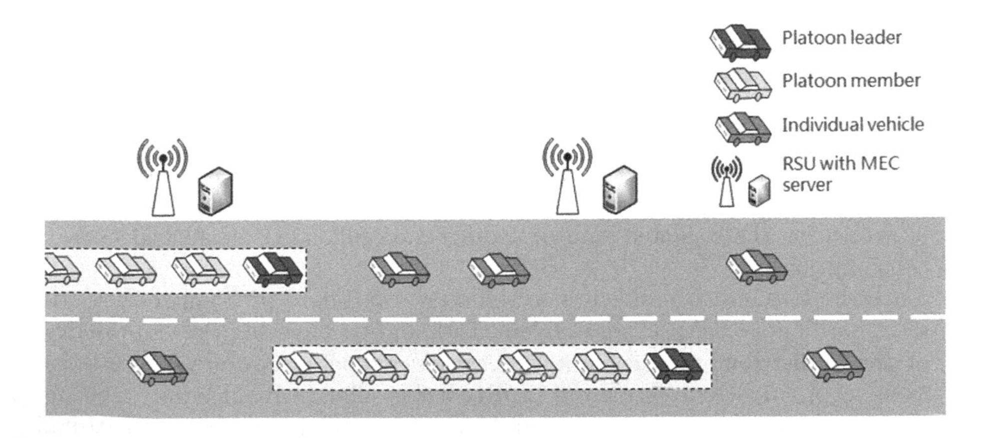

Fig. 1. The multi-lane scenario.

We consider the task offloading procedure for IVs. We assume that the vehicles running on the road follows a Poisson process with rate λ per unit time interval. At each time slot there are $U(t)$ vehicles which have to offload, defined as $\mathcal{U}^t = \left\{V_1^t, V_2^t, ..., V_{U(t)}^t\right\}, U(t) < v$. The coordinate of the vehicle u at the time slot t is (a_u^t, b_u^t). Suppose the vehicle's velocity in each time slot remains unchanged, we use v_u^t to represent the velocity of V_u^t and use $v_i^{t,P}$ to represent the velocity of platoon i. The velocity is assumed to follow constrained Gaussian distribution independently and identically with probability distribution function (PDF)[12]

$$\hat{f}(v) = \frac{1}{\sqrt{2\pi}\sigma} e^{-\frac{(v-\mu)^2}{2\sigma^2}}, v \in [v_{\min}, v_{\max}], \tag{1}$$

where $v_{\min}$ and $v_{\max}$ are lower and upper bound of velocity, respectively; μ and σ respectively represent mean and standard deviation of velocity.

Let r_i^L and r_i^{Mem} denote the transmission ranges of the platoon leaders and members, respectively. Assume that in each platoon the member's transmission range is the same, and $r_i^{Mem} < r_i^L$. Platoon leaders are trucks with higher-placed antennas in order to cover all the members. Since the transmitting power of each RSU is different, the wireless coverage range of each RSU is different. Let r_i denote the coverage range of RSU i.

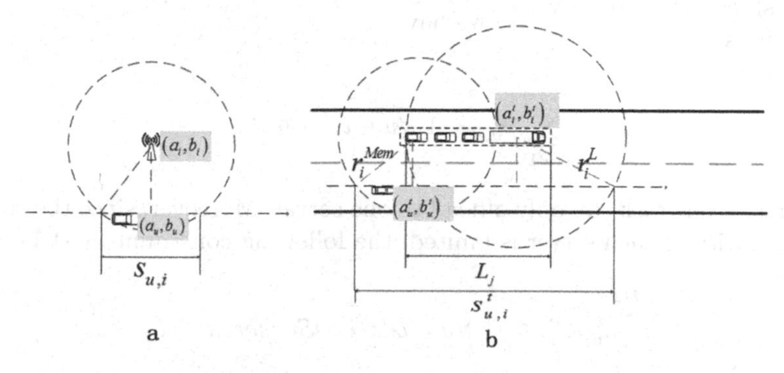

Fig. 2. A topology of the vehicle within the range of the RSU and the platoon.

As shown in Fig. 2a, at time slot t, the distance traveled by a vehicle within the coverage of RSU i is $s_{u,i}^t = 2\sqrt{r_i^2 - (b_u^t - b_i^t)^2}$. The duration of IV staying within the coverage is

$$t_{u,i,t}^{stay} = \begin{cases} \frac{\frac{s_{u,i}^t}{2} + (a_i^t - a_u^t)}{v_u^t}, & |a_i^t - a_u^t| \le \frac{s_{u,i}^t}{2} \\ 0, & |a_i^t - a_u^t| > \frac{s_{u,i}^t}{2} \end{cases}. \tag{2}$$

As shown in Fig. 2b, at time slot t, the distance traveled by a vehicle within the coverage of platoon i is $s_{u,i}^t = L_i + 2\left(\sqrt{(r_i^L)^2 - (b_u^t - b_i^t)^2} + \sqrt{(r_i^{Mem})^2 - (b_u^t - b_i^t)^2}\right)$, where $v_{rel} = \left|v_i^{t,P} - v_u^t\right|$ is the relative velocity. The duration of IV staying within the coverage is

$$t_{x,i,t}^{stay} = \begin{cases} 0, & |a_i^t - a_u^t| > \sqrt{(r_i^L)^2 - (b_u^t - b_i^t)^2} \,\&\, |a_i - a_x| < L + \sqrt{(r_i^{Mem})^2 - (b_u^t - b_i^t)^2} \\ \frac{s_{u,i} - \left(a_i^t - a_u^t + \sqrt{(r_i^L)^2 - (b_u^t - b_i^t)^2}\right)}{v_{rel}}, & v_i^{t,P} \geq v_u^t \\ \frac{a_i^t - a_u^t + \sqrt{(r_i^L)^2 - (b_u^t - b_i^t)^2}}{v_{rel}}, & v_i^{t,P} < v_u^t \end{cases}. \quad (3)$$

We assume that when a vehicle enters the communication range of the platoon, it is deemed that it can establish a connection with the leader.

2.2 Computation Model

We focus on the widely used task model $Q_u^t = \{I_u^t, C_u^t\}$, where I_u^t and C_u^t stand for the size of computation input data and the total number of CPU cycles needed to accomplish the task, respectively. In this paper, we consider these two parts can be split proportionally.

In general, a computation task can be executed at an RSU or a platoon. Let $\rho_{u,i}^t \in (0,1)$ denote the offloading decision, $\rho_{u,i}^t = \begin{cases} 1, & \text{task } u \text{ offloads to server } i \\ 0, & \text{others} \end{cases}$, we have

$$\sum_{i=1}^{m+n} \rho_{u,i}^t = 1, \forall u \in \mathcal{U}^t, t \in \mathcal{T}, \quad (4)$$

which means one task can only offload to one server. Moreover, since the computation capacity of each server is limited, the following constraint must be held,

$$\sum_{u=1}^{u(t)} \rho_{u,i}^t C_u^t \leq f_i^t, \forall u \in \mathcal{U}^t,\ i \in \mathcal{S}erver,\ t \in \mathcal{T}. \quad (5)$$

Let $\theta_{u,i,j}^t \in (0,1)$ denote the task allocation proportion for the j-th vehicle in platoon i when offloading task u, and the corresponding offloading decision of task u is denoted by $\bar{\theta}_{u,i}^t = \left\{\theta_{u,i,1}^t, \theta_{u,i,2}^t, ..., \theta_{u,i,N_i}^t\right\}$,

$$\sum_{j=1}^{N_i} \theta_{u,i,j}^t = 1, \forall u \in \mathcal{U}^t, t \in \mathcal{T}, i \in [m+1, m+n] \quad (6)$$

Also, the computation capacity of each vehicle is limited,

$$\sum_{u=1}^{u(t)} \rho_{u,i}^t \theta_{u,i,j}^t C_u^t \leq f_{i,j}^t, \forall u \in \mathcal{U}^t, i \in [m+1, m+n], j \in \{1, 2, ..., N_i\}, t \in \mathcal{T}. \quad (7)$$

After time slot t, the computation capacity of each vehicle and each RSU becomes

$$\begin{cases} f_{i,j}^{t} = f_{i,j}^{t-1} - \sum\limits_{u=1}^{u_{t-1}} \sum\limits_{j=1}^{N_i} \varphi_{u,i}^{t} \rho_{u,i}^{t} \theta_{u,i,j}^{t} C_u^{t} \\ f_i^{t} = f_i^{t-1} - \sum\limits_{u=1}^{u_{t-1}} \sum\limits_{j=1}^{N_i} \varphi_{u,i}^{t} \rho_{u,i}^{t} C_u^{t} \end{cases}, \tag{8}$$

where $\varphi_{u,i}^{t} = 1$ indicates that task u offloaded to server i can not be completed in this time slot.

2.3 Communication Model

We consider that one IV only accesses to one server in a time slot for the data transmission. Let $\left|h_{u,i}^{t}\right|^2$ and $d_{u,i}^{t}$ denote the coefficient of the effective channel power gain and the distance from vehicle u to server i, respectively. In this paper, we consider the scenario that the users move very slowly during the data offloading, so $\left|h_{u,i}^{t}\right|^2$ can be seen as a constant in a time slot and can change over different time slots (i.e., block fading channel). Hence, the corresponding channel power gain can be given as $\left|h_{u,i}^{t}\right|^2 = G \cdot \left(d_{u,i}^{t}\right)^{-\alpha} \cdot \left|h_0\right|^2$, where G is the power gain constant introduced for the amplifier and antenna, and $h_0 \sim CN(0,1)$ represents the complex Gaussian variable that represents Rayleigh fading [13].

Without loss of generality, $d_{u,i}^{t}$ denotes the distance between the server (the RSU center or the vehicle leader) and vehicle at the beginning of each time slot and $d_{u,i}^{t}$ is $d_{u,i}^{t}=\sqrt{\left(a_i^t - a_u^t\right)^2 + \left(b_u^t - b_i^t\right)^2}$. After a time slot the distance changes to

$$d_{u,i}^{t+1} = \begin{cases} \sqrt{\left(a_i^t - a_u^t - v_u^t \Delta t\right)^2 + \left(b_u^t - b_i^t\right)^2}, & i \in (1, m) \\ \sqrt{\left(a_i^t - a_u^t - v_{rel} \Delta t\right)^2 + \left(b_u^t - b_i^t\right)^2}, & i \in (m+1, m+n) \end{cases}. \tag{9}$$

The signal noise ratio (SNR) of task u transmission from V_u^t to server i at time slot t is expressed as

$$\gamma_{u,i}^{t} = \frac{P_u \left|h_{u,i}^{t}\right|^2}{\sigma^2}. \tag{10}$$

where P_u, σ^2 are the uplink transmit power of vehicle u, the noise power, respectively. The uplink data rate of a vehicle that chooses to offload its task to the server via a wireless link can be expressed as

$$R_{u,i}^{t}=B\log_2\left(1 + \gamma_{u,i}^{t}\right), \tag{11}$$

where B is the available spectrum bandwidth.

On the occasion of a vehicle decide to offload the task to a platoon, the task will first be offloaded to the leader, then the leader completes the secondary offloading of the task according to the characteristics of the platoon and the allocation of computing and communication resources in the platoon.

Let $\omega_{u,i,j}^{t} \in (0,1)$ denote the spectrum allocation proportion for the j-th vehicle in platoon i when transferring task u, and the corresponding offloading decision of task u is denoted by $\bar{\omega}_{u,i}^{t} = \{\omega_{u,i,1}^{t}, \omega_{u,i,2}^{t}, ..., \omega_{u,i,N_i}^{t}\}$,

$$\sum_{j=1}^{N_i} \omega_{u,i,j}^{t} = 1, \forall u \in \mathcal{U}^t, i \in \mathcal{Server}, t \in \mathcal{T}. \tag{12}$$

According to the above, when the leader sends task u to v_i^j, the SNR can be expressed as

$$\gamma_{u,i,j}^{t} = \frac{P_i^L \left|h_{u,i,j}^{t}\right|^2}{\sigma^2}. \tag{13}$$

The data rate can be expressed as

$$R_{u,i,j}^{t} = \omega_{i,j}^{t} B \log_2\left(1 + \gamma_{u,i,j}^{t}\right), \tag{14}$$

where P_i^L, $\left|h_{u,i,j}^{t}\right|^2$ and B are the transmit power of the leader of platoon i, the effective channel power gain from leader to V_i^j and the available spectrum bandwidth, respectively.

2.4 Problem Formulation

In this subsection, the optimal problem formulation will be described in detail. The server needs to make a joint optimization of the offloading proportion, communication resource and computation resource allocation. For each task generated by a vehicle, it can either be computed by a platoon or an RSU. According to different computation and resource allocation strategies, the latency and energy consumption of the task may be different.

For in-vehicle applications, latency must be taken into consideration seriously. On the other hand, due to the limited battery of the vehicles energy consumption should also be taken into account. However, minimize both energy consumption and the latency are conflicting. For example, the vehicle can save the energy by setting the lowest frequency all the time, but this will certainly increase the computing time. Therefore, we consider a trade-off analysis between the energy consumption and the execution delay for the offloading decision. Next we will introduce the formulas for calculating energy consumption and time delay.

When the task is offloaded to RSU, the communication time between the vehicle task generator and the RSU can be expressed by

$$T_{u,i}^{t,R,com} = \frac{I_u^t}{B \log_2\left(1 + \gamma_{u,i}^{t,R}\right)}. \tag{15}$$

The computation execution time of task u completed by RSU can be expressed as

$$T_{u,i}^{t,R,cmp} = \frac{C_u^t}{f_i^t}. \tag{16}$$

Then, the communication and computation energy consumption of task u by accomplishing at RSU can be calculated as

$$E_{u,i}^{t,R,com} = P_u T_{u,i}^{t,R,com}. \tag{17}$$

$$E_{u,i}^{t,R,cmp} = P_i^{t,R} T_{u,i}^{t,R,cmp}, \tag{18}$$

respectively, where $P_i^{t,R}$ indicate the computation power of RSU i.

When the task choose to offload at platoon, the communication time between the vehicle which has task and the leader is defined as follows,

$$T_{u,i}^{t,P,com1}(\omega) = \frac{I_u^t}{B\log_2\left(1+\gamma_{u,i}^{t,P}\right)}. \tag{19}$$

And the communication time within the platoon, i.e. the transmission between the leader and the member, can be expressed as

$$T_{u,i,j}^{t,P,com2}(\theta,\omega) = \frac{\theta_{u,i,j}^t I_u^t}{\omega_{u,i,j}^t B\log_2\left(1+\gamma_{u,i,j}^t\right)}. \tag{20}$$

The computation execution time of task u completed by each vehicle assigned to the task is

$$T_{u,i,j}^{t,P,cmp}(\theta) = \frac{\theta_{u,i,j}^t C_u^t}{f_{i,j}^t}. \tag{21}$$

Then, the communication energy consumption of task offloaded to the leader can be calculated as

$$E_{u,i}^{t,R,com1}(\omega) = P_u T_{u,i}^{t,P,com1}(\omega), \tag{22}$$

and the intra-platoon communication and computation energy consumption are

$$E_{u,i,j}^{t,R,com2}(\theta,\omega) = P_i^L T_{u,i,j}^{t,P,com2}(\theta,\omega), \tag{23}$$

$$E_{u,i,j}^{t,P,cmp}(\theta,\omega) = P_{i,j}^t T_{u,i,j}^{t,P,cmp}(\theta) \tag{24}$$

respectively, where $P_{i,j}^t$ indicates the computation power of member j in platoon i.

Generally, the amount of data returned are very small, so the time delay of the results back to the vehicle can be ignored.

The total energy consumption and time delay are shown to be

$$E_{u,i}^t = \begin{cases} E_{u,i}^{t,R,com} + E_{u,i}^{t,R,cmp}, & i \in (1,m) \\ E_{u,i}^{t,R,com1} + \sum\limits_{j=1}^{N_i}\left[E_{u,i,j}^{t,R,com2} + E_{u,i,j}^{t,P,cmp}\right], & i \in (m+1, m+n) \end{cases}, \tag{25}$$

$$D_{u,i}^{t}=\begin{cases}T_{u,i}^{t,R,com}+T_{u,i}^{t,R,cmp}+\varphi_{u,i}^{t-1}T_{wait}, & i\in(1,m)\\ T_{u,i}^{t,P,com1}+\max\limits_{j=1}^{N_i}\left[T_{u,i,j}^{t,P,com2}+T_{u,i,j}^{t,P,cmp}\right]+\varphi_{u,i,j}^{t-1}T_{wait}, & i\in(m+1,m+n)\end{cases}, \tag{26}$$

If the task can't be completed in this time slot, as for RSU, $\varphi_{u,i}^{t}=1$ means $T_{u,i}^{t,R,cmp}>\Delta t-T_{u,i}^{t,R,com}$. As for platoon, $\varphi_{u,i,j}^{t}=1$ when $T_{u,i}^{t,P,cmp}>\Delta t-T_{u,i}^{t,P,com1}-T_{u,i}^{t,P,com2}$. So the time delay to complete the remaining tasks in the previous slot is

$$T_{wait}=\left(C_{u}^{t-1}-f_{u,i}^{t-1}T_{u,i}^{t-1,R,cmp}\right)/f_{u,i}^{t}, \tag{27}$$

$$T_{wait}=\left(\theta_{u,i,j}^{t-1}C_{u}^{t-1}-f_{u,i}^{t-1}T_{u,i}^{t-1,P,cmp}\right)/f_{u,i}^{t}, \tag{28}$$

on the occasion of offloading to RSU and platoon respectively.

Then we consider an optimization problem about the offloading decision, communication resource and computation resource allocation. The aim is to provide optimal computation offloading decision $\rho_{u,i}^{t}$, task allocation proportion $\bar{\theta}_{u,i}^{t}$ and spectrum allocation proportion $\bar{\omega}_{u,i}^{t}$ for all vehicles such that the energy consumption and the maximal task completion time is minimized. The corresponding optimization problem can be formulated as follows,

$$\min_{\rho_{u,i}^{t},\theta_{u,i,j}^{t},\omega_{u,i,j}^{t}}\ \frac{1}{T}\sum_{t=1}^{T}\sum_{u=1}^{u(t)}\sum_{i=1}^{m+n}\left\{\rho_{x,i}^{t}\left[\alpha E_{u,i}^{t}(\theta,\omega)+\beta\min\left[D_{u,i}^{t}(\theta,\omega),\Delta t\right]\right]\right\} \tag{29}$$

$$\begin{aligned}s.t.\ & C1:(4)\\ & C2:(6)\\ & C3:(5),(7)\\ & C4:(12)\\ & C5:\left\lceil\theta_{u,i,j}^{t}\right\rceil\odot\left\lceil\omega_{u,i,j}^{t}\right\rceil=1,\forall u\in\mathcal{U}^{t},\ i\in\mathcal{S},\ j\in\{1,2,...,N_i\},t\in\mathcal{T}\\ & C6:P_u\ge P_{th},\rho_{u,i}^{t}P_i^{L}\ge P_{th},\forall u\in\mathcal{U}^{t},i\in\mathcal{S},t\in\mathcal{T}\\ & C7:\gamma_{u,i}^{t}\ge\gamma_{th},\rho_{u,i}^{t}\theta_{u,i,j}^{t}\gamma_{u,i,j}^{t}\ge\gamma_{th},\forall u\in\mathcal{U}^{t},i\in\mathcal{S},j\in\{1,2,...,N_i\},t\in\mathcal{T}.\end{aligned}$$

where α, β are weighting factor.

Constraint C5 means that only platoon members with tasks will have the allocated spectrum. Constraint C6 ensures that the transmission power of each vehicle is lower than the threshold P_{th}. C7 is the SNR requirement for successful transmission.

3 Deep Reinforcement Learning for Task Offloading and Resource Allocation

Deep reinforcement learning combines the perception ability of deep learning with the decision-making ability of reinforcement learning, which can be controlled directly according to the input information. In our study, we used DDPG to learn optimal task offloading and resource allocation strategies.

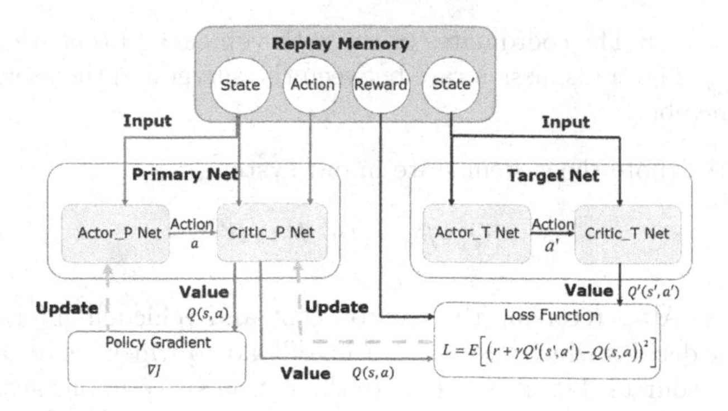

Fig. 3. The framework of deep deterministic policy gradient learning algorithm.

3.1 DDPG-Based Learning Model

The DDPG algorithm is a combination of Deterministic Policy-Gradient [14] Algorithms, the Actor-Critic [15] Methods, and the Deep Q-Network (DQN). It follows the target network and experience replay technology in the DQN algorithm, using two deep-Q Networks in the algorithm: one is Actor network used to approximate the policy function and the other is Critic network used to approximate the value function. The Actor realizes the output of continuous action values, and the Critic evaluates the execution effect of the action. The framework of DDPG algorithm is demonstrated in Fig. 3.

In reinforcement learning, agents find optimal strategies through interaction with the environment and trial-and-error learning. States, actions and rewards are three key factors of reinforcement learning and we formulate the task as a Markov decision process (MDP). We define a 4-tuple $\langle \mathcal{S}, \mathcal{A}, \mathcal{R}, \mathcal{P} \rangle$, where they represent the agent state space, action space, reward function $r = \mathcal{R}(s, a, s')$ and the transition function $\mathcal{P}^a_{ss'}$, respectively. And $s, s' \in \mathcal{S}$, $a \in \mathcal{A}$, $\mathcal{P}^a_{ss'}$ is the probability of a transition from state s to state s' when taking action a.

Apply it to the task offloading and resource allocation problem, we define that each IV and each platoon is an agent, but the decisions are made from a central server. Generally, an agent observes a state $s^t \in \mathcal{S}$ at each time t, then accordingly takes action $a^t \in \mathcal{A}$ based on the policy π. By taking the action a^t, the agent receives a reward r^t and the environment transits to the next state s^{t+1}. Then we define the state spaces, action spaces and reward function of the joint optimization of task offloading and resource allocation as follows:

State Spaces: At the beginning of each time period, the agent obtains environmental information and vehicular information. Specifically, the system state space contains:

- $Q^t_u = \{I^t_u, C^t_u\}$: Current information of task u.
- f^t_i, $f^t_{i,j}$: The maximum computation capability of the i-th server and the j-th vehicle in platoon i.
- v^t_u, $v^{t,P}_i$: The velocity of the u-th vehicle and the i-th platoon.

- (a_u^t, b_u^t), (a_i^t, b_i^t): The coordinates of the u-th vehicle and the i-th platoon.
- $R_{u,i}^t$, $R_{u,i,j}^t$: The transmission rate between the server and the user, the leader and the member.

Let $s^t \in \mathcal{S}$ denote the system state in our system, i.e.,

$$s^t = \left[Q_u^t, f^t, v^t, \left(a^t, b^t\right), R^t\right]. \tag{30}$$

Action Spaces: After receiving the environment and vehicular information, the central server decides which task should be offload to which server and how to allocate the resources. Let $a^t \in \mathcal{A}$ denote the action space in our system, i.e.,

$$a^t = \left[\rho_{u,i}^t, \bar{\theta}_{u,i}^t, \bar{\omega}_{u,i}^t\right] \tag{31}$$

Reward Function: Our objective is to minimize the total delay and energy consumption of the network by interacting with the environments. Thus, we design a reward function $\mathcal{R}_t$ to get immediate return by executing action a^t as

$$\mathcal{R}_t = \sum_{u=1}^{u(t)} \sum_{i=1}^{m+n} \left\{\rho_{x,i}^t \left[\alpha E_{u,i}^t\left(\theta, \omega\right) + \beta \min\left[D_{u,i}^t\left(\theta, \omega\right), \Delta t\right]\right]\right\} \tag{32}$$

The agent can achieve the optimal results by adjusting Q value according to the updated rule

$$Q\left(s, a\right) \leftarrow Q\left(s, a\right) + \alpha \left[r + \gamma \max_{a' \in \mathcal{A}} Q\left(s', a'\right) - Q\left(s, a\right)\right] \tag{33}$$

where α and γ are the learning rate and the discount factor, respectively, $\alpha, \gamma \in [0, 1]$.

The policy gradient can be computed as (32) to update the actor's primary network,

$$\begin{aligned} \nabla_{\theta^\mu} J &\approx \frac{1}{|\mathbb{D}|} \sum_t \left[\nabla_{\theta^\mu} Q\left(s, a|\theta^Q\right)|_{s,a=\mu(s|\theta_\mu)}\right] \\ &= \frac{1}{|\mathbb{D}|} \sum_t \left[\nabla_a Q\left(s, a|\theta^Q\right)|_{s,a=\mu(s)} \nabla_{\theta^\mu} \mu\left(s|\theta_\mu\right)|_s\right]. \end{aligned} \tag{34}$$

The value function is donated as $V^\pi\left(s\right)$. There is an optimal value $V^*\left(s\right)$ corresponding to an optimal policy $\pi*$ of all the possibility of the value function $V^\pi\left(s\right)$. By choosing the action to maximize the reward, the optimal policy $\pi*$ can be retrieved from optimal value function $V^*\left(s\right)$. So we can define $V^*\left(s\right)$ and the optimal policy as

$$V^*\left(s\right) = V^{\pi*}\left(s\right) = \max_{a \in \mathcal{A}} \left\{r\left(s, a\right) + \gamma \sum_{s' \in \mathcal{S}} p\left(s'|s, a\right) V^*\left(s'\right)\right\} \tag{35}$$

$$\pi*\left(s\right) = \arg\max_{a \in \mathcal{A}} \left\{r\left(s, a\right) + \gamma \sum_{s' \in \mathcal{S}} p\left(s'|s, a\right) V^*\left(s'\right)\right\} \tag{36}$$

We use the primary deep neural network with the parameters θ^{μ} and θ^{Q} Besides, the parameters of the target policy network and the target value network are respectively represented as $\theta^{\mu'}$ and $\theta^{Q'}$ which are approached to the primary network parameters with a small amount periodically by (36),

$$\begin{cases} \theta^{\mu'} \leftarrow \tau\theta^{\mu} + (1-\tau)\,\theta^{\mu} \\ \theta^{Q'} \leftarrow \tau\theta^{Q} + (1-\tau)\,\theta^{Q} \end{cases}, \tag{37}$$

where τ is the update coefficient, $\tau \in [0, 1]$.

3.2 DDPG-Based Joint Task Offloading Decision Algorithm (JTO)

The process of DDPG-based JTO Algorithm is summarized in Algorithm 1. We construct a replay memory to store a series of historical experiences. By randomly choosing a mini-batch sample from replay memory, the network parameters will be updated.

The algorithm parameters include the replay memory $\mathbb{D}$, the total time slot T, the number of servers and vehicles which have task to offload, discount factor and learning rates. After initializing the parameters, the decision of task offloading and resource allocation is executed at the beginning of each episode t. The system will choose an action with random noise N_t and then receive the reward and turn to the next state. Furthermore, the parameters of the network will be updated.

Algorithm 1: DDPG-based JTO Algorithm

Input: $\mathbb{D}$, K, S, T, V, γ, σ, s_0

1 **initialize** replay memory $\mathbb{D}$ to capacity D^*, $\mathbb{D}(K) = \emptyset$;
2 **initialize** networks θ_{μ} and θ_Q with random weight, target networks $\theta_{\mu'} = \theta_{\mu}$ and $\theta_{Q'} = \theta_Q$;
3 **for** *episode* $= 0 \rightarrow T-1$ **do**
4 initialize multi-lane scenario,task assignment and resource allocation process as the state $s_0 = \{Q, F, v, (a, b), R\}$;
5 **for** $t = 0 \rightarrow l-1$ **do**
6   **perform** action $a = \theta_{\mu} + N_t$;
7   **observe** reward r and s', **execute** action a, based on (32), (33) and (36);
8   **Store** transition (s, a, r, s') in $\mathbb{D}$;
9   **if** *replay memory* $\mathbb{D}$ *is full* (D^*) **then**
10   sample a random batch of K transitions (s, a, r, s') from $\mathbb{D}$;
11   Set $\Delta_t = r + \upsilon Q(s', a'|\theta_Q)$;
12   Update the parameter of critic network θ_Q by minimizing the loss: $L = E^2[\Delta_t - Q(s, a|\theta_{Q'})]$;
13   Update the parameter of actor network θ_{μ} by using the sampled policy gradient in (34) ;
14   Update target network $\theta_{\mu'}$ and $\theta_{Q'}$ using (37);
15 return θ_{μ} and θ_Q;

4 Performance Evaluation

In this section, we use Python to build a simulation environment for the multi-lane platoon system. Furthermore, we use Tensorflow platform to implement the DDPG-based JTO scheme. Our implementation is based on the open-source package DDPG [16].

For performance comparison, we present two benchmark schemes: RSU task offloading scheme (RTO) and other vehicles task offloading scheme (VTO),

1) *RTO:* Each IV only chooses to offload the task to RSU by V2I.
2) *VTO:* Each IV only chooses to offload the task to other vehicles by V2V.

The simulation parameters are given in Table 1.

Table 1. Simulation parameters

System parameter	Value/description
Number of IVs (tasks), u	10
Number of RSUs, m	4
Number of platoons, n	5
Number of vehicles in a platoon, N_i	4
Bandwith, B	20 MHz
Computation power P_i^t	1 W
Transmission power P_u	125 mW
Time slot Δt	90 s
Size of computation input data I_u^t	[250, 600] MB
Total number of CPU cycles C_u^t	[20, 30] cycle
Computation frequency f_i^t	[5, 20] cycles/s
Transmission rate R_i^t	[5, 20] Mbps
α, β	1

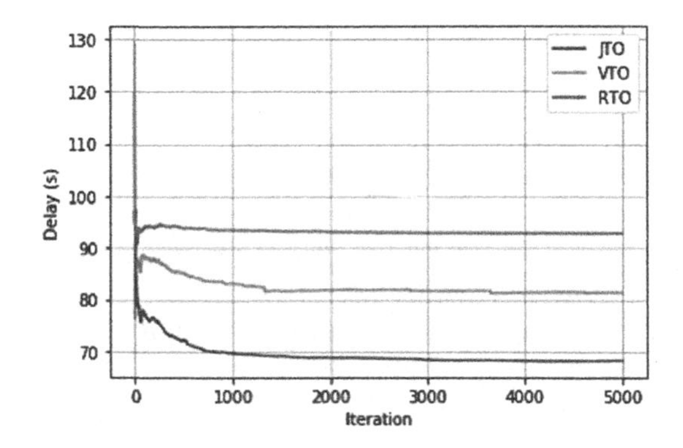

Fig. 4. The task execution delay achieved by different task offloading schemes.

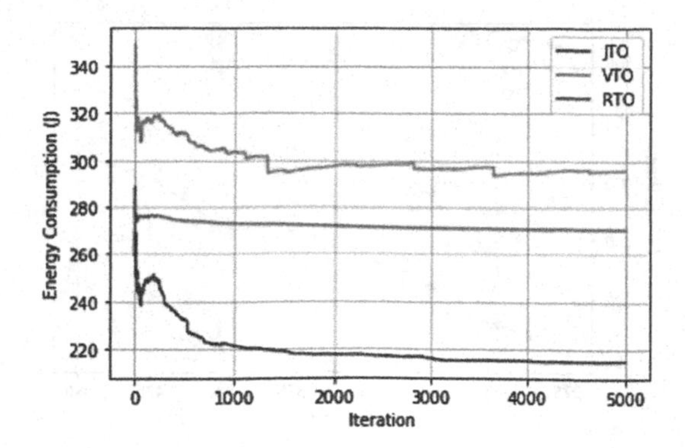

Fig. 5. The energy consumption achieved by different task offloading schemes.

In Fig. 4, we compare the performance of task execution delay of different task offloading schemes based on the DDPG learning algorithm. It can be easily seen that all the three task offloading schemes can approach their stable point as the number of episodes increases. We can draw the following observations from Fig. 5. Firstly, RTO has the highest delay because the RSU layout on the road may be sparse, and the distance is longer than that between vehicles. The channel quality is more easily affected too. Secondly, VTO can reduce latency since the possibility of connection interruption during transmission is less due to the relative speed between vehicles. Moreover, our proposed JTO schemes can yield the lowest delay as compared to the other benchmark schemes.

Figure 5 shows the comparison of the three schemes' energy consumption. Different to the task execution delay, the highest energy consumption is caused by VTO. The reason is that the computation capability of the vehicle is much smaller than that of RSU, so only using the V2V would require sending tasks to multiple vehicles, which would consume a lot of energy. However, the proposed JTO scheme combines the advantages of the other two schemes, so both the delay and the energy consumption are the lowest. As for the sudden rise for JTO when iteration is between 100 and 200, it is because the system tentatively assigning more tasks to the IVs, and then learning other schemes after increasing energy consumption in the learning process.

Figure 6 depicts the overall cost of those three proposed schemes, including the delay and the energy consumption with different amounts of tasks in the system. As the number of tasks increases, the energy consumption and delay of the whole system increase under these three scenarios. When there are fewer tasks, VTO must be lower than RTO because it uses vehicle collaboration. However, as the number of tasks increases, many tasks in V2V have to be offloaded to vehicles with poor communication quality, so its consumption increases rapidly. However, the algorithm proposed by us can reduce the consumption because it can choose RSUs and platoons adaptively.

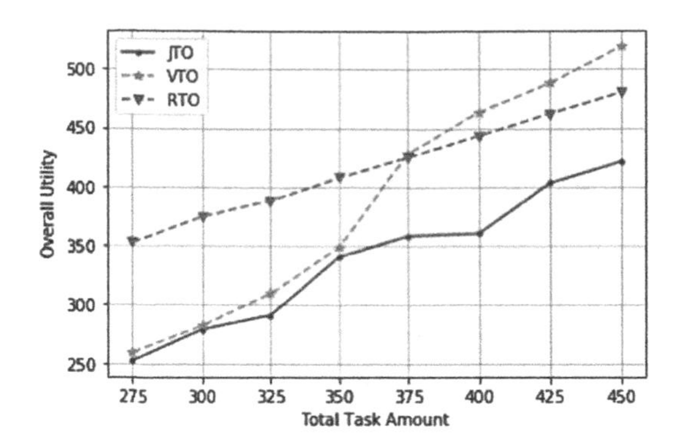

Fig. 6. The overall cost of different task offloading schemes.

5 Conclusion

In this paper, we have introduced a platoon-based VEC mechanism, which makes use of computing capabilities of both platoons and edge computing enabled RSU. Then we make a jointly optimization of task offloading target selection and inner-platoon resource allocation. We aims at minimizing the offloading cost, including task execution delay and energy consumption. In order to solve the problem, a DDPG-based algorithm has been proposed. Compared with other benchmark schemes, our proposed scheme significantly reduces task offloading latency and energy consumption, obtaining the optimized resource scheduling strategy, and especially in the case with a large number of tasks.

References

1. Li, W., Zhu, C., Leung, V.C.M., Yang, L.T., Ma, Y.: Performance comparison of cognitive radio sensor networks for industrial IoT with different deployment patterns. IEEE Syst. J. **11**(3), 1456–1466 (2017). https://doi.org/10.1109/JSYST.2015.2500518
2. Alam, K.M., Saini, M., Saddik, A.E.: Toward social internet of vehicles: concept, architecture, and applications. IEEE Access **3**, 343–357 (2015)
3. Huang, C., Chiang, M., Dao, D., Su, W., Xu, S., Zhou, H.: V2V data offloading for cellular network based on the software defined network (SDN) inside mobile edge computing (MEC) architecture. IEEE Access **6**, 17741–17755 (2018). https://doi.org/10.1109/ACCESS.2018.2820679
4. Zhou, H., et al.: TV white space enabled connected vehicle networks: challenges and solutions. IEEE Netw. **31**(3), 6–13 (2017). https://doi.org/10.1109/MNET.2017.1600049NM
5. Choo, S., Kim, J., Pack, S.: Optimal task offloading and resource allocation in software-defined vehicular edge computing. In: 2018 International Conference on Information and Communication Technology Convergence (ICTC), Jeju, pp. 251–256 (2018). https://doi.org/10.1109/ICTC.2018.8539726

6. Liu, P., Li, J., Sun, Z.: Matching-based task offloading for vehicular edge computing. IEEE Access **7**, 27628–27640 (2019). https://doi.org/10.1109/ACCESS.2019.2896000
7. Wang, H., Li, X., Ji, H., Zhang, H.: Federated offloading scheme to minimize latency in MEC-enabled vehicular networks. In: 2018 IEEE Globecom Workshops (GC Wkshps), Abu Dhabi, United Arab Emirates, 2018, pp. 1–6 (2018). https://doi.org/10.1109/GLOCOMW.2018.8644315
8. Xu, X., Xue, Y., Li, X., Qi, L., Wan, S.: A computation offloading method for edge computing with vehicle-to-everything. IEEE Access **7**, 131068–131077 (2019). https://doi.org/10.1109/ACCESS.2019.2940295
9. Nguyen, K.K., Duong, T.Q., Vien, N.A., Le-Khac, N., Nguyen, L.D.: Distributed deep deterministic policy gradient for power allocation control in D2D-based V2V communications. IEEE Access **7**, 164533–164543 (2019). https://doi.org/10.1109/ACCESS.2019.2952411
10. Qi, Q., et al.: Knowledge-driven service offloading decision for vehicular edge computing: a deep reinforcement learning approach. IEEE Trans. Veh. Technol. **68**(5), 4192–4203 (2019). https://doi.org/10.1109/TVT.2019.2894437
11. Loh, K.H., et al.: Solving the maximum cardinality bin packing problem with a weight annealing-based algorithm. Oper. Res. Cyber Infrastruct. **47**, 147–164 (2009)
12. Roess, R.P., Prassas, E.S., McShane, W.R.: Traffic Engineering, 3rd edn. Pearson/Prentice Hall, New Jersey (2004)
13. Rappaport, T.S.: Wireless Communications: Principles and Practice, 2nd edn. Prentice Hall: Englewood Cliffs, New Jersey (1996)
14. Jan, P., Stefan, S.: Reinforcement learning of motor skills with policy gradients. Neural Netw. **21**(4), 682–697 (2008). https://doi.org/10.1016/j.neunet.2008.02.003
15. Konda, V.R., Tsitsiklis, J.N.: Actor-critic algorithm. In: NIPS, vol. 13, pp. 1008–1014 (1999)
16. Reimplementation of DDPG (Continuous Control with Deep Reinforcement Learning) based on OpenAI Gym + Tensorflow. https://github.com/songrotek/DDPG. Accessed Dec 2019

Joint Power Allocation and Splitting Control in SWIPT-Aided Multi-carrier NOMA System

Jie Tang[1(✉)], Jingci Luo[1], Junhui Ou[1], Xiuyin Zhang[1], Nan Zhao[2], Danial So[3], and Kai-Kit Wong[4]

[1] South China University of Technology, Guangzhou, China
eejtang@scut.edu.cn, 201720110615@mail.scut.edu.cn, oujunhui_1990@foxmail.com, zhangxiuyin@scut.edu.cn

[2] Dalian University of Technology, Dalian, China
zhaonan@dlut.edu.cn

[3] The University of Manchester, Manchester, UK
d.so@manchester.ar.uk

[4] University College London, London, UK
kai-kit.wong@ucl.ac.uk

Abstract. The combination of non-orthogonal multiple access (NOMA) and simultaneous wireless information and power transfer (SWIPT) contributes to improve the spectral efficiency (SE) and the energy efficiency (EE) at the same time. In this paper, we investigate the throughput maximization problem for the downlink multi-carrier NOMA (MC-NOMA) system with the application of power splitting (PS)-based SWIPT, in which power allocation and splitting are jointly optimized with the constraints of maximum transmit power supply as well as the minimum demand for energy harvesting (EH). To tackle the non-convex problem, a dual-layer approach is developed, in which the power allocation and splitting control are separated and the corresponding sub-problems are respectively solved through Lagrangian duality method. Simulation results validate the theoretical findings and demonstrate the superiority of the application of PS-based SWIPT to MC-NOMA over SWIPT-aided single-carrier NOMA (SC-NOMA) and SWIPT-aided orthogonal multiple access (OMA).

Keywords: Multi-carrier non-orthogonal multiple access (MC-NOMA) · Simultaneous wireless information and power transfer (SWIPT) · Deep learning

1 Introduction

With the rapid development of fifth generation (5G) and its advanced application scenarios, the limited spectrum and energy resources are increasingly difficult to meet the requirements for the communication system. Hence, it is

X. Wang et al. (Eds.): 6GN 2020, LNICST 337, pp. 86–100, 2020.
https://doi.org/10.1007/978-3-030-63941-9_7

considerably significant to improve the spectrum efficiency (SE) and energy efficiency (EE). The non-orthogonal multiple access (NOMA) scheme has been considered as a significant technique to achieve a higher SE for 5G and the future communication system due to the elimination of channel orthogonality [1]. Besides, NOMA technology enables the communication system to provide higher data rate, lower latency, greater reliability and larger connectivity, etc. [2]. Thus, NOMA has aroused great attention and the application of NOMA to other advanced techniques has also been investigated, including multiple-input multiple-output (MIMO) [3], cognitive radio [4], multi-point cooperative relaying [5], etc. On the other hand, simultaneous wireless information and power transfer (SWIPT) [6], which makes it possible to collect energy and receive information parallelly, is viewed as an energy-efficient solution to the green communication system. Therefore, it has attracted extensive concern in both academic and industry.

Previous studies in [7,8] have investigated the performance comparison between NOMA and the conventional orthogonal multiple access (OMA) with the application of SWIPT. However, most of the existing works considered the single-carrier NOMA (SC-NOMA) systems and the performance of SWIPT-aided MC-NOMA is still an open topic. Motivated by this conversation, we considered a novel system which combines the spectrum-efficient MC-NOMA and the energy-efficient SWIPT, where the total throughput maximization problem is investigated with the constraints of transmit power supply and energy harvesting (EH) requirement.

2 System Model and Problem Formulation

2.1 System Model

In this section, we focus on the downlink of MC-NOMA system with the application of PS-based SWIPT, in which one BS communicates with K MUs via N subcarriers (SCs). Denote the set of all MUs' indexes and the set of all SCs' indexes as $\mathcal{K} = \{1, 2, \cdots, K\}$ and $\mathcal{N} = \{1, 2, \cdots, N\}$, respectively. The available bandwidth BW is equally divided into N orthogonal SCs and hence the bandwidth of each SC is $BW_n = BW/N$. Thus, the received signal of the k-th MU via the n-th SC is given by

$$y_{n,k} = h_{n,k}\left(\sqrt{p_{n,k}}s_{n,k} + \sum_{j\in\mathcal{K},j\neq k}\sqrt{p_{n,j}}s_{n,j}\right) + z_{n,k}, \qquad (1)$$

where $h_{n,k}$ represents the channel coefficient from the BS to the k-th MU over the n-th SC; $s_{n,k}(s_{n,j})$ indicates the data symbol transmitted from the BS to the k-th (j-th) MU over the n-th SC, which is a random signal with the energy of $\mathbb{E}[|s_{n,k}|^2](\mathbb{E}[|s_{n,k}|^2]) = 1$; $z_{n,k} \sim \mathcal{CN}(0, \sigma_{n,k}^2)$ denotes the additive white Gaussian noise (AWGN) to the k-th MU on the n-th SC.

At the receiving end, the received signal of the k-th ($k \in \mathcal{K}$) MU is split into two parts by a PS-based SWIPT scheme, where $\sqrt{\rho_k}$ and $\sqrt{1-\rho_k}$ are the

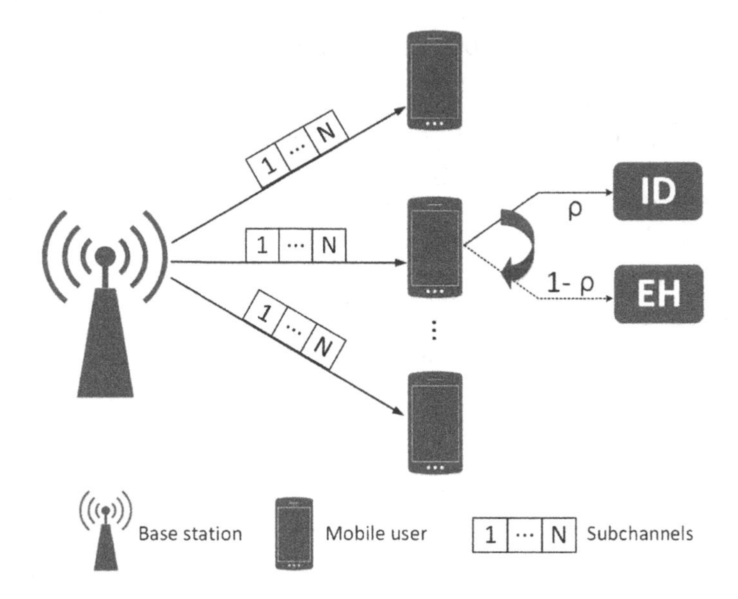

Fig. 1. The system model of the downlink SWIPT-aided MC-NOMA with PS-based receivers.

ratios of the received signal for information decoding (ID) and EH, respectively. Thereby, the received signal for ID and EH can be respectively written as (Fig. 1)

$$y_{n,k}^{\mathrm{ID}} = \underbrace{h_{n,k}\sqrt{\rho_k p_{n,k}}s_{n,k}}_{\text{Intended signal}} + \underbrace{h_{n,k}\sum_{\substack{j\in\mathcal{K},\\ j\neq k}}\sqrt{\rho_k p_{n,j}}s_{n,j}}_{\text{Interference signal}} + \underbrace{\sqrt{\rho_k}z_{n,k} + z_{n,k}^{\mathrm{ID}}}_{\text{Noise}}, \tag{2}$$

$$y_{n,k}^{\mathrm{EH}} = h_{n,k}\sum_{j=1}^{j=K}\sqrt{(1-\rho_j)p_{n,j}}s_{n,j} + \sqrt{1-\rho_k}z_{n,k}, \tag{3}$$

in which $z_{n,k}^{\mathrm{ID}} \sim \mathcal{CN}(0,(\sigma_{n,k}^{\mathrm{ID}})^2)$ refers to the noise generated during the PS process.

In order to reduce the interference during the ID process, the successive interference cancellation (SIC) technique is applied by the ID receivers. Let $\tilde{h}_{n,k} = h_{n,k}^2/\sigma_{n,k}^2$ denote the channel to noise ratio (CNR) for the k-th MU over the n-th SC. In practice, the order of ID in the downlink NOMA is usually the same as the order of the CNR. Therefore, the interference for the k-th MU on the n-th SC can be reduced as

$$I_{n,k} = \rho_k h_{n,k}^2 \sum_{\substack{j\in\mathcal{K},\\ \tilde{h}_{n,j}>\tilde{h}_{n,k}}} p_{n,j}. \tag{4}$$

Accordingly, the signal to interference plus noise ratio (SINR) and the available data rate of the k-th MU on the n-th SC can be respectively given by

$$\mathrm{SINR}_{n,k} = \frac{h_{n,k}^2 \rho_k p_{n,k}}{I_{n,k} + \rho_k \sigma_{n,k}^2 + (\sigma_{n,k}^{\mathrm{ID}})^2}, \tag{5}$$

$$R_{n,k} = BW_n \log_2(1 + \mathrm{SINR}_{n,k}). \tag{6}$$

Moreover, the harvested energy of the k-th MU on the n-th SC can be written as

$$E_{n,k} = \eta(1-\rho_k)\left(h_{n,k}^2 \sum_{j=1}^{K} p_{n,j} + \sigma_{n,k}^2\right), \tag{7}$$

where η corresponds to the efficiency of the EH receivers when harvesting energy.

Hence, the achievable data rate and the available harvested energy for the k-th MU can be respectively written as

$$R_k = \sum_{n=1}^{N} R_{n,k}, \quad E_k = \sum_{n=1}^{N} E_{n,k}. \tag{8}$$

Consequently, the total throughput of the considered MC-NOMA system with the application of PS-based SWIPT can be expressed as

$$R_{\mathrm{sum}} = \sum_{n=1}^{N}\sum_{k=1}^{K} R_{n,k} = \sum_{k=1}^{K} R_k. \tag{9}$$

2.2 Problem Statement

In this study, we focus on the total throughput maximization problem for our considered PS-SWIPT aided MC-NOMA system by jointly optimizing the power allocation and splitting control with the constraints of maximum transmit power supply as well as the minimum EH requirement. Thus, the optimization problem can be formulated as follows

$$\max_{\boldsymbol{\rho},\mathbf{p}} \ R_{\mathrm{sum}}(\boldsymbol{\rho}, \mathbf{p}) \tag{10}$$

$$\text{s.t.} \ E_k \geq E_{\mathrm{req}}, \ \forall\, k \in \mathcal{K}, \tag{11}$$

$$0 < \rho_k < 1, \ \forall\, k \in \mathcal{K}, \tag{12}$$

$$p_{n,k} \geq 0, \ \forall\, n \in \mathcal{N}, \ \forall\, k \in \mathcal{K}, \tag{13}$$

$$\sum_{n=1}^{N} p_{n,k} \leq p_k^{\max}, \ \forall\, k \in \mathcal{K}, \tag{14}$$

in which $\boldsymbol{\rho} = [\rho_1, \rho_2, \cdots, \rho_K]^T$ and $\mathbf{p} = [\mathbf{p}_1, \mathbf{p}_2, \cdots, \mathbf{p}_N]^T$ with the component $\mathbf{p}_n = [p_{n,1}, p_{n,2}, \cdots, p_{n,K}]^T$ $(1 \leq n \leq N)$. The inequality in (11) corresponds to the minimum requirement for EH of each MU, i.e., E_{req}W. The inequality in (12) indicates that the PS ratio for each MU should be within $(0, 1)$. The constraint (13) ensures the non-negativity of the power allocation for the k-th MU through the n-th SC and the constraint (14) limits the power allocation for the k-th MU (i.e., $\sum_{n=1}^{N} p_{n,k}$) not to exceed $p_k^{\max}$. Moreover, the maximum power supply of the BS can be implied to be $\sum_{k=1}^{K} P_k^{\max}$ according to (14).

The throughput maximization problem formulated in (10)–(14) is non-convex owing to the coupled multiple variables (i.e., $\boldsymbol{\rho}, \mathbf{p}$) and the co-channel interference. Additionally, the aforementioned maximization problem is a widely-known NP-hard problem, and hence it is difficult to obtain the solution directly. In the following section, we propose a dual-layer iterative approach to tackle the problem given in (10)–(14).

3 Algorithm Based on Lagrangian Duality

In this section, we develop the power allocation and PS control strategy for the involved PS-SWIPT aided MC-NOMA system. Since the coupled variables $\boldsymbol{\rho}$ and $\mathbf{p}$ make the original problem (10)–(14) non-convex, it is extremely tough to derive the optimal solution directly. According to [9], for any optimization problem involving multiple variables, it is practicable to deal with the sub-problem over part of variables while considering the remainder as constants, and next turn to handle the sub-problem over the remaining variables. As a result, $\mathbf{p}$ and $\boldsymbol{\rho}$ are separated to develop the practical and effective solution for the considered optimization problem.

3.1 PS Control with Fixed Power Allocation

We first consider the case where all the components of the power allocation matrix $\mathbf{p}$ are constants. In this case, we focus on optimizing the PS ratios under the fixed power allocation. Hence, the corresponding sub-problem can be simplified as

$$\max_{\boldsymbol{\rho}} \; R_{\text{sum}}(\boldsymbol{\rho}) \tag{15}$$

$$\text{s.t.} \;\; 0 < \rho_k < 1, \;\; \forall\, k \in \mathcal{K}, \tag{16}$$

$$E_k \geq E_{\text{req}}, \;\; \forall\, k \in \mathcal{K}. \tag{17}$$

According to (7), (8) and (17), ρ_k ($\forall\, k \in \mathcal{K}$) is required to satisfy the following condition

$$\rho_k \leq 1 - \frac{E_{\text{req}}}{\eta \sum_{n=1}^{N} h_{n,k}^2 \sum_{j=1}^{K} p_{n,j} + \sigma_{n,k}^2} \triangleq \rho_k^{\text{UB}}. \tag{18}$$

Considering (16) and (18) together, the optimization problem (15)–(17) is infeasible unless

$$\rho_k^{\text{UB}} > 0, \;\; \forall\, k \in \mathcal{K}. \tag{19}$$

Proposition 1: Under the fixed power allocation $\overline{\mathbf{p}}$ satisfying (13), (14) and (19), the throughput maximization sub-problem given in (15)–(17) is strictly convex with regard to $\boldsymbol{\rho}$.

Consequently, strong duality holds between the sub-problem (15)–(17) and its corresponding dual problem, which makes it possible to solve (15)–(17) optimally by employing the Lagrangian duality based method [9]. The corresponding Lagrangian function is formulated as

$$\begin{aligned}&\mathcal{L}(\boldsymbol{\rho},\boldsymbol{\mu},\boldsymbol{\nu},\boldsymbol{\omega}) = \\ &\sum_{n=1}^{N}\sum_{k=1}^{K} BW_n \log_2\left(1+\frac{\rho_k h_{n,k}^2 \overline{p}_{n,k}}{\rho_k(h_{n,k}^2\sum_{j\in\mathcal{K},\tilde{h}_{n,j}>\tilde{h}_{n,k}}\overline{p}_{n,j}+\sigma_{n,k}^2)+C_{n,k}}\right)+\sum_{k=1}^{K}\mu_k\rho_k \\ &+\sum_{k=1}^{K}\nu_k(1-\rho_k)+\sum_{k=1}^{K}\omega_k\left(\sum_{n=1}^{N}\eta(1-\rho_k)\left(h_{n,k}^2\sum_{j=1}^{K}\overline{p}_{n,j}+\sigma_{n,k}^2\right)-E_{\text{req}}\right),\end{aligned} \tag{20}$$

in which $\boldsymbol{\mu}=[\mu_1,\cdots,\mu_K]^T$, $\boldsymbol{\nu}=[\nu_1,\cdots,\nu_K]^T$ and $\boldsymbol{\omega}=[\omega_1,\cdots,\omega_K]^T$ are non-negative Lagrange multipliers. More specifically, $\boldsymbol{\mu}$ and $\boldsymbol{\omega}$ are corresponding to the constraint (16) while $\boldsymbol{\omega}$ is pertaining to the constraint (17).

Then the Lagrange dual objective function can be accordingly written as

$$g(\boldsymbol{\mu},\boldsymbol{\nu},\boldsymbol{\omega}) = \max_{\boldsymbol{\rho}} \ \mathcal{L}(\boldsymbol{\rho},\boldsymbol{\mu},\boldsymbol{\nu},\boldsymbol{\omega}). \tag{21}$$

Thus, the Lagrange dual problem can be modelled as

$$\min_{\boldsymbol{\mu},\boldsymbol{\nu},\boldsymbol{\omega}} \ g(\boldsymbol{\mu},\boldsymbol{\nu},\boldsymbol{\omega}) \tag{22}$$

$$\text{s.t.} \ \boldsymbol{\mu}\succeq\mathbf{0},\boldsymbol{\nu}\succeq\mathbf{0},\boldsymbol{\omega}\succeq\mathbf{0}. \tag{23}$$

To solve the Lagrange dual problem, we first optimize the PS ratio $\boldsymbol{\rho}$ with the given dual variables $\{\boldsymbol{\mu},\boldsymbol{\nu},\boldsymbol{\omega}\}$ through gradient ascent method, and then update the dual variables $\{\boldsymbol{\mu},\boldsymbol{\nu},\boldsymbol{\omega}\}$ with the optimized $\boldsymbol{\rho}$ through well-known sub-gradient scheme [10].

Optimizing $\boldsymbol{\rho}$ with Given Dual Variables $\{\boldsymbol{\mu},\boldsymbol{\nu},\boldsymbol{\omega}\}$. We first calculate the gradient direction of the Lagrangian function (20) regarding the PS ratio ρ_k ($\forall k\in\mathcal{K}$), which is given as

$$\begin{aligned}\nabla_{\rho_k}\mathcal{L} = &\sum_{n=1}^{N}\frac{BW_n}{\ln 2}\cdot\frac{A_{n,k}C_{n,k}}{(A_{n,k}\rho_k+B_{n,k}\rho_k+C_{n,k})(B_{n,k}\rho_k+C_{n,k})} \\ &+\mu_k-\nu_k-\omega_k\sum_{n=1}^{N}\left(\eta h_{n,k}^2\sum_{j=1}^{K}\overline{p}_{n,j}+\sigma_{n,k}^2\right).\end{aligned} \tag{24}$$

Particularly, ρ_k can be sequentially updated according to the following formula

$$\rho_k(n+1) = \rho_k(n)+\varepsilon(n)\nabla_{\rho_k(n)}\mathcal{L}, \tag{25}$$

where $\rho_k(n)$ and $\rho_k(n+1)$ denote the ρ_k in the n-th and $(n+1)$-th iteration respectively, and $\varepsilon(n)$ represents the updating step for ρ_k in the n-th iteration, which is required to satisfy the following condition

$$\varepsilon(n) = \arg\max_{\varepsilon} \ \mathcal{L}(\boldsymbol{\rho}(n+1), \boldsymbol{\mu}, \boldsymbol{\nu}, \boldsymbol{\omega})|_{\boldsymbol{\rho}(n+1)=\boldsymbol{\rho}(n)+\varepsilon\nabla_{\boldsymbol{\rho}(n)}\mathcal{L}}. \tag{26}$$

Process in (25) is repeated until $|\nabla_{\rho_k(n)}\mathcal{L}| \leq \epsilon_1$ for any $k \in \mathcal{K}$, and the optimal PS ratio is denoted as $\boldsymbol{\rho}^*$. Therefore, the Lagrange dual objective function in (21) is further determined as

$$g(\boldsymbol{\mu}, \boldsymbol{\nu}, \boldsymbol{\omega}) = \mathcal{L}(\boldsymbol{\rho}^*, \boldsymbol{\mu}, \boldsymbol{\nu}, \boldsymbol{\omega}). \tag{27}$$

Updating $\{\boldsymbol{\mu}, \boldsymbol{\nu}, \boldsymbol{\omega}\}$ with the Optimized $\boldsymbol{\rho}^*$. With the obtained PS ratio $\boldsymbol{\rho}^*$, the corresponding optimal Lagrange multipliers $\{\boldsymbol{\mu}, \boldsymbol{\nu}, \boldsymbol{\omega}\}$ can be determined accordingly through solving the Lagrange dual problem in (22)–(23).

Obviously, the dual problem is convex on the Lagrange multipliers $\{\boldsymbol{\mu}, \boldsymbol{\nu}, \boldsymbol{\omega}\}$. Therefore, one-dimensional search scheme can be adopted to optimize the dual variables. Nevertheless, the objective function (22) is not necessarily differentiable and thus this gradient-based approach is not always feasible. Otherwise, we apply the widely-used sub-gradient method to determine the dual variables $\{\boldsymbol{\mu}, \boldsymbol{\nu}, \boldsymbol{\omega}\}$, for which the sub-gradient directions are given in *Proposition* 2.

Proposition 2: The sub-gradient of the Lagrange dual function regarding the Lagrange multipliers can be respectively calculated by

$$\nabla_{\mu_k} g = \rho_k^*, \tag{28}$$

$$\nabla_{\nu_k} g = 1 - \rho_k^*, \tag{29}$$

$$\nabla_{\omega_k} g = \sum_{n=1}^{N} \eta(1-\rho_k^*) \left(h_{n,k}^2 \sum_{j=1}^{K} \overline{p}_{n,j} + \sigma_{n,k}^2 \right) - E_{\text{req}}. \tag{30}$$

Proof: Please refer to [10] for more details. ■

According to Proposition 2, the value of μ_k (ν_k, ω_k) should decrease if $\nabla_{\mu_k} g > 0$ ($\nabla_{\nu_k} g > 0$, $\nabla_{\omega_k} g > 0$), and vice versa. Based on this observation, we apply the binary search algorithm [10] to determine the optimal Lagrange multipliers (denoted as $\{\boldsymbol{\mu}^*, \boldsymbol{\nu}^*, \boldsymbol{\omega}^*\}$).

The algorithms developed in *1)* and *2)* operate alternately until the strong duality holds, i.e.,

$$R_{\text{sum}}(\boldsymbol{\rho}^*) = g(\boldsymbol{\mu}^*, \boldsymbol{\nu}^*, \boldsymbol{\omega}^*). \tag{31}$$

3.2 Power Allocation with Fixed PS Ratio

After obtaining the optimal solution of the PS ratio $\boldsymbol{\rho}^*$, now we aimed at optimizing the power allocation $\mathbf{p}$ under the optimized $\boldsymbol{\rho}^*$. Correspondingly, the original optimization problem in (10)–(14) is predigested into the following sub-problem

$$\max_{\mathbf{p}} \; R_{\text{sum}}(\mathbf{p}) \tag{32}$$

$$\text{s.t.}\; E_k \geq E_{\text{req}}, \;\; \forall\, k \in \mathcal{K}, \tag{33}$$

$$p_{n,k} \geq 0, \;\; \forall\, n \in \mathcal{N}, \;\; \forall\, k \in \mathcal{K}, \tag{34}$$

$$\sum_{n=1}^{N} p_{n,k} \leq p_k^{\max}, \;\; \forall\, k \in \mathcal{K}. \tag{35}$$

Proposition 3: Suppose that the process of PS in the receiving ends is almost idealized and the noise power for all MUs on the n-th SC is equal, i.e., $(\sigma_{n,k}^{\text{ID}})^2 \to 0$ and $\sigma_{n,k}^2 = \sigma_{n,j}^2 = \sigma_n^2 (\forall k, j \in \mathcal{K})$, the sub-problem (32)–(35) is convex if the feasible domain is non-empty.

Similar to the previous section III.A, strong duality can also be guaranteed between the sub-problem (32)–(35) and its dual problem, and thus the Lagrangian duality based algorithm is also employed here to optimize the power allocation $\mathbf{p}$.

Specifically, we define the relationship between the k-th MU and its decoding order as $k = \pi(i)$. The corresponding Lagrangian function for the sub-problem (32)–(35) can be written as

$$\begin{aligned}\widetilde{\mathcal{L}}(\mathbf{p}, \boldsymbol{\alpha}, \boldsymbol{\beta}, \boldsymbol{\gamma}) = &\sum_{n=1}^{N}\sum_{i=1}^{K} BW_n \log_2\left(1 + \frac{h_{n,\pi(i)}^2 p_{n,\pi(i)}}{h_{n,\pi(i)}^2 \sum_{j=i+1}^{K} p_{n,\pi(j)} + \sigma_n^2}\right) \\ &+ \sum_{i=1}^{K} \alpha_i \left(\sum_{n=1}^{N} \eta(1-\rho_i^*)\left(h_{n,\pi(i)}^2 \sum_{j=1}^{K} p_{n,\pi(j)} + \sigma_n^2\right) - E_{\text{req}}\right) \\ &+ \sum_{n=1}^{N}\sum_{i=1}^{K} \beta_{n,i} p_{n,\pi(i)} + \sum_{i=1}^{K} \gamma_i \left(p_k^{\max} - \sum_{n=1}^{N} p_{n,\pi(i)}\right),\end{aligned} \tag{36}$$

in which $\boldsymbol{\alpha} = [\alpha_1, \cdots, \alpha_K]^T$, $\boldsymbol{\beta} = [\boldsymbol{\beta}_1, \cdots, \boldsymbol{\beta}_N]^T$ with $\boldsymbol{\beta}_n = [\beta_{n,1}, \cdots, \beta_{n,K}]^T$ and $\boldsymbol{\gamma} = [\gamma_1, \cdots, \gamma_K]^T$ are non-negative multipliers with respect to (33), (34) and (35), respectively.

Then, the Lagrange dual objective function is given by

$$\widetilde{g}(\boldsymbol{\alpha}, \boldsymbol{\beta}, \boldsymbol{\gamma}) = \max_{\mathbf{p}} \; \widetilde{\mathcal{L}}(\mathbf{p}, \boldsymbol{\alpha}, \boldsymbol{\beta}, \boldsymbol{\gamma}), \tag{37}$$

Thus, the corresponding dual optimization problem can be formulated as follows

$$\min_{\boldsymbol{\alpha},\boldsymbol{\beta},\boldsymbol{\gamma}} \ \widetilde{g}(\boldsymbol{\alpha},\boldsymbol{\beta},\boldsymbol{\gamma}) \tag{38}$$

$$\text{s.t.} \ \boldsymbol{\alpha} \succeq \mathbf{0}, \boldsymbol{\beta} \succeq \mathbf{0}, \boldsymbol{\gamma} \succeq \mathbf{0}. \tag{39}$$

The proposed algorithm to solve the aforementioned problems consists of two steps, and more specific details are developed as follows.

Optimizing p Under Fixed Lagrange Multipliers $\{\boldsymbol{\alpha},\boldsymbol{\beta},\boldsymbol{\gamma}\}$. The gradient ascent method is employed to determine the optimal power allocation $\mathbf{p}^*$. Firstly, we analyze the gradient direction of the Lagrangian function given in (36) with regard to the power allocation component $p_{n,\pi(i)}$, which is calculated as

$$\begin{aligned}
&\nabla_{p_{n,\pi(i)}}\widetilde{\mathcal{L}} = \frac{BW_n}{\ln 2}\cdot \\
&\left(\frac{h^2_{n,\pi(1)}}{h^2_{n,\pi(1)}\Theta_{n,\pi(1)}+\sigma_n^2} + \sum_{i'=2}^{i}\left(\frac{h^2_{n,\pi(i')}}{h^2_{n,\pi(i')}\Theta_{n,\pi(i')}+\sigma_n^2} - \frac{h^2_{n,\pi(i'-1)}}{h^2_{n,\pi(i'-1)}\Theta_{n,\pi(i')}+\sigma_n^2}\right)\right) \\
&+ \beta_{n,i} - \gamma_i + \sum_{j=1}^{K}\alpha_j\eta\left(1-\rho_j^*\right)h^2_{n,\pi(j)}.
\end{aligned} \tag{40}$$

On the n-th ($1 \le n \le N$) SC, the power allocation for each MU can be successively updated through the following expressions

$$\begin{aligned}
&\underbrace{p_{n,\pi(1)}(1) \dashrightarrow p_{n,\pi(K)}(1)}_{\text{The 1-st iteration}} \dashrightarrow \underbrace{p_{n,\pi(1)}(t) \dashrightarrow p_{n,\pi(K)}(t)}_{\text{The } t\text{-th iteration}} \\
&\qquad \rightarrow \underbrace{p_{n,\pi(1)}(t+1) \dashrightarrow p_{n,\pi(K)}(t+1)}_{\text{The } (t+1)\text{-th iteration}},
\end{aligned} \tag{41}$$

$$p_{n,\pi(i)}(t+1) = p_{n,\pi(i)}(t) + \widetilde{\varepsilon}(t)\nabla_{p_{n,\pi(i)}(t)}\widetilde{\mathcal{L}}, \tag{42}$$

where t and $t+1$ indicate the number of iterations and $\widetilde{\varepsilon}(t)$ represents the updating step in the t-th iteration.

The process (41)–(42) for the power allocation on the n-th SC proceeds until $|\nabla_{p_{n,\pi(i)}}\widetilde{\mathcal{L}}| \le \epsilon_3$ for any $1 \le i \le K$. Correspondingly, the optimal power allocation on the n-th SC is expressed as p_n^* and thus $\mathbf{p}^* = [p_1*, \cdots, p_N^*]^T$. Then, the dual objective function in (37) can be reformulated as

$$\widetilde{g}(\boldsymbol{\alpha},\boldsymbol{\beta},\boldsymbol{\gamma}) = \widetilde{\mathcal{L}}(\mathbf{p}^*,\boldsymbol{\alpha},\boldsymbol{\beta},\boldsymbol{\gamma}). \tag{43}$$

Optimizing $\{\boldsymbol{\alpha},\boldsymbol{\beta},\boldsymbol{\gamma}\}$ Under the Obtained $\mathbf{p}^*$. Similar to the section III.A, sub-gradient approach is employed here to tackle the optimization of Lagrange multipliers $\{\boldsymbol{\alpha}, \boldsymbol{\beta}, \boldsymbol{\gamma}\}$, for which the sub-gradient directions are respectively denoted as follows

$$\nabla_{\alpha_i} \widetilde{g} = \sum_{n=1}^{N} \eta(1-\rho_i^*) \left(h_{n,\pi(i)}^2 \sum_{j=1}^{K} p_{n,\pi(j)} + \sigma_n^2 \right) - E_{\text{req}}, \tag{44}$$

$$\nabla_{\beta_{n,i}} \widetilde{g} = p_{n,\pi(i)}, \tag{45}$$

$$\nabla_{\gamma_i} \widetilde{g} = p_k^{\max} - \sum_{n=1}^{N} p_{n,\pi(i)}. \tag{46}$$

It is worth noting that the binary search method is also applicable to determine the optimal solution of the Lagrange multipliers here, which are denoted as $\{\boldsymbol{\alpha}^*, \boldsymbol{\beta}^*, \boldsymbol{\gamma}^*\}$.

The algorithms developed in *1)* and *2)* are repeated alternately until the zero duality gap is achieved, i.e.,

$$R_{\text{sum}}(\mathbf{p}^*) = \widetilde{g}(\boldsymbol{\alpha}^*, \boldsymbol{\beta}^*, \boldsymbol{\gamma}^*). \tag{47}$$

3.3 Complete Solution for Joint Power Allocation and Splitting

Up to now, the solutions to the sub-problems for optimizing $\boldsymbol{\rho}$ and $\mathbf{p}$ have been proposed in the sections III.*A* and III.*B*, respectively. Now we develop the complete solution for jointly optimizing the original problem (10)–(14), which is summarized in Algorithm 1.

Algorithm 1. Complete Solution for Joint Power Allocation and Splitting Control

1: Initialize $\overline{\mathbf{p}}$ and stop criteria ϵ_1, ϵ_2, ϵ_3, ϵ_4,.
2: **repeat**
3: *Step 1: optimize the PS ratio under fixed power allocation:*
4: **repeat**
5: initialize dual variables $\{\boldsymbol{\mu}, \boldsymbol{\nu}, \boldsymbol{\omega}\}$;
6: solve the problem (21) to obtain the optimal $\boldsymbol{\rho}^*$ according to (24)-(25) until $|\nabla_{\rho_{k}(n)} \mathcal{L}| \leq \epsilon_1 (\forall k \in \mathcal{K})$;
7: determine the optimal dual variables $\{\boldsymbol{\mu}^*, \boldsymbol{\nu}^*, \boldsymbol{\omega}^*\}$ according to *Proposition 2*;
8: **until** $R_{\text{sum}}(\boldsymbol{\rho}^*) = g(\boldsymbol{\mu}^*, \boldsymbol{\nu}^*, \boldsymbol{\omega}^*)$.
9: *Step 2: optimize the power allocation with fixed PS ratio:*
10: **repeat**
11: initialize the PS ratio assignment as $\boldsymbol{\rho}^*$;
12: solve the problem (37) to acquire the optimal $\mathbf{p}^*$ according to (40)-(42) until $|\nabla_{p_{n,\pi(i)}} \widetilde{\mathcal{L}}| \leq \epsilon_3 (\forall i \in \mathcal{K})$;
13: determine the optimal dual variables $\{\boldsymbol{\alpha}^*, \boldsymbol{\beta}^*, \boldsymbol{\gamma}^*\}$ according to (44)-(46);
14: **until** $R_{\text{sum}}(\mathbf{p}^*) = \widetilde{g}(\boldsymbol{\alpha}^*, \boldsymbol{\beta}^*, \boldsymbol{\gamma}^*)$.
15: **until** $R_{\text{sum}}(\boldsymbol{\rho}^*) = R_{\text{sum}}(\mathbf{p}^*)$.

Remark 1: The complete algorithm can be regarded as a dual-layer process. In the inner-layer, the complexity of the gradient decent algorithm is $\mathcal{O}(K)$ and the number of this loop iteration is approximately $\mathcal{O}\log(1/\epsilon_1^2)$ [11]; and the complexity of the binary search method with error tolerance ϵ_2 is $\mathcal{O}\log(1/\epsilon_2^2)$. Similarly, in the outer-layer, the computational complexity of the gradient algorithm method is $\mathcal{O}(NK)$ and the number of this loop iteration is approximately $\mathcal{O}\log(1/\epsilon_3^2)$; and the complexity of the binary search method is $\mathcal{O}\log(1/\epsilon_4^2)$. To summarize, the computational complexity of the complete solution is $\mathcal{O}(NK^2\log(1/\epsilon_1^2)\log(1/\epsilon_2^2)\log(1/\epsilon_3^2\log(1/\epsilon_4^2))$.

4 Numerical Results

In this section, numerical results are provided to evaluate the convergence performance of our proposed dual-layer iterative approach and the superiority of our considered MC-NOMA system with the application of PS-based SWIPT in terms of throughput. Assume that the BS is located at the center of a circular cell with a radius of 300 m, within which all MUs are randomly and independently located. The available bandwidth of the system is assumed to BW = 100 MHz. Referring to the typical 3GPP propagation setting, the channel from the BS to the MU includes three parts, i.e., i.i.d Rayleigh block fading, Log-Normal shadowing with standard deviation of 8 dB and path loss given by $(\frac{d_0}{d})^v$. In particular, d, $d_0 = 2.5$ and $v = 3.76$ indicates the propagation distance, the reference distance and the path-loss exponent, respectively. Moreover, the power spectrum density (PSD) of the channel noise and the additional noise generated during PS process are set to −96 dBm/Hz and −192 dBm/Hz, respectively. The efficiency of the EH circuits is supposed to $\eta = 38\%$.

Firstly, we investigate the convergence performance of the developed dual-layer iterative approach. We take a SWIPT-based MC-NOMA system with two SCs and two MUs as an example, where the maximum power supply of the BS and the minimum demand for EH are set to 4 W and 0.01 W respectively. As shown in Fig. 2, it is evident that the proposed Lagrangian duality-based approach is gradually converged to the optimal value acquired by the exhaustive search algorithm. This confirms our convergence analysis.

Then, performance in terms of total throughput of the proposed approach with various constraints is investigated. We taken $N = 2, K = 2$ and $N = 4, K = 4$ for comparison. We firstly evaluate the throughput performance under different minimum transmit power supplies. Assume that the minimum requirement for EH of each MU is $E_{\text{req}} = 0.1$ W and the transmit power budget varies from 2 W to 20 W. It is obviously shown in Fig. 3 that the total throughput is monotonically non-decreasing with the increase of the transmit power budget. This is because that with the growth in the transmit power budget, the received signal is more likely to be split to ID once the requirement for EH of each MU is satisfied, eventually leading to an increase in the throughput. Then we evaluate the throughput performance under various minimum demands for EH. In particular, it is supposed that the maximum transmit power supply is 10 W and

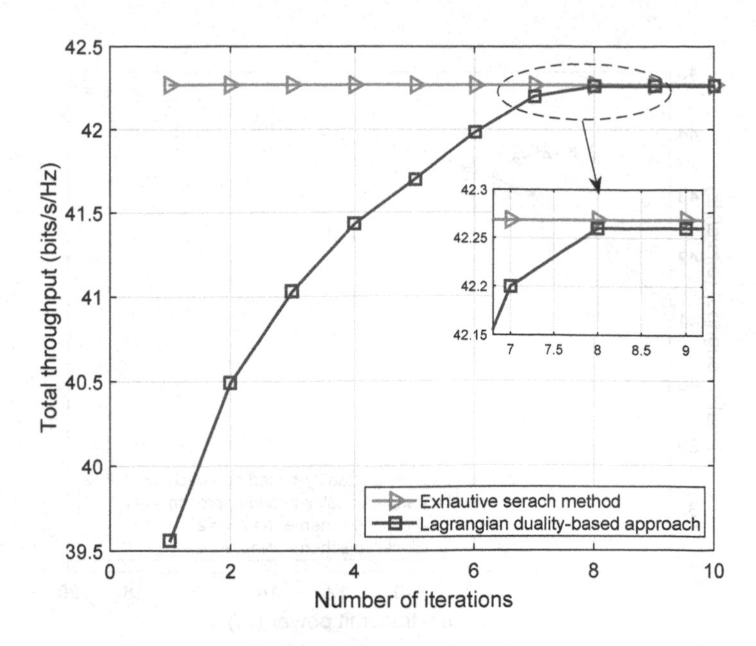

Fig. 2. The convergence behavior of the proposed Lagrangian duality-based approach.

the minimum requirement for EH varies from 0.1 W to 1 W. It is shown in Fig. 4 that the total throughput is monotonically decreasing as the minimum demand for EH grows, resulting from the fact that the received signal is more likely to be split to EH to fulfill the EH requirement and thereby the signal split to ID is cut off. Additionally, we can conclude from Fig. 3 and Fig. 4 that our developed dual-layer iterative approach outperforms the equal power allocation scheme in terms of throughput performance.

Lastly, we examine the performance comparison in terms of total throughput among our considered MC-NOMA with PS-based SIWPT and other schemes in the existing studies, including MC-NOMA with TS-based SWIPT, SC-NOMA with PS-based and OMA with PS-based. In particular, the maximum transmit power supply and the requirement for EH of each MU are respectively supposed to 10 W and 0.1 W, the number of SCs for two MC-NOMA schemes is set to $N = 3$, and the number of MUs is assumed to $K = 4$. It is evidently depicted in Fig. 5 that both SWIPT-aided MC-NOMA and SWIPT-aided SC-NOMA is always superior to the SWIPT-aided OMA, which further confirms that the NOMA scheme is more spectrum-efficient than the conventional OMA scheme. More significantly, our considered PS-based MC-NOMA system outperforms either the TS-based MC-NOMA or the PS-based SC-NOMA. This result demonstrates the superiority of our developed joint power allocation and splitting approach for the considered MC-NOMA system with the application of PS-based SWIPT, and accordingly provides a significant direction for practical communication system design.

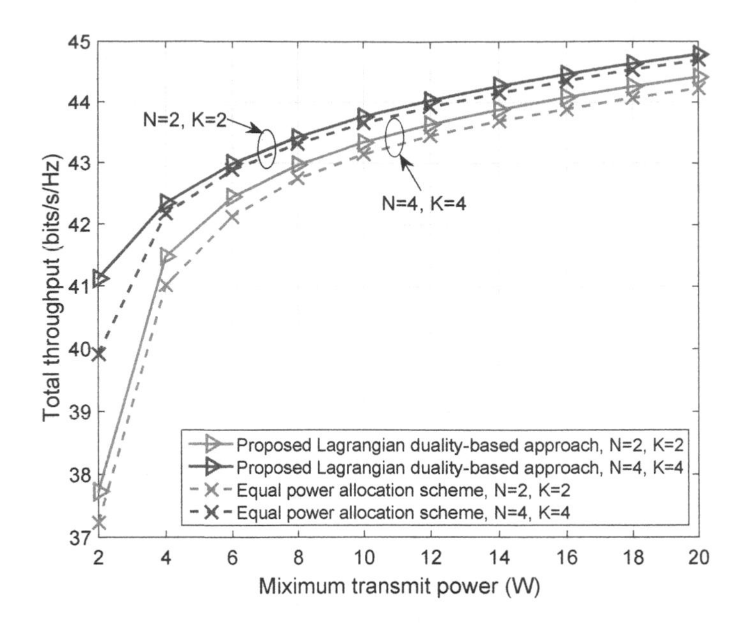

Fig. 3. Throughput performance of the proposed Lagrangian duality-based approach (total throughput vs maximum transmit power).

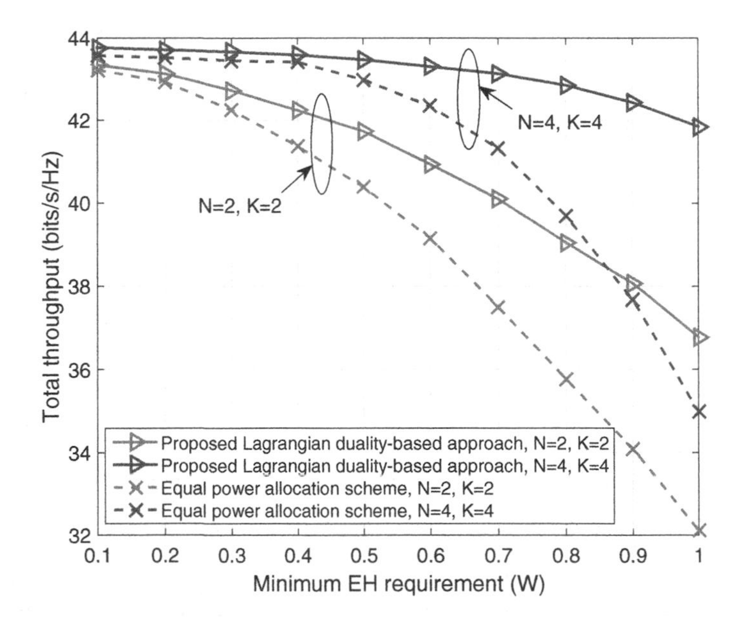

Fig. 4. Throughput performance of the proposed Lagrangian duality-based approach (total throughput vs minimum EH requirement).

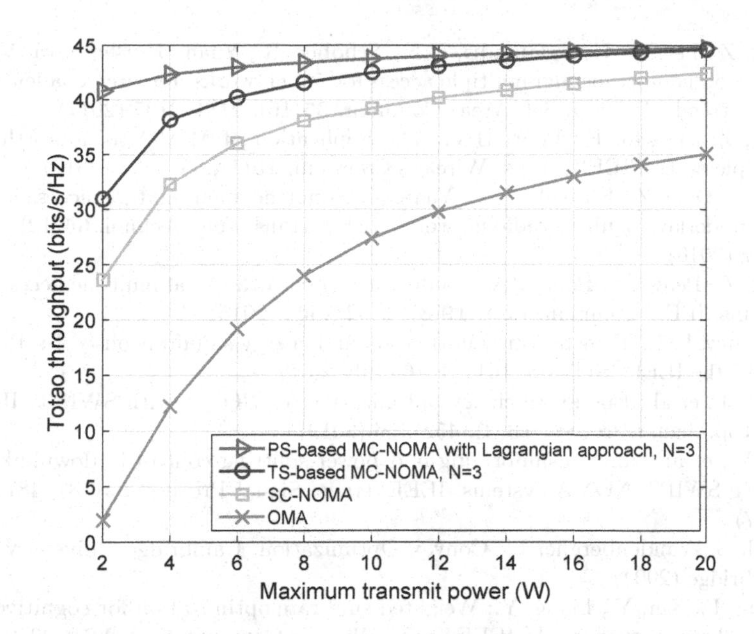

Fig. 5. Performance comparison in terms of throughput among different system model - OMA, SC-NOMA, TS-based MC-NOMA and PS-based MC-NOMA, K = 4

5 Conclusions

In this work, we have studied the total throughput maximization problem for the downlink MC-NOMA system with the application of PS-based SWIPT under the constraints of the maximum transmit power supply and the minimum demand for EH. The formulated optimization problem was non-convex owing to the coupled variables as well as the multi-user interference, and thus it was challenging to obtain the optimal solution directly. To solve this problem, we proposed a dual-layer iterative approach in which the coupled variables, i.e., the power allocation and the PS ratio assignment, were separated. Then the corresponding sub-problems were solved by employing the Lagrangian duality-based method. Simulation results verified the theoretical analysis of the convergence performance. More importantly, it was confirmed that the considered MC-NOMA system with the application of PS-based SWIPT outperformed other existing schemes in terms of throughput, including MC-NOMA with TS-SWIPT, SC-NOMA with PS-SWIPT and OMA with PS-SWIPT.

References

1. Benjebbour, A., Saito, Y., Kishiyama, Y., Li, A., Harada, A., Nakamura, T.: Concept and practical considerations of non-orthogonal multiple access (NOMA) for future radio access. In: Proceedings of the IEEE Intelligent Signal Processing and Communication Systems (IEEE ISPACS), pp. 770–774, November 2013

2. Ding, Z., Lei, X., Karagiannidis, G.K., Schober, R., Yuan, J., Bhargava, V.K.: A survey on non-orthogonal multiple access for 5G networks: research challenges and future trends. IEEE J. Sel. Areas Commun. **35**(10), 2181–2195 (2017)
3. Ding, Z., Adachi, F., Poor, H.V.: The application of MIMO to non-orthogonal multiple access. IEEE Trans. Wireless Commun. **15**(1), 537–552 (2016)
4. Liu, Y., Ding, Z., Elkashlan, M., Yuan, J.: Nonorthogonal multiple access in large-scale underlay cognitive radio networks. IEEE Trans. Veh. Technol. **65**(12), 10152–10157 (2016)
5. Ding, Z., Peng, M., Poor, H.V.: Cooperative non-orthogonal multiple access in 5G systems. IEEE Commun. Lett. **19**(8), 1462–1465 (2015)
6. Varshney, L.R.: Transporting information and energy simultaneously. In: Proceedings of the IEEE ISIT, pp. 1612–1616, July 2008
7. Tang, J., et al.: Energy efficiency optimization for NOMA with SWIPT. IEEE J. Sel. Top. Signal Process. **13**(3), 452–466 (2019)
8. Xu, Y., et al.: Joint beamforming and power-splitting control in downlink cooperative SWIPT NOMA systems. IEEE Trans. Signal Process. **65**(18), 4874–4886 (2017)
9. Boyd, S., Vandenberghe, L.: Convex Optimization. Cambridge University Press, Cambridge (2004)
10. Zhang, L., Xin, Y., Liang, Y.: Weighted sum rate optimization for cognitive radio MIMO broadcast channels. IEEE Trans. Wireless Commun. **8**(6), 2950–2959 (2009)
11. More, J.J., Calamai, P.H.: Projected gradient methods for linearly constrained problems. Math. Program. **39**(1), 93–116 (1987)

Decentralized Resource Sharing Platform for Mobile Edge Computing

Hongbo Zhang[1,2], Sizheng Fan[1,2], and Wei Cai[1,2](✉)

[1] The Chinese University of Hong Kong, Shenzhen, Guangdong, China
[2] Shenzhen Institute of Artificial Intelligence and Robotics for Society, Shenzhen, China
{hongbozhang,sizhengfan}@link.cuhk.edu.cn, caiwei@cuhk.edu.cn

Abstract. Recently, the Internet of Things (IoT) technology is booming in the industrial field. More and more industrial devices begin to connect to the internet. Compared with cloud computing, edge computing can well shorten the delay time on information transmission and improve the Quality of Service (QoS) of task computing, which promotes the development of the industrial Internet of things (IIoT) to some extent. The state-of-the-art edge computing service providers are specifically designed for customized applications. In our previous work, we proposed a blockchain-based toll collection system for edge resource sharing to improve the utility of these Edge Nodes (ENs). We provide a transparent, quick, and cost-efficient solution to encourage the participation of edge service providers. However, there exists a debatable issue since the system contains a centralized proxy. In this paper, we introduce the consortium blockchain to record the results of the service matching process in order to solve the issue. Besides, we propose a service matching algorithm for IIoT devices to select the optimal node and implement it using smart contract.

Keywords: Industrial Internet of Things · Mobile Edge Computing · Blockchain

1 Introduction

Internet of Things (IoT) can be regarded as a global network that consists of various connected devices that rely on sensing, communication, networking, and information processing technologies. It has made significant progress in recent decades [17]. IoT devices are widely used in industrial control, network equipment systems, public safety equipment, environmental monitoring, and many other fields. In order to satisfy the requirements of smart city, smart factory, and medical system, there are also a large-scale of IoT devices deployed to perform

This work was supported by Project 61902333 supported by National Natural Science Foundation of China, by the Shenzhen Institute of Artificial Intelligence and Robotics for Society (AIRS).

X. Wang et al. (Eds.): 6GN 2020, LNICST 337, pp. 101–113, 2020.
https://doi.org/10.1007/978-3-030-63941-9_8

tasks such as monitoring, sensing, pre-processing, and real-time decision-making. More and more scientists hope to apply the IoT to the industry, which will help them achieve industry 4.0 [13].

Driven by the development of the 5G network, the industrial Internet of things (IIoT) is attracting growing attention all over the world [9]. In industry, IIoT devices are often used to monitor the regular operations on factory equipment. The future IoT system combined with 5G can monitor the vehicle data in real-time, and the data of the vehicle can be calculated at the edge to give the vehicle control instructions [14]. The IIoT paradigm in healthcare enables users to interact with various types of sensors via secure wireless medical sensor networks (WMSNs) [1]. In these application scenarios, IoT devices handle tasks with large amounts of data. However, those IoT devices are relatively weak in performance, and they are heterogeneous. So, they are not feasible to directly support the intensive computing load brought by the large-scale IoT data.

In order to handle the mentioned real-time data processing scenarios which require low latency and high Quality of Service (QoS), we introduce Mobile Edge Computing (MEC). MEC is a novel paradigm that extends the computing capabilities and storage resources from cloud computing to the edge of the mobile network [8]. It can reduce the significant delay in delivering the computing tasks to the cloud. Due to the dense geographical distribution, support for high mobility, and open platform [4], users can upload their computing tasks to Edge Nodes (ENs) no matter when and where. With the mentioned features, MEC can support applications and services with lower latency and higher QoS, which significantly promotes the development of IoT applications.

In our previous work, we designed and implemented EdgeToll, a blockchain-based toll collection system for heterogeneous edge resource sharing [15]. By leveraging the payment channel technique, EdgeToll provides a transparent, quick, and cost-efficient solution to encourage the participation of edge service providers. The payment channel is an efficient way to trade for multiple frequent transactions between two stakeholders. It requires stakeholders to deposit tokens and set up the receiver in this channel first. We set the payment channel as uni-directional, which means that only the receiver of the channel can withdraw coins. Instead of building a payment channel directly between users and edges, we introduce a proxy to handle payment delivery. In the payment stage, the proxy receives a signature on the agreement of splitting coins from users and then sign the same amount signature to edge computing services providers based on the address in the public blockchain network. Because the verification of signature and the delivery of payment signature are all operations with nearly no cost, the payment channel can reduce the cost of public transactions on the public blockchain.

However, it is quite controversial to introduce a centralized proxy in our system, which violates the decentralization spirit of the blockchain. As a third party, the proxy is responsible for the service matching process. The system might be vulnerable to have a centralized proxy. It might cause significant collusion if the proxy is unsupervised. It is possible that proxy colludes with one of the edge service providers and prefers to recommend that provider's ENs to users. To solve the mentioned issue, we construct a consortium blockchain to record the service matching results on the consortium blockchain. We combine different edge service providers and proxy in this consortium.

Compared with the public blockchain, the consortium blockchain has many advantages, such as higher efficiency, higher scalability, and more transaction privacy. Rather than having all nodes joining the consensus process, the Practical Byzantine Fault Tolerance (PBFT) consensus process of the consortium blockchain is controlled by a set of selected nodes. At least 10 out of 15 nodes in the consortium need to sign and approve the block for it to be valid [2]. In our case, we select the proxy and edge service providers as the consensus nodes. In other words, every matching result needs to be signed and approved by at least two-thirds of the consortium.

To better attract IIoT devices and edge service providers to use our system, we deploy a service matching smart contract on the consortium blockchain, which is convenient for both IIoT devices and edge service providers. As buyers, the IIoT devices call the smart contract to search for the recommended ENs to handle computational tasks. As sellers, the edge service providers call the smart contract to record their ENs' location information. After receiving the information on the task, the proxy recommends the nearest EN according to the location of the IIoT device, which is an excellent way to reduce the time-consuming. Moreover, the smart contract can handle the above situations automatically, and the service matching results are recorded as transactions on the consortium blockchain.

The rest of the paper is organized as follows. We review related work in Sect. 2 and illustrate the system model in Sect. 3. Then, we present the technical design of the blockchain framework in Sect. 4. The test-bed implementation of the proposed system is shown in Sect. 5 to validate our system. In Sect. 6, we conclude our work and have a discussion about future work.

2 Related Work

2.1 Cloud and Edge Integration

Integrating edge to cloud platform involves a series of research topics in data and computational offloading. Traditional approach offloading schemes adopt virtualization techniques to host multiple copies of virtual machines in both clouds and edges [12]. At the same time, another group of researchers has investigated the possibility of dynamic code partitioning [3,6]. However, despite the form of offloading, the ENs intrinsically provide resource services for end-users through direct network connectivity. In this work, we assume the end-users are requesting micro-services installed in the ENs to simplify our model.

2.2 Blockchain in Edge Computing

Many existing works in MEC adopts blockchain for various purposes. For example, [11] presents an in-home therapy management framework, which leverages blockchain to preserve the therapeutic data privacy, ownership, generation, storage, and sharing. [7] proposes a blockchain-based framework for video streaming with MEC, which uses blockchain to build decentralized peer-to-peer networks with flexible monetization. To incentivize users with no mutual trust and different interests, [16] uses the reward-penalty model to align incentives in the ecosystem on MEC. Meanwhile, they implement the model using the blockchain smart contract to solve the high centralization problem in the ecosystem.

2.3 Payment Channel

The payment channel [10] is designed for "off-chain" transactions to overcome the long response latency and the monetary costs introduced by frequent transactions. It allows users to exchange tokens for multiple times with a minimum number of smart contract invocations. The state-of-the-art payment channels can be classified into two types: uni-directional payment channel and bi-directional payment channel. A uni-directional payment channel only allows single directional transactions, while a bidirectional payment channel [5] allows both parties to send transactions. The duplex payment channel is composed of two uni-directional payment channels, which allows transactions to be sent from both directions.

3 System Model

3.1 System Overview

In this section, we start by basically introduce the previous system. The system contains three types of users, including proxy, users, and ENs from different companies. Before the previous system showed up, users need to register different companies' accounts to use their ENs' computing resources, which is quite inconvenient for users. Hence, the previous system builds up payment channels between proxy and users and between the proxy and ENs. In this way, users only need to register one public blockchain address and pay the bills to proxy through the payment channel. The proxy then sends the tokens to the selected ENs through the payment channel. Besides, the payment channel is implemented in a smart contract. In this way, the previous system can provide a convenient, low cost and transparent payment platform for edge computing.

Based on the existing system, we use consortium blockchain to solve the centralized proxy problem and introduce a new type of system user, which is the company. Companies take control of different ENs. One EN belongs to only one company. As illustrated in Fig. 1, ENs, proxy, companies, and users all have their specific address on the consortium blockchain. In our case, users are the IIoT devices, and companies are the edge service providers. The consortium blockchain has several functions.

First, we can implement the service matching through the smart contract and deploy it on the consortium blockchain. After being deployed on the blockchain,

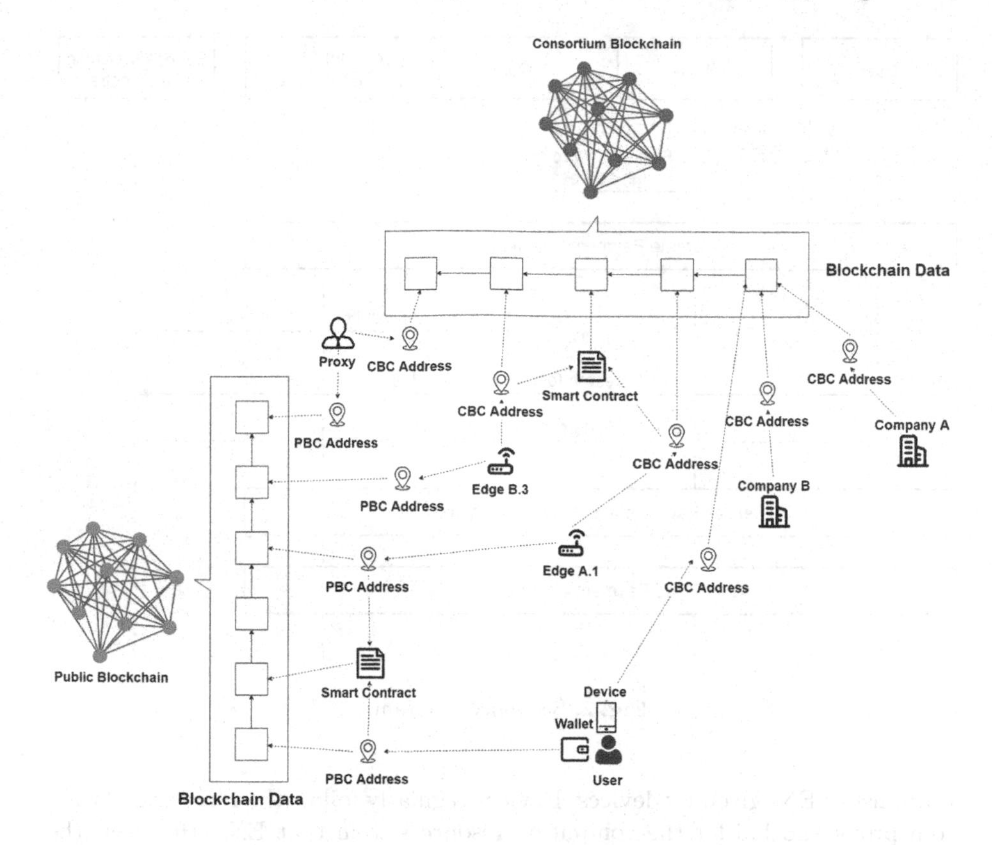

Fig. 1. System architecture

the smart contract has a specific address. Through the address, ENs can call the smart contract to record their locations to rent their surplus computing resources to devices that need computing resources. Devices can send their location to the smart contract to find the recommend EN and rent for the computing resources from the recommend node. Through the smart contract, the whole procedure of services matching can be done automatically. The result of an auction will be sent back to the ENs and devices which join the auction. To better demonstrate the procedure of using our system, we draw the sequence diagram in Fig. 2. Compare to the public blockchain, the performance of consortium blockchain is much better, and the cost of calling or deploying smart contract is zero, which save the costs of paying the gas fee. Meanwhile, the consortium blockchain reserves the characteristics of transparent, traceable, and unalterable. All the transactions, such as calling and deploying the smart contract, will be recorded on the consortium blockchain.

Second, the system can provide a more decentralized proxy with the consortium blockchain. In other words, the result of the service matching process needs to be supervised. As a third-party platform, proxy recommends ENs according

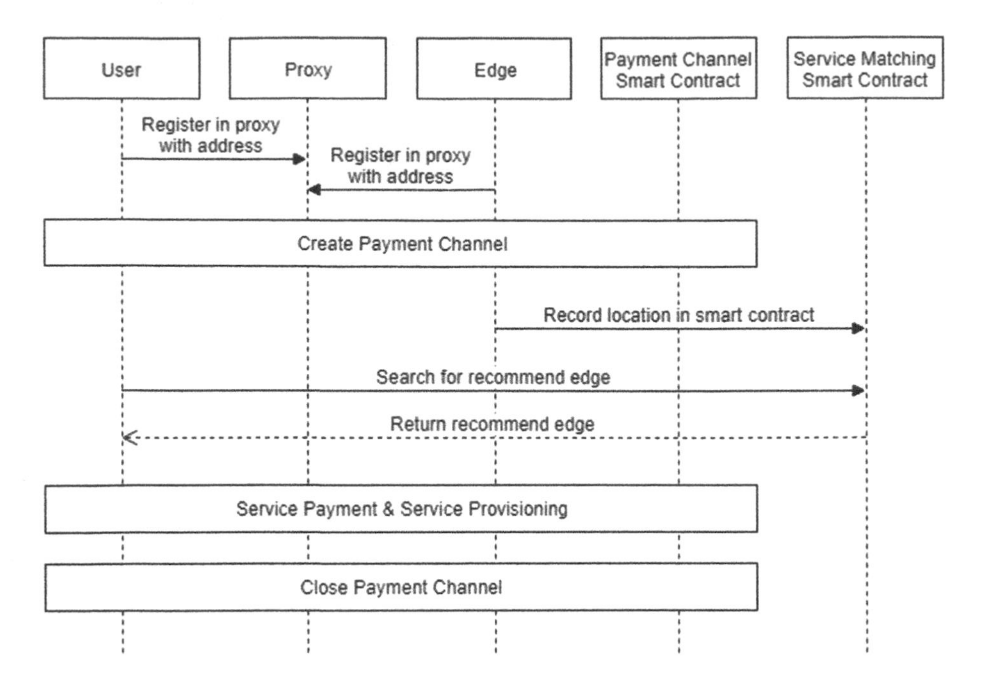

Fig. 2. Sequence diagram

to the list of ENs given by devices. Devices regularly follow the recommendation from proxy and bid for the computing resources from that ENs. However, the fairness of recommendation results can not be guaranteed since the whole procedure is entirely decided by proxy. If we do not deal with the situation, it will lead to a centralized proxy in our system, which violates the decentralization spirit of blockchain. In order to handle this centralized part of our system, we introduce consortium blockchain with the PBFT consensus algorithm.

4 The Blockchain Framework

We implement a decentralized edge computing resource sharing platform combining the advantages of two types of blockchain in our system design.

4.1 Software Architecture

Figure 3 illustrates the software architecture for the system.

4.2 Consortium Blockchain

Group Member Management. In order to better manage the system, the nodes on the blockchain are divided into two groups, which are sealers and

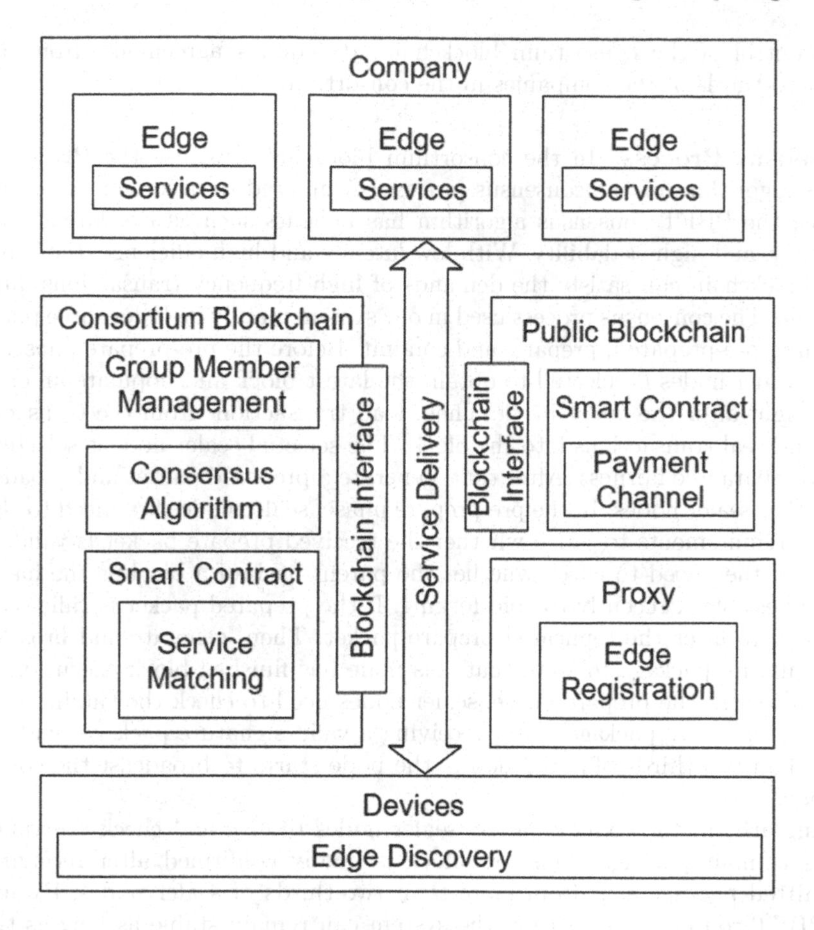

Fig. 3. Software architecture

observers. For both observer nodes and sealer nodes, they can send transactions in the consortium blockchain. The transactions are waiting in the transaction pool to be sealed into blocks. Compared to observer nodes, sealer nodes take part in the consensus process while sealing transactions into blocks. In our system, there are four types of system users, including proxy, devices, companies, and ENs. Different ENs belong to different companies. They are divided into two groups according to their types. For proxy and companies, they are the sealer nodes in the consortium blockchain. While for devices and ENs, they are the observer nodes in the consortium blockchain. Under this kind of classification, it is easier to manage the system when newcomers are accessing the system. For new devices and ENs, they can be directly added to the consortium blockchain after registration because they do not join the consensus process. They need to provide identifying information during registration to prevent DDoS attacks. However, the registration of the sealer node is stricter. Because sealer nodes take control of the consortium blockchain. It requires agreements from more than two-thirds of the companies in the consortium.

Consensus Process. In the consortium blockchain, we use the PBFT consensus algorithm for the consensus process. Compared to other consensus algorithms, the PBFT consensus algorithm has benefits such as low latency, high efficiency, and high scalability. With low latency and high efficiency, the consortium blockchain can satisfy the demands of high-frequency transactions during auctions. The consensus process used in our system mainly includes three phases, including pre-prepared, prepare, and commit. Before the pre-prepare phase, one of the sealer nodes is selected to obtain the latest block and populate an empty block right after the latest block. Then, load transactions from the transaction pool and seal transactions into the block. The selected sealer node is selected in turn to guarantee fairness. After that, generate a prepared packet and broadcast it to other sealer nodes. In the pre-prepare phase, sealer nodes first need to check several requirements to judge whether the received prepare packet is valid. For example, they need to check whether the parent hash of a block is the hash of the highest block recently to void forking. If the prepared packet is valid, cache it locally to filter the replicated prepare packet. Then, generate and broadcast the signature package to state that this node has finished block execution and verification. In the prepare phase, sealer nodes need to check the validity of the received signature package. After receiving a valid signature package send from more than two-thirds of sealer nodes, the node starts to broadcast the commit package.

Similarly, in the commit phase, sealer nodes receive and check the validity of the commit package. The new block is finally confirmed after receiving a committed package sent from more than two-thirds of sealer nodes. By using the PBFT consensus algorithm, the system can remain stable as long as there are more than two-thirds of non-malicious nodes.

Service Matching. Similarly, we implement the service matching algorithm by using the smart contract and deploy it on the consortium blockchain. The smart contract also has a unique address and can be called by any other nodes in the consortium blockchain. Devices can call the smart contract according to the address and receive a recommended ENs according to the service matching algorithm. Since proxy and companies are responsible for the consensus process, the results of service matching are supervised by them. It can prevent the proxy from colluding with any other company. Companies will not allow the situation happened because it is related to their benefits. As a result, companies and proxy will supervise each other, and the fairness of service matching can be guaranteed.

5 Test-Bed Implementation

In order to better demonstrate our system, we implement a prototype and conduct several experiments on it. In this section, we introduce the enabling technologies, the system deployment of the prototype, and demonstrate it with several shortcuts.

5.1 Test-Bed Specification

The edge computing server in our test-bed is Dell Precision 3630 Tower Workstation equipped with 16 GB RAM, Intel i7-9700 CPU, and NVIDIA GeForce GTX 1660. The edge computing server is also equipped with three wireless access points. The first one is TP-LINK WR886N, which adopts IEEE 802.11b/g/n standard with up to 450 Mbps data rate and 2.4 GHz radiofrequency. The second one is NanoPi R1, which adopts IEEE 802.11b/g/n standard with up to 450 Mbps data rate and 2.4 GHz radiofrequency. The third one is Phicomm K2P, which adopts IEEE 802.11b/g/n/ac standard with up to 1267 Mbps data rate and 2.4/5 GHz radiofrequency.

Figure 4 illustrates the test-bed implementation of our system. We use the workstation to work as an edge computing server. We equipped the server with three wireless access points to work as the ENs in MEC. The IIoT devices can submit their computational tasks through the wireless network. Then, edge computing server runs the computational tasks and return the results to IIoT devices.

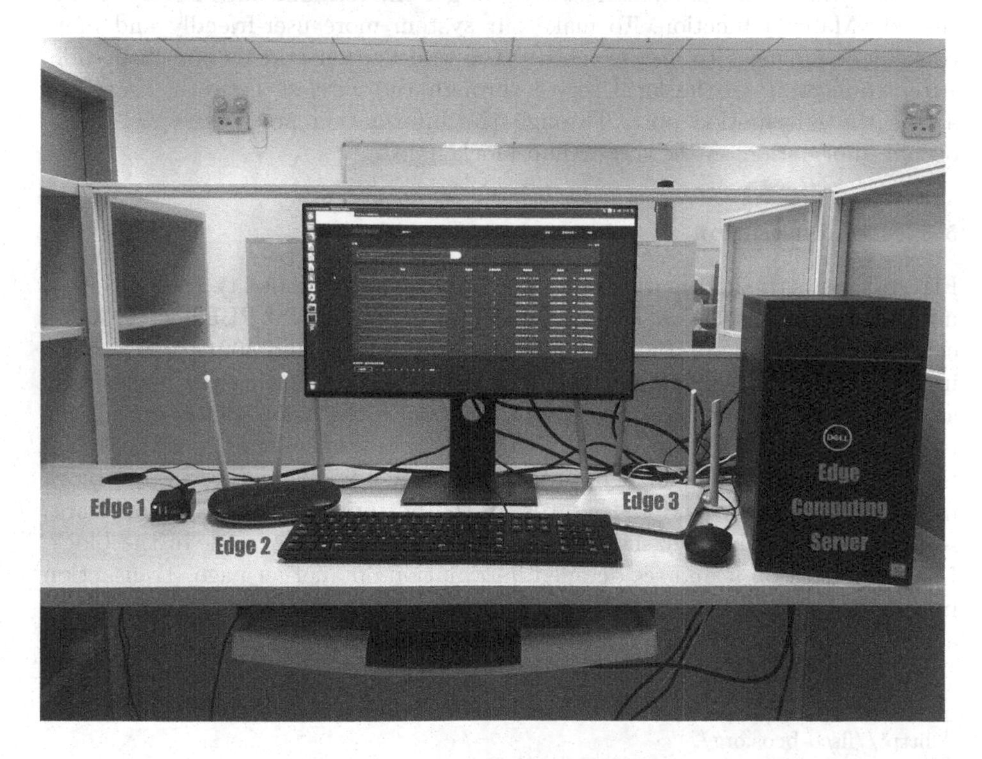

Fig. 4. Test-bed implementation

5.2 Enabling Technologies

To implement the prototype, we select various platforms and technologies. For the consortium blockchain platform, we choose FISCO-BCOS[1], which is an open-source consortium blockchain platform. For the programming language, we use Solidity[2] to write the service matching smart contract. Besides, FISCO-BCOS also provides an information port. With the information port, all system users can check the information of transactions and the smart contract. We choose information port as our client-side to demonstrate the information of the consortium blockchain. As for the interaction with consortium blockchain, we use the python-SDK[3] provided by the FISCO-BCOS.

5.3 Blockchain Deployment

We deploy the consortium blockchain in our local server. Initially, proxy deploys the service matching smart contract on the consortium blockchain. The smart contract has a unique address on the consortium blockchain. Both edge service providers and users can access the smart contract through the unique address. Edge service providers can record the information of their ENs by calling the addNote() function. As for users, they can get the recommended EN by calling the edgeMatch() function. To make our system more user-friendly and more convenient for users to use, we choose the information port as our client-side. After connecting to the local server through the wireless network, users can access the information port. Through the information port, users can easily obtain information on the consortium blockchain.

5.4 Demonstration

Figure 5 shows the information port. The upper-left part shows the current block number, total transactions, dealing transactions, and current PBFT view. As we can see from the figure, there are already 40012 blocks and 40012 transactions in the consortium blockchain. Next to it is the curve showing the transaction amount in the last 15 days. Since we do not have any transactions during the last 15 days, the curve stays flat. Below is the information of some nodes on the consortium blockchain. The information includes node ID, current block number, PBFT view, and node status. In the bottom-left part, it demonstrates several block information, including created time and sealer node of the block. Users can click on it and see detailed information on another page. Transaction information is shown in the bottom-right part. Similarly, users can click on it to achieve more information.

---

[1] http://fisco-bcos.org/.

[2] https://github.com/ethereum/solidity.

[3] https://github.com/FISCO-BCOS/FISCO-BCOS-DOC/tree/release-2/docs/sdk/python_sdk.

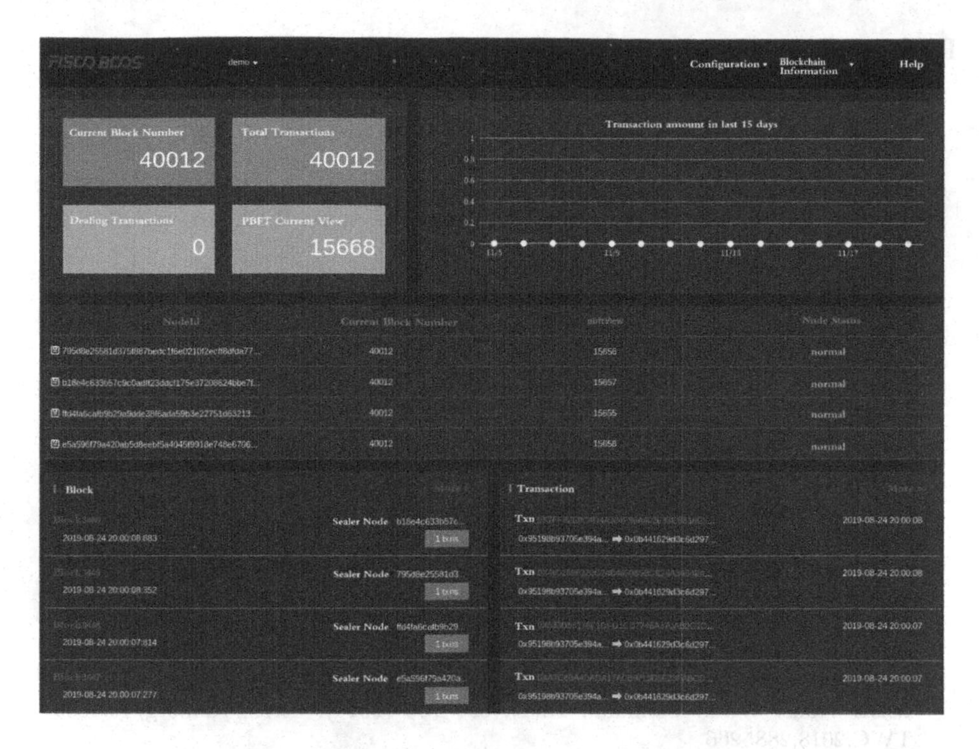

Fig. 5. Shortcut

6 Conclusion and Future Work

In our previous work, we provide a low-latency and cost-efficient solution for a decentralized, transparent, and auditable toll collection system by leveraging the payment channel technique. In this work, we demonstrate a decentralized toll collection system which solves the centralized proxy problem in our previous work. By adding consortium blockchain, we introduce a low-cost solution to solve out the centralized proxy problem. The decisions used to made by a centralized proxy can be implemented in the smart contract, and the smart contract will be deployed on the consortium blockchain. In this way, the results will be supervised by edge computing companies and proxy.

In the future, a more efficient and rational service matching model will be considered, and the proxy will consider the type of both ENs and tasks. In particular, the task allocation process will focus on high efficiency and low cost. In order to attract IIoT devices and edge service providers, a new dynamic pricing strategy will be proposed, which not only focuses on the incentive mechanism but also aims to improve the utility of ENs.

References

1. Al-Turjman, F., Alturjman, S.: Context-sensitive access in industrial internet of things (IIoT) healthcare applications. IEEE Trans. Ind. Inform. **14**(6), 2736–2744 (2018). https://doi.org/10.1109/TII.2018.2808190
2. Cai, W., Wang, Z., Ernst, J.B., Hong, Z., Feng, C., Leung, V.C.: Decentralized applications: the blockchain-empowered software system. IEEE Access **6**, 53019–53033 (2018)
3. Chun, B.G., Ihm, S., Maniatis, P., Naik, M., Patti, A.: CloneCloud: elastic execution between mobile device and cloud. In: Proceedings of The Sixth Conference on Computer Systems, pp. 301–314. ACM (2011)
4. Corcoran, P., Datta, S.K.: Mobile-edge computing and the internet of things for consumers: extending cloud computing and services to the edge of the network. IEEE Consum. Electron. Mag. **5**(4), 73–74 (2016)
5. Decker, C., Wattenhofer, R.: A fast and scalable payment network with bitcoin duplex micropayment channels. In: Pelc, A., Schwarzmann, A.A. (eds.) SSS 2015. LNCS, vol. 9212, pp. 3–18. Springer, Cham (2015). https://doi.org/10.1007/978-3-319-21741-3_1
6. Kosta, S., Aucinas, A., Hui, P., Mortier, R., Zhang, X.: ThinkAir: dynamic resource allocation and parallel execution in the cloud for mobile code offloading. In: 2012 Proceedings IEEE INFOCOM, pp. 945–953. IEEE (2012)
7. Liu, M., Yu, F.R., Teng, Y., Leung, V.C.M., Song, M.: Distributed resource allocation in blockchain-based video streaming systems with mobile edge computing. IEEE Trans. Wirel. Commun. **18**(1), 695–708 (2019). https://doi.org/10.1109/TWC.2018.2885266
8. Mach, P., Becvar, Z.: Mobile edge computing: a survey on architecture and computation offloading. IEEE Commun. Surv. Tutor. **19**(3), 1628–1656 (2017)
9. Palattella, M.R., et al.: Internet of things in the 5G era: enablers, architecture, and business models. IEEE J. Sel. Areas Commun. **34**(3), 510–527 (2016). https://doi.org/10.1109/JSAC.2016.2525418
10. Poon, J., Dryja, T.: The bitcoin lightning network: scalable off-chain instant payments (2016)
11. Rahman, M.A., et al.: Blockchain-based mobile edge computing framework for secure therapy applications. IEEE Access **6**, 72469–72478 (2018). https://doi.org/10.1109/ACCESS.2018.2881246
12. Satyanarayanan, M., Bahl, V., Caceres, R., Davies, N.: The case for VM-based cloudlets in mobile computing. IEEE Pervasive Comput. **8**, 14–23 (2009)
13. Shrouf, F., Ordieres, J., Miragliotta, G.: Smart factories in industry 4.0: a review of the concept and of energy management approached in production based on the internet of things paradigm. In: 2014 IEEE International Conference on Industrial Engineering and Engineering Management, pp. 697–701, December 2014. https://doi.org/10.1109/IEEM.2014.7058728
14. Utsunomiya, H., Kobayashi, N., Yamamoto, S.: A safety knowledge representation of the automatic driving system. Procedia Comput. Sci. **96**, 869–878 (2016). https://doi.org/10.1016/j.procs.2016.08.265. http://www.sciencedirect.com/science/article/pii/S1877050916320816. Knowledge-Based and Intelligent Information & Engineering Systems: Proceedings of the 20th International Conference KES-2016
15. Xiao, B., Fan, X., Gao, S., Cai, W.: EdgeToll: a blockchain-based toll collection system for public sharing of heterogeneous edges. In: 2019 IEEE Conference on Computer Communications Workshops (INFOCOM 2019 WKSHPS) (2019)

16. Xu, J., Wang, S., Bhargava, B.K., Yang, F.: A blockchain-enabled trustless crowd-intelligence ecosystem on mobile edge computing. IEEE Trans. Ind. Inform. **15**(6), 3538–3547 (2019). https://doi.org/10.1109/TII.2019.2896965
17. Xu, L.D., He, W., Li, S.: Internet of things in industries: a survey. IEEE Trans. Ind. Inform. **10**(4), 2233–2243 (2014). https://doi.org/10.1109/TII.2014.2300753

Research on D2D Resource Fair Allocation Algorithm in Graph Coloring

Yanliang Ge[1], Qiannan Zhao[1], Guanghua Zhang[1](✉), and Weidang Lu[2]

[1] Northeast Petroleum University, Daqing 163318, China
dqzgh@nepu.edu.cn
[2] Zhejiang University of Technology, Hangzhou 310014, China

Abstract. In a heterogeneous network composed of cellular users and device-to-device (D2D) users, D2D users multiplex the spectrum resources of cellular users in heterogeneous networks, which improves the shortage of spectrum resources. But this will bring a series of interference problems, which will greatly affect the throughput of the system and the service rate of users. On the premise of ensuring the service quality and throughput of users in the system in a heterogeneous network, to improve the user service rate, a fair distribution algorithm of D2D resources in graph coloring is proposed. First, in a heterogeneous network system, allowing multiple D2D users to share the resources of the same cellular user at the same time can improve the utilization of spectrum resources; Secondly, the interference graph is constructed by users and the interference between users in the heterogeneous network, and then the resource allocation colored by the D2D graph is added with a priority factor so that the fairness of user resource acquisition in the system is improved. Finally, it is verified by simulation, the algorithm improves the fairness of D2D users' access to resources while maintaining stable system throughput. It also reduces the system's packet loss rate and improves the user's service quality.

Keywords: Heterogeneous network · Resource allocation · Graph coloring · Throughput · Fairness

1 Introduction

At present, the continuous development of mobile communication technology and Internet technology has promoted the popularization of various intelligent terminals and the increase in the number of users, which ultimately led to a serious shortage of spectrum resources [1]. Device-to-device (D2D) technology is a communication method under the control of a cellular system, which can communicate directly by sharing the resources of cellular users. At the same time, D2D users will share the spectrum resources of cellular users, so that the utilization of spectrum resources is improved, and the problem of shortage of spectrum resources for wireless communications is solved. Therefore, D2D communication technology is listed as one of the key technologies in the new generation mobile communication system [2]. However, when D2D communication is introduced

X. Wang et al. (Eds.): 6GN 2020, LNICST 337, pp. 114–123, 2020.
https://doi.org/10.1007/978-3-030-63941-9_9

into a cellular network, it will cause serious interference [3, 4]. It is of great practical significance and research value on how to allocate resources reasonably to improve the spectrum utilization rate of the system and enhance the fairness of users receiving services.

In [5], when ensuring the user's service quality requirements, a pair of D2D users multiplex multiple channel resource blocks for communication, and a resource allocation algorithm based on Kuhn-Munkres optimal matching is proposed, thereby improving the overall system throughput. In [6], by analyzing the mathematical characteristics of the uplink and downlink interference area of D2D communication under the cellular network, the D2D communication system is designed and optimized. In [7], the problem of maximizing the system and the rate is transformed into an integer programming problem. On this basis, a resource allocation algorithm based on a bipartite hypergraph is proposed. In [8], through the proposed channel assignment scheme of hypergraph theory, the interference coordination between D2D users and cellular users are studied. In [5–8], they only consider the situation that one D2D user can only reuse the channel spectrum resources of one or more cellular users. However, when there is more available channel spectrum in the system, the remaining spectrum resources cause a waste of resources, thereby reducing the spectrum utilization rate of the system.

In [9], the capacity-oriented restricted (core) area is introduced into the traditional Stackelberg game method, which reduces the interference suffered by cellular users, and a resource allocation scheme for fair and safe capacity is proposed. In [10], by proposing graph theory coloring and QoS clustering method to allocate channel resources to D2D users, the satisfaction of D2D users, and the fairness of the system are solved. In [11], by using polynomial time proportional fair resource allocation schemes, it meets the requirement of the user's rate. However, it is considered that multiple resource blocks are allocated to one D2D user, which will cause a waste of resources. In [12], it is proposed to design a heuristic D2D resource allocation scheme based on the proportional fair scheduling algorithm. This scheme guarantees the data rate requirement of D2D communication improves the system throughput, and at the same time, the fairness of user scheduling is also taken into account. In [11, 12], on the premise of meeting the requirements of the cellular user signal-to-noise ratio (SINR), a resource allocation algorithm is proposed to achieve proportional fairness. In [13], in order to improve the overall performance of the cellular network system and reduce the fairness of D2D users to the cellular network system, improved proportional fairness (IPF) interference control scheme is proposed, but the scheme does not consider the fairness of D2D users' access to resources.

In response to the above problems, this paper uses multiple D2D users to reuse the same cellular user resource allocation scheme, so that the spectrum utilization rate and throughput of the system are improved.

2 System Model

2.1 Network Scenario

In a single-cell cellular network, when multiplexing uplink spectrum resources is compared with multiplexing downlink spectrum resources, the base station can control the allocation of system spectrum resources through its own interference, and the interference received by D2D users is also less. Therefore, this article considers the scenario where the uplink spectrum resources are multiplexed by D2D users in a network scenario where cellular users and D2D users share resources [14]. As shown in Fig. 1, assuming that the base station (BS) is located in the center of the cell, M cellular users and N D2D users are randomly distributed in the cell. Cellular users in the cell network are expressed as $CUE = \{C_1, C_2, \cdots, C_m, \cdots, C_M\}$, D2D users expressed as $DUE = \{D_1, D_2, \cdots, D_n, \cdots, D_N\}$, and the number of D2D users in the system is greater than the number of cellular users (M < N). Each D2D user exists in pairs, consisting of a transmitter (DT) and a receiving terminal (DR), and it is assumed that the initial transmission power of the data transmission of the D2D users is the same.

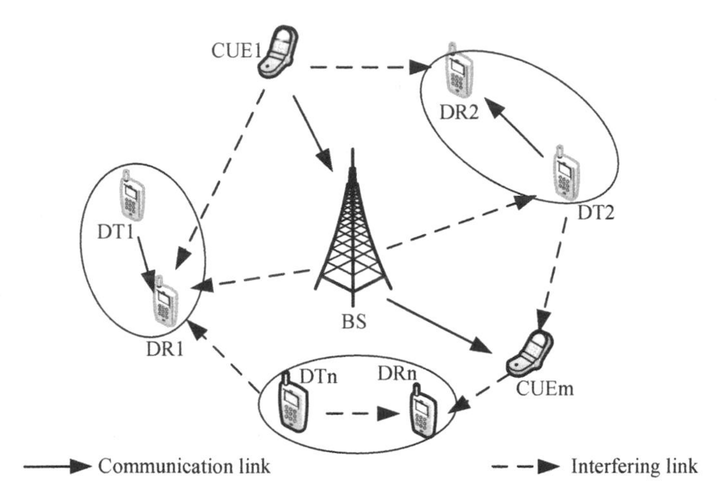

Fig. 1. Single-cell network system model

2.2 Mathematical Model

As shown in Fig. 1, when the uplink channel of the cellular user is multiplexed by the D2D user, there will be interference from D2D users and other D2D users' DR who reuse the same resources, between D2D users' DR and cellular users, and between D2D users' DT and base station. It can be seen that the interference caused by D2D users multiplexing the channel resources of cellular users is more complex. In order to avoid that the quality of service of users in the system can't be guaranteed when improving the spectrum utilization and throughput of the system, it is necessary to coordinate the complex interference in the cell network through reasonable resource allocation. In addition, it can prevent users with poor channel quality from being unable to allocate resources for a long time, which makes resource scheduling unfair. Therefore, the optimization goal of this paper is to find an optimal matching matrix for resource allocation and to ensure the user quality of service of the system's throughput level, as much as possible to improve the fairness of user resource acquisition.

$$R = nax(\sum_{m=1}^{M} B_0 \log(1 + SINR_m^c) + \sum_{m=1}^{M} \sum_{n=1}^{N} \alpha_{nm} B_0 \log(1 + SINR_n^d)) \tag{1}$$

When the spectrum resources of cellular users are multiplexed and shared by multiple D2D users at the same time, the signal-to-noise ratio (SINR) of the cellular users is:

$$SINR_m^c = \frac{p_m^c G_{c,b}}{\sum p_n^d G_{d,b} + N_0} \tag{2}$$

Among them, p_m^c represents the transmit power of the mth cellular user; $p_{n_m}^d$ represents the power sent from the DT of the nth D2D user to the mth cellular user; $G_{c,b}$ represents the channel gain between the cellular user and the base station; $G_{d,b}$ represents the channel gain between the DT of the D2D user and the base station; N_0 represents the noise power in the system.

The interference received by the D2D user DR in the system includes interference from cellular users and interference from other D2D users that multiplex the same resource at the same time, so the SINR of the D2D user's DR is:

$$SINR_n^d = \frac{p_n^d G_{dt,dr}}{\sum p_{n'}^d G_{dt',dr} + p_m^c G_{c,d} + N_0} \tag{3}$$

Among them, $G_{dt,dr}$ represents the channel gain between the DT and DR of the nth D2D user; $G_{dt',dr}$, represents the channel gain between DTs of other D2D users multiplexing the same resources as the DR of the nth D2D user; $G_{c,d}$ represents the channel gain between the DR of the cellular user and the D2D user. In order to ensure the user's communication quality requirements, the default minimum SINR threshold for cellular users is $SINR_{th}^c$, the SINR threshold of normal communication for D2D users is $SINR_{th}^d$, and under the conditions of fairness, the throughput should be improved as much as possible. Then:

$$\begin{aligned} s.t. SINR_m^c &\geq SINR_{th}^c, \forall m \\ SINR_n^d &\geq SINR_{th}^d, \forall n \\ p_m^c &\leq p_{\max}^d, \forall m \\ \sum_{m=1}^{M} p_{n_m}^d &\leq p_{\max}^d, \forall n \\ \sum_{m=1}^{M} \alpha_{n,m} &\geq 1, \forall n \\ \sum_{n=1}^{N} \alpha_{n,m} &\geq 1, \forall m \end{aligned} \tag{4}$$

Among them, B_0 is the bandwidth of the channel resource, $p_{\max}^c$ and $p_{\max}^d$ represent the maximum transmission power of cellular users and D2D users, $\alpha_{n,m} \in \{0, 1\}$ It indicates whether the same channel resources can be shared simultaneously cellular users and user

D2D, $\sum_{m=1}^{M} \alpha_{n,m} \geq 1, \forall n$ Indicates that the spectrum resources of multiple cellular users can be reused by one D2D user, $\sum_{n=1}^{N} \alpha_{n,m} \geq 1, \forall m$. It means that the spectrum resource of one cellular user can be multiplexed by multiple cellular D2D users. This objective function is the resource allocation problem of interference control, which can be solved by graph coloring theory.

3 Fair Color Allocation Based on Graph Coloring

According to the actual situation, in a single-cell network system in which D2D users and cellular users coexist, D2D users exist in the system as an auxiliary communication method, which causes a decrease in the service rate of D2D users. In order to avoid this situation and a certain occupied channel resource is excessive, a priority factor is added to the resource allocation to adjust the user coloring order, which is used to improve the fairness of user resource acquisition. In the fair resource allocation based on graph coloring, all users in the network system are abstracted as "vertices" in the graph and the interference suffered by the users is abstracted as "edges" in the graph, so a piece of interference can be constructed Figure G.

3.1 Construct an Interference Graph

In a network system where cellular users and D2D users coexist, a specific interference graph G can be established through the interference relationship between users and users. Among them, the vertices abstracted by all users in the system, the set is V, which includes the set of vertices composed of cellular users and the set of vertices composed of D2D users, then V is expressed as $V = \{v_i, i = 1, 2, \cdots, M, \cdots, N+M\}$, it is a vector of length $N+M$. The first M elements represent the vertices of the cellular user, and the last N elements represent the vertices of the D2D user. If the mutual interference edge matrix between vertices using the same spectrum resource is E, then the set of matrix E is represented as $E = \{e_{i,j}, i = 1, 2, \cdots, N+M, j = 1, 2, \cdots, N+M\}, e_{i,j} \in \{0, 1\}$, which is a (N + M) × (N + M) matrix. When $e_{i,j} = 1$, it means that vertex i and vertex j are both cellular users, or it can be compared by the distance $d_{i,j}$ between vertex i and vertex j and the interference distance threshold d_{th} between vertices, if it is less than the interference distance threshold $(d_{i,j} < d_{th})$, it indicates that there is interference between vertex i and vertex j, so user i and user j cannot reuse the same spectrum resource; When $e_{i,j} = 0$ indicates that there is no interference between vertex i and vertex j, then user i and user j can reuse the resources of the same spectrum. Thus, the interference matrix $E(G)$ between the vertices of the interference graph G can be obtained. By using C to indicate that the set of available subchannel resources in the heterogeneous network where the cellular and D2D coexist is the color set $C(G)$ of the interference graph G, then $C = \{c_1, c_2, \cdots, c_N\}$. If the cellular users in the system are fully loaded with $|C| = N$, it means that each cellular user can be assigned a subchannel. Therefore, the interference graph is constructed as G = (V, E, C), and the sub-channel allocation problem of cellular users and D2D users in the system is transformed into the problem of coloring the interference graph G.

3.2 Improve Resource Allocation Coloring Process

In order to avoid the problem that users with poor channel quality cannot be allocated resources for a long time and improve the service rate of D2D users, this paper proposes a graph coloring fair resource allocation strategy. In order to improve the fairness of users' access to resources during the graph coloring process, it avoids the problem of node coloring due to excessive user occupation of channel resources, so the priority factor of user coloring can be set during resource allocation.

Suppose that the sub-channels assigned by the cellular user $C_1, C_2, \cdots, C_m, \cdots, C_M$ to the vertex $v_1, v_2, \cdots, v_m, \cdots, v_M$, that is, the color of the graph coloring, respectively correspond to $c_1, c_2, \cdots, c_m, \cdots, c_M$. Therefore, the resource allocation problem of system throughput can be transformed into the resource allocation problem of D2D users. When a D2D user whose vertex is v_{M+n} reuses the color c_m corresponding to cellular user C_m, the utility value of vertex v_{M+n} to the network is:

$$R_{v_{M+n},c_m} = \frac{\log(1 + SINR^c_{m,c_n}) + \log(1 + SINR^d_{n,c_n})}{\deg(v_{M+n})} \tag{5}$$

Among them, $SINR^c_{m,c_m}$ represents the SINR of the cellular user C_m when the user assigns the color c_m; $SINR^d_{n,c_m}$ represents the SINR of the D2D user D_n when the user assigns the color c_m; $\deg(v_{M+n})$ represents the degree of the vertex v_{M+n}.

Assuming that the color c_m corresponding to the cellular user C_m is multiplexed for the D2D user vertex v_{M+n}, the priority factor of the colored c_m is w_{v_{M+n},c_m}. The priority factor is the utility value R_{v_{M+n},c_m} of the vertex v_{M+n} to the network and the number of colors assigned to the value of v_{M+n} which is:

$$w_{v_{M+n},c_m} = \frac{R_{v_{M+n},c_m}}{\sum_{m=1}^{M} A_{MN} + 1} \tag{6}$$

Among them, A_{MN} is a D2D user spectrum allocation matrix composed of $\alpha_{n,m}$ elements; adding 1 to the denominator is to avoid the problem of infinite priority factor. An improved resource allocation algorithm based on graph coloring, the implementation steps are as follows:

(1) Construct the interference graph and initialize the graph G = (V, E, C);
(2) for i = 1: M, coloring cycle;
(3) According to whether the vertices in E(G) interfere with each other, obtain a D2D user set K that can use the color c_m occupied by the vertex v_m;
(4) According to the set K, determine whether this vertex can be colored;
(5) According to formula (5), the vertex corresponding to the maximum network utility value is obtained and colored;
(6) According to formula (6), the coloring order, that is, the order of resource allocation, is determined.

4 Analysis of Simulation Results

4.1 Simulation Parameters

In order to verify the effect of D2D resource fairness allocation algorithm (IFCA) on the graph coloring proposed in this paper on the system performance. The three aspects of system fairness, system throughput, and system packet loss rate are respectively compared with the random coloring allocation algorithm (RCA), the graph coloring resource allocation algorithm (DVCA) considering the delay and saturation, the traditional graph coloring allocation algorithm (TGCA) for simulation comparison. Assuming that the cell radius is 500 m, the base station BS is located in the center of the cell, and other main simulation parameters are shown in Table 1.

Table 1. System simulation parameters.

Parameter	Value
Cell radius/m	500
Total system bandwidth/MHz	10
Maximum transmit power of cellular users/dBm	24
Distance between D2D users/m	10–25
Maximum transmit power of D2D users/dBm	20
Noise power spectral density/dB/HZ	−174
Path loss model for D2D link/dB	$148 + 40\lg d$
Path loss model for cellular link/dB	$128.1 + 37.6\lg d$
SINR threshold/dB for normal communication of D2D users	30
SINR threshold for normal communication for cellular users/dB	20

4.2 Analysis of Simulation Results

Figure 2 shows the relationship between the fairness of system D2D resource allocation under different graph coloring resource allocation algorithms. It can be seen from Fig. 2 that the greater the number of D2D users accessed in the system, the worse the fairness of resource allocation in the system. Among them, when the number of D2D users accessed in the system is within 30 pairs, because the number of D2D users accessed in the cell network is not large, there is not much interference, so the fairness of resource allocation of the four algorithms is particularly different small. However, when the number of D2D users connected in the system is between 80 and 100 pairs, the IFCA algorithm proposed in this paper can be improved by about 4% compared with the DVCA algorithm. This is because a priority factor is added on the basis of the DVCA algorithm so that users who cannot obtain channel resources due to poor channel communication quality in the system can obtain resources preferentially. Therefore, the fairness of resource allocation in the system is improved.

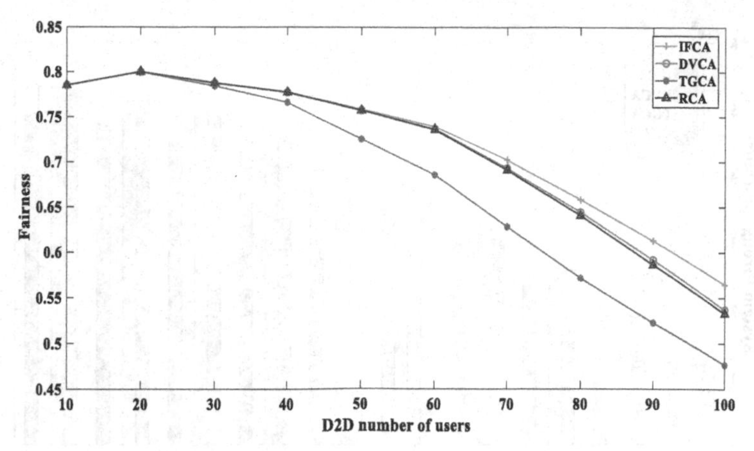

Fig. 2. Comparison of the fairness of different algorithms

Figure 3 shows the relationship between the number of D2D users accessing the system and the throughput. It is known from the simulation results that the three algorithms of IFCA, DVCA, and TGCA are based on ensuring normal communication between users and allow multiple pairs of D2D users to reuse the resources of the same cellular user to achieve maximum utilization of channel resources. However, when more D2D users are connected to the system, the higher the throughput of the system, the worse the communication service quality of the users. It can be seen from Figs. 3 and 4 that IFCA compares with DVCA and TGCA, the throughput of the system will be slightly different, but when more and more D2D users in the system, the service quality of IFCA users is better. Although the TGCA algorithm has improved the system throughput compared with the IFCA algorithm, it seriously sacrifices the fairness of user resource scheduling. Except for the RCA algorithm, the other three algorithms have no significant difference in system throughput.

Figure 4 compares the system packet loss rate of the number of D2D users connected to the system under different algorithms. It can be seen from the figure that as the number of D2D users in the system increases, the data packets in the system are more likely to be lost. In the TGCA algorithm, only the saturation of user vertices is considered, and no delay is considered. Therefore, some users occupy the channel resources too much, which causes an increase in the rate of packet loss. The algorithm IFCA considers the delay and adds a priority factor for the fairness of user resource scheduling, which avoids some users from occupying too many resources, improves the service rate of D2D users, and reduces the system's packet loss rate.

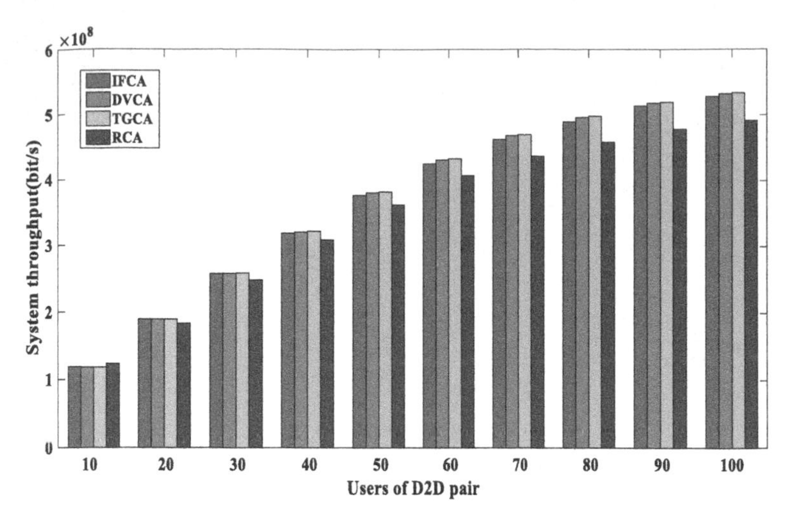

Fig. 3. Comparison of system throughput under different.

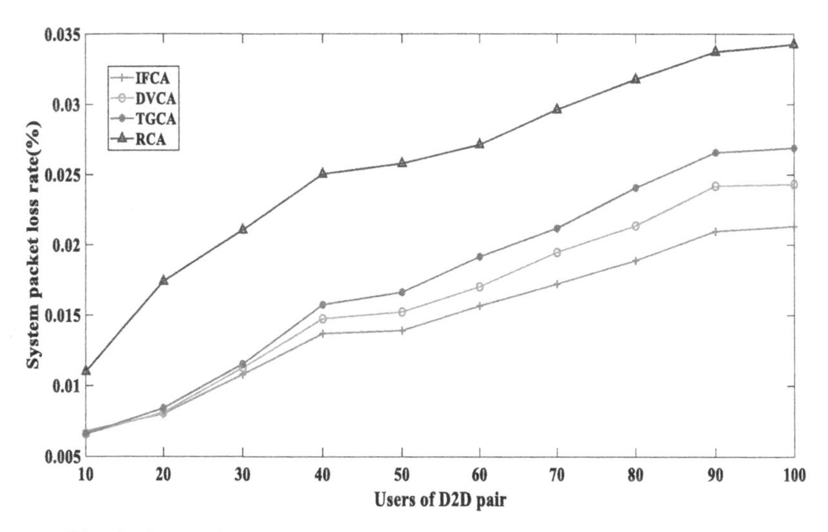

Fig. 4. Comparison of system packet loss rate under different algorithms.

5 Conclusion

In this paper, through the communication model of a heterogeneous network, the problem of resource allocation in the environment where D2D users are much larger than cellular users is studied, so that the spectrum utilization rate has been effectively improved. In order to ensure the throughput of the system and the fairness of users' access to resources, a resource fairness allocation algorithm based on graph coloring theory is proposed. Simulation results show that the proposed algorithm can improve the fairness of D2D users to obtain resources and reduce the packet loss rate on the premise of ensuring the system throughput, which can better guarantee the quality of service of users in the system. However, in this paper, we only consider the interference problem when D2D users and cellular users coexist in a single cell. While ignoring the interference between users in multiple cells, further research can be done in this regard in the future.

References

1. Zhou, Y., Pan, Z., Zhai, G., et al.: Research on the standardization outlook and key technology of the 5G mobile communication system. J. Data Acquis. Process. **30**(4), 714–724 (2015)
2. Qian, Z., Wang, X.: Review of D2D technology for 5G communication network. J. Acta Commun. Sin. **37**(7), 1–14 (2016)
3. Wang, H., Leung, S.H., Song, R.: Uplink area spectral efficiency analysis for multichannel heterogeneous cellular networks with interference coordination. J. IEEE Access **6**(1), 14485–14497 (2018)
4. Wang, Q.: An interference control scheme for D2D communication underlaying LTE network. J. Radio Eng. (09), 13–17+61 (2015)
5. Huang, J., Liu, X.: Design of resource allocation scheme for D2D based on Kuhn-Munkres optimal matching. J. Comput. Appl. Res. **32**(03), 827–829 (2015)
6. Li, X., Wang, Z., Sun, Y., Gu, Y., Hu, J.: Mathematical characteristics of uplink and downlink interference regions in D2D communications underlaying cellular networks. Wirel. Pers. Commun. **93**(4), 917–932 (2017). https://doi.org/10.1007/s11277-016-3936-y
7. Wang, Z., Zhao, Y., Xue, W., et al.: Resource allocation for D2D communication underlaid cellular networks using bipartite hypergraph. J. Comput. Sci. **44**(08), 82–85+94 (2017)
8. Zhang, H., Song, L., Han, Z.: Radio resource allocation for device-to-device underlay communication using hypergraph theory. J. IEEE Trans. Wirel. Commun. **15**(7), 4852–4861 (2016)
9. Chang, W., Lin, Y.C., Lee, Y., et al.: Fairness and safety capacity oriented resource allocation scheme for D2D communications. In: Wireless Communications & Networking Conference, San Francisco, CA, pp. 1–6 (2017)
10. Shi, C., Zhu, Q.: QoS-based clustering channel allocation algorithm in D2D communication system. J. Signal Process. **33**(07), 953–960 (2017)
11. Mondal, I., Neogi, A., Chaporkar, P., et al.: Bipartite graph based proportional fair resource allocation for D2D communication. In: Wireless Communications & Networking Conference, San Francisco, CA, pp. 1–6 (2017)
12. Wang, J., Li, H., Bu, Z.: A heuristic D2D resource allocation scheme based on proportional fairness. J. Comput. Eng. **43**(12), 78–82 (2017)
13. Wang, B., Meng, W., Zhou, W., Wang, Z., et al.: A scheme to improve fairness in D2D communications underlaying cellular networks. J. Beijing Univ. Posts Telecommun. **38**(02), 21–26 (2015)
14. Liu, H., Yan, B., Chen, Y.: Many – to many resource allocation scheme for D2D based on hypergraph. J. Comput. Eng. Des. **39**(12), 3605–3609+3621 (2018)

Future Wireless Communications

On Relay Selection for Relay-Assisted D2D Communications with Adaptive Modulation and Coding in Cellular Systems

Xixi Bi[1], Ruofei Ma[1,2](✉), Zhiliang Qin[2], and Gongliang Liu[1]

[1] Harbin Institute of Technology, Weihai 264209, Shandong, China
19S130290@stu.hit.edu.cn, {maruofei,liugl}@hit.edu.cn
[2] Beiyang Electric Group Co. Ltd., Weihai, Shandong, China
qinzhiliang@beiyang.com

Abstract. Developing device-to-device (D2D) communications can improve spectrum efficiency and traffic offloading capability of cellular systems. But only considering direct D2D communication mode may limit these benefits brought in by D2D communications, for direct D2D mode may not be available due to the long separation distance and poor link quality. Hence, relay-assisted D2D communication is presented to expand the coverage of D2D communications. One of the key points in developing relay-assisted D2D communication is to find out the optimal relay user equipment (UE) to assist data transmission between the source and destination D2D-capable UEs. In this paper, a cross-layer relay selection scheme is researched, which considers the end-to-end data rate, end-to-end transmission delay, and remaining battery time of the relay-capable UEs. Specially, we propose a method to estimate the end-to-end transmission delay for the relay-involved D2D path when adaptive modulation and coding (AMC) is taken into account. Performances of the proposed scheme are also evaluated.

Keywords: Device-to-device (D2D) communication · Relay selection · Transmission delay · Adaptive modulation and coding

1 Introduction

With increasing of wireless applications, it is important to develop new technologies to improve the spectrum efficiency. Amongst these technologies, device-to-device (D2D) communication is considered as one of the most promising way. D2D communication enables reducing the burden of cellular network by

This work was supported partially by National Natural Science Foundation of China (Grant No. 61801144, 61971156), Shandong Provincial Natural Science Foundation, China (Grant No. ZR2019QF003, ZR2019MF035), and the Fundamental Research Funds for the Central Universities, China (Grant No. HIT.NSRIF.2019081).

X. Wang et al. (Eds.): 6GN 2020, LNICST 337, pp. 127–143, 2020.
https://doi.org/10.1007/978-3-030-63941-9_10

allowing two proximity user equipments (UEs) to communicate directly bypassing the base station (BS) [1]. In addition, D2D communications can improve the performances on energy efficiency and system-wise throughput [2]. Hence, corresponding topics of D2D communications have attracted many scholars to research and investigate. According to whether or not using relay node during data transmissions, D2D communications can be commonly divided into direct and relay-assisted D2D communications. At present, most works of D2D communications focus on direct D2D communications, which commonly considers two types of D2D modes. One is dedicated D2D mode and the other is reuse D2D mode. Direct D2D pair in dedicated mode requires dedicated cellular radio resources [3], whereas the one in reuse mode needs to reuse the channel that is being used by a general cellular UEs (CUE), i.e., a CUE shares the same cellular spectrum resources with the D2D pair. Involving reuse mode can improve spectrum efficiency, while it also causes interference between D2D UEs and CUEs. Thus, there are many works concentrating on how to choose reuse channels and how to assign powers for D2D UEs and CUEs to limit the interference in the reuse mode [4–7]. However, it might not be enough while only considering direct D2D communications, because direct D2D communication probably can not be performed due to long separation distance between the source and destination D2D-capable UEs (DUEs) [8]. Hence, to expand the communication range of D2D communications and further enhance the system capacity and spectrum efficiency, including relay-assisted D2D mode may be a promising way [9,10].

A crucial problem on developing relay-assisted D2D communications is to find out the optimal relay-capable UE (RUE) for the DUEs in relay-assisted D2D mode under the BS scheduling. Thus, many criterias of choosing relay for relay-assisted D2D communications have been proposed, e.g., throughput, data rate, link outage probability, and power consumption, which are all designed on physical layer (PHY) [11,12]. To get an overall good performance via selecting the optimal relay, high layer criteria and remaining battery of the candidate RUEs need also to be considered. In [13], the transmission delay combining the packet buffer of relay is proposed, whereas, ignoring the packet buffer of the source UEs. In [14], authors proposes a cross-layer relay selection scheme combining the queue state information (QSI) of buffer for relay UEs, also ignoring the buffer of the source.

Base on the above analysis, cross-layer scheme should be considered more for its better overall performance. However, on the aspect of the delay estimate, most works only concentrate on the buffer of the relay, and lack comprehensive end-to-end cross-layer design. Also, the battery of the relay should also be considered as it has influences on the operation time of relay-assisted D2D communications, thus we proposed a cross-layer relay selection scheme which combines the end-to-end data rate, end-to-end delay and RUE remaining battery time in our previous work [15]. But the previous research we did ignores the time-varying of channels. To overcome this shortage, in this paper, on the aspect of the end-to-end delay, we investigate a new scheme combining the end-to-end delay with adaptive modulation and coding (AMC). Accordingly, in this paper, we only talk about derivation of the end-to-end delay with AMC, and the derivation of the end-to-end data rate and RUE remaining battery time can be found in the aforementioned work [15]. But on the performance evaluation, we will jointly assess the performance of the three criterias.

The rest of the paper is organized as follows. In Sect. 2, a single-cell system model for D2D communications is described. Section 3 introduces the problem formulation for relay selection. AMC technology is briefly introduced in Sect. 4. Section 5 provides the derivation process of end-to-end transmission delay for the relay-assisted D2D path, while Sect. 6 presents performance analysis of the relay selection scheme when taking AMC into account. Section 7 concludes the paper.

2 System Model

The considered system model is a single-cell scenario in a cellular system involving relay-assisted D2D communications. As illustrated in Fig. 1, we assume that there are N CUEs, M candidate RUEs, and a pair of DUEs intending to perform D2D communications in the cell, and all of them are controlled by the BS. The M RUEs are candidate relay nodes to aid the two DUEs to form relay-assisted D2D path. Also assume that N CUEs communicate with the BS via N orthogonal channels and the N orthogonal channels are with the same bandwidth. For ease of the expression in the follow sections, we use $\{C_1, C_2, \cdots, C_j, \cdots, C_N\}$ and $\{R_1, R_2, \cdots, R_i, \cdots, R_M\}$ to denote the N CUEs and M RUEs, respectively. Use S and D to denote the source DUE and the destination DUE, respectively. We also assume that there is no idle cellular channel for DUEs, i.e., DUEs must reuse an uplink channel of CUEs to access the system and communicate with each other. If DUEs work on direct D2D mode, the two DUEs reuse only one uplink channel of the CUEs to perform data transmission. If the two DUEs work on relay-assisted D2D mode, the source-to-relay and relay-to-destination transmissions can reuse the same uplink channel of a CUE or two different uplink channels of two CUEs. In the relay-assisted D2D mode, the first-hop and the second-hop (i.e., source-to relay and relay-to-destination) transmissions were performed on two different time slots sequentially. In addition, we further assume that the BS can obtain perfect channel state information (CSI) of all links, including the D2D source-to-relay link, relay-to-destination link, CUE-to-relay link and CUE-to-destination link.

3 Problem Formulation for Relay Selection

In this paper, three criterias, i.e., end-to-end data rate, end-to-end transmission delay, and candidate RUEs' remaining battery time, will be considered in the relay selection procedure of the relay-assisted D2D mode. The relay-assisted D2D mode is used as a supplement the direct D2D mode, which means that only when the direct D2D mode is not available, can the relay-assisted D2D mode be used. The relay selection is a part of the joint scheduling performed at

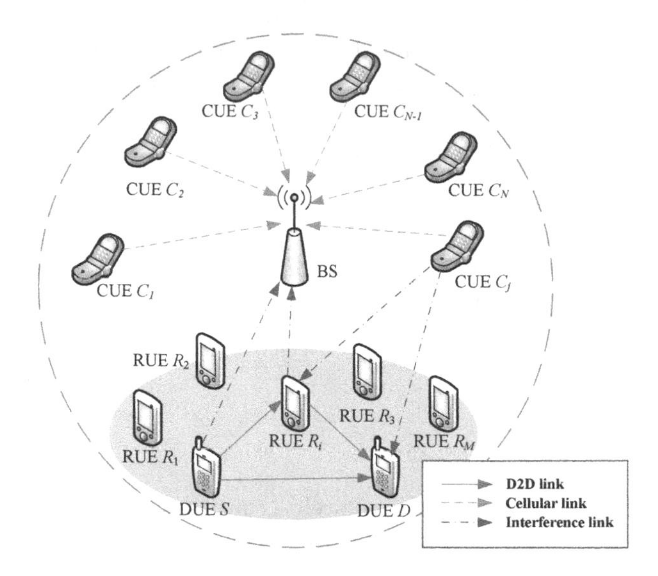

Fig. 1. A single-cell system model for relay-involved D2D communications underlying cellular network, where the uplink channel of CUE C_j is reused by a D2D communication pair in relay-assisted mode (where DUE S and DUE D are the source and destination D2D UEs, respectively, and RUE R_i is the selected relay UE).

the BS side. Once the availability of the relay-assisted D2D mode for two DUEs is being assessed by the BS, it will first try to select an optimal RUE as the relay UE for the two DUEs according to three criterias mentioned above. Hence, the most important thing to do first is to derive the mathematical expressions for the three criterias, based on which the relay selection process can be formulated into a optimization problem, as we have done in our previous work [15]. We have derived the mathematical expressions for the end-to-end data rate and the candidate RUEs' remaining battery time in the previous work. Thus, in this paper, we will put our focus on building an end-to-end transmission delay estimation model while taking the AMC strategy into account. The optimization problem formation for the relay selection process is definitely similar to that in our previous works, so we just make a brief demonstration here. The derivation process of end-to-end transmission delay involving AMC will be presented in Sect. 5. Next we will first provide some knowledge of AMC.

4 Adaptive Modulation and Coding

4.1 Principle of AMC

At physical layer, the channel state is time-varying. To support transmissions of high-speed multimedia services on limited spectrum resources, lots of adaptive technologies have been proposed, link adaptation and AMC. By adjusting transmission parameters according to the available CSI, AMC can improve spectral efficiency and link reliability to maximize data rate.

The schematic diagram for realization of AMC in a wireless communication link is depicted in Fig. 2, in which the AMC selector makes decisions on the mode of modulation and coding according to acquired CSI (e.g., signal to interference and noise ratio, SINR) from the channel estimator, and the decisions are then returned to the transmitter through a feedback channel. The transmitter adjusts modulation and coding scheme (MCS) to adapt changes of the channel. Modulation determines number of data bits that every modulation symbol can transmit, whereas coding provides the function of error correction at the receiver side via pre-adding redundant bits to the data bits before transmission. This is definitely the principle of AMC.

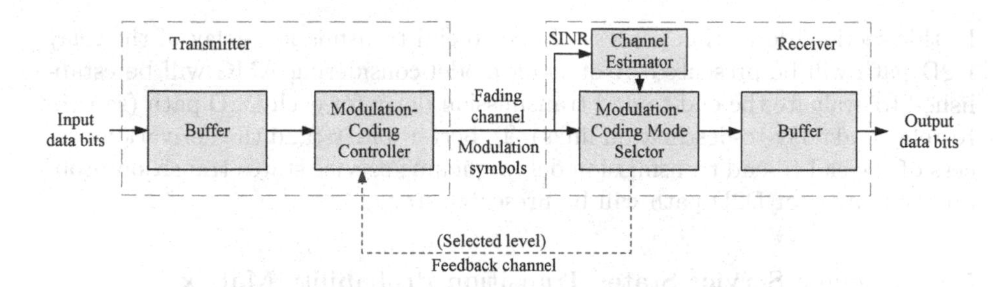

Fig. 2. Schematic diagram of realizing AMC in wireless links

4.2 SINR Boundaries in AMC

As different ranges of SINR corresponds to different MCS levels. In order to select the optimal MCS for a transmission, the SINR boundaries for each MCS level must be decided first. This is also the basis for deriving the end-to-end transmission delay of the relay D2D path in Sect. 5. Let L denote the total number of MCS levels, and we assume that the transmit power is constant. Then divide the entire SINR range into L non-overlapping consecutive intervals, with SINR boundary points denoted as $\{\Gamma_1, \Gamma_2, \cdots, \Gamma_{L+1}\}$. When

$$\gamma \in [\Gamma_l, \Gamma_{l+1}),\ l \in \{1, 2, \cdots, L+1\}, \tag{1}$$

the MCS level l is chosen, where γ is the receive SINR and l is the MCS level index. For the relay-assisted D2D path, γ can be expressed as

$$\gamma_{S,R_i}^{j} = \frac{P_S |\ h_{S,R_i}|^2}{P_{C_j} |\ h_{C_j,R_i}|^2 + \sigma^2}, \tag{2}$$

and

$$\gamma_{R_i,D}^{k} = \frac{P_{R_i} |\ h_{R_i,D}|^2}{P_{C_k} |\ h_{C_k,D}|^2 + \sigma^2}, \tag{3}$$

where γ_{S,R_i}^j represents the SINR of S-R_i link (reuse the uplink channel of CUE C_j) and $\gamma_{R_i,D}^k$ represents the SINR of R_i-D link (reuse the uplink channel of CUE C_k). P_S, P_{R_i} and P_{C_j} represent transmit power of the source DUE S, relay UE R_i, and CUE C_j, respectively. $|h_{S,R_i}|^2$, $|h_{C_j,R_i}|^2$, $|h_{R_i,D}|^2$, and $|h_{C_k,D}|^2$ denote the channel gains of $S - R_i$, $C_j - R_i$, $R_i - D$, and $C_k - D$ links, respectively. σ^2 is the variance of additive white Gaussian noise (AWGN). Based on the relationship between SINR and packet error rate (PER), the minimum required SINR for the MCS level l can be obtained for a given target PER [16]. And then we can set the boundary Γ_l for the MCS level l equalling to the minimum SINR. With the obtained Γ_l, the SINR boundaries for different MCS levels can be set.

5 End-to-End Transmission Delay Analysis with AMC

In this section, derivation process of end-to-end transmission delay of the relay D2D path will be presented. A queuing model considering AMC will be established to evaluate the end-to-end transmission delay for each D2D path (source-to-relay and relay-to-destination links). To better understand the derivation process of the end-to-end transmission delay, queuing service states transition probability matrix for D2D path will be presented first.

5.1 Queuing Service States Transition Probability Matrix

For a D2D path, the channel is time-varying and different channel states correspond to the different MCS levels. We have assumed that there are L MCS modes, and the L MCS levels available for each D2D link. For each hop of the relay-assisted D2D path, we use c_l to denote the number of data packets that can be transmitted at a time slot (we call it queuing service state) and let $C = \{c_1, c_2, \cdots, c_L\}$ denote the set of queuing service states. Obviously, queuing service state reflects the channel states and is determined by the MCS adopted. In such a case, the set of queuing service state can be described by a finite state Markov chain (FSMC) model. Use $p_{l,m}$ to denote the queuing service state transition probability from c_l to c_m, $(l,m) \in \{1, 2, \cdots, L\}$. It is also assumed that the state transitions only happen between adjacent states, i.e.,

$$p_{l,m} = 0, \ \forall\, |l - m| > 1, \ (l,m) \in \{1, 2, \cdots, L\} \tag{4}$$

Then the adjacent-state transition probability can be determined by [17]

$$p_{l,l+1} = \frac{\chi(\Gamma_{l+1})\Delta T}{\pi_l}, l = 1, \cdots, L-1, \tag{5}$$

and

$$p_{l,l-1} = \frac{\chi(\Gamma_l)\Delta T}{\pi_l}, l = 2, \cdots, L, \tag{6}$$

where π_l is the steady-state probability of state c_l, and ΔT is the duration of a time slot which is set as 1 ms. $\chi(\Gamma_l)$ represents the average number of times per unit interval that a fading signal crosses a given signal level Γ_l. $\chi(\Gamma_l)$ and π_l can be acquired referring the method in [17]. Jointly combining (4), (5), and (6) the state transition probability matrix can be obtained, which will be used in following derivation.

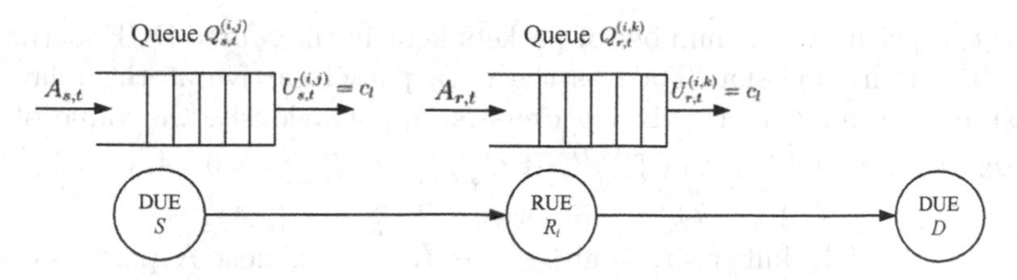

Fig. 3. End-to-end queuing model involving AMC for relay-assisted D2D path, where each UE (except for destination DUE) along the route maintains a fixed-size buffer to store arrived packets.

5.2 A Queuing Model

A queuing model for end-to-end delay estimation of a relay-assisted D2D path combining AMC is illustrated in Fig. 3, where i ($i \in \{1, 2, \cdots, M\}$) denotes the index of RUE, j ($j \in \{1, 2, \cdots, N\}$) denotes the index of CUE whose uplink channel is reused by S-R_i link, and k ($k \in \{1, 2, \cdots, N\}$) denotes the index of CUE whose uplink channel is reused by R_i-D link. To better depict the model, the time axis is evenly divided into time slots, each of which is with a duration of ΔT. When there are new data packets arriving to the source DUE in time slot t, use $A_{s,t}$ to represent the number of data packets arrived, and assume that the packet arrival process is stationary and follows Poisson distribution, i.e.,

$$\mathbb{P}(A_{s,t} = n) = \frac{(\lambda_s \Delta T)^n}{n!} \exp(-\lambda_s \Delta T), \ n \in \{0, 1, 2, \cdots\}, \tag{7}$$

where $\lambda_s \Delta T$ indicates the average number of packets that arrive at the source DUE in a time slot with an arrival rate of λ_s, i.e., the mean of $A_{s,t}$ is $\lambda_S \Delta T$. The arrived packets are stored in a buffer of the source DUE. Assume that the buffer size is K, i.e., the source DUE can store K data packets at most. $Q_{s,t}^{(i,j)}$ denotes the number of packet kept in the source DUE at the end of the tth time slot, and $U_{s,t}^{(i,j)}$ denotes the maximal number of packets that can be transmitted on S-R_i

link in the tth time slot. Obviously, $U_{s,t}^{(i,j)}$ is equal to c_l (number of transmitted data packets determined by AMC at the tth time slot), thus the queuing state transition relationship on S-R_i link can be expressed as

$$Q_{s,t}^{(i,j)} = \min\{K, \max\{0, Q_{s,t-1}^{(i,j)} - U_{s,t}^{(i,j)}\} + A_{s,t}\}, \tag{8}$$

where $Q_{s,t-1}^{(i,j)}$ denotes the number of packets kept in the source DUE at the end of the $(t-1)$th time slot. Then as the data packets arrive at the relay UE, we assume the number of arrival packets is $A_{r,t}$. Obviously, the value of $A_{r,t}$ is correlated with $Q_{s,t-1}^{(i,j)}$ and $U_{s,t}^{(i,j)}$. If $Q_{s,t-1}^{(i,j)} \geq U_{s,t}^{(i,j)} \geq 0$, $A_{r,t} = U_{s,t}^{(i,j)}$; if $U_{s,t}^{(i,j)} \geq Q_{s,t-1}^{(i,j)} \geq 0$, $A_{r,t} = Q_{s,t-1}^{(i,j)}$; otherwise $(U_{s,t}^{(i,j)} = 0)$, $A_{r,t} = 0$.

At the relay UE, buffer size is also set as K, i.e., at most K packets can be stored at a relay UE. In the tth time slot, the number of packets queued in the relay UE's buffer is denoted as $Q_{r,t}^{(i,k)}$ and the maximal number of packets that can be transmitted on the $R_i - D$ link is denoted as $U_{r,t}^{(i,k)}$. Accordingly, the queuing state transition of $R_i - D$ link can be expressed as

$$Q_{r,t}^{(i,k)} = \min\{K, \max\{0, Q_{r,t-1}^{(i,k)} - U_{r,t}^{(i,k)}\} + A_{r,t}\} \tag{9}$$

According to (8), $Q_{s,t}^{(i,j)}$ is closely related to $Q_{s,t-1}^{(i,j)}$, $U_{s,t}^{(i,j)}$, and $A_{s,t}$. As $A_{s,t}$ follows Poisson distribution and is independent of $Q_{s,t-1}^{(i,j)}$ and $U_{s,t}^{(i,j)}$ according to (7), it can be isolated from state $\{Q_{s,t-1}^{(i,j)}, U_{s,t}^{(i,j)}\}$, where $Q_{s,t-1}^{(i,j)}$ and $U_{s,t}^{(i,j)}$ are closely related. To analyze the system behavior, we use a two-dimension state $\{Q_{s,t-1}^{(i,j)}, U_{s,t}^{(i,j)}\}$ to depict both queuing and servicing states. Generally, the queuing and servicing states transition can be described by a FSMC model, i.e., $Q_{s,t}^{(i,j)}$ is only determined by its former state and $U_{s,t}^{(i,j)}$ (or c_l) is determined by its adjacent-state which has been stated in the previous part.

The queuing and AMC behaviors on $S-R_i$ link can be described by a varying-rate queuing system. To do this, the joint queuing and service state transition probability matix should be first formed. The transition probability from states $(Q_{s,t-1}^{(i,j)} = \varphi_s, U_{s,t}^{(i,j)} = c)$ to state $(Q_{s,t}^{(i,j)} = \theta_s, U_{s,t+1}^{(i,j)} = d)$ can be expressed as

$$p_{(\varphi_s,c),(\theta_s,d)}^{(i,j)} = \mathbb{P}(Q_{s,t}^{(i,j)} = \theta_s, U_{s,t+1}^{(i,j)} = d | Q_{s,t-1}^{(i,j)} = \varphi_s, U_{s,t}^{(i,j)} = c), \tag{10}$$

where $\varphi_s, \theta_s \in [0, 1, 2, \cdots, K]$ and $c, d \in [c_1, c_2, \cdots, c_L]$.

Due to the fact that $U_{s,t+1}^{(i,j)}$ is only dependent on $U_{s,t}^{(i,j)}$, (10) can be further simplified as

$$p_{(\varphi_s,c),(\theta_s,d)}^{(i,j)} = \mathbb{P}(U_{s,t+1}^{(i,j)} = d | U_{s,t}^{(i,j)} = c)\mathbb{P}(Q_{s,t}^{(i,j)} = \theta_s | Q_{s,t-1}^{(i,j)} = \varphi_s, U_{s,t}^{(i,j)} = c), \tag{11}$$

where $\mathbb{P}(U_{s,t+1}^{(i,j)} = d | U_{s,t}^{(i,j)} = c)$ is the element in the channel service state transmission probability matrix which has been obtained in the previous subsection. Thus the key work is to compute $\mathbb{P}(Q_{s,t}^{(i,j)} = \theta_s | Q_{s,t-1}^{(i,j)} = \varphi_s, U_{s,t}^{(i,j)} = c)$ to get the joint state transition probability.

Based on (8), we can obtain

$$\mathbb{P}\left(Q_{s,t}^{(i,j)} = \theta_s | Q_{s,t-1}^{(i,j)} = \varphi_s, U_{s,t}^{(i,j)} = c\right) = \begin{cases} 0, & \text{if } 0 \le \theta_s < K, \theta_s < \max\{0, \varphi_s - c\} \\ \mathbb{P}\left(A_{s,t} = \theta_s - \max\{0, \varphi_s - c\}\right), & \text{if } 0 \le \theta_s < K, \theta_s \ge \max\{0, \varphi_s - c\} \\ 1 - \sum\limits_{\theta_s=0}^{K-1} \mathbb{P}\left(A_{s,t} = \theta_s - \max\{0, \varphi_s - c\}\right), & \text{if } \theta_s = K \end{cases} \tag{12}$$

and then the transition probability from state $(Q_{s,t-1}^{(i,j)} = \varphi_s, U_{s,t}^{(i,j)} = c)$ to state $(Q_{s,t}^{(i,j)} = \theta_s, U_{s,t+1}^{(i,j)} = d)$ can be obtained.

As the queuing process can be modeled as a FSMC, the queuing state will be steady after a sufficiently long time, under which each queuing state $Q_{s,t}^{(i,j)}$ corresponds to a stationary probability. Let $\boldsymbol{\Omega}_s^{(i,j)}$ denote the stationary probability vector for the queuing process on S-R_i link. It can be obtained by

$$\begin{cases} \boldsymbol{\Omega}_s^{(i,j)} = \boldsymbol{\Omega}_s^{(i,j)} \boldsymbol{P}_s^{(i,j)}, \\ \sum\limits_{c=c_1}^{c_L} \sum\limits_{\theta_s=0}^{K} \Omega_{s,(\theta_s,c)}^{(i,j)} = 1, \end{cases} \tag{13}$$

where $\boldsymbol{P}_s^{(i,j)}$ is the queuing and servicing joint states transition probability matrix, whose element can be derived by (11). Let $\Omega_{s,(\theta_s,c)}^{(i,j)}$ denote the stationary probability for the queuing state of the source DUE's buffer, i.e., it corresponds to the situation that queue length at the source DUE buffer is θ_s while the channel state is c. $\Omega_{s,(\theta_s,c)}^{(i,j)}$ then can be expressed as

$$\Omega_{s,(\theta_s,c)}^{(i,j)} = \lim_{t \to \infty} \mathbb{P}(Q_{s,t-1}^{(i,j)} = \theta_s, U_{s,t}^{(i,j)} = c) \tag{14}$$

Apparently, $\Omega_{s,(\theta_s,c)}^{(i,j)}$ is an element of $\boldsymbol{\Omega}_s^{(i,j)}$.

Based on the above, we can further get the average packet queue length $\overline{Q_s}^{(i,j)}$, and its mathematical expression is

$$\overline{Q_s}^{(i,j)} = \sum_{c=c_1}^{c_L} \sum_{\theta_s=0}^{K} \left(\theta_s \times \Omega_{s,(\theta_s,c)}^{(i,j)}\right) \tag{15}$$

Similarly, the average packet queue length $\overline{Q_r}^{(i,k)}$ of R_i-D link can be further derived. The derivation process is depicted in Appendix A.

5.3 End-to-End Delay

There are two main factors leading to the transmission delay on S-R_i link and R_i-D link. One is the buffer queues on devices along the transmission route and the other is the packet retransmissions caused by packet losses and packet errors on the links. In this paper, we only consider the finite-length of the buffers at DUE and RUEs as the causes for packet losses. As the length of buffers are finite, the data packets will be dropped when a buffer is full. Use $\nu_s^{(i,j)}$ to denote the average transmission delay for the packet queuing on the S-R_i link. Then according to the Littles law, $\nu_s^{(i,j)}$ can be expressed as

$$\nu_s^{(i,j)} = \frac{\overline{Q_s}^{(i,j)}}{T_s^{(i,j)}} = \frac{\overline{Q_s}^{(i,j)}}{\mathbb{E}\left[A_{s,t}\right] \times \left(1 - p_s^{(i,j)}\right)}, \tag{16}$$

where $\overline{Q_s}^{(i,j)}$ denotes the average packet queue length of the source DUEs which has been obtained via (15). $\mathbb{E}\left[A_{s,t}\right]$ is the average number of arrived data packets at the tth time slot and equals to $\lambda_S \Delta T$. $p_s^{(i,j)}$ is the packet loss rate, which can be expressed as

$$p_s^{(i,j)} = \lim_{T \to \infty} \frac{\sum\limits_{t=1}^{T} D_{s,t}^{(i,j)}}{\sum\limits_{t=1}^{T} A_{s,t}}, \tag{17}$$

where $A_{s,t}$ represents the number of new arrived data packets and $D_{s,t}^{(i,j)}$ denotes the dropped packets at the tth time slot. $D_{s,t}^{(i,j)}$ can be expressed as

$$D_{s,t}^{(i,j)} = \max\left[0, A_{s,t} - K + \max\left(0, Q_{s,t-1}^{(i,j)} - U_{s,t}^{(i,j)}\right)\right] \tag{18}$$

According to [15], stationary distribution for $A_{s,t}$, $Q_{s,t-1}^{(i,j)}$, and $U_{s,t}^{(i,j)}$ all exist, and here we use A_s, $Q_s^{(i,j)}$, and $U_s^{(i,j)}$ to denote them respectively. Then we have

$$D_s^{(i,j)} = \max\left[0, A_s - K + \max\left(0, Q_s^{(i,j)} - U_s^{(i,j)}\right)\right], \tag{19}$$

where $A_s = n$, $n \in [0, 1, 2, \cdots]$, $Q_s^{(i,j)} = \theta_s$, $\theta_s \in [0, 1, 2, \cdots, K]$, and $U_s^{(i,j)}$ is determined according to the selected MCS level.

Also referring to our previous work [15], packet loss rate can be obtained as

$$p_s^{(i,j)} = \lim_{T \to \infty} \frac{\sum\limits_{t=1}^{T} D_{s,t}^{(i,j)}}{\sum\limits_{t=1}^{T} A_{s,t}} = \frac{\mathbb{E}\left[D_s^{(i,j)}\right]}{\mathbb{E}\left[A_s\right]} = \frac{\mathbb{E}\left[D_s^{(i,j)}\right]}{\lambda_s \Delta T}, \tag{20}$$

where $E\left[D_s^{(i,j)}\right]$ can be obtained as

$$\mathbb{E}\left[D_s^{(i,j)}\right] = \sum_{n=0}^{\infty} \sum_{\theta_s=0}^{K} D_s^{(i,j)} \times \mathbb{P}\left(A_s = n\right) \times \mathbb{P}\left(Q_s^{(i,j)} = \theta_s\right), \tag{21}$$

in which $\mathbb{P}(A_s = n) = \mathbb{P}(A_{s,t} = n)$. As the stationary probability for the queuing states have been obtained above, combining (19)–(21), the transmission delay on S-R_i link can be obtained. Similarly, the transmission delay $\nu_r^{(i,k)}$ on the R_i-D link can be derived, which is depicted in Appendix B.

Accordingly, the end-to-end transmission delay ν of the entire relay-assisted D2D path can be acquired, i.e.,

$$\nu = \nu_s^{(i,j)} + \nu_r^{(i,k)} \tag{22}$$

If BS captures CSI of all links, it selects the optimal relay UE that corresponds to longer operation time, higher end-to-end data rate, and lower end-to-end transmission delay. The format of the objective function combining these three criterias can be found in [15].

6 Performance Evaluation

In this section, we will assess the end-to-end delay performance of the relay selection scheme when taking into account AMC. A single-cell scenario is established in Matlab environment and the cell radius is set as 500 m. All CUEs are randomly distributed in the cell and only one D2D pair is included, where the source DUE randomly locates at least half of the cell radius away from BS, and the destination DUE locates randomly on a circle centering at the source DUE. All RUEs are randomly distributed in the circular area with the source DUE as its center. The number of data bits in a packet is set as 512. The parameters related to ACM is presented in Table 1, and the maximum SINR boundary is ∞ which is not presented in the table. Other major parameters used in the simulation can be found in Table I of our previous work [15].

Table 1. Major parameters related to AMC.

Level l	R_l (bits/sym)	SINR boundary Γ_l
1	0	0
2	2	2.17
3	4	6.82
4	4	16.94
5	6	40.7

Figure 4 shows the delay performance with different average packet arrival rates at source DUE under two relay selection strategies and different channel bandwidths. The delay performances on the relay selection strategies with and without AMC are compared. It is seen that with a fixed channel bandwidth and relay selection strategy, increasing the average packet arrival rate leads to an increase of the transmission delay. Obviously, when other conditions do not

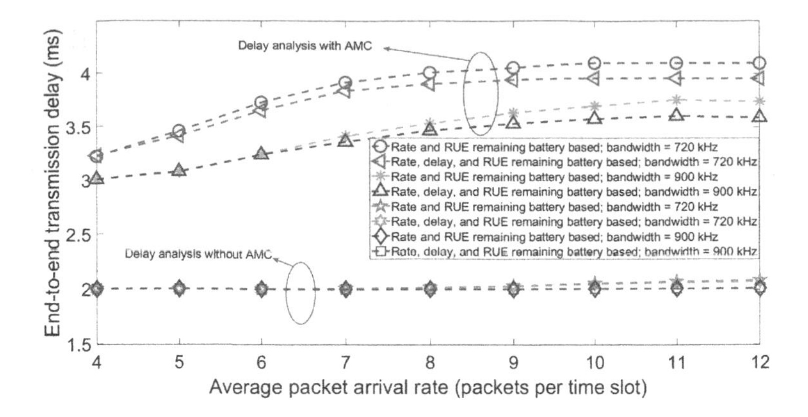

Fig. 4. End-to-end transmission delay versus a varying packet arrival rate at source DUE under different relay selection strategies and channel bandwidths. Channel bandwidths are 720 kHz and 900 kHz, numbers of CUEs and RUEs are 20 and 10, respectively, the distance between source and destination DUEs is 100 m and the buffer size of each UE is 15.

change, increasing arrival rate at DUE S means increasing packet dropping probability at source DUE and the selected RUE due to their finite-length buffers, so the transmission delay on each link will increase. No matter which relay selection strategy is used (either the rate and RUE remaining battery based relay selection strategy or the rate, delay, and RUE remaining battery based relay selection strategy), it is seen that delay will dramatically increase while reducing channel bandwidth, because a smaller bandwidth corresponds to a lower data rate on each link which will decrease the service rate of the queuing system established on each link. The rate and RUE remaining battery based strategy has worse delay performance, while the rate, delay and RUE remaining battery based strategy offers a better delay performance (both in analysis with and without AMC), validating the effectiveness of taking into account the end-to-end delay criteria in relay selection.

In Fig. 4, it is seen that under the same conditions the scheme with AMC corresponds to higher transmission delay than that without AMC. This is reasonable, because the latter always evaluates the performance based on the data rate calculated by Shannon formula, that is, the calculated data rate is always maximal under the current channel condition. The situation with AMC (the data rate is determined by the selected MCS level) shows a more realistic result. In addition, it is seen that with same packet arrival rate and channel bandwidth, the end-to-end delay performance of the rate, delay, and RUE remaining battery based relay selection strategy is better than that of the rate and RUE remaining battery based relay selection strategy.

Figure 5 shows the delay performance with varying distances between DUEs S and D. Increasing separation distance between source and destination DUEs will enlarge the path loss, and result in a decreasing service rate, so the delay

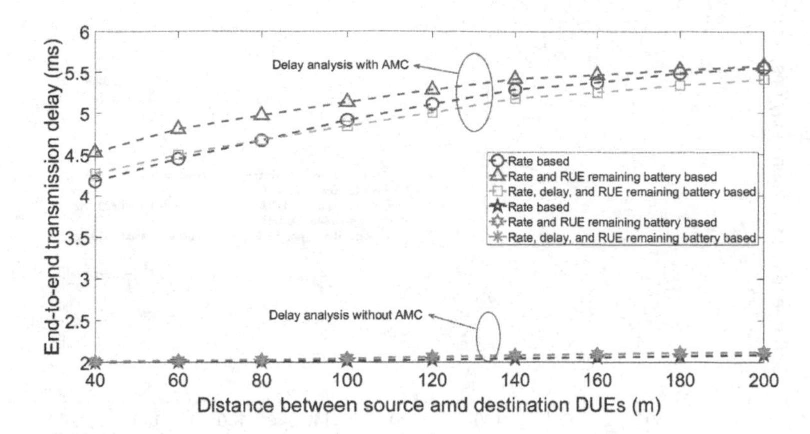

Fig. 5. End-to-end transmission delay versus distance between source and destination DUEs under different relay selection strategies. Channel bandwidth is 720 kHz, the numbers of CUEs and RUEs are 20 and 10, respectively, the average number of packets arrived in each time slot is 10, and the buffer size is 15.

will increase no matter which relay selection strategy is used. It is seen that the strategies with AMC show higher transmission delay than those without AMC due to difference of data rate calculations. In addition, for the strategies without AMC, with same system parameters, the rate based relay selection strategy offers a much better delay performance, compared to other two relay selection strategies. Although the rate based relay selection strategy gives the best delay performance, the fact that RUE remaining battery time is not considered in relay selection may cause that a RUE with short remaining operation time is selected as the relay UE. The rate and RUE remaining battery based strategy has the worst delay performance. While the rate, delay, and RUE remaining battery based strategy offers a delay performance between the other two strategies.

As shown in Fig. 5, for the delay performance with AMC, the rate and RUE remaining battery based strategy still has the worst delay performance. We can notice that the rate based strategy has the best delay performance under certain distance condition, but the rate, delay and RUE remaining battery based strategy will be with the best delay performance when distance is larger than a threshold. The SINR of link is larger while the distance between source and destination nodes is shorter. SINR impacts on the MCS level selection, and higher SINR and corresponding MCS level corresponds to smaller delay, hence the data rate based selection has the best performance while the distance between source and destination DUEs is short. For the relay-assisted D2D mode, although SINR of each link on the relay path may decrease as the distance becomes larger, the entire relay-assisted D2D path may still corresponds to the same MCS level due to the selection cross layer selection criteria.

The delay performance with varying distances between source and destination DUEs under different channel bandwidths is shown in Fig. 6, which only illustrates delay performance with AMC. Only two relay selection strategies are

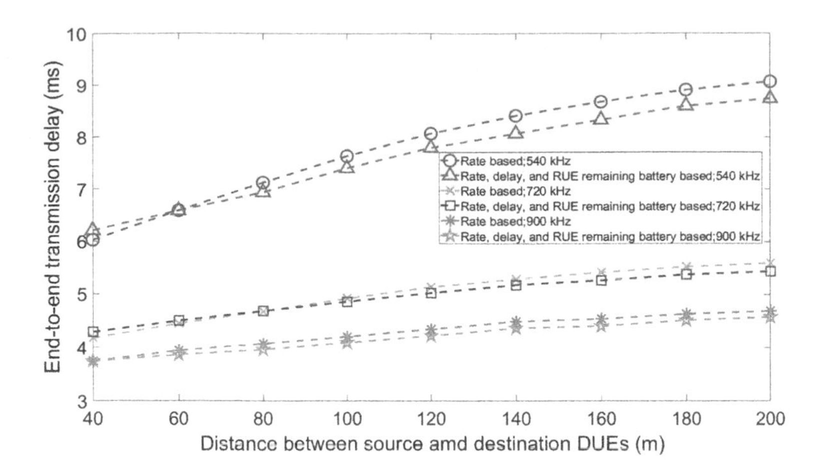

Fig. 6. End-to-end transmission delay versus distance between source and destination DUEs under different relay selection strategies. The numbers of CUEs and RUEs are 20 and 10, respectively, the average number of packets arrived in each time slot is set to 10, and the buffer size of each UEs is 15.

simulated, and the bandwidth' s influence on the delay performance is straightforward.

7 Conclusions

In this paper, we considered the influence of the channels' time-varying property on the relay selection scheme for relay-assisted D2D communication in cellular system. We took into account the delay criteria based on a queuing model involving AMC mechanism. The derivation process for the end-to-end transmission delay of the relay-assisted D2D path was provided. The delay performances were finally evaluated via simulations to validate the effectiveness of the derivation.

A Appendix

We have obtained the queuing states transition relationship of the R_i-D link in (9). We can formulate the transition probability from state $(Q_{r,t-1}^{(i,k)} = \varphi_r, U_{r,t}^{(i,k)} = e)$ to state $(Q_{r,t}^{(i,k)} = \theta_r, U_{r,t+1}^{(i,k)} = f)$ as

$$p_{(\varphi_r,e),(\theta_r,f)}^{(i,k)} = \mathbb{P}(Q_{r,t}^{(i,k)} = \theta_r, U_{r,t+1}^{(i,k)} = f | Q_{r,t-1}^{(i,k)} = \varphi_r, U_{r,t}^{(i,k)} = e), \quad (23)$$

which can also be further expressed as

$$p_{(\varphi_r,e),(\theta_r,f)}^{(i,k)} = \mathbb{P}(U_{r,t+1}^{(i,k)} = f | U_{r,t}^{(i,k)} = e)\mathbb{P}(Q_{r,t}^{(i,k)} = \theta_r | Q_{r,t-1}^{(i,k)} = \varphi_r, U_{r,t}^{(i,k)} = e), \quad (24)$$

where φ_r, $\theta_r \in [0, 1, 2, \cdots, K]$ and e, $f \in [c_1, c_2, \cdots, c_L]$. Similar to (12), we also need acquire the $\mathbb{P}(Q_{r,t}^{(i,k)} = \theta_r | Q_{r,t-1}^{(i,k)} = \varphi_r, U_{r,t}^{(i,k)} = e)$ by

$$\mathbb{P}\left(Q_{r,t}^{(i,k)} = \theta_r | Q_{r,t-1}^{(i,k)} = \varphi_r, U_{r,t}^{(i,k)} = e\right) = \begin{cases} 0, & \text{if } 0 \leq \theta_r < K, \theta_r < \max\{0, \varphi_r - e\}, \\ \mathbb{P}\left(A_{r,t} = \theta_r - \max\{0, \varphi_r - e\}\right), & \text{if } 0 \leq \theta_r < K, \theta_r \geq \max\{0, \varphi_r - e\}, \\ 1 - \sum\limits_{\theta_r=0}^{K-1} \mathbb{P}\left(A_{r,t} = \theta_r - \max\{0, \varphi_r - e\}\right), & \text{if } \theta_r = K. \end{cases} \tag{25}$$

Assume that the number of the arrival packets is ξ. To acquire transition probability, we need to first get the probability of the $A_{r,t}$, i.e., $\mathbb{P}(A_{r,t} = \xi)$. $\mathbb{P}(A_{r,t})$ is related to $U_{s,t}^{(i,j)}$ and $Q_{s,t-1}^{(i,j)}$, which have been discussed in the Subsect. 5.2. We can use $\mathbb{P}\left(Q_{s,t-1}^{(i,j)} = \theta_s, U_{s,t}^{(i,j)} = c\right)$ to replace $\mathbb{P}(A_{r,t})$. Theoretically, we can use stationary probability to replace the transient probability for a specific queue length, i.e., we can use $\lim_{t\to\infty}(Q_{s,t-1}^{(i,j)} = \theta_s, U_{s,t}^{(i,j)} = c)$ replace $\mathrm{P}\left(Q_{s,t-1}^{(i,j)} = \theta_s, U_{s,t}^{(i,j)} = c\right)$. Accordingly, $\mathbb{P}(A_{r,t} = \xi)$ can be expressed as

$$\mathbb{P}\left(A_{r,t} = \xi\right) = \sum_{\theta_s \in [0,1,2,\ldots,K]} \Omega_{s,(\theta_s,c)}^{(i,j)}, \text{ for } (c = \xi, \theta_s > \xi) \text{ or } (c \geq \xi, \theta_s = \xi) \tag{26}$$

Based on the above calculation, the transition probability for queuing state of R_i-D link can be obtained. Similar to S-R_i link, there also exists a stationary probability vector $\boldsymbol{\Omega}_r^{(i,k)}$ for the queuing states of R_i-D link. The calculation of the $\boldsymbol{\Omega}_r^{(i,k)}$ is as

$$\begin{cases} \boldsymbol{\Omega}_r^{(i,k)} = \boldsymbol{\Omega}_r^{(i,k)} \boldsymbol{P}_r^{(i,k)}, \\ \sum\limits_{e=c_1}^{c_L} \sum\limits_{\theta_s=0}^{K} \Omega_{r,(\theta_r,e)}^{(i,k)} = 1, \end{cases} \tag{27}$$

where $\Omega_{r,(\theta_r,e)}^{(i,k)}$ is the element of $\boldsymbol{\Omega}_r^{(i,k)}$, which represents the stationary probability of the state that θ_r packets are queued in the buffer of R_i corresponding to the service state e, i.e.,

$$\Omega_{r,(\theta_r,\mathrm{e})}^{(i,\mathrm{k})} = \lim_{t\to\infty} \mathbb{P}\left(Q_{r,t-1}^{(i,k)} = \theta_r, U_{r,t}^{(i,k)} = e\right). \tag{28}$$

Then the average packet queue length at R_i can be obtained as

$$\bar{Q}_r^{(i,k)} = \sum_{e=c_1}^{c_L} \sum_{\theta_r=0}^{K} \left(\theta_r \times \Omega_{r,(\theta_r,e)}^{(i,k)}\right). \tag{29}$$

B Appendix

Similar to the transmission delay of the S-R_i link, the transmission delay $\nu_r^{(i,k)}$ on the R_i-D link caused by queuing at the buffer of R_i can be expressed as

$$\nu_r^{(i,k)} = \frac{\overline{Q_r}^{(i,k)}}{T_r^{(i,k)}} = \frac{\bar{Q}_r^{(i,k)}}{\mathbb{E}\left[A_{r,t}\right] \times \left(1 - p_r^{(i,k)}\right)}, \tag{30}$$

where $p_r^{(i,k)}$ denotes the packet loss rate on R_i-D link due to the buffer overflow, $T_r^{(i,k)} = \mathbb{E}\left[A_{r,t}\right] \times \left(1 - p_r^{(i,k)}\right)$ denotes the average throughput of the link, and $\mathbb{E}\left[A_{r,t}\right]$ is the average number of data packets that arrive to R_i in a time slot. $\mathbb{E}\left[A_{r,t}\right]$ can be calculated by

$$\mathbb{E}\left[A_{r,t}\right] = \sum_{\xi=0}^{\min\{K, U_s^{(i,j)}\}} \xi \times \mathbb{P}\left(A_{r,t} = \xi\right), \tag{31}$$

where $\mathbb{P}\left(A_{r,t} = \xi\right)$ is given by (26).

The packet loss rate $p_r^{(i,k)}$ can be calculated as

$$p_r^{(i,k)} = \lim_{T \to \infty} \frac{\sum_{t=1}^{T} D_{r,t}^{(i,k)}}{\sum_{t=1}^{T} A_{r,t}^{(i,k)}} = \frac{\mathbb{E}\left[D_r^{(i,k)}\right]}{\mathbb{E}\left[A_r\right]}, \tag{32}$$

where $D_r^{(i,k)}$ represents the number of dropped packets on the R_i-D link in the tth time slot. According to [15], the stationary distribution $D_r^{(i,k)}$ of $D_{r,t}^{(i,k)}$ can be acquired by

$$\mathbb{E}\left[D_r^{(i,k)}\right] = \sum_{\xi=0}^{\min\{K, U_s^{(i,j)}\}} \sum_{\theta_r=0}^{K} D_r^{(i,k)} \times \mathbb{P}\left(A_r = \xi\right) \times \mathbb{P}\left(Q_r = \theta_r\right), \tag{33}$$

and then $\mathbb{E}\left[D_r^{(i,k)}\right]$ can be further acquired as

$$\mathbb{E}\left[D_r^{(i,k)}\right] = \sum_{\xi=0}^{\min\{K, U_s^{(i,j)}\}} \sum_{\theta_r=0}^{K} \max\left[0, \xi - K + \max\left(0, \theta_r - U_r^{(i,k)}\right)\right] \mathbb{P}(A_r = \xi) \mathbb{P}\left(Q_r^{(i,k)} = \theta_r\right), \tag{34}$$

where $Q_r^{(i,k)} = \theta_r$, $\theta_r \in [0, 1, 2, \ldots, K]$ is the stationary distribution for the queuing states, and $U_r^{(i,k)}$ is determined by the selected MCS level.

By inserting the obtained values of $\mathbb{E}\left[D_r^{(i,k)}\right]$ and $\mathbb{E}[A_r]$ into (32), we can get the transmission delay of R_i-D link.

References

1. Liu, J., Kato, N., Ma, J., Kadowaki, N.: Device-to-device communication in LTE-advanced networks: a survey. IEEE Commun. Surv. Tutor. **17**(4), 1923–1940 (2015)
2. Asadi, A., Wang, Q., Mancuso, V.: A survey on device-to-device communication in cellular networks. IEEE Commun. Surv. Tutor. **16**(4), 1801–1819 (2014)
3. Yu, G., Xu, L., Feng, D., Yin, R., Li, G.Y., Jiang, Y.: Joint mode selection and resource allocation for device-to-device communications. IEEE Trans. Commun. **62**(11), 3814–3824 (2014)
4. Feng, D., Lu, L., Yuan-Wu, Y., Li, G.Y., Feng, G., Li, S.: Device-to-device communications underlaying cellular networks. IEEE Trans. Commun. **61**(8), 3541–3551 (2013)
5. Kim, H., Na, J., Cho, E.: Resource allocation policy to avoid interference between cellular and D2D links/and D2D links in mobile networks. In: The International Conference on Information Networking (ICOIN), pp. 588–591 (2014)
6. Guo, B., Sun, S., Gao, Q.: Power optimization of D2D communications underlying cellular networks in multi-user scenario. In: IEEE International Conference on Communication Systems 2014, pp. 212–216 (2014)
7. Tang, H., Ding, Z.: Mixed mode transmission and resource allocation for D2D communication. IEEE Trans. Wirel. Commun. **15**(1), 162–175 (2016)
8. Hasan, M., Hossain, E., Kim, D.I.: Resource allocation under channel uncertainties for relay-aided device-to-device communication underlaying LTE-A cellular networks. IEEE Trans. Wirel. Commun. **13**(4), 2322–2338 (2014)
9. Hasan, M., Hossain, E.: Distributed resource allocation for relay-aided device-to-device communication: a message passing approach. IEEE Trans. Wirel. Commun. **13**(11), 6326–6341 (2014)
10. Zhou, B., Hu, H., Huang, S., Chen, H.: Intra cluster device-to-device relay algorithm with optimal resource utilization. IEEE Trans. Veh. Technol. **62**(5), 2315–2326 (2013)
11. Nomikos, N., et al.: A survey on buffer-aided relay selection. IEEE Commun. Surv. Tutor. **18**(2), 1073–1097 (2016)
12. Ibrahim, A.S., Sadek, A.K., Su, W., Liug, K.J.R.: Cooperative communications with relay-selection: when to cooperate and whom to cooperate with? IEEE Trans. Wirel. Commun. **7**(7), 2814–2827 (2008)
13. Poulimeneas, D., Charalambous, T., Nomikos, N., Krikidis, I., Vouyioukas, D., Johansson, M.: Delay- and diversity-aware buffer-aided relay selection policies in cooperative networks. In: IEEE Wireless Communications and Networking Conference, pp. 1–6 (2016)
14. Miao, M., Sun, J., Shao, S.: A cross-layer relay selection algorithm for D2D communication system. In: International Conference on Wireless Communication and Sensor Network (WCSN), pp. 448–543 (2014)
15. Ma, R., Chang, Y., Chen, H., Chiu, C.: On relay selection schemes for relay-assisted D2D communications in LTE-a systems. IEEE Trans. Veh. Technol. **66**(9), 8303–8314 (2017)
16. Liu, Q., Zhou, S., Giannakis, G.B.: Queuing with adaptive modulation and coding over wireless links: cross-layer analysis and design. IEEE Trans. Wirel. Commun. **4**(3), 1142–1153 (2005)
17. Zhang, Q., Kassam, S.A.: Finite state Markov model for Rayleigh fading channels. IEEE Trans. Commun. **47**(11), 1688–1692 (1999)

An Outage Probability Based Channel Bonding Algorithm for 6G Network

Weifeng Sun(✉), Guanghao Zhang, Yiming Zhou, and Rui Guo

School of Software, Dalian University of Technology, Dalian, China
wfsun@dlut.edu.cn, zguanghao@163.com, hakura_zym@qq.com,
329000842@qq.com

Abstract. Multiple input multiple output (MIMO) can greatly improve the throughput and frequency utilization of wireless transmission systems. In the IEEE 802.11ax protocol, 4G, 5G, and even future 6G communications, MIMO has greatly increased the throughput of transmissions. Proper channel bonding strategies can improve channel utilization and transmission throughput. This paper proposes an outage probability based channel bonding algorithm (OP-CB). OP-CB calculates the outage probability according to the average signal-to-noise ratio (SNR) and the threshold signal-to-noise ratio. SNR is affected by transmission power and transmission distance. And then OP-CB performs channel bonding according to the outage of data transmission. The algorithm proposed in this paper can be used in massive MIMO (mMIMO) in 5G and future 6G communications. Combing the outage probability with the channel bonding technique can improve the success rate of data transmissions. In addition, we use software defined network (SDN) to improve the efficiency of the algorithm in 6G network. The simulation results show that OP-CB can improve the throughput of data transmission by about 40% when the transmission power is low or the transmission distance is large. OP-BC can also reduce the bit error rate (BER).

Keywords: mMIMO · Outage probability · Channel bonding · 6G communication

1 Introduction

With the popularity of 5G communication, 6G communication has gradually become a hot topic. Compared with 4G and 5G communication, the channel frequency used in 6G communication will be higher, and the number of channels will be much larger. Therefore, 6G communication will support more users to transmit at a higher rate. In addition, the combination of IPv6 and 6G networks brings many new features to 6G communications, which can be used not only in daily communication, but also in industrial fields, such as the Industrial Internet of Things (IIoT). Studying the reasonable utilization of channel resources in 6G networks is crucial to improving the overall performance of 6G networks.

X. Wang et al. (Eds.): 6GN 2020, LNICST 337, pp. 144–159, 2020.
https://doi.org/10.1007/978-3-030-63941-9_11

MIMO builds multiple channels between the sender and the receiver by using multiple antennas, which greatly improves the channel capacity [1]. Multi-User Multiple-Input Multiple-Output (MU-MIMO) in IEEE 802.11ax protocol and 8-antenna MIMO in 4G communication are both typical MIMO systems. The 256–1024 antenna mMIMO used in 5G communication significantly increases system capacity and realizes reliable transmission through spatial multi-grading and multiplexing technology, which plays a key role in mobile communication [2]. In MIMO systems, a good channel binding strategy can maintain good channel utilization while ensuring good transmission quality. 6G communication is expected to use 10,000-antenna MIMO technology in the terahertz (THz) frequency band [3]. As shown in Fig. 1, the frequency band used by 6G communication is much higher than 4G and 5G. In such MIMO systems with more channels, a reasonable channel bonding strategy is very important.

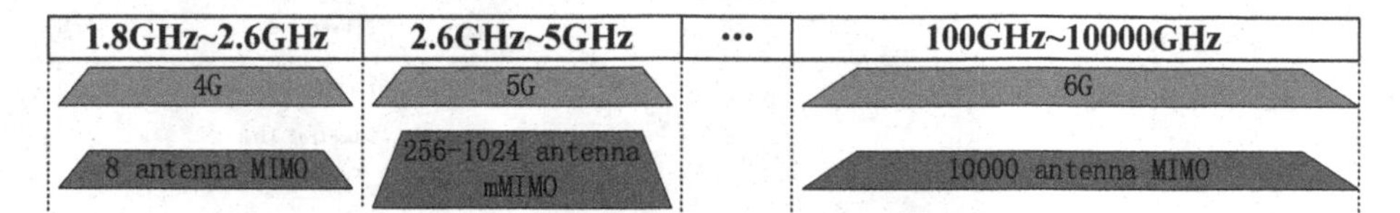

Fig. 1. Comparison of 4G, 5G, 6G bands.

D2D communication is also one of the hotspot technologies in 5G at present, and it is also an attractive research direction in 6G in the future. In the condition of many densely distributed devices in IIoT, D2D is an efficient communication method. D2D is a new technology that allows users to communicate directly through shared cellular resources with or without the control of a cellular system. In this way, the channel resource utilization can be greatly improved. The mMIMO-D2D system model is shown in Fig. 2.

Base stations (BS) and users have a large number of available channels. In this model u1 communicates with other users through the base station, u2 and u3 communicate directly (D2D) with each other under the control of the base station, and u4 and u5 directly perform D2D communication. In such a communication system with small transmission power, short transmission distance, and large number of nodes, a suitable dynamic channel bonding strategy is even more needed.

The IEEE 802.11ax is a typical protocol using MIMO technology. The channel bandwidth of this protocol supports 20 MHz, 40 MHz, and 80 MHz, and even can reach to 160 MHz through the channel bonding mechanism. Focusing on the IEEE 802.11ax protocol, this paper studies a dynamic channel bonding algorithm. By analyzing the outage probability and its influencing factors, we proposed an outage probability based channel bonding (OP-CB) algorithm for channel bonding in MIMO system. The outage probability is determined by the average SNR of the channel. Transmission power and transmission distance are important factors in received average SNR. OP-CB gets the average SNR based on the collected channel information, and then calculates the outage probability, and then dynamically bonds the channel according to the outage probability. The algorithm can be used in massive MIMO (mMIMO) in 5G and future 6G communication.

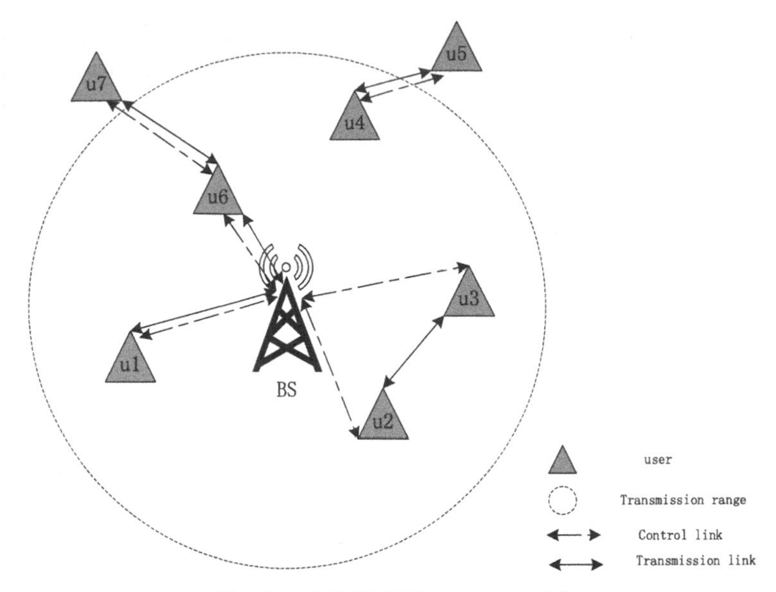

Fig. 2. mMIMO-D2D system model.

The rest parts of this paper are arranged as follows: the second part is related work. The third part introduces the concept of outage probability and its influencing factors. The fourth part introduces the OP-CB algorithm. The fifth part introduces the simulation of OP-CB in MATLAB and NS3 and the simulation results are analyzed. The last part concludes this paper.

2 Related Work

Channel bonding methods in a wireless network can be divided into static channel allocation, adaptive channel allocation, and variable bandwidth channel allocation. Static channel allocation allocates channels for each user when the network is deployed and each user will reduce the channel utilization due to load unbalance. Adaptive channel allocation considers the change of network states to make a quick respond and then adjust the channel allocation of each user. In paper [4], Yuan Zhao et al. proposed a channel bonding scheme based on packet loss mechanism and introduced a packet discarding mechanism to help the controller reduce the average delay of the system. In paper [5], YuKi Takei et al. studied the multi-interface parallel data transmission introduced into the 802.11n protocol to improve the performance of wireless communication in small pipes. The results showed that when the communication distance is less than 4 m, using multi-channel bonding could improve the communication quality. In paper [6], Aslam et al. studied the signal Rayleigh fading in massive MIMO and found that the distribution of the antenna array had a large impact on the throughput of the fading transmission. In paper [7], authors analyzed variable channels using game theory in IEEE 802.11ac networks and proposed a payment-based incentive method to make the system fairer in channel bandwidth allocation. In paper [8], authors used a power control algorithm. The base station adjusted the transmit power of the signal according to the carrier sensing threshold to allocate a larger transmit power to the busy node. In paper [9], Sergio et al. discussed dynamic channel bonding (DCB) strategies in spatially dense WLANs. The result showed that choosing the maximum bandwidth for communication could improve the throughput for a single transmission. However, it was the unfair competition for other data transmissions. Therefore, transmission with random bandwidth could improve fairness, in the case of high-density WLANs.

The static channel allocation and the packet loss-based channel bonding mechanism will cause problems of poor fairness and low channel resource utilization. Although the random channel bonding strategy has good fairness, it is difficult to meet all data transmission requirements. This problem is solved in this paper by using a channel bonding algorithm based on outage probability. According to the calculated outage probability, combined with the bonding mechanism, different numbers of channels are bonded to data transmission to achieve efficient data transmission.

3 Description of Outage Probability

3.1 Outage Probability

Definition: Outage probability is an expression of link capacity. When the link capacity cannot meet the required user rate (the information quantity I is less than the transmission rate R), an interrupt event will occur. The probability of an outage event depends on the average SNR of the link and the channel fading distribution model. In a noise-limited system, the outage probability is defined as the probability that the instantaneous SNR is lower than a predetermined threshold SNR [10], which is shown in (1).

$$P_{out} = P[r \leq r_{th}] \tag{1}$$

In Eq. (1), r represents the instantaneous SNR at the receiver, and r_{th} refers to the minimum SNR required to meet the transmission requirements, that is, the threshold SNR. If the instantaneous SNR is higher than the threshold SNR, the quality of service will be guaranteed. The r_{th} depends on the requirements of the receiver, the type of modulation, and other factors. We can derive the outage probability based on the average SNR of the channel and the noise distribution of the channel. From the paper [10], in the absence of a relay node, the data is sent directly from the source node s to the destination node d, and the outage probability is shown in Eq. (2).

$$P_{out} = 1 - \frac{2r_{th}}{\sqrt{\bar{r}}}.K_1(\frac{2r_{th}}{\sqrt{\bar{r}}})e^{-r_{th}(\frac{1}{\bar{r}})} \tag{2}$$

Where $\bar{r}$ is the average SNR of the channel. K_1 is a first-order modified Bessel function. When node s communicates with node d through relay node m, the outage probability is shown in Eq. (3).

$$P_{out} = 1 - \frac{2r_{th}}{\sqrt{\bar{r}_1.\bar{r}_2}}.K_1(\frac{2r_{th}}{\sqrt{\bar{r}_1.\bar{r}_2}})e^{-r_{th}(\frac{1}{\bar{r}_1}+\frac{1}{\bar{r}_2})} \tag{3}$$

Where $\bar{r}_1$ and $\bar{r}_2$ respectively are average SNR from the source node s to the relay node m and from the relay node m to the destination node d.

To validate Eq. (3), a Monte Carlo simulation is done. Where SNR is exponentially distributed, and $\bar{r}_1 = \bar{r}_2$, $r_{th} = 15$, We repeat 1000 times and count the probability that the instantaneous SNR is less than the threshold SNR under different average SNR. The result is shown in Fig. 3. It can be seen that the outage probability obtained by the Monte Carlo algorithm after repeating 1000 times is similar to that calculated by Eq. (3). When there is no relay node, the conclusion is similar according to Eq. (2).

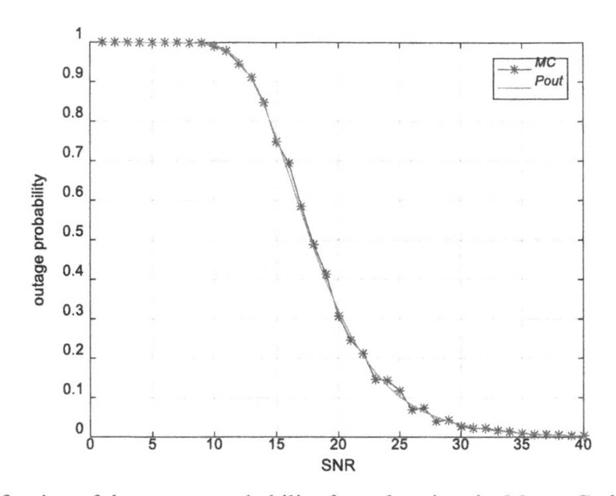

Fig. 3. Verification of the outage probability formula using the Monte Carlo simulation.

3.2 Factors Affecting Outage Probability

Assuming that the noise power at the receiver remains basically the same as N_0, the SNR at the receiver is defined as Eq. (4).

$$SNR = \frac{P_r}{N_0} \tag{4}$$

In wireless transmissions, the relationship between the transmit power, the transmission gain and loss is given by Eq. (5) [11].

$$P_r(dBm) = P_t(dBm) + G_t(dBi) + G_r(dBi) - L_{pf}(dB) \tag{5}$$

Where P_r refers to the received power; P_t refers to the transmit power; G_t refers to the gain of the transmit antenna; G_r is the gain of the receiving antenna; Typical values for G_t and G_r are 3 and 0, respectively. L_{pf} is the free space loss. L_{pf} in 5 GHz band is given by Eq. (6) [12].

$$L_{pf}(dB) - 20\lg f(MHz) + N * \lg d(m) + P_{f(n)} - 28 \tag{6}$$

where f is the channel center frequency, and here f is about 5000 MHz, d is the transmission distance, $P_{f(n)}$ refers to the influence of the n-story interval on the fading, the typical value here is 16, the typical value of the coefficient N is 30.

According to the above equations, bringing the typical value into the equations, the relationship between received power P_r, transmission distance d and the transmit power P_t is as shown in Eq. (7):

$$\mathrm{P}_r(dBm) = P_t(dBm) - 30\lg d(m) - 59 \tag{7}$$

According to Eqs. (4) and (7), the relationship between SNR, transmit power and distance can be known: the SNR is positively correlated with the transmit power and inversely related to the transmission distance.

It can be known from Eqs. (3), (4), and (7) that the outage probability is related to the SNR, which in turn is related to the transmission power and the transmission distance. The outage probability is inversely correlated with the transmit power and positively related to the transmission distance. We use MATLAB to simulate the relationship known from Eqs. (3), (4) and (7). The result is shown in Fig. 4, where $\bar{r}_1 = \bar{r}_2,\quad r_{th} = 15$, and the average SNR is calculated from transmit power and transmission distance.

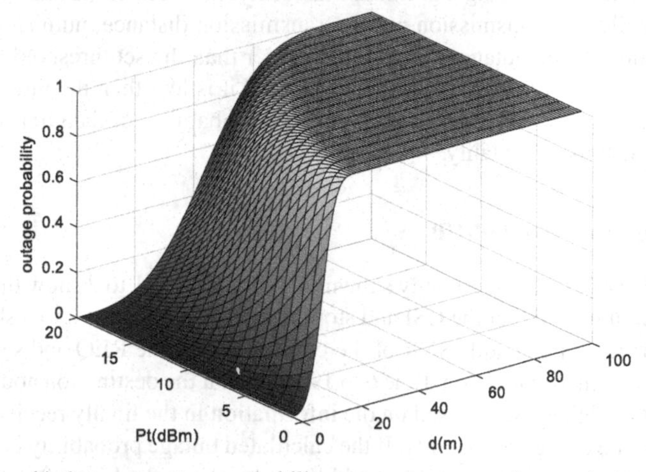

Fig. 4. Relationship between outage probability, transmit power and transmission distance when OP-CB algorithm is not applied.

As can be seen from Fig. 4, when the transmission power gradually decreases from 20 dBm or the transmission distance gradually increases from 1 m to 100 m, the outage probability rapidly becomes large and approaches 1. When there is no relay node, the conclusion is similar according to Eq. (2).

According to Eqs. (2) or (3), We can increase the channel SNR or reduce the threshold SNR to reduce the outage probability.

According to Shannon formula shown in Eq. (8), If the channel capacity C is kept constant and the channel bandwidth B increases, the required SNR (i.e., threshold SNR) can be reduced.

$$C = B\log(1 + SNR) \tag{8}$$

Therefore, when the transmission distance is large or the transmission power is low, the SNR at the receiver is relatively low. We can increase the transmission bandwidth by bonding more channels to reduce the outage probability. In addition, we have simulated different numbers of transmit and receive antennas in our previous work [13]. The relationship between SNR and channel capacity proves that under the same SNR condition, the increase of the number of bonded channels will increase the channel capacity and increase the data transmission rate.

4 OP-CB: Outage Probability Based Channel Bonding

OP-CB can be used in single-hop networks and multi-hop networks (This article only considers the case where there are 0 or 1 relay nodes for explanation) in MIMO systems. It is a dynamic channel bonding algorithm. The receiver calculates the outage probability based on the collected transmission power, transmission distance, number of hops and other information. If the outage probability is larger than the set threshold (10% in this paper) outage probability, bond more channels until it is less than the threshold outage probability. In this way, we can reduce the outage probability of data transmission and improve the transmission quality.

4.1 Working Process of OP-CB

The pseudocode of OP-CB is roughly shown in Table 1. Line 1 to 2 show that the nodes first scan the channels, to get the CSI and store it in matrix C_I. Line 3 to 5 show that the source node or relay node adds SNR of the current hop to the REQ and sends it to the next hop to establish a connection. Line 6 to 19 show that the destination node selects the channel with the highest SNR based on the information in the finally received REQ and then calculates the outage probability. If the calculated outage probability is greater than the threshold outage probability (20%), add bonding channels. Line 20 to 22 describes the process of updating the system after data transmission is completed.

In order to implement the above algorithm, the base stations in 6G network need to exchange CSI and coordinate work. Because of the high frequency band used for 6G communication, the single-hop transmission distance is relatively short according to Eq. (6). Therefore, compared to 4G and 5G communication, more base stations need to be set. What's more, in the 6G era, people's range of activities has expanded from the ground to the air. In order to ensure the complete coverage of the network, 6G network will integrate the ground network and the air network, and the network topology will expand from two dimensions to three dimensions. To make a large number of base stations work in coordination, we use SDN to coordinate and control the base stations

Table 1. pseudocode of OP-CB.

```
1: Nodes scan all channels to get channel state information (CSI);
2: Nodes store the information in matrix C_I;
3: //For relay node or source node:
4: Add current_SNR to REQ;
5: Send REQ to next-hop node;
6: //For destination node:
7: While non-solved REQs exist:
8:      If free channels exist:
9:          Choose channel with the highest SNR;
10:         Do compute outage probability;
11:         While P_out > 0.2:
12:             Add bonding channels to decrease outage probability;
13:         End While
14:         Update C_I;
15:     End If
16:     Else:
17:         Return No free channel;
18:     End Else
19: End While
20: Transmit data;
21: If data transmission finished: Release channels;
22: Update C_I;
```

in a certain area, which makes the data forwarding between base stations more efficient. The three-dimensional 6G network using SDN controllers is shown in Fig. 5. Under the control of the SDN controller, CSI is transmitted from the user to the base station.

After determining the number of bound channels, the base station sends channel binding information to the user.

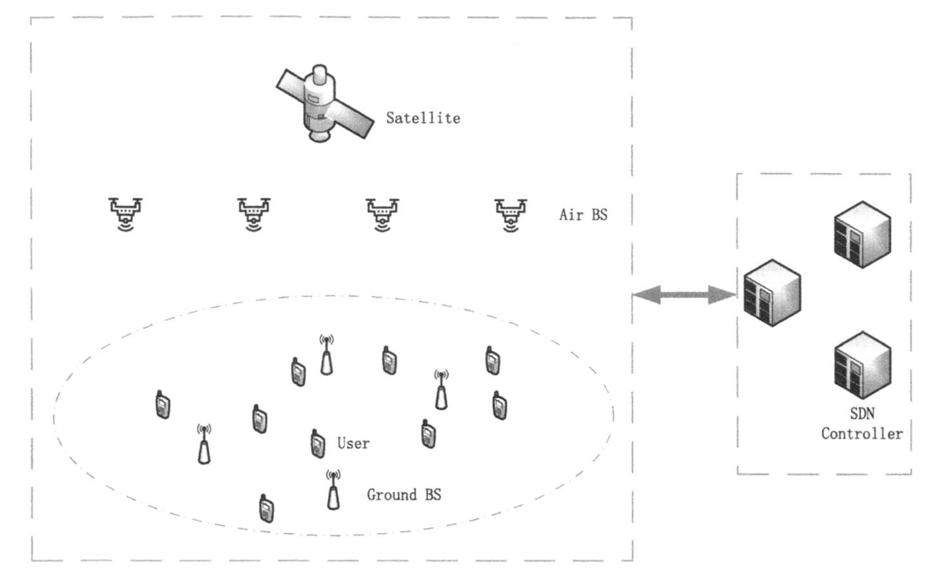

Fig. 5. 6G network with SDN controllers.

4.2 Analyze of OP-CB

Combining the outage probability with the channel bonding algorithm, we can control the bandwidth to adjust the threshold SNR of the data transmission. If the success rate of data transmission is to be guaranteed, it is necessary to ensure that the average SNR of data transmission is greater than the threshold SNR. We can bond multiple channels for transmission and increase the transmission bandwidth to reduce threshold SNR.

We use MATLAB to show the working process of OP-CB and evaluate it. Figure 6 is a comparison of the results of the outage probability at different SNRs using and not using the OP-CB algorithm. The number of channels bonded in OP-CB is shown in Fig. 7. The average SNR ranges from 1 to 40, and the SNR is exponentially distributed. The threshold SNR is set to 15 dB. The threshold of the outage probability set here is 0.2.

It can be seen that in the process of changing the SNR from 40 dB to 1 dB, when the OP-CB algorithm is not used, the outage probability is rapidly increased. When the SNR is about 23 dB, it exceeds 0.2. With the SNR gradually decreasing, the outage probability increases rapidly.

When using OP-CB, as the SNR is gradually reduced from 40 dB to 1 dB, the outage probability gradually become larger. But when the outage probability reaches 0.2 (SNR is about 16 dB), multiple channels are bonded for data transmission, and outage probability is reduced. As SNR decreases further, the outage probability continues to

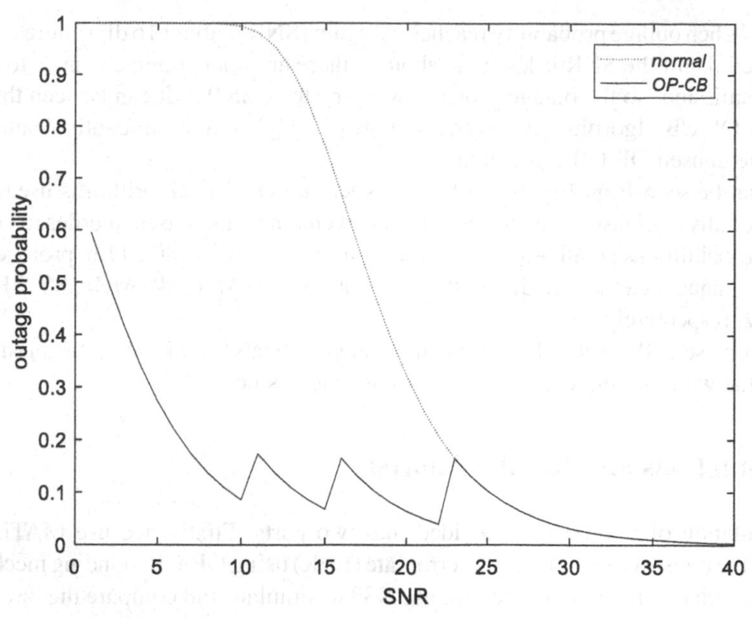

Fig. 6. Comparison of outage probability results with different SNRs with OP-CB algorithm and normal channel bonding.

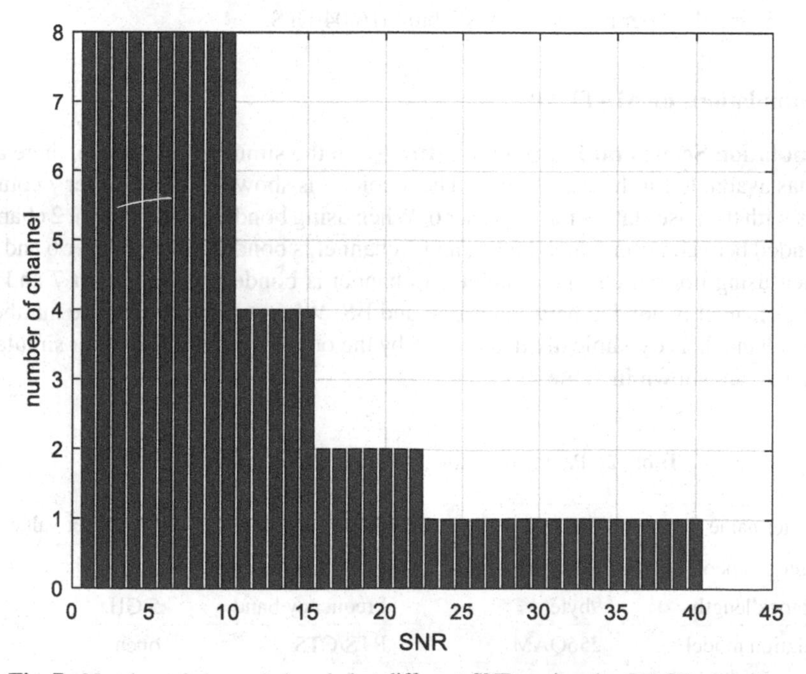

Fig. 7. Number of channels bonded at different SNRs using the OP-CB algorithm.

increase. When outage probability reaches 0.2 again (SNR is about 16 dB), more channels are bonded. Until the SNR is less than about 5, there are not enough channels to use for data transmission, so the outage probability is greater than 0.2. It can be seen that after using the OP-CB algorithm, the overall outage probability is significantly smaller than that of the unused OP-CB algorithm.

As can be seen from Fig. 6 and Fig. 7, when the OP-CB algorithm is used, as the SNR gradually decreases, the number of bonded channels increases in order to keep the outage probability as small as possible. According to the IEEE 802.11ax protocol, 1, 2, 4, and 8 channels can be bonded with bandwidths of 20 MHz, 40 MHz, 80 MHz, and 160 MHz, respectively.

It can be seen that OP-CB controls the outage probability of data transmission to a small value while saving channel resources as much as possible.

5 Simulations and Results Analysis

The simulation of this paper is divided into two parts. Firstly, we use MATLAB to compare the transmission rate and bit error rate (BER) using OP-CB, bonding mechanism and normal channel bonding. Then we use NS3 to simulate and compare the throughput with and without the OP-BC algorithm.

The MATLAB simulation uses the simulation tool MATLAB2016a and the operating system is the window 10 home version. The simulation tool used in the NS3 simulation is NS3.26, and the operating system is Ubuntu16.04 LTS.

5.1 Simulations on MATLAB

A. Simulation Scene and Parameter Settings. In the simulation scenario, there are 8 antennas available for the base station. The topology is shown in Fig. 2. User 7 communicates with the base station through user 6. When using bonding mechanism, 2 channels are bonded between user 6 and the BS, and 1 channel is bonded between user 6 and user 7. When using normal channel bonding, 1 channel is bonded between user 7 and user 6, and 1 channel is bonded between user 6 and BS. When using OP-CB, the number of bonded channels is dynamically determined by the outage probability. Other simulation parameters are shown in Table 2.

Table 2. Parameter settings of simulation on MATLAB.

Parameter name	Parameter value	Parameter name	Parameter value
Transfer protocol	IEEE802.11ax	Number of users	<4
Data block length	9byte	Frequency band	5 GHz
Modulation model	256QAM	RTS/CTS	open
Number of antennas	8		

B. Simulations Results and Analysis. Figure 8 is a comparison of the data transmission rate as a function of SNR when different channel bonding methods are used. When using the OP-CB algorithm, the channel bonded is determined according to the outage probability, and the outage probability is calculated by comparing the current SNR with the threshold SNR (10% in the simulation). When the outage probability is too large, the base station allocates multiple channels for the current data transmission, reducing the outage probability. The simulation results show that using OP-CB algorithm can obtain higher data transmission rate, and with the increase of SNR, the rate of increase is faster than using other methods.

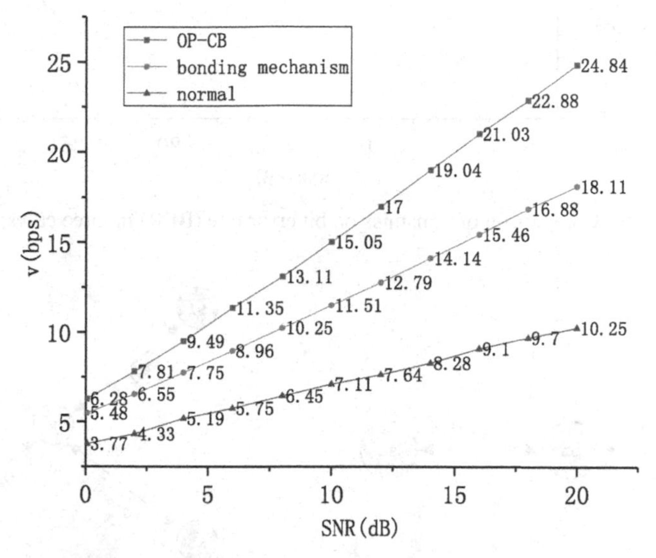

Fig. 8. Comparison of transmission rates in three cases.

Figure 9 shows the error bit rate comparison of data at the receiver when different channel allocation mechanisms are used. When the OP-CB algorithm is adopted, the final result obtained is optimal among the three channel allocation methods. When the SNR exceeds 13, the BER is first reduced to near 0.

5.2 Simulations on NS3

A. Simulation Scene and Parameter Settings. Use NS3 to simulate both single-hop and two-hop transmissions. The network topologies are respectively shown in Fig. 10a and Fig. 10b.

In the single-hop scenario in Fig. 10a, the source node s1 and the source node s2 communicate with the node BS using channels 36 and 52, respectively. In the two-hop scenario in Fig. 10b. The source node s1 communicates with the destination node d1 through the relay node n using channel 36, and the source node s2 also communicates with the destination node d2 through the relay node n using channel 52.

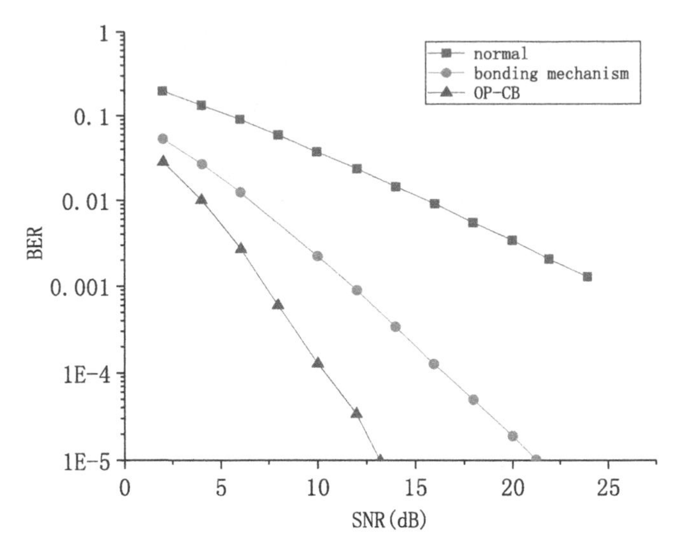

Fig. 9. Comparison of transmission bit error rate (BER) in three cases.

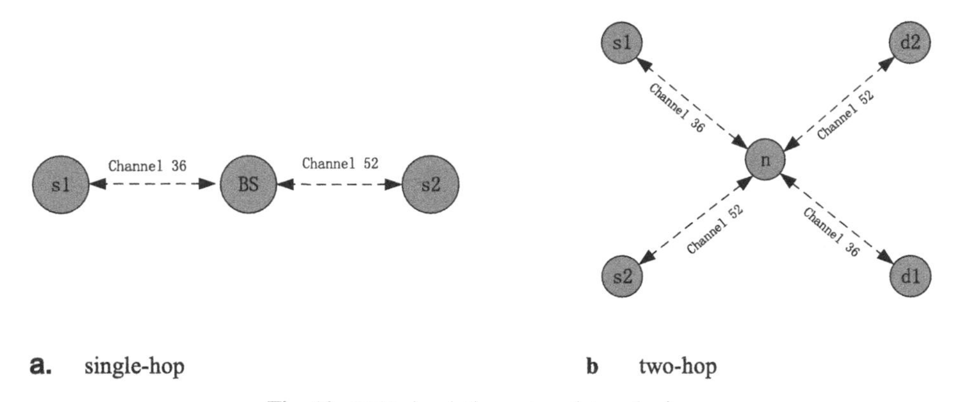

Fig. 10. NS3 simulation network topologies.

The distance between the source nodes s1, s2 and the BS node in the single-hop network remain equal and changes synchronously. In the two-hop network, the distance between the relay node and the source node s1, s2 and the destination node d1, d2 also remains equal and changes synchronously. In both network topologies, the transmit power remains the same. Other simulation parameters are shown in Table 3.

B. simulation Results and Analysis. Figure 11 shows the comparison of the throughput of the used and unused OP-CB algorithms with the data transmission distance in a single-hop network topology. Figure 11a shows that the source nodes s1 and s2 do not use the OP-CB algorithm. Since the transmit power and transmission distance are the same, the throughput of the two transmissions is almost the same, and both decrease rapidly with increasing distance. Figure 11b shows that s2 uses the OP-CB algorithm and s1 is not used. It can be seen that before 50 m, the distance is not large enough, s2 does

Table 3. Parameter settings of simulation on NS3.

Parameter name	Parameter value	Parameter name	Parameter value
Transfer protocol	IEEE802.11ac	Number of antennas	8
Application layer protocol	UDP	Number of users	2 or 4
Packet size	1472byte	Frequency band	5 GHz
Modulation model	256QAM	RTS/CTS	Open

not start to bond more channels, and the throughputs of the two transmissions remain almost identical. When the distance is greater than 50 m, since the outage probability is greater than the threshold designed in this paper, s2 begins to bond more channels. As the distance increases, the throughput of both transmissions decreases, and the throughput of s2 falls slower than that of s1. More, the overall throughput of s2 is 40%–50% higher than that of s1. However, when the distance is large, the gap of throughput between s2 and s1 is not very large.

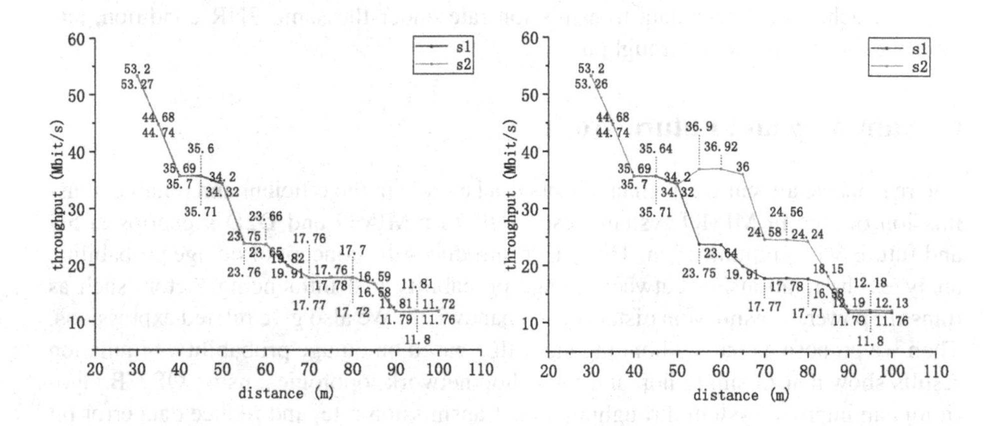

a. s1 not uses, s2 not uses **b.** s1 not uses, s2 uses

Fig. 11. Comparison of throughput between unused and unused OP-CB algorithms in a single-hop network topology.

Similarly, in the result of the two-hop network topology shown in Fig. 12, while s2 using the OP-CB algorithm and s1 not, and the distance is 60 m, the outage probability exceeds the set threshold, and s2 begins to bond more channels. The throughput of s2 is about 40% higher than that of s1. Because of the relay node, the transmission distance is about twice that of the single-hop network in Fig. 11, when the throughput is basically the same.

The simulation results show that the OP-CB algorithm considers the outage probability of data transmission, which will increase the information volume of the transmission

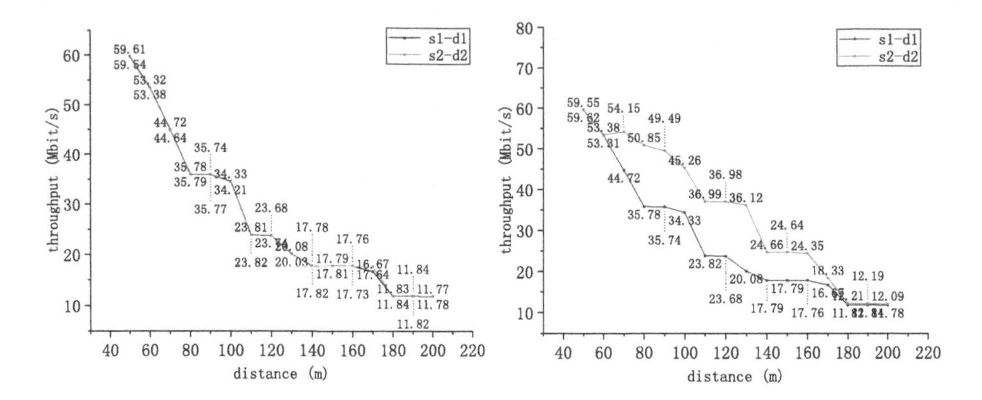

a. s1 not uses, s2 not uses

b. s1 not uses, s2 uses

Fig. 12. Comparison of throughput between unused and unused OP-CB algorithms in a two-hop network topology.

channel, achieve a larger data transmission rate under the same SNR condition, and improve the transmission throughput.

6 Summary and Future Work

The reasonable allocation of channels plays a key role in the efficient and reliable transmission of data in MIMO systems, especially in mMIMO and D2D scenarios in 5G and future 6G communication. This paper introduces the concept of outage probability, analyzes the relationship between outage probability and influencing factors such as transmit power, transmission distance, and bandwidth. We also give related expressions. Then we propose a channel bonding algorithm based on outage probability. Simulation results show that in single-hop and multi-hop network topologies, using OP-CB algorithm can improve system throughput, data transmission rate, and reduce data error bit rate.

For the future work, we will further study the use of deep learning in 6G networks (broader spectrum and greater number of channels) to optimize channel allocation methods based on the various characteristics of channels.

Acknowledgment. This research was funded by CERNET Innovation Project (NGII20190801), the Fundamental Scientific Research Project of Dalian University of Technology (DUT18JC28) and National Key R&D Program of China (2018YFB1700100).

References

1. Raja, M., Muthuchidambaranathan, P.: Joint precoding and decoding in MU-MIMO downlink systems with perfect channel state information (CSI). Procedia Technol. **10**(6), 708–715 (2012)

2. Ahmadinejad, A., Talebi, S.: Performance evaluation of chaotic spreading sequences in a multi-user MIMO-OFDM system. Phys. Commun. **1**(19), 11–17 (2016)
3. Yang, P., Xiao, Y., Xiao, M., LI, S.Q.: 6G wireless communications: vision and potential techniques. IEEE Netw. **33**(4), 70–75 (2019)
4. Zhao, Y., Peng, M., Li, H.: A channel bonding scheme with packet dropping mechanism in centralized cognitive radio networks. In: The 30th China Conference on Control and Decision Making, China, pp. 1548–1551. Institute of Electrical and Electronics Engineers Inc. (2018)
5. Takei, Y., Liu, Z., Ishihara, S.: Effect of channel bonding and parallel data transmission with IEEE802.11n wireless LAN in a small sewer pipe. In: 2018 International Conference on Information Networking (ICOIN), Chiang Mai, pp. 223–228. IEEE Computer Society (2018)
6. Aslam, M.Z., Corre, Y., Bjornson, E., Larsson, E.G.: Performance of a dense urban massive MIMO network from a simulated ray-based channel. EURASIP J. Wirel. Commun. Netw. **2019**(1) (2019)
7. Chiti, F., Fantacci, R., Duzgun, F.G., Vespri, V.: A game theory-inspired channel allocation policy for multi-radio pervasive ad hoc communications. Wirel. Commun. Mob. Comput. **16**(13), 1826–1836 (2015)
8. Feng, M.J., Mao, S.W., Jiang, T.: Joint duplex mode selection, channel allocation, and power control for full-duplex cognitive femtocell networks. Digit. Commun. Netw. **1**(1), 30–44 (2015)
9. Barrachina-Munoz, S., Wilhelmi, F., Bellalta, B.: Dynamic channel bonding in spatially distributed high-density WLANs. IEEE Trans. Mob. Comput. **19**(4), 821–835 (2019)
10. Hasna, M.O., Alouini, M.S.: Performance analysis of two-hop relayed transmissions over Rayleigh fading channels. In: Proceedings IEEE 56th Vehicular Technology Conference, Canada, pp. 1992–1996. Institute of Electrical and Electronics Engineers Inc. (2002)
11. Izumi, S., Edo, A., Abe, T., Suganuma, T.: An adaptive multipath routing scheme based on SDN for disaster-resistant storage systems. In: 2015 10th International Conference on Broadband and Wireless Computing, Communication and Applications, Poland, pp. 478–483. Institute of Electrical and Electronics Engineers Inc. (2015)
12. Wan, Y., Chen, X., Lu, J.: Fair and bandwidth-efficient broadcast link rate adaption in wireless LANs. Tsinghua Sci. Technol. **17**(1), 60–66 (2012)
13. Piao, X., Sun, W., Ji, Z., Yuan, X.: An energy efficiency channel binding mechanism for multi-user MIMO in 802.11ac. In: 2015 IEEE International Conference on Data Science and Data Intensive System, Sydney, pp. 420–426. Institute of Electrical and Electronics Engineers Inc. (2015). https://ieeexplore.ieee.org/document/7396535

Adaptive Collaborative Computing in Edge Computing Environment

Jianji Ren, Haichao Wang, and Xiaohong Zhang(✉)

Henan Polytechnic University, Jiaozuo 454000, Henan, China
xhzhpuedu@163.com

Abstract. The rapid development of 5th generation mobile networks (5G) and Internet of Things (IoT) technologies will generate a large amount of data, the processing and analysis requirements of big data will challenge existing networks and processing platforms. As the most promising technology in 5G networks, edge computing will greatly ease the pressure on network and data processing analysis on the edge. In this paper, we consider the coordination between compute and cache resources between multi-level edge computing nodes (ENs), users under this system can offload computing tasks to ENs to improve quality of service (QoS). We aim to maximize the long-term profit on the edge, while satisfying the low-latency computing of the users, and jointly optimize the edge-side node offloading strategy and resource allocation. However, it is challenging to obtain an optimal strategy in such a dynamic and complex system. Therefore, we use double deep Q-learning (DDQN) to make decisions to solve the complex resource allocation problem on the edge and make edge have certain adaptation and cooperation. Ability to maximize long-term gains while making quick decisions. The simulation results prove the effectiveness of DDQN in maximizing revenue when allocation resources on the edge.

Keywords: Collaborative computing · Edge computing · Optimization strategy

1 Introduction

As mobile communication and IoT technologies advances, smart cities, health care systems, etc. are deeply integrated with IoT technologies, and a large amount of data generated will pose challenges to data analysis and processing. Although the cloud computing [1] platform provides an efficient computing platform for big data processing, high bandwidth and high latency are unacceptable for scenarios with low latency requirements such as industrial control and real-time analysis.

In recent years, edge computing [2] as a new computing paltform attracted the attention of researchers, although edge computing does not have a uniform definition, in essence it is by deploying computing resources at the edge of the Internet,

X. Wang et al. (Eds.): 6GN 2020, LNICST 337, pp. 160–172, 2020.
https://doi.org/10.1007/978-3-030-63941-9_12

thereby reducing service delays, mitigating traffic pressure on the backhaul link and meeting the computational requirements of low latency applications.

Edge computing and cloud computing, distributed computing, parallel computing, etc. provide the necessary technical means to achieve accurate and fast integrated computing analysis. Cloud computing is based on platform virtualization, distributed storage, and parallel computing, flexible computing resources are allocated. Edge computing can be used as an extension of cloud computing [2,3]. It provides ubiquitous and low latency and reliable computing.

However, edge computing has no powerful computing power of cloud computing. When a single computing node has many computing tasks, it is prone to high latency caused by long task queues. Therefore, edge computing still has great challenges in deployment and application.

(1) Firstly, the uncertainty of the computing task, due to the uncertainty of factors such as the size of the computing task, the length of computing time, and the delay of the task, the workload between edge computing nodes may vary greatly; (2) Secondly, the workload scheduling of a single node, the task dynamic scheduling and computing resource allocation between nodes when collaborative computing between multiple nodes; (3) Finally, the time interval, the most valuable part of edge computing is the computing of low latency, so collaborative computing should meet the requirements of low latency.

To solve the above problems, in this work, we use deep reinforcement learning agents to determine the relevant nodes of collaborative computing. Specifically, we are using double deep Q-learning [4,5] to minimize the delay of collaborative computing and ensure load balancing between nodes.

The rest of the paper is organized as follows: The second part summarizes the related work of collaborative computing in edge computing. The third part describes the dynamic system model of edge collaborative computing. We provide the results of simulation experiments and experiments in the fourth part. Finally, the fifth part summarizes our work and discusses the direction of future work.

2 Related Work

In recent years, edge computing networks based on multiple access have received extensive attention from academia and industry. Edge computing eliminates latency by providing a large number of computing resources for application services that require low latency and high computational demands. Although cloud computing has become very popular due to its powerful computing and flexible resource allocation strategies. However, because the distance between the cloud and the end device is typically large, cloud computing services may not provide assurance for low latency applications in the edge network.

To solve these problems, Edge Computing (EC) [2,3,6] has been studied to deploy computing resources closer to the user device, which can effectively improve applications Quality of Service (QoS) that require large amounts of computation and low latency.

The computing of the task at the edge is complicated by the complex factors of computing, storage, caching, network, energy consumption, etc., it is difficult to make an offload strategy under low latency calculation limits, therefore, researchers use game theory to solve such problems. Zheng et al. [7] expressed the dynamic offloading decision process of mobile users as a random game and proposed an efficient multi-agent random learning algorithm to solve the multi-user computing offloading problem. Chen et al. [8] proposed a game-based computational offloading algorithm as a solution for MEC multi-user computing offloading in multi-channel wireless interference environments, with excellent performance in terms of energy consumption and computational execution time.

In addition, heuristic algorithms or dynamic programming methods can also be used to solve computational offloading problems. Dinh et al. [9] proposed a joint optimization computational offloading framework that can improve task allocation decisions and adjusts the CPU frequency of mobile devices. Mao et al. [10] proposed a dynamic computational offloading algorithm based on Lyapunov optimization, which can jointly determine the offloading strategy and CPU frequency for the MECO problem of energy harvesting equipment.

Recently, researchers have begun to use machine learning or deep learning to optimize the computational offload strategy for edge computing. Zhang et al. [11] proposed an intermittent connection cloudlet system based on Markov decision process for the dynamic offloading problem of mobile users. But in the literature [12], the author studies the dynamic service migration problem in the mobile edge cloud and proposes a sequential offloading decision framework based on the Markov decision process.

Li et al. [13] proposed an RL-based optimization framework to solve the resource allocation problem in wireless MEC. The framework optimizes the offloading decision and computing resource allocation by optimizing the total cost of delay and energy consumption of all UEs. Yang et al. [14] proposed a computing resource allocation strategy based on deep reinforcement learning for URLLC edge computing networks with multiple users.

Wang et al. [15] combined deep reinforcement learning and federated learning frameworks with mobile edge systems to optimize mobile analysis edge computing, caching and communication. Ren et al. [16] considering the dynamic workload and complex radio environment in the IoT environment, indicate the decision of the IoT device through multiple Deep Reinforcement Learning (DRL) agents deployed on multiple edge nodes, and use the federated learning (FL) to train DRL agents in a distributed manner.

When we consider communication, computing resource allocation, delay constraints, etc., the complexity of the edge computing system will be very high, it is challenging to obtain an optimal strategy in such a dynamic and complex system. Deep reinforcement learning is an advanced reinforcement learning algorithm that uses a deep Q network to approximate the Q-value action function [4] and has been utilized in wireless networks to achieve automatic resource allocation.

Therefore, we proposed that edge nodes use deep reinforcement learning agents to determine the allocation of computing resources and the maximum long-term benefits. Specifically, we use DDQN to make decisions to solve the complex resource allocation problem on the edge and make edge have certain adaptation and cooperation. Ability to maximize long-term gains while making quick decisions.

3 System Model

This paper uses the nodes with computing ability in the edge computing environment to analyze, as shown in Fig. 1. Overall, the system is divided into four levels. The first is the device layer where the user device is located, including various networked devices of the user, such as mobile phones, IoT, VR, PC, etc., which establish connections with the Internet through a wireless network access point or 5G.

Secondly, the base station, cellular network, wireless network access point, etc., which the user device is connected. These equipments are located at the edge of the Internet, connecting users and the Internet, and closest to the user device. Placing the edge computing nodes here will greatly reduce the delay and improve the users experience, this paper assumes that the base station has an edge computing node which user connected to, and the node is marked as a level 1 computing node.

Then there is a level 2 compute node. The level 2 compute node is located between the level 1 compute node and the cloud computing platform of the core network. It acts as a collaborator for the compute node of the level 1 compute node. A level 2 compute node can coordinate a cache, calculation, etc. There are several levels 1 compute nodes in the area, and the level 2 compute nodes are close to the user, but not as close as the level 1 node.

Finally, the cloud computing platform is located in the core network. The cloud computing platform stores the running environment of the user application and the latest data and can release the file image to the computing node near the user when needed.

3.1 Communication Model

The system includes M level 1 computing nodes and N level 2 computing nodes, wherein the level 1 computing node m belongs to $M, M = \{1, 2, ..., M\}$, and the level 2 computing node n belongs to $N, N = \{1, 2, ..., N\}$. There are a total of $D, D = \{1, 2, ..., D\}$ user devices, and user device d belong to D. Among them, D devices are divided into M groups, and user devices in each group are connected to M. For quantitative analysis, time horizon is discretized into time epochs indexed by i with equivalent duration as ϕ (in seconds).

We described the network model using a single base station m and the user device d connected to it. When the device d establishes a connection with the base station m, the base station allocates W Hz spectrum resources to the device,

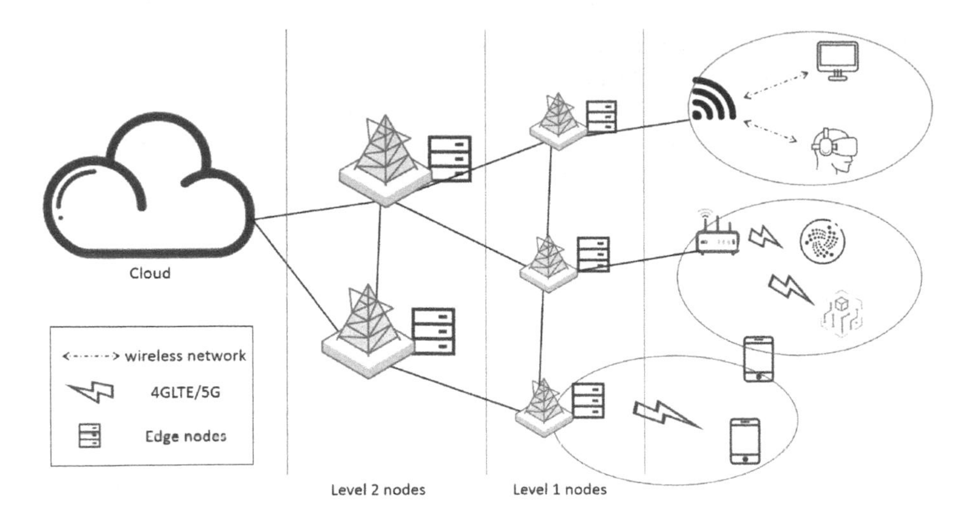

Fig. 1. System architecture diagram.

but the base station channel experiences a time-space change of rayleigh fading, and following the flat fading model.

We denote s_i^m as the channel gain during the epoch i between the device and an EN $m \in M$, which is assumed static and independently taken from a finite state space S_e. The f_i^{tr} is the transmit power with maximum limitation f_{max}^{tr}, which $A$is the power of interference plus noise. The transmission speed v of the user device is calculated as follows:

$$v = W * log_2(1 + s_i^m \cdot f_i^{tr}/A) \tag{1}$$

3.2 Computing Model

We assume that every computing node in the system has the ability to be virtualization, and the application running environment image of the user device can be found in the cloud. Each level 1 compute node has an IP address and a remaining computing resource table of other level 1 nodes connected to it and a level 2 computing node n of the upper level. Within a certain period of time i, the node connects to the surrounding node and one level 2 node n. The level 1 node m broadcasts its own remaining computing resource $p_m, p_m = (C_m, S_m)$ and accepts the remaining computing resource p broadcasts from other nodes, updating the local resource table based on the broadcast content.

We treat D_m as a set of service requesters, where each requester d belongs to D_m, connects to the nearest base station m according to its signal strength, and then sends a request to the compute node m where the base station is located, where $R_m^d, R_m^d = (D_s, T_l, C_r, S_r)$ is computing request send from user d to the node m. Where D_s includes the task data and image file globally unique identifier (GUID), T_l is the computational delay limit of node m, C_r is the computing resource size required, and S_r is the storage resource size required.

After node m receives the task request R_m^d, First check the remaining computing resources $p_m = (C_m, S_m)$ at the node m, and if the remaining computing resource p_m meets the user computing requirement $C_r < C_m \& S_r < S_m$, then the service is started. During the service start phase, the node m searches for the image cache resource image required for the user task computing from the local cache area. If the image resource is in the local cache area, the image is loaded from the local cache area and the computing service is started. If there is no cache image locally, the node download the image file image from the cloud platform to the node m and start the computing service. When the calculation ends, the node m returns the computing result to the user device, completes the computing task of this period, waits for the user to compute the task for the next period or the user ends the computing command and pay for the task R_m^d.

If the remaining computing resource p_m at node m cannot satisfy the user computing requirement, $Cr > C_m \| S_r > S_m$, the service cannot be opened locally, and node m will forward the user request to the node in the computing resource table that meets the user's computing requirements, if the node x accepts the computing task, the user's computing service will be completed by the node x, and the base station m performs the transfer function here.

If there is no node in the calculation resource table that meets user requirement, the length of the queue determines whether the task is placed in the task queue or offloaded to the cloud. If the queue at node m is not full, $l_m < l^{max}$, the computing task is placed in the local task queue, and the task queue is a first in first out FIFO model; if the task queue at node m is full $l_m = l^{max}$, the computing task is offloaded to the cloud. In summary, the user computing task R_m^d offload policy p_m^d is:

$$p_m^d = \begin{cases} 0 & \text{if } C_r < C_m, S_r < S_m \text{ and } l_m = 0, \text{ Computed at node } m; \\ 1 & \text{if } C_r < C_x, S_r < S_x \text{ and } l_x = 0, \text{ Computed at node } x; \\ 2 & \text{if } C_r < C_x, S_r < S_m \text{ and } l_m < l^{max}, \text{ Waiting in node } m; \\ 3 & \text{Offload task } R_m^d \text{ to the cloud.} \end{cases} \tag{2}$$

3.3 Payment Strategy

After the user's computing request R_m^d is completed by node m and the computing result is returned to the user device, the user device will pay to the node m according to the delay of the computing completion del_t^m. If the computing is completed within the limited time T_l, the user pays according to the actual delay time. If the edge node times out to complete the computing task, users will not pay it.

$$fe_m^d = \begin{cases} \pi * del_t^m & \text{if } del_t^m < T_l, \pi \text{ is the price of edge;} \\ 0 & \text{if } del_t^m > T_l; \\ \eta & \text{if task } R_m^d \text{ failed, and } \eta < 0. \end{cases} \tag{3}$$

The delay del_t^m in this paper is defined as: the time interval between when the user equipment initiates the calculation request and when the device receives the node calculation result. If the computing task is placed at the base station m to which the user device is connected, the computing delay del_t^m can be expressed as:

$$del_t^m = tr_d + tr_{res} + h_m + ct_m \tag{4}$$

Where tr_d is the time spent on the transmit task R_m^d, and tr_{res} is the time it takes to transmit the result. h_m is the time taken by the computing node m to switch the computing task at the base station, which is a small fixed value.

$$tr_d = D_s/v \tag{5}$$

ct_m is the time required for the computing node to complete the computing task. It is usually related to the CPU frequency fr_{CPU}, the size of CPU cache B_{ca}, and the data size D_s of the task data.

$$ct_m = D_s/(fr_{CPU} * B_{ca}) \tag{6}$$

If the computing task is placed at the neighboring base station x, the delay del_t^x is:

$$del_t^x = tr_d + tr_{res} + h_x + ct_x + 2 * dis_m^x/c \tag{7}$$

Where dis_m^x is the fixed distance between the base station m and the neighbor node x, c is the speed of light. Finally, our objective function is:

$$\begin{aligned} &\max \sum_{m\in\{M,N\}} \sum_i fe_m^d \\ &s.t. \forall d \in D, \forall m \in \{M, N\} \\ &C_r >= 0, S_r >= 0 \end{aligned} \tag{8}$$

4 Collaborative Computing Strategy Based on Double Deep Q-Learning

In order to better understand the DDQN agent, we briefly introduce DDQN in this paper. First, we introduce reinforcement learning. Reinforcement learning is an important branch of machine learning, agents in reinforcement learning can learn the actions that maximize the return through interaction with the environment. Unlike supervised learning, reinforcement learning does not learn from the samples provided. Instead, act and learn from their own experience in an uncertain environment.

Algorithm 1. Collaborative computing framework

1: **Initialize:**
2: Each edge node loads DDQN model as agent;
3: The compute node m broadcasts its remaining computing power $P(C_m, S_m)$ to other nodes;
4: Node m receives the remaining computing power broadcast and adds it to the calculation table C;
5: **If:** there is computing task $R_m^d = (Ds, Tl, Cr, Sr)$;
6: The agent takes the node m status s and the task R_m^d as input, $s = C$;
7: Generate a decision policy according to action a ;
8: **Switch**(a): ;
9: case 0: Immediately allocate computing resources and perform computing tasks;
10: case 1: Put tasks into the local task queue q and wait to allocate computing resources;
11: case 2: Send task to the node in the computing resource table;
12: case 3: Send tasks to the cloud computing platform;
13: **Return:** Send the results to the user device ;
14: The user pays the node according to the calculation delay and the task resource allocation amount ;

Reinforcement learning has two salient features: multiple trials and delayed rewards. Trial testing means weighing trade-offs between exploration and development. Agents tend to use the effective actions they have tried in the past to generate rewards, but it must also explore better new actions to generate higher returns in the future. Agents must take a variety of actions and gradually get the most out of it. Another feature of reinforcement learning is that agents should look at the global, not only considering immediate rewards, but also long-term cumulative rewards, which are designated as reward functions.

Model-free reinforcement learning has been successfully applied to the processing of deep neural networks and value functions [4]. It can directly use the original state representation as a neural network input to learn the strategy of difficult tasks [5]. Q-Learning is a model-free reinforcement learning algorithm. The most important component of the Q-learning algorithm is a method for correctly and effectively estimating the Q value. Q-functions can be implemented simply by look-up tables or function approximators, sometimes by nonlinear approximators, such as neural networks or even more complex deep neural networks. Q-learning is combined with deep neural networks, so-called Deep learning Q-learning (DQN). The formula for Q-learning is:

$$Q_\pi(s,a) = E[R_1 + \gamma R_2 + \cdots | S_0 = s, A_0 = a, \pi] \tag{9}$$

The parameter update formula is:

$$\theta_{t+1} = \theta_t + \alpha(Y_t^Q - Q(S_t, A_t; \theta_t)) \bigtriangledown_{\theta_t} Q(S_t, A_t; \theta_t) \tag{10}$$

Which Y_t^Q is defined as:

$$Y_t^Q = R_{t+1} + \gamma \cdot \underset{a}{max} Q(S_{t+1}, A_t; \theta_t) \tag{11}$$

The formula of deep Q-learning is:

$$Y_t^{DQN} = R_{t+1} + \gamma \cdot \max_a Q(S_{t+1}, a; \theta_t^{'}) \tag{12}$$

Improved DQN: double deep Q-learning. In conventional DQN, selecting an action and evaluating the selected action uses a maximum value that exceeds the Q value, which results in an overly optimistic estimate of the Q value. In order to alleviate the problem of overestimation, the target value in DDQN is designed and updated to

$$Y_t^{Q} = R_{t+1} + \gamma \cdot Q(S_{t+1}, \underset{a}{argmax} Q(S_{t+1}, a; \theta_t); \theta_t) \tag{13}$$

The error function in DDQN is rewritten as:

$$Y_t^{DoubleQ} = R_{t+1} + \gamma \cdot Q(S_{t+1}, \underset{a}{argmax} Q(S_{t+1}, a; \theta_t); \theta_t^{'}) \tag{14}$$

Among them, the action selection is separated from the target Q value generation. This simple technique makes the overestimation significantly reduced and the training process runs faster and more reliably.

Algorithm 2. Collaborative edge computing strategy based on double deep Q-learning

---

1: **Initialization:**
2: Initialize replay memory: R and the memory capacity: M;
3: Main deep-Q network with random weights: θ;
4: Target deep-Q network with weights: $\theta^- = \theta$;
5: **For** epoch i in I:
6: Input the system state s into the generated Q-network;
7: Compute the Q-value $Q(s, a; \theta)$;
8: Input the system state s' into the generated Q-network;
9: Compute the Q-value $Q(s', a; \theta)$;
10: Input the system state s' into the target Q-network;
11: Compute the Q-value $Q(s', a; \theta^-)$;
12: Compute the target Q-value: ;
13: $Y = p(s, a) + \gamma Q(s', argmax(s', a; \theta), \theta^-)$;
14: **Output:** action a ;
15: Record the changed status s'' and reward f after action a to memory R ;
16: **End For**
17: **Save model.**

---

5 Simulation Results

5.1 Experiment Setup

This paper uses a simulation experiment method to instantiate user device and edge computing nodes for simulation through Python programming. The operating system used in the experiment was CentOS7, the processor was Intel E5-2650

V4, the hard disk size was 480G SSD + 4T enterprise hard disk, and the memory was 32G. The code interpreter is Python, version 3.6, and the code runtime dependencies include Tensorflow, Keras, Numpy, Scipy, Matplotlib, CUDA, etc.

The experimental data includes the computational task of the user offloading to the edge node in the time period i, which is randomly generated by calling the bernoulli and poisson functions in the Scipy library. The experiment assumes that the user device has the ability to connect to the network and can offload computing tasks and receive computing results. Experiments in this paper compared Double deep Q-learning (DDQN), Deep Q-learning (DQN), Dueling deep Q-learning and Natural Q-learning, where the learning rate is set to 0.001, the replay memory size is 200, and the total training steps is 12000.

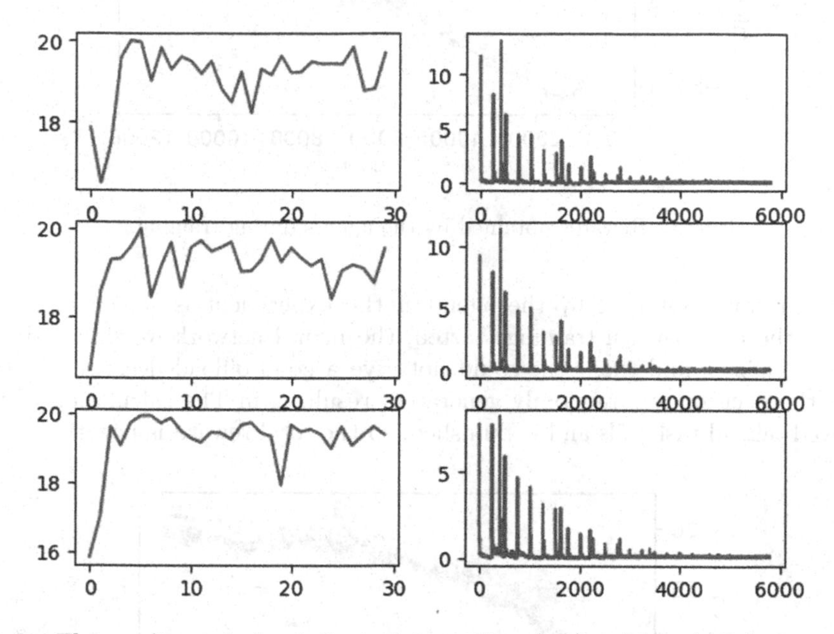

Fig. 2. The result graph shows the training utility (left) and loss (right) changes of random three nodes.

5.2 Result Analysis

In Fig. 2, the learning curve of agents are shown above, the probability of a user offloading a task is 0.5, the utility and loss of three nodes are randomly selected as the display. The left side is the utility of the node, the abscissa is the training period, a total of 30 cycles are trained, each cycle includes 200 training steps, and each step of the training will get a utility. After averaging the utility of each cycle, the utility curve of the agents is got. The utility graph shows that with the training of the agent, the edge node's profit gradually increases in a certain period of time, but as the training increases, the growth rate of utility begins to decrease. On the right is the training loss map. The loss is defined as the mean square error between the target value and the predicted value. A total of 6000 steps are trained, each step will have some fluctuation due to the update of the weight of the neural network. However, from the trend point of view, the loss is gradually reduced.

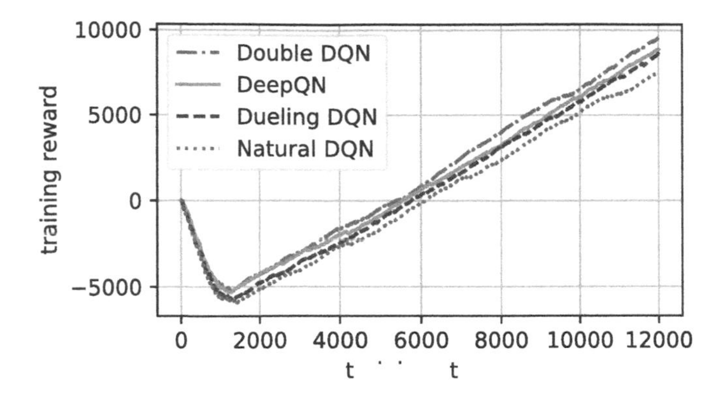

Fig. 3. Rewards obtained by the agents during training.

The reward obtained by the agent in the experiment is shown in Fig. 3. During the period when training started, the neural network weights had just been initialized, and the agent could not give a good offload decision. At this time, the decision was randomly generated, resulting in The calculation of the received offload task fails and is punished, so the total reward is negative.

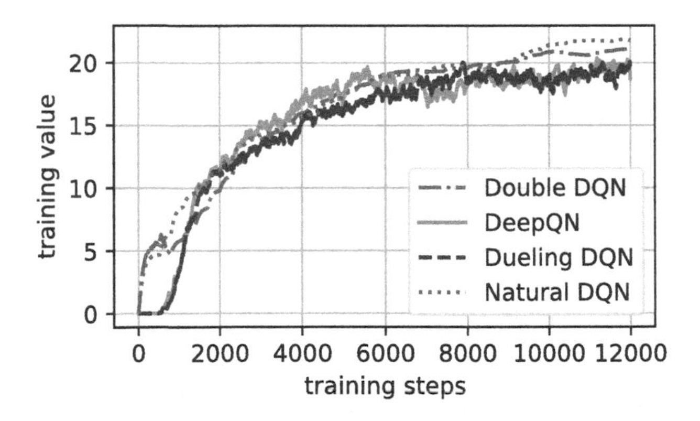

Fig. 4. Value changes in agents.

Figure 4 shows the value of the value calculated by the value function in the agent. At the beginning of the training, the value cannot be well estimated, but as the training progresses, the value estimation continues to approach the true level. And reached a stable level in the end.

6 Conclusion and Future Work

In this paper, we consider the bandwidth, computing and cache resources of the ENs, benefit from the deep learning and powerful learning ability and decision-making characteristics, maximize the edge while satisfying the low-latency computing of users at the edge. In addition, it also considers the horizontal and vertical coordination cache and computing at the edge, which has certain adaptability and can fully coordinate the computing resources at the edge to maximize the value. However, the DDQN on the EN has a long training period and the effect is unstable. It needs to train for a while to make better decisions. In addition, when multiple ENs in the same group perform collaborative computing, we have not studied the prioritization strategy of computing resources or the computing resource bidding strategy. Future work we will focus on competitive bidding and allocation priorities. In addition, the security of users on the edge is also the focus of research.

References

1. Josep, A.D., Katz, R.A.D., Konwinski, A.D., et al.: A view of cloud computing. Commun. ACM **53**(4), 50–58 (2010)
2. Shi, W., Cao, J., Zhang, Q., et al.: Edge computing: vision and challenges. IEEE Internet Things J. **3**(5), 637–646 (2016)
3. Satyanarayanan, M.: The emergence of edge computing. Computer **50**(1), 30–39 (2017)
4. Van Hasselt, H., Guez, A., Silver, D.: Deep reinforcement learning with double Q-learning. In: Thirtieth AAAI Conference on Artificial Intelligence (2016)
5. Mnih, V., Kavukcuoglu, K., Silver, D., et al.: Human-level control through deep reinforcement learning. Nature **518**(7540), 529 (2015)
6. Hu, Y.C., Patel, M., Sabella, D., et al.: Mobile edge computing? A key technology towards 5G. ETSI White Pap. **11**(11), 1–16 (2015)
7. Zheng, J., Cai, Y., Wu, Y., et al.: Stochastic computation offloading game for mobile cloud computing. In: 2016 IEEE/CIC International Conference on Communications in China (ICCC), pp. 1–6. IEEE (2016)
8. Chen, X., Jiao, L., Li, W., et al.: Efficient multi-user computation offloading for mobile-edge cloud computing. IEEE/ACM Trans. Netw. **24**(5), 2795–2808 (2015)
9. Dinh, T.Q., Tang, J., La, Q.D., et al.: Offloading in mobile edge computing: task allocation and computational frequency scaling. IEEE Trans. Commun. **65**(8), 3571–3584 (2017)
10. Mao, Y., Zhang, J., Letaief, K.B.: Dynamic computation offloading for mobile-edge computing with energy harvesting devices. IEEE J. Sel. Areas Commun. **34**(12), 3590–3605 (2016)
11. Zhang, Y., Niyato, D., Wang, P.: Offloading in mobile cloudlet systems with intermittent connectivity. IEEE Trans. Mob. Comput. **14**(12), 2516–2529 (2015)
12. Wang, S., Urgaonkar, R., Zafer, M., et al.: Dynamic service migration in mobile edge-clouds. In: 2015 IFIP Networking Conference (IFIP Networking), pp. 1–9. IEEE (2015)
13. Li, J., Gao, H., Lv, T., et al.: Deep reinforcement learning based computation offloading and resource allocation for MEC. In: 2018 IEEE Wireless Communications and Networking Conference (WCNC), pp. 1–6. IEEE (2018)

14. Yang, T., Hu, Y., Gursoy, M.C., et al.: Deep reinforcement learning based resource allocation in low latency edge computing networks. In: 2018 15th International Symposium on Wireless Communication Systems (ISWCS), pp. 1–5. IEEE (2018)
15. Wang, X., Han, Y., Wang, C., et al.: In-edge AI: intelligentizing mobile edge computing, caching and communication by federated learning. IEEE Netw. **33**(5), 156–165 (2019)
16. Ren, J., Wang, H., Hou, T., et al.: Federated learning-based computation offloading optimization in edge computing-supported internet of things. IEEE Access **7**, 69194–69201 (2019)

Towards the Future Data Market: Reward Optimization in Mobile Data Subsidization

Zehui Xiong[1,2], Jun Zhao[2], Jiawen Kang[3(✉)], Dusit Niyato[2], Ruilong Deng[4], and Shengli Xie[5,6]

[1] Alibaba-NTU Joint Research Institute, NTU, Singapore, Singapore
[2] School of Computer Science and Engineering, NTU, Singapore, Singapore
[3] Energy Research Institute, NTU, Singapore, Singapore
kavinkang@ntu.edu.sg
[4] College of Control Science and Engineering, School of Cyber Science and Technology, Zhejiang University, Hangzhou, China
[5] Guangdong-Hong Kong-Macao Joint Laboratory for Smart Discrete Manufacturing, Guangzhou, China
[6] Joint International Research Laboratory of Intelligent Information Processing and System Integration of IoT, Ministry of Education, Guangzhou, China

Abstract. Mobile data subsidization launched by network operators is a promising business model to provide some economic insights on the evolving direction of the 4G/5G and beyond mobile data market. The scheme allows content providers to partly subsidize mobile data consumption of mobile users in exchange for displaying a certain amount of advertisements. The users are motivated to access and consume more content without being concerned about overage charges, yielding higher revenue to the data subsidization ecosystem. For each content provider, how to provide appropriate data subsidization (reward) competing with others to earn more revenue and gain higher profit naturally becomes the key concern in such a ecosystem. In this paper, we adopt a hierarchical game approach to model the reward optimization process for the content providers. We formulate an Equilibrium Programs with Equilibrium Constraints (EPEC) problem to characterize the many-to-many interactions among multiple providers and multiple users. Considering the inherent high complexities of the EPEC problem, we propose to utilize the distributed Alternating Direction Method of Multipliers (ADMM) algorithm to obtain the optimum solutions with fast-convergence and decomposition properties of ADMM.

Keywords: Data subsidization · Next-generation mobile data · Network economics · Game theory

Corresponding author: Jiawen Kang

X. Wang et al. (Eds.): 6GN 2020, LNICST 337, pp. 173–189, 2020.
https://doi.org/10.1007/978-3-030-63941-9_13

1 Introduction

With the rapidly growing demand of network data volume in mobile 5G and beyond, *mobile data subsidization* is envisioned as a promising scheme to bring in additional revenue for network operator in wireless networks. This scheme allows users to consume content (e.g., watch video content from HBO, Hulu, and Netflix) subsidized by content providers without eating into their data allowances. Therefore, the scheme motivates users to access and consume more content without being concerned about overage charges, yielding higher revenue to the data subsidization ecosystem. Data subsidization has become an appealing area of research since its emergence. There are numerous application examples of data subsidization programs. Some network operators in the US such as AT&T and Verizon have proposed the *Data Sponsoring Plan* and *FreeBee Sponsored Data Plan*, respectively, that permit content providers to partly pay for the data consumption fees instead of mobile users themselves [1]. Under the data subsidization scheme, the mobile users, i.e., the subscribers of network content services can offload their data usage fees to content providers. In particular, the data usage of mobile users is allowed to be partly subsidized, but in return the mobile users need to receive and view a certain amount of advertisements. Although the content providers bear the cost of subsidization, the advertisement earnings increases consequently. Clearly, the data subsidization potentially leads to a mutual benefit for both participants in wireless networks. Meanwhile, this scheme will guarantee the continuity and efficiency of cellular data service in 5G and beyond, thereby improving the connectivity of massive number of mobile devices to heterogeneous cellular networks [6].

The subsidization can be treated as the reward offered to users, and how to provide appropriate subsidization, i.e., the reward optimization, is naturally a key concern for the content providers. Furthermore, the data subsidization scheme can foster greater reward competition among the content providers. However, the competition for earning from the users among content providers has not been formally studied in the literature. Moreover, most of the existing works [2,14,15,17,21] on mobile data subsidization consider that the users can only consume the content with advertisement (i.e., *sponsored content*) and passively accept the corresponding reward. Therein, the only active strategies of users are to choose how much sponsored content that they demand. However, the users are able to reject the sponsored content and still consume the *normal content* paying full data usage fees without advertisement. The users aim to maximize their individual payoff in a self-interest manner by finding the balance between sponsored content and normal content. The strategic behaviors of users further make it more challenging to explore the reward competition among providers. Essentially, when there are tremendous numbers of providers and users in future mobile data market, the constrained optimization [7] and traditional game techniques [19,21] are practically inapplicable in terms of the complexity and scalability. Nevertheless, this has not been well-addressed in existing literature that motivates the study of this paper.

In this paper, we study the reward optimization for content providers in the framework of mobile data subsidization by analyzing the rational behaviors of both content providers and users. We discern the fact that game theory is a suitable analytical tool to address such a two-sided interaction problem [9,10,18]. Therefore, we investigate the interactions among the content providers and mobile users by formulating a hierarchical Stackelberg game model. The game model is developed to jointly maximize the profits of content providers, and the utilities of mobile users. In the game, the content providers are the leaders that determine the reward offered to users first. Then, the users are the followers that decide on how much sponsored content and normal content to consume based on the reward claimed by the leaders. Specifically, the major contributions of this paper are summarized as follows:

1. A hierarchical Stackelberg game is established to model the strategic interactions among multiple content providers and multiple users in the data subsidization system, where the profits of content providers and the utilities of users are jointly maximized.
2. To explore the reward competition among providers, we then formulate an Equilibrium Programs with Equilibrium Constraints (EPEC) problem to characterize the many-to-many interactions among multiple providers and multiple users. Considering the inherent high complexities of EPEC problem, we propose to employ the distributed Alternating Direction Method of Multipliers (ADMM) algorithm to tackle the EPEC problem. Taking advantage of the fast-convergence property and high scalability of ADMM, we derive the optimum solutions with reasonable complexity in a distributed manner.
3. Numerical simulations are conducted to demonstrate the analytical results and evaluate the system performance in the proposed Stackelberg game-based schemes. The results confirm that with the proposed scheme, the optimization of the profits of content providers and the optimization of the utility of users can be jointly attained.

The rest of the paper is structured as follows. Section 2 shows the system description characterizing the mobile data subsidization system and develops a hierarchical game framework. We formulate and study the multi-provider multi-user Stackelberg game in Sect. 3. Section 4 presents the performance evaluation, and Sect. 5 concludes the paper.

2 Problem Formulation

In this section, we propose the model of mobile data subsidization, and we utilize a hierarchical game approach to characterize the model and analyze the reward optimization for content providers therein.

2.1 Mobile Data Subsidization Model

As illustrated in Fig. 1, we consider a 4G/5G system with mobile data subsidization in which there are M heterogeneous Content Providers (CPs) labeled as $o_1, o_2, \ldots, o_M$, such as the video CPs: Youtube, Netflix, Hulu, and Vimeo. The CPs can subsidize the mobile data usage of N heterogeneous Mobile Users

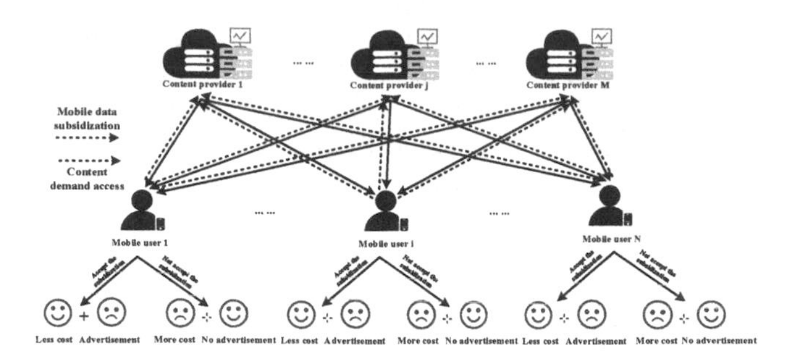

Fig. 1. Schematics of mobile data subsidization model.

(MUs) labeled as $s_1, s_2, \ldots, s_N$. The MUs can access and consume the content from different CPs, and the content can be downloaded directly through the network infrastructure such as 4G/5G base stations. The data usage of MUs can be partly subsidized by CPs via displaying a certain amount of advertisements. The CP earns the advertisement revenue when the MU accepts the offered data subsidization, i.e., the reward. Meanwhile, the MUs are able to reject viewing the advertisements, and still access and consume the content paying the full data usage fees without subsidization.

We denote y as the content (volume) demand from an MU, and $\sigma g(y)$ as its utility derived from enjoying the content, where $\sigma > 0$ is a factor representing the utility coefficient of MU, e.g., a particular valuation between MU and content. Similar to [5,16], we first define the following function, $g(y) = \frac{1}{1-\alpha} y^{1-\alpha}$, where $0 < \alpha < 1$ is a given coefficient. Specifically, $g(\cdot)$ is a non-decreasing and concave function with decreasing marginal satisfaction, representing the decreasing marginal preference of MUs to content. In traditional content consumption, the usage-based pricing is popular, i.e., the network operator charges MU s_i a certain unit price p_i for the volume of content downloaded. Here, the unit price is the same for all MUs for fairness, i.e., $p_i = p, \forall i \in \{1, 2, \ldots, N\}$. Therefore, the utility formulation of the MU with content demand y is expressed by $v(y) = \sigma g(y) - py$.

With the data subsidization scheme, the data usage fees for accessing the content can be partly subsidized by the CP. Denote $\theta \in [0, 1]$ as a reward factor of content subsidized by the CP, i.e., θ units of the content is subsidized. Therefore, if the MU accepts the subsidization, it pays for the rest $(1 - \theta)y$ units of content, with the cost $(1 - \theta)py$ incurred to the MU [16]. However, the MU's utility of enjoying the content is discounted as it needs to view a certain amount of advertisement displayed by the CP. We denote l_a as the amount of advertisement imposed by the CP per volume of content. In what follows, we assume that l_a is constant for all content. For example, Hulu plays the same amount of advertisement regularly between videos for all of its subscribers (users). We consider a normalized $l_a \in [0, 1]$ since the CPs have the amount of advertisements strictly less than that of the provided content. For the ease of derivation, we define an auxiliary variable as $\tau = \frac{1}{1+l_a}, \tau \in \left[\frac{1}{2}, 1\right]$, representing a discounting factor in terms of viewing advertisement. The larger the length of the advertisement per content, the smaller of τ. Therefore, the utility of the MU which has the content demand y that accepts the subsidization from the CP is formulated as follows:

$$\hat{u}(y) = \tau\sigma g(y) - (1-\theta)py. \quad (1)$$

2.2 Game Formulation

Without loss of generality, we first assume that the MU, i.e., the buyer, decides on a fraction of the content demand to access that accepts the data subsidization (*sponsored content*), denoted by x, where $x \in [0, 1]$. Then, the fraction of the content demand of that not accepts the data subsidization, i.e., the *normal content* demand of the MU is $1-x$ accordingly. Note here that the MUs' actions (the content volume to be purchased) are normalized to sum up to one, e.g., one video content [16]. Nevertheless, the analytical results will not structurally change even if the content demand is not normalized [16].

Mobile Users in Stage II. Let $x_{i,j}$ and $\theta_{i,j}$ denote the content demand of MU s_i that accepts the data subsidization from CP o_j, and the reward factor of CP o_j for MU s_i, respectively. Given the reward factors $\boldsymbol{\theta}_j = \{\theta_{1,j}, \theta_{2,j}, \ldots, \theta_{N,j}\}$ determined by each CP $o_j, j \in \{1, 2, \ldots, M\}$, each rational MU $s_i, i \in \{1, 2, \ldots, N\}$ decides the strategies of content demand from different CPs that maximizes its utility. Specifically, each MU aim to maximize its utility by finding the balance between the content with subsidization (sponsored content) and that without subsidization (normal content). Let $\boldsymbol{x}_i = \{x_{i,1}, x_{i,2}, \ldots, x_{i,M}\}$ be the content demand strategies of MU s_i from all CPs, $\boldsymbol{x}_{-i}$ be the strategies of all other MUs except MU s_i, and $\boldsymbol{\Theta} = \{\boldsymbol{\theta}_1, \boldsymbol{\theta}_2, \ldots, \boldsymbol{\theta}_M\}$ be the reward strategies of all CPs, i.e., the reward factors. Therefore, the sum utility of MU s_i obtained from all CPs is formulated as follows:

$$\begin{aligned} u_i(\boldsymbol{x}_i; \boldsymbol{x}_{-i}, \boldsymbol{\Theta}) = \sum_{j=1}^{M} \omega_{i,j} \Big(\tau\sigma_i f(x_{i,j}) - (1-\theta_{i,j})x_{i,j}p \\ + \sigma_i f(1-x_{i,j}) - (1-x_{i,j})p \Big), \\ \forall i \in \{1, 2, \ldots, N\}, \end{aligned} \quad (2)$$

where

$$f(x_{i,j}) = \frac{1}{1-\alpha} x_{i,j}^{1-\alpha}, \quad (3)$$

and $\omega_{i,j}$ is the probability for MU s_i to choose CP o_j to access the content, and $\sum_{j=1}^{M} \omega_{i,j} = 1, \forall i = \{1, 2, \ldots, N\}$. For example, $\omega_{i,j} = 1$ (or $\omega_{i,j} = 0$) means that MU s_i only access (or not access) the content from CP o_j.

Consequently, each MU s_i needs to choose its optimal strategies $\boldsymbol{x}_i$ based on the strategies of all other MUs except MU s_i (i.e., $\boldsymbol{x}_{-i}$) and the reward factors announced by all CPs (i.e., $\boldsymbol{\Theta}$), by solving the following optimization problem:

Problem 1. (The MU s_i sub-game)

$$\begin{aligned} \underset{\boldsymbol{x}_i}{\text{maximize}} \quad & u_i(\boldsymbol{x}_i; \boldsymbol{x}_{-i}, \boldsymbol{\Theta}), \forall i \in \{1, 2, \dots, N\}, \\ \text{subject to} \quad & x_{i,j} \in [0, 1]. \end{aligned} \tag{4}$$

Content Providers in Stage I. Let μ_j denote the advertisement revenue coefficient of CP o_j, and hence $\mu_j h(x)$ represents the advertisement revenue obtained from MUs that watch the advertisements, in which $h(x)$ is defined by using the α-fair function [16,19], as follows:

$$h(x) = \frac{1}{1-\gamma} x^{1-\gamma}, \tag{5}$$

where $0 < \gamma < 1$ which is a coefficient. Each CP o_j aims to maximize the obtained total profit, i.e., the gained advertisement revenue minus the cost of subsidization offered to MUs, which can be formulated as follows:

$$\Pi_j(\boldsymbol{\theta}_j; \boldsymbol{\theta}_{-j}, \boldsymbol{X}, \boldsymbol{\omega}) = \sum_{i=1}^{N} \omega_{i,j} \Big(\mu_j h(x_{i,j}) - \theta_{i,j} p x_{i,j} \Big), \tag{6}$$

$\forall j \in \{1, 2, \dots, M\}$, where $\boldsymbol{\omega} = \langle \omega_{i,j} \rangle, i \in \{1, 2, \dots, N\}, j \in \{1, 2, \dots, M\}$ denotes the pairing probability between CPs and MUs, $\boldsymbol{\theta}_{-j}$ denotes the optimal strategies of all other CPs except CP o_j ($\boldsymbol{\theta}_{-j} = \boldsymbol{\Theta} \backslash \boldsymbol{\theta}_j$), and $\boldsymbol{X}$ denotes the optimal strategies of all MUs in terms of content demand, i.e., $\boldsymbol{X} = \{\boldsymbol{x}_1, \boldsymbol{x}_2, \dots, \boldsymbol{x}_N\}$.

By adopting the incentive mechanism method in [20], we can set the probability of each MU s_i choosing CP o_j as

$$\omega_{i,j} = \frac{\theta_{i,j}}{\sum_{k=1}^{M} \theta_{i,k}}. \tag{7}$$

By observing from (7), we know that the value of $\omega_{i,j}$ is larger on condition that the reward factor offered by CP o_j increases given the fixed reward strategies of other CPs. This indicates that MU s_i obtains the greater data subsidization from CP o_j. As such, in order to attract more MUs, each CP o_j ($j \in \{1, 2, \dots, M\}$), tends to provide greater subsidization (i.e., higher reward factors) to MUs. The reason is that MUs are more likely to access and consume the content from CP o_j when $\omega_{i,j}$ increases. However, each CP o_j, $\forall j \in \{1, 2, \dots, M\}$, cannot maintain the value of reward factor $\theta_{i,j}$ that is too high to reduce the subsidization cost. Moreover, when each CP o_j determine its reward factors $\boldsymbol{\theta}_j$ for different MUs, the CP needs to consider the reward factors offered by other CPs (i.e., $\boldsymbol{\theta}_{-j}$) as well as the strategies of all MUs (i.e., $\boldsymbol{X}$). This thereby leads to the reward competition among CPs. Therefore, the optimization problem for each CP is defined as follows:

Problem 2. (The CP o_j sub-game)

$$\begin{aligned} &\underset{\boldsymbol{\theta}_j}{\text{maximize}} \quad \Pi_j(\boldsymbol{\theta}_j; \boldsymbol{X}, \boldsymbol{\theta}_{-j}, \boldsymbol{\omega}), \forall j \in \{1, 2, \ldots, M\}, \\ &\text{subject to} \quad \theta_{i,j} \in [0, 1]. \end{aligned} \tag{8}$$

Considering the inherent leader-follower relations among the CPs and MUs, we hence model the two-sided interaction problem as a hierarchical game. In particular, we model the problem as a multi-leader multi-follower Stackelberg game, in which the CPs are leaders and the MUs are followers. Consequently, **Problems 1** and **2** jointly form a Stackelberg game with the objective of finding the Stackelberg equilibrium. The Stackelberg equilibrium is defined as a point where the payoffs of the leaders are maximized given that the followers adopt their best responses [3]. In the following, we define the Stackeberg game.

Definition 1. *Let $\boldsymbol{X}^*$ and $\boldsymbol{\Theta}^*$ denote the optimal content demand strategies of MUs (followers) and the optimal reward strategies of CPs (leaders), respectively. Let $\boldsymbol{x}_i$ be the strategy of MU s_i, $\boldsymbol{x}_{-i}$ be the strategies of all other MUs except MU s_i, $\boldsymbol{\theta}_j$ be the strategy of CP o_j, and $\boldsymbol{\theta}_{-j}$ be the strategies of all other CPs except CP o_j. Then, the point $(\boldsymbol{X}^*, \boldsymbol{\Theta}^*)$ is the Stackelberg equilibrium of the multi-leader multi-follower game provided that the following conditions,*

$$\Pi_j(\boldsymbol{\theta}_j^*, \boldsymbol{\theta}_{-j}^*, \boldsymbol{X}^*) \geq \Pi_j(\boldsymbol{\theta}_j, \boldsymbol{\theta}_{-j}^*, \boldsymbol{X}^*), \forall j, \tag{9}$$

and

$$u_i(\boldsymbol{x}_i^*, \boldsymbol{x}_{-i}^*, \boldsymbol{\Theta}^*) \geq u_i(\boldsymbol{x}_i, \boldsymbol{x}_{-i}^*, \boldsymbol{\Theta}^*), \forall i, \tag{10}$$

are satisfied, where $\boldsymbol{X} = \{\boldsymbol{x}_1, \boldsymbol{x}_2, \ldots, \boldsymbol{x}_N\}$ and $\boldsymbol{\Theta} = \{\boldsymbol{\theta}_1, \boldsymbol{\theta}_2, \ldots, \boldsymbol{\theta}_M\}$.

In the context of game theory, each of the leaders (CPs) or the followers (MUs) is rational and autonomous making the decision in a distributed manner [8]. In the following sections, we investigate the Stackelberg equilibrium by analyzing the optimal strategies of the followers and leaders in the game.

3 Multi-CP Multi-MU Game as EPEC

In this section, we investigate the general multi-leader multi-follower game that incorporates M CPs and N MUs, i.e., the many-to-many interaction. In such a multi-CP and multi-MU scenario, each MU has multiple CP choices to access the content, and each CP is able to provide subsidization to multiple MUs to earn more advertisement revenue. Accordingly, each individual CP competes with others for the equilibrium, which is constrained by the lower equilibrium among the MUs. This leads to the Equilibrium Programs with Equilibrium Constraints (EPEC) problem formulation.

For the EPEC problem, we also follow the basic idea of backward induction and consider the sub-game problem among MUs with the fixed CP strategies first. Clearly, we have the following lower equilibrium condition among MUs:

$$\begin{aligned}\boldsymbol{x}_i^* &= \arg\max_{\boldsymbol{x}_i} u_i(\boldsymbol{x}_i; \boldsymbol{x}_{-i}, \boldsymbol{\Theta}) \\ &= \arg\max_{\boldsymbol{x}_i} \sum_{j=1}^{M} \omega_{i,j} \Big(\tau\sigma_i f(x_{i,j}) - (1-\theta_{i,j}) x_{i,j} p \\ &\quad + \sigma_i f(1 - x_{i,j}) - (1 - x_{i,j}) p \Big), \forall i \in \{1, 2, \ldots, N\}. \end{aligned} \tag{11}$$

With the anticipation of all MUs' behaviors as indicated in (11), each CP o_j, $j \in \{1, 2, \ldots, M\}$, aims to set its reward factors so as to receive the optimal profit, which is given by

$$\Pi_j = \sum_{i=1}^{N} \omega_{i,j} \Big(\mu_j h(x_{i,j}^*) - \theta_{i,j} p x_{i,j}^* \Big), \forall j \in \{1, 2, \ldots, M\}, \tag{12}$$

where $\omega_{i,j}$ is related to the setting reward factors of all CPs, as indicated in (7). Therefore, in order to obtain the maximum profit, each CP also needs to consider the strategies of all other CPs. Since each CP can provide the data subsidization for multiple MUs simultaneously, we include the following constraint to indicate the limited total budget in terms of data subsidization held by each CP o_j:

$$\sum_{i=1}^{N} P_{i,j}\theta_{i,j} - Q_j \leq 0, j \in \{1, 2, \ldots, M\}, \tag{13}$$

where all $\{P_{i,j} \,|i = 1, 2, \ldots, N\}$ and Q_j are real, scalar constants. In summary, the CPs' optimization problems are formulated as the following EPEC problems:

$$\begin{aligned} &\underset{\boldsymbol{\theta}_j}{\text{maximize}} \quad \Pi_j = \sum_{i=1}^{N} \omega_{i,j} \Big(\mu_j h(x_{i,j}^*) - \theta_{i,j} p x_{i,j}^* \Big), \\ &\text{subject to} \quad \begin{cases} \sum_{i=1}^{N} P_{i,j}\theta_{i,j} - Q_j \leq 0, \\ 0 \leq \theta_{i,j} \leq 1, \\ \boldsymbol{x}_i^* = \arg\max_{\boldsymbol{x}_i} u_i(\boldsymbol{x}_i; \boldsymbol{x}_{-i}, \boldsymbol{\Theta}), \\ \text{subject to} \begin{cases} x_{i,j} \geq 0, \\ x_{i,j} \leq 1, \end{cases} \end{cases} \end{aligned} \tag{14}$$

for all $i \in \{1, 2, \ldots, N\}$, and $j \in \{1, 2, \ldots, M\}$.

The EPEC describes the hierarchical optimization problems that contain equilibrium problems at both the upper and lower levels [11,23]. As aforementioned, the CPs are independent and rational entities, which aim to maximize their individual profit. However, maximizing Π_j for CP o_j affects the profits of other CPs and the utilities of all MUs. Likewise, the utility maximization of MU s_i affects the profits of all CPs. In practice, when the number of CPs and the number of MUs are large, the centralized optimization in terms of the profits and the utilities of all CPs and all MUs, respectively, is difficult to achieve the optimal solutions simultaneously. Furthermore, in the multi-CP multi-MU scenario, the coordination of multiple conflicting payoffs leads to high complexity to achieve the optimal result. As such, we propose to utilize the distributed Alternating Direction Method of Multipliers (ADMM) algorithm with the fast convergence property for the above large-scale optimization problem, which is guaranteed to converge to the optimum results [12,13,23]. ADMM is an efficient large-scale optimization tool for solving convex or even nonconvex functions.

3.1 ADMM-Form Optimization Concepts

Before presenting the ADMM implementation for solving the EPEC problem, we briefly introduce a typical ADMM-form optimization problem. To facilitate the narrative, we first focus on a simple system with a single provider and N users. Therein, the objective of the provider is expressed as follows:

$$\begin{aligned} \underset{\boldsymbol{y}}{\text{minimize}} \quad & L(\boldsymbol{m}) = \sum_{i=1}^{N} l_i(m_i) \\ \text{subject to} \quad & \sum_{i=1}^{N} G_i m_i - T_i = 0, \end{aligned} \tag{15}$$

where $\boldsymbol{m} = \{m_1, \ldots m_i, \ldots, m_N\}$, and $l_i(m_i)$ represents the cost of provider j if its strategy is m_i. Specifically, G_i and T_i are real scalar constants, and m_i is a real scalar variable. $l_i(m_i)$ is convex on m_i.

With t denoted as the iteration index, the provider iteratively updates the value of $\boldsymbol{m}$ such that

$$\boldsymbol{m}(t+1) = \arg\min(H(\boldsymbol{m})) - \sum_{i=1}^{N} \lambda_i(t) G_i m_i + \Psi, \tag{16}$$

where

$$\Psi = \frac{\rho}{2} \sum_{i=1}^{N} \|G_i m_i - T_i\|_2^2. \tag{17}$$

$\rho > 0$ is a damping factor, and $\|\cdot\|_2^2$ represents the Frobenius-2 norm. Likewise, the dual variable λ is iteratively updated by

$$\lambda_i(t+1) = \lambda_i(t) - \rho \left(\sum_{i=1}^{N} G_i m_i(t+1) - Q_i \right). \tag{18}$$

If $l_i(m_i)$ is separable and convex, the ADMM algorithm will eventually converge to the set of stationary solutions [12,13,22,23]. It is worth noting that for the case of non-convex objective functions, the convergence of ADMM can still be ensured in certain cases [4].

3.2 Multi-CP Multi-MU Based ADMM for Solving EPEC

In what follows, we elaborate the iteration process of the Multi-CP Multi-MU based ADMM that is leveraged to optimize the profits of CPs and the utilities of MUs in the framework of mobile data subsidization. Specifically, each iteration is composed of two-layer optimization as follows:

(1) **Utility Optimization in lower layer:** In the lower layer, each MU s_i ($i = 1, 2, \ldots, N$) observes the announced reward factors $\theta_{i,j}^{(q)}$ from CPs at the start of each iteration q, and decides on the content demand strategies towards different CP o_j, $x_{i,j}$ (within the strategy space $[0, 1]$), maximizing its utility $u_i(\boldsymbol{x}_i)$. Note that the superscript (q) represents the qth iteration of the ADMM in the external loop. The objective of each MU s_i is to maximize its individual utility $u_i(\boldsymbol{x}_i)$, and obtain the optimal values of $\boldsymbol{x}_i$. This forms the internal loop of the ADMM. Hence, the value of $\boldsymbol{x}_i$ is updated by each MU s_i at each iteration of the internal loop as follows:

$$\boldsymbol{x}_i^{(q)}(t+1) = \arg\max \left(u_i(\boldsymbol{x}_i^{(q)}) \right). \tag{19}$$

During each iteration of the external loop q, the MUs are able to derive a set of values of content demand, $x_{i,j}^{(q)}$, which maximize their utilities at the end of the internal loop. t is the index of iteration in the internal loop. At the same time, these values can be predicted by all CPs, which will be employed to update the values of $\theta_{i,j}$ in the higher layer.

(2) **Profit Optimization in higher layer:** In this layer, the CPs are aware of the behaviors of MUs due to the first-moving advantage and hence can predict the content traffic to be transferred to MUs. Specifically, each CP o_j ($j = 1, 2, \ldots, M$) controls the values of $\boldsymbol{\theta}_j$ within the strategy space $[0, 1]$ to maximize its profit by invoking ADMM as follows:

$$\boldsymbol{\theta}_j^{(q)}(t+1) = \arg\max \left(\Pi_j(\boldsymbol{\theta}_j^{(q)}) \right) + \sum_{i=1}^{N} \lambda_i^{(q)}(t) P_{i,j}\theta_{i,j} + \Psi, \tag{20}$$

where

$$\Psi = \frac{\rho}{2} \sum_{i=1}^{N} \left\| \sum_{m=1, m\neq j}^{M} x_{i,m}^{(q)}(\tau) + P_{i,j}\theta_{i,j} - Q_j \right\|_2^2, \tag{21}$$

and $\tau = t$ if $m > j$, and $\tau = t + 1$ if $m < j$. Specifically, t is the index of iteration in the internal loop. $\rho > 0$ is the damping factor, and λ is the dual variable which will be updated as follows:

$$\lambda_i^{(q)}(t+1) = \lambda_i^{(q)}(t) + \rho \left(\sum_{i=1}^{N} P_{i,j}\theta_{i,j}^{(q)}(t+1) - Q_j \right). \tag{22}$$

The updated reward factors $\theta_{i,j}^{(q+1)}$ are then broadcasted to the MUs for the next iteration, i.e., the $(q+1)^{\text{th}}$ iteration. This forms the external loop of the algorithm. The external loop will not terminate until the condition

$$\left\|\sum_{j=1}^{M} \Pi_j(\boldsymbol{p}_j^{(q)}) - \sum_{j=1}^{M} \Pi_j(\boldsymbol{p}_j^{(q-1)})\right\| \leq \epsilon \tag{23}$$

holds, where ϵ is a pre-determined small-valued threshold.

The detailed steps of the above ADMM algorithm is presented in Algorithm 1. Moreover, we can obtain the following theorem after analyzing the utility and profit functions of MU and CP, respectively.

Theorem 1. *The utility function of each MU s_i in (2), and the profit function of each CP o_j in (6) are strictly concave, where $i = 1, 2, \ldots, N$ and $j = 1, 2, \ldots, M$.*

Proof. The first-order and second-order derivatives of (2) with respect to $x_{i,j}$ can be expressed by

$$\frac{\partial u_i}{\partial x_{i,j}} = \sum_{j=1}^{M} \omega_{i,j} \Big(\tau\sigma x_{i,j}^{-\alpha} - \sigma(1 - x_{i,j})^{-\alpha} + \theta p\Big), \tag{24}$$

$$\frac{\partial^2 u_i}{\partial x_{i,j}^2} = \sum_{j=1}^{M} \omega_{i,j} \Big(- \alpha\tau\sigma x_{i,j}^{-\alpha-1} - \alpha\sigma(1 - x_{i,j})^{-\alpha-1}\Big) < 0. \tag{25}$$

We can hence easily conclude with the negativity of (25).

Moreover, we have the profit function of CP o_j

$$\begin{aligned}
\Pi_j &= \sum_{i=1}^{N} \omega_{i,j} \Big(\mu_j h(x_{i,j}) - \theta_{i,j} p x_{i,j}\Big) \\
&= \sum_{i=1}^{N} \frac{\theta_{i,j}}{\sum_{k=1}^{M} \theta_{i,k}} \Big(\mu_j h(x_{i,j}) - \theta_{i,j} p x_{i,j}\Big) \\
&= \sum_{i=1}^{N} \left(\frac{\theta_{i,j}}{\sum_{k=1}^{M} \theta_{i,k}} \mu_j h(x_{i,j}) - \frac{\theta_{i,j}^2}{\sum_{k=1}^{M} \theta_{i,k}} p x_{i,j} \right),
\end{aligned} \tag{26}$$

$\forall j \in \{1, 2, \ldots, M\}$. To demonstrate the concavity of Π_j on $\theta_{i,j}$, we need to ensure the negativity of $\frac{\partial^2 \Pi_j}{\partial \theta_{i,j}^2}$. We expand the first-order and second-order derivatives of (26) with respect to $\theta_{i,j}$ in (27), and (28), respectively:

$$\frac{\partial \Pi_j}{\partial \theta_{i,j}} = \sum_{i=1}^{N} \left(\frac{\mu_j h(x_{i,j}) \sum\limits_{k=1,k\neq j}^{M} \theta_{i,k}}{\left(\sum\limits_{k=1,k\neq j}^{M} \theta_{i,k} + \theta_{i,j} \right)^2} - px_{i,j} \frac{2\theta_{i,j} \left(\sum\limits_{k=1,k\neq j}^{M} \theta_{i,k} + \theta_{i,j} \right) - \theta_{i,j}^2}{\left(\sum\limits_{k=1,k\neq j}^{M} \theta_{i,k} + \theta_{i,j} \right)^2} \right)$$

$$= \sum_{i=1}^{N} \left(\frac{\mu_j h(x_{i,j}) \sum\limits_{k=1,k\neq j}^{M} \theta_{i,k} - 2\theta_{i,j} px_{i,j} \sum\limits_{k=1,k\neq j}^{M} \theta_{i,k} - \theta_{i,j}^2 px_{i,j}}{\left(\sum\limits_{k=1,k\neq j}^{M} \theta_{i,k} + \theta_{i,j} \right)^2} \right), \quad (27)$$

and

$$\frac{\partial^2 \Pi_j}{\partial \theta_{i,j}^2} = \sum_{i=1}^{N} \left(\frac{\left(-2px_{i,j} \sum\limits_{k=1,k\neq j}^{M} \theta_{i,k} - 2\theta_{i,j} px_{i,j} \right) \left(\sum\limits_{k=1,k\neq j}^{M} \theta_{i,k} + \theta_{i,j} \right)^2}{\left(\sum\limits_{k=1,k\neq j}^{M} \theta_{i,k} + \theta_{i,j} \right)^4} \right)$$

$$- \sum_{i=1}^{N} \left(\frac{\left(2 \sum\limits_{k=1,k\neq j}^{M} \theta_{i,k} + 2\theta_{i,j} \right) \mathrm{A}}{\left(\sum\limits_{k=1,k\neq j}^{M} \theta_{i,k} + \theta_{i,j} \right)^4} \right)$$

$$= -2 \sum_{i=1}^{N} \left(\frac{px_{i,j} \left(\sum\limits_{k=1,k\neq j}^{M} \theta_{i,k} + \theta_{i,j} \right)^2 + \mathrm{B}}{\left(\sum\limits_{k=1,k\neq j}^{M} \theta_{i,k} + \theta_{i,j} \right)^3} \right)$$

$$= -2 \sum_{i=1}^{N} \left(\frac{px_{i,j} \left(\sum\limits_{k=1,k\neq j}^{M} \theta_{i,k} \right)^2 + \mu_j h(x_{i,j}) \sum\limits_{k=1,k\neq j}^{M} \theta_{i,k}}{\left(\sum\limits_{k=1,k\neq j}^{M} \theta_{i,k} + \theta_{i,j} \right)^3} \right) < 0, \quad (28)$$

where

$$\mathrm{A} = \mu_j h(x_{i,j}) \sum_{k=1,k\neq j}^{M} \theta_{i,k} - 2\theta_{i,j} px_{i,j} \sum_{k=1,k\neq j}^{M} \theta_{i,k} - \theta_{i,j}^2 px_{i,j}, \quad (29)$$

and

$$\mathrm{B} = \mu_j h(x_{i,j}) \sum_{k=1,k\neq j}^{M} \theta_{i,k} - 2\theta_{i,j} px_{i,j} \sum_{k=1,k\neq j}^{M} \theta_{i,k} - \theta_{i,j}^2 px_{i,j}. \quad (30)$$

We can then deduce that $\frac{\partial^2 \Pi_j}{\partial \theta_{i,j}^2}$ is negative, and hence validate the concavity of Π_j on $\theta_{i,j}$.

Algorithm 1. Multi-CP Multi-MU Based ADMM for Solving EPEC problem

1: **Input:**
Initial input $\theta_{i,j} \in [0,1]$, where $i = 1,2,\ldots,N$, $j = 1,2,\ldots,M$, a pre-determined small-valued threshold ϵ, $q = 1$;
2: **repeat**
3: Utility optimization for MUs using ADMM (Internal Loop): MUs observe the announced reward factors $\theta_{i,j}$, and decide on their content demand strategies $x_{i,j}^{(q)}$ by evaluating their derived utilities, $u_i(\boldsymbol{x}_i)$;
4: Profit optimization for CPs using ADMM (External Loop): CPs predict the MU behaviors $x_{i,j}$, and invoke ADMM to perform the maximization for their individual profit Π_j. The optimal reward factor $\theta_{i,j}^{(q)}$ is obtained by maximizing their profits;
5: $q = q + 1$;
6: **until** $\left\| \sum_{j=1}^{M} \Pi_j(\boldsymbol{\theta}_j^{(q)}) - \sum_{j=1}^{M} \Pi_j(\boldsymbol{\theta}_j^{(q-1)}) \right\| \le \epsilon$
Output: The optimal strategies of content demand $\boldsymbol{x}_i^* = \boldsymbol{x}_i^{(q)}$, where $i = 1,2,\ldots,N$; The optimal reward factors $\boldsymbol{\theta}_j^* = \boldsymbol{\theta}_j^{(q)}$, where $j = 1,2,\ldots,M$.

According to [12,13,23], if the optimization problems faced by MUs and CPs are both convex, the ADMM can converge to the optimum results, i.e., $\boldsymbol{x}_i^*$, $i \in \{1,2,\ldots,N\}$ and $\boldsymbol{p}_j^*$, $j \in \{1,2,\ldots,M\}$ in a distributed manner. We further confirm the convergence of the ADMM in the next section.

4 Performance Evaluation

In this section, we employ numerical simulations to justify the analytical results and evaluate the system performance metrics in the mobile data subsidization model, with default network parameters set as follows: $\alpha = 0.8$, $\gamma = 0.9$, $l_a = 0.5$, $\tau = 1/(1+0.5)$, $p = 100$, $M = 3$, and $N = 3$.

Before evaluating the system performance with the proposed scheme, we first confirm the convergence of the distributed ADMM algorithm in a data subsidization system with 3 CPs ($\mu_1 = 10, \mu_2 = 20, \mu_3 = 60$) with 3 MUs ($\sigma_1 = 5, \sigma_2 = 15, \sigma_3 = 30$). The results are presented in Figs. 2 and 3, where the EPEC problem is solved in an iterative manner. In particular, the results in Fig. 2 show the convergence of the competition among MUs to achieve the lower-layer equilibrium, and the results in Fig. 3 show the convergence of the competition among CPs to achieve the higher-layer equilibrium. Different MUs and CPs finally achieve different payoffs at the convergence point. We find that the MU with the higher value of σ_i will obtain the higher utility as it derives more benefit from viewing the content. Moreover, the CP with the higher value of μ_j has greater competitiveness as it extracts more advertisement revenue from content traffic, and hence it is able to offer more data subsidization to attract MUs.

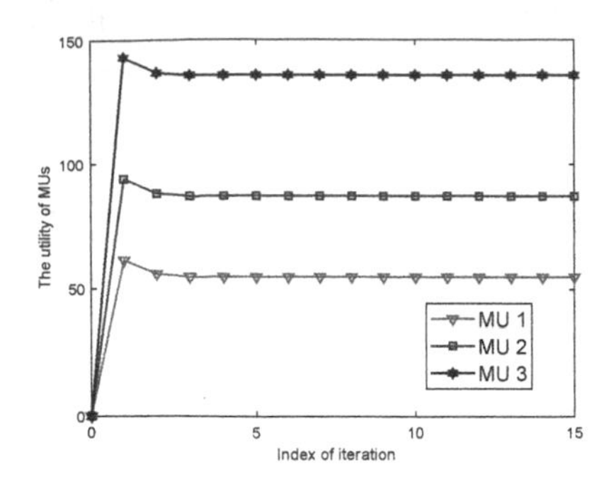

Fig. 2. The utilities of MUs vs. the index of iteration.

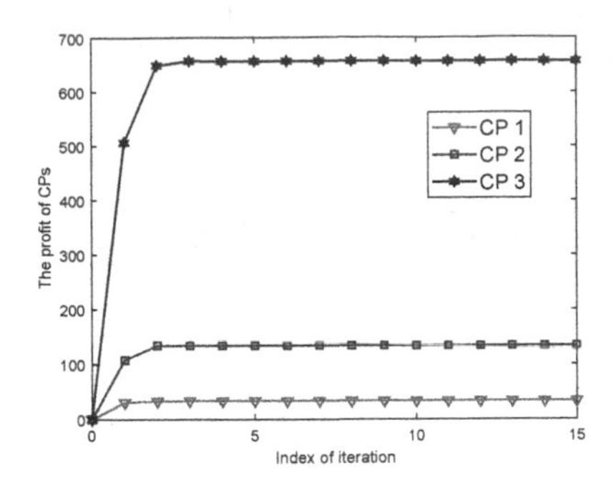

Fig. 3. The profits of CPs vs. the index of iteration.

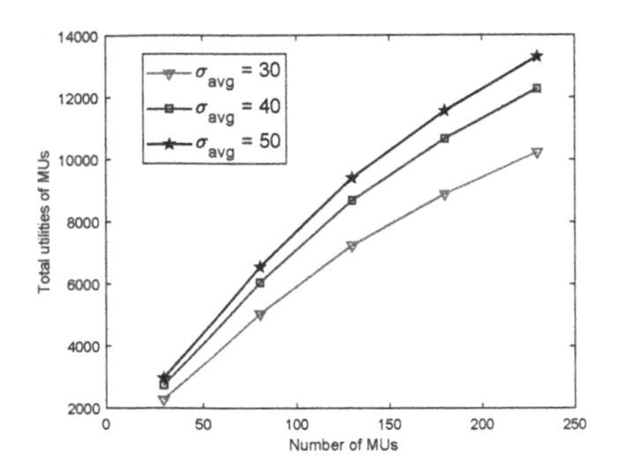

Fig. 4. Total utilities of MUs vs. the number of MUs

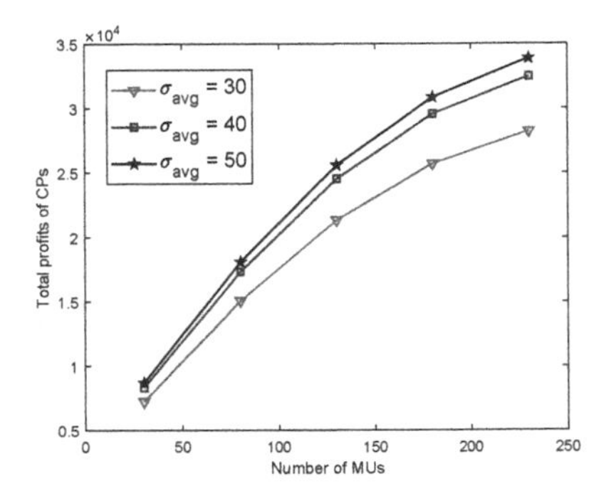

Fig. 5. Total profits of CPs vs. the number of MUs

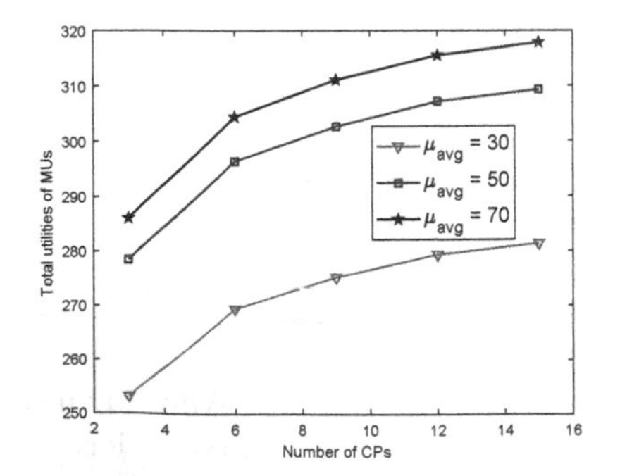

Fig. 6. Total utilities of MUs vs. the number of CPs.

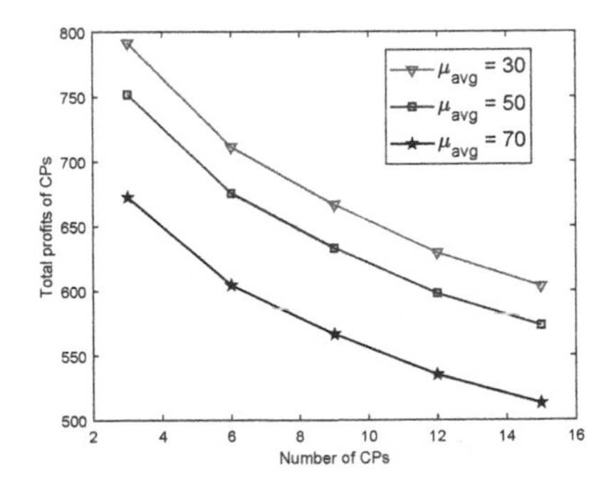

Fig. 7. Total profits of CPs vs. the number of CPs.

Then, we study the system performance of a general data subsidization system with M CPs and N MUs under the proposed scheme. We assume the utility coefficients of MUs σ_i and advertisement revenue coefficients of CPs μ_j follow the normal distribution $\mathcal{N}(\sigma_{\mathrm{avg}}, 2)$, and $\mathcal{N}(\mu_{\mathrm{avg}}, 2)$, respectively. Specifically, we evaluate the performance with different numbers of MUs and CPs, as depicted in Figs. 4, 5, 6 and 7. From Figs. 4 and 5, we can observe that the total utilities of MUs and the total profits of CPs both increase with the increase of the number of MUs. The reason is that the more MUs will consume more "sponsored content", which results in more profits of CPs. Nevertheless, as the number of MUs increases, the marginal increase of the total utilities of MUs and total profits of CPs both decrease. This is due to the limited data subsidization constraint. As the total reward is limited, the CP cannot provide enough subsidization for all MUs when the number of MUs is high. Therefore, the increasing rate of the total utilities of MUs decreases as the number of MUs increases. The restrained subsidization from CPs cannot continuously achieve greater profits for themselves. Consequently, the increasing rate of total profit of CPs decreases as well. Furthermore, by comparing different values of σ_{avg}, we find that the increase of the average utility coefficients leads to the increase of total utilities of MUs and the increase of total profits of CPs. The increase of σ_{avg} improves the valuation derived from viewing the content, which hence promotes the higher willingness of MUs to access and enjoy the content. This in turn increases the total profits of CPs.

In addition, we observe from Fig. 6 that the total utilities of MUs increase but the total profits of CPs decrease as the number of CPs increases. The reason is that the competition among CPs becomes sharper when there are more CPs in the system. Each CP competes with other CPs to promote its content traffic by providing more subsidization for MUs. Therefore, the total profits of CPs decrease in presence of sharper competition. Meanwhile, the competition among CPs results in the greater subsidization to MUs, leading to the increase of total utilities of MUs. Moreover, we find that the increase of the average advertisement revenue coefficients μ_{avg} improves the total utilities of MUs but reduces the total profits of CPs. The reason is that the higher value of μ_{avg} intensifies the reward competition among CPs in data subsidization system since the CPs can obtain greater advertisement revenue from the given "sponsored content" traffic. On one hand, the CPs provide greater data subsidization that incurs more cost, and hence the total profits are reduced. On the other hand, the greater subsidzation improve the total utilities of MUs.

5 Conclusion

In this work, we have established a hierarchical Stackelberg game to model the interactions among content providers and mobile users in the framework of mobile data subsidization scheme. We have characterized the many-to-many interactions among multiple providers and multiple users by formulating an Equilibrium Programs with Equilibrium Constraints (EPEC) problem. Moreover, we have employed the distributed Alternating Direction Method of Multipliers (ADMM) algorithm to tackle the EPEC problem by utilizing the fast-convergence properties of ADMM. Numerical results have been presented to confirm the analytical solutions and evaluate the system performance with the proposed schemes.

Acknowledgment. The research of Zehui Xiong was supported by Alibaba Group through Alibaba Innovative Research (AIR) Program and Alibaba-NTU Singapore Joint Research Institute (JRI), Nanyang Technological University, Singapore. The research of Jun Zhao was supported by 1) Nanyang Technological University (NTU) Startup Grant, 2) Alibaba-NTU Singapore Joint Research Institute (JRI), 3) Singapore Ministry of Education Academic Research Fund Tier 1 RG128/18, Tier 1 RG115/19, Tier 1 RT07/19, Tier 1 RT01/19, and Tier 2 MOE2019-T2-1-176, 4) NTU-WASP Joint Project, 5) Singapore National Research Foundation (NRF) under its Strategic Capability Research Centres Funding Initiative: Strategic Centre for Research in Privacy-Preserving Technologies & Systems (SCRIPTS), 6) Energy Research Institute @NTU (ERIAN), 7) Singapore NRF National Satellite of Excellence, Design Science and Technology for Secure Critical Infrastructure NSoE DeST-SCI2019-0012, and 8) AI Singapore (AISG) 100 Experiments (100E) programme. The research of Dusit Niyato was supported by the National Research Foundation (NRF), Singapore, under Singapore Energy Market Authority (EMA), Energy Resilience, NRF2017EWT-EP003-041, Singapore NRF2015-NRF-ISF001-2277, Singapore NRF National Satellite of Excellence, Design Science and Technology for Secure Critical Infrastructure NSoE DeST-SCI2019-0007, A*STAR-NTU-SUTD Joint Research Grant on Artificial Intelligence for the Future of Manufacturing RGANS1906, Wallenberg AI, Autonomous Systems and Software Program and Nanyang Technological University (WASP/NTU) under grant M4082187 (4080), Singapore MOE Tier 2 MOE2014-T2-2-015 ARC4/15, and MOE Tier 1 2017-T1-002-007 RG122/17. The work of Ruilong Deng was supported in part by the National Natural Science Foundation of China (NSFC) under Grant 61873106 and 62061130220. The work was also supported by NSFC under grant No. 61973087 and U1911401.

References

1. AT&T sponsored plan. https://www.att.com/
2. Andrews, M., Özen, U., Reiman, M.I., Wang, Q.: Economic models of sponsored content in wireless networks with uncertain demand. In: Proceedings of IEEE INFOCOM Workshops, Turin, Italy, April 2013
3. Han, Z., Niyato, D., Saad, W., Baar, T., Hjrungnes, A.: Game Theory in Wireless and Communication Networks: Theory, Models, and Applications. Cambridge University Press, Cambridge (2012)
4. Hong, M., Luo, Z.Q., Razaviyayn, M.: Convergence analysis of alternating direction method of multipliers for a family of nonconvex problems. SIAM J. Opt. **26**(1), 337–364 (2016)
5. Joe-Wong, C., Ha, S., Chiang, M.: Sponsoring mobile data: an economic analysis of the impact on users and content providers. In: Proceedings of IEEE INFOCOM, Hong Kong, China, April 2015
6. Joe-Wong, C., Zheng, L., Ha, S., Sen, S., Tan, C.W., Chiang, M.: Smart data pricing in 5G systems. In: Key Technologies for 5G Wireless Systems, p. 478 (2017)

7. Ma, R.T.: Subsidization competition: vitalizing the neutral internet. In: Proceedings of ACM CoNEXT, Sydney, Australia, December 2014
8. Nie, J., Luo, J., Xiong, Z., Niyato, D., Wang, P.: A Stackelberg game approach toward socially-aware incentive mechanisms for mobile crowdsensing. IEEE Trans. Wirel. Commun. **18**(1), 724–738 (2018)
9. Nie, J., Luo, J., Xiong, Z., Niyato, D., Wang, P., Guizani, M.: An incentive mechanism design for socially aware crowdsensing services with incomplete information. IEEE Commun. Mag. **57**(4), 74–80 (2019)
10. Nie, J., Xiong, Z., Niyato, D., Wang, P., Luo, J.: A socially-aware incentive mechanism for mobile crowdsensing service market. In: 2018 IEEE Global Communications Conference (GLOBECOM), pp. 1–7. IEEE (2018)
11. Outrata, J., Kocvara, M., Zowe, J.: Nonsmooth approach to optimization problems with equilibrium constraints: theory, applications and numerical results, vol. 28. Springer, Cham (2013). https://doi.org/10.1007/978-1-4757-2825-5
12. Raveendran, N., Zhang, H., Niyato, D., Yang, F., Song, J., Han, Z.: VLC and D2D heterogeneous network optimization: a reinforcement learning approach based on equilibrium problems with equilibrium constraints. IEEE Trans. Wirel. Commun. **18**(2), 1115–1127 (2019)
13. Raveendran, N., Zhang, H., Zheng, Z., Song, L., Han, Z.: Large-scale fog computing optimization using equilibrium problem with equilibrium constraints. In: Proceedings of IEEE GLOBECOM, Singapore (December 2017)
14. Wang, W., Xiong, Z., Niyato, D., Wang, P., Han, Z.: A hierarchical game with strategy evolution for mobile sponsored content and service markets. IEEE Trans. Commun. **67**(1), 472–488 (2018)
15. Xiong, Z., Kang, J., Niyato, D., Wang, P., Poor, H.V., Xie, S.: A multi-dimensional contract approach for data rewarding in mobile networks. IEEE Trans. Wirel. Commun. **19**, 5779–5793 (2020)
16. Xiong, Z., Feng, S., Niyato, D., Wang, P., Leshem, A., Han, Z.: Joint sponsored and edge caching content service market: a game-theoretic approach. IEEE Trans. Wirel. Commu. **18**(2), 1166–1181 (2019)
17. Xiong, Z., Feng, S., Niyato, D., Wang, P., Zhang, Y.: Economic analysis of network effects on sponsored content: a hierarchical game theoretic approach. In: Proceedings of IEEE GLOBECOM, Singapore, December 2017
18. Xiong, Z., Feng, S., Niyato, D., Wang, P., Zhang, Y., Lin, B.: A stackelberg game approach for sponsored content management in mobile data market with network effects. IEEE Internet Things J. **7**, 5184–5201 (2020)
19. Xiong, Z., Zhao, J., Yang, Z., Niyato, D., Zhang, J.: Contract design inhierarchical game for sponsored content service market. IEEE Trans. Mob. Comput. (2020, early access)
20. Zhang, H., Xiao, Y., Bu, S., Yu, R., Niyato, D., Han, Z.: Distributed resource allocation for data center networks: a hierarchical game approach. IEEE Trans. Cloud Comput. **8**, 778–789 (2018)
21. Zhang, L., Wang, D.: Sponsoring content: motivation and pitfalls for content service providers. In: Proceedings of IEEE INFOCOM Workshops, Toronto, Canada, April 2014
22. Zheng, Z., Song, L., Han, Z., Li, G.Y., Poor, H.V.: Game theoretic approaches to massive data processing in wireless networks. IEEE Wirel. Commun. **25**(1), 98–104 (2018)
23. Zheng, Z., Song, L., Han, Z., Li, G.Y., Poor, H.V.: Game theory for big data processing: multi-leader multi-follower game-based ADMM. IEEE Trans. Signal Process. **6**, 3933–3945 (2018)

Research on SDN Enabled by Machine Learning: An Overview

Pu Zhao, Wentao Zhao, and Qiang Liu(✉)

College of Computer, National University of Defense Technology, Changsha 410005, Hunan, China
qiangliu06@nudt.edu.cn

Abstract. Network abstraction brings the birth of Software Defined Network (SDN). SDN is a promising network architecture that separates the control logic the network from the underlying forwarding elements. SDN gives network centralized control ability and provides developers with programmable ability. In this review, the latest advances in the field of artificial intelligence (AI) have provided SDN with learning capabilities and superior decision-making capabilities. In this study, we focus on a sub-field of artificial intelligence: machine learning (ML) and give a brief review of recent researches on introducing ML into SDN. Firstly, we introduce the backgrounds of SDN and ML. Then, we conduct a brief review on existing works about how to apply several typical ML algorithms to SDN. Finally, we give conclusion towards integrating SDN with ML.

Keywords: Software Defined Networking · Machine learning · Artificial intelligence

1 Introduction

The Internet has led to the birth of a digital society in which almost everything is connected and accessible from anywhere. The network usually involves different devices, runs different protocols, and supports different applications. With new network devices, resources and protocols deployed in the network, the network is becoming more and more complex and heterogeneous. Heterogeneous network infrastructure enhances the complexity of the network and brings many challenges in effectively organizing, managing and optimizing network resources [47].

Fortunately, Software Defined Networking (SDN) based on the idea of logically centralized management proposes a simplified solution for complex tasks, such as traffic engineering [36], network optimization [21], orchestration, and so on. In SDN network paradigm, a logic centralized SDN controller manages network devices and arranges network resources. The SDN controller has a overall perspective of the network by monitoring and gathering the timely status of the network and configuration information of the network, and supports a stream

X. Wang et al. (Eds.): 6GN 2020, LNICST 337, pp. 190–203, 2020.
https://doi.org/10.1007/978-3-030-63941-9_14

level resource scheduling of the underlying layers. This kind of creation leads to a huge transformation in the way of networks construction, operation and maintenance. Therefore, SDN framework can be regarded as a means to solve multifarious problems in the network from another perspective, and can also be used to meet the demand of new technologies, such as the IoT and the fifth generation (5G) [8]. As a promising way to rebuild the network, SDN has become the frontier of innovation in industry and academia. The Open Networking Foundation (ONF) is a leader in SDN standardization, and it has the support of more than 100 companies that together accelerate the creation of standards, products, and applications, such as NEC, Google, IBM, and VMware [26].

However, the key to the success of SDN is whether it can effectively solve the problems that can not be well solved in the traditional networking architectures, such as scalability, network awareness, on-demand quality assurance, intelligent traffic scheduling, and so on. Noteworthily, machine learning (ML) provides great potential for SDN innovation. The researches shows that ML technology has been widely used to solve various problems in the network, such as resource allocation, network routing, load balancing, traffic classification, traffic clustering, intrusion detection, fault detection, quality of service (QoS) and quality of experience (QoE) optimization, and so on [25]. Meanwhile, SDN creates conditions for the smooth deployment of ML in the network because of the unique advantages of SDN, such as programmability, global view, centralized control, and so on. Firstly, a mass of data is the key point to implement a data-driven ML algorithm. The SDN controller maintains a overall network view, as well as can monitor and collect all kinds of network data, which can provide a lot of timely and historical data for ML algorithms. Secondly, the optimized solution (e.g., configuration and resource allocation) can be easily deployed in the network due to the programmability of SDN [47]. Therefore, as a subset of artificial intelligence, ML technology has gained more and more the interest of the researchers in the application of SDN technology.

From the view of how to use ML technology to solve the problems faced by SDN, Xie et al. [47] have reviewed the ML technology which can better solve the key problems in the development of SDN, and discussed the research of using ML technology to improve the performance, intelligence, efficiency and security of SDN. On the other hand, from the standpoint of ML key algorithms, we further discuss the methods and characteristics of several typical ML technologies in the application of SDN paradigm. We argue that our work is a complement to research of Xie et al., to better reveal the important role and broad prospects of ML in SDN paradigm. In the paper, we first introduce the backgrounds of SDN and ML. Then, we conduct a brief review on existing works about how to apply widely-used ML algorithms to SDN. Finally, we give conclusion towards integrating SDN with ML.

2 Overview of Software Defined Networking

In this chapter, we will briefly introduce the background of SDN. Firstly, we discusses the background of the birth of SDN, points out the inevitability of the emergence of SDN, then introduces the framework of SDN.

2.1 The Origin and Development of SDN

The distributed control and transport network protocols deployed by the distributed forwarding device are the key point to make traditional Internet successful. However, with the rapid development of network, traditional networks are complex and hard to manage [22]. Therefore, the traditional network architecture needs to be reformed. The related research of programmable network provides a theoretical basis for the generation of SDN [45]. Active network [44,45] allows data packets to carry user programs and can be automatically executed by network devices. Users can dynamically configure the network by programming, which facilitates the management of the network. However, due to the low demand and poor compatibility of protocol, it has not been deployed in the industry. The 4D architecture [18,48] separates the programmable decision plane (i.e. the control plane) from the data plane, centralizes and automates the decision plane. Its design idea generates the rudiment of SDN controller [19]. The term SDN was originally used to describe Stanford's ideas and work around OpenFlow. According to the original definition, SDN refers to a centralized network architecture, in which the data forwarding plane is separated from the distributed control and controlled by a remote centralized controller.

In addition, many standardization organizations have joined in the formulation of SDN standards. The Open Networking Foundation (ONF) is a famous organization specializing in SDN interface standards. The OpenFlow protocol formulated by this organization has become the mainstream standard of SDN interface. Many operators and manufacturers have developed according to this standard. The ForCES Working Group of the Internet Engineering Task Force (IETF), the SDN Research Group of the Internet Research Task Force (IRTF) and several working groups of the International Telecommunication Standardization Sector (ITU-T) also aim at the new methods and new technologies of SDN [12]. The follow-up of standardization organization has promoted the rapid development of SDN market. With the development and application of 5G communication technology, SDN has become an important enabling technology for 5G. Thus, SDN has broad prospects for development and great research value.

2.2 Network Architecture

The design idea of SDN is to separate the control plane of the network from the data forwarding plane to realize a centralized network control and provide a programmable network for developers. Referring to the structure of computer system, there will be three kinds of virtualization concepts in the SDN architecture: forwarding abstraction, distributed state abstraction and configuration abstraction. According to the original design intention of SDN, the forwarding abstraction should be able to support any forwarding behavior required by network applications, and hide the implementation information of the underlying hardware. Openflow is a practical implementation according to this design idea. Compared with the traditional computer operating system, it can be regarded as the "device driver" in an operating system. At the same time, the SDN applications are not affected by the distributed state of forwarding plane, and they enjoy a unified network view. The unified network view is provided by the distributed state abstraction. The control plane can gather the distribution state information of devices and construct an overall network view so that the applications can set the network uniformly through the whole network state. Configuration abstraction can provide users with a more simplified network model. The users can automatically complete the unified deployment of forwarding devices along the path, through the application interface provided by the control layer. Therefore, network abstraction is the decisive factor for generation of SDN architecture decoupling between data and control planes and providing unified interface.

According to different requirements, many organizations have proposed corresponding SDN reference architectures. SDN Architecture was first proposed by ONF and has been widely accepted in academia and industry. The typical SDN architecture is shown in Fig. 1. SDN consists of three parts: data plane, control plane and application plane. The data plane contains a series of forwarding devices interconnected through wireless channels or wired cables. They are responsible for data processing, forwarding, and status collection based on flow tables. The forwarding plane communicate with the control plane by the southbound interface (SI). The SI defined the communication protocol between the forwarding elements and the controllers, such as OpenFlow protocol. The protocol formalize the way that the controllers and the forwarding elements interact. The control plane includes a series of logically centralized controllers regarded as the brain of the network. The controller is mainly responsible for the arrangement of data plane resources, maintenance of network topology, status information, and so on. The SDN controller can offer the APIs to application developers. The APIs represent the northbound interface (NI), i.e., a common interface for developing applications. The application plane includes a variety of businesses and applications such as load balancers, network routing, firewalls, monitoring, and so on. The network application program communicates with SDN controller by NI to control the network reasonably, so as to realize the business logic of the application program itself.

3 Overview of Machine Learning

The general definition of ML is that intelligent machines learn from experience (i.e. from available data in the environment) and use learned methods to improve overall performance [33,38]. In the case, ML technology can be divided into four groups: supervised, unsupervised, semi-supervised, and reinforcement learning. In this section, Each category is briefly explained to help the reader understand what follows. A more in-depth discussion of ML technology and the basic concepts about ML, please refer to [33,38].

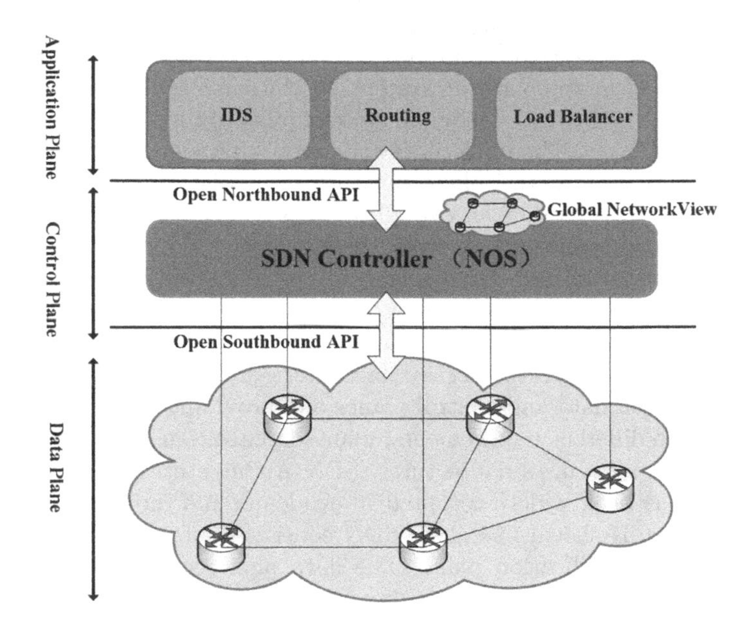

Fig. 1. Illustration of SDN architecture.

3.1 Supervised Learning

Supervised learning methods require predefined knowledge. For example, a training data set consisting of "input-output combinations" in which the model learns a function that maps a given input to an corresponding output [38]. The method needs a test data set that represents the best performance of the current research system. The test data set can be used to evaluate the performance of the final learning method [31].

3.2 Unsupervised Learning

Unsupervised learning is carried out without pre-defined knowledge (that is, only unlabeled data) [38]. Therefore, the system mainly focuses on finding rules or knowledge in the input data. A common use case of unsupervised learning is clustering algorithm, which is used to distinguish meaningful groups in input data according to similar attributes defined by appropriate distance measures (such as Euclidean distance and cosine distance measure) [34,38].

3.3 Semi-supervised Learning

The semi-supervised learning model learns from labeled and unlabeled data. Labeled and unlabeled data may contain random noise in supervised learning and unsupervised learning [38]. As in many practical applications, since the data is labeled manually by experts, it is more realistic to collect many labeled data, while it is easier to collect a large number of unlabeled data [16]. Semi-supervised learning is superior to unsupervised learning because it contains some small labeled data [16].

3.4 Reinforcement Learning

The reinforcement learning (RL) model is based on a set of "reinforcement" in the environment to learn a superior behavior. For example, the system is rewarded or punished according to whether it works well [38]. Every time the system interacts with the environment, it gets feedback information, and it will make full use of the feedback to update its performance [33]. An important property of reinforcement learning is Markov property, because of this property, the subsequent state of reinforcement learning system is determined by the current state [3].

4 Discussion of Applying Machine Learning to SDN

Due to the great efforts of industry and academia, the role of ML in the network has been significantly enhanced. ML technology has been widely used to solve various network-related problems, such as network routing, load balancing, traffic classification and clustering, fault detection, intrusion detection, QoS and QoE optimization, and so on. In this section, we will investigate the application areas of several typical ML methods in SDN.

4.1 Application of Neural Network in SDN

The advantage of neural network (NN) is that it can approximate any function, but because of the need to adjust a large number of parameters, the computational cost is very high. Neural network method is used mainly for intrusion detection [1,14,17], traffic classification [4,28,32], load balancing [15], performance prediction [13,39,41], service level agreement (SLA) execution [6,7], solving the problems of controller placement [2,20] and optimal virtual machine (VM) placement [30], etc.

In this paper, six application examples are discussed. Sander et al. [41] presented the design and performance of DeePCCI, a passive congestion control identification method based on deep learning which only needs to train the traffic of congestion control variables. Compared with the traditional methods, it can be directly applied to encrypted traffic and easier to expand, because it only needs the time of arrival information of the packets. To solve the problem of weighted controller configuration, He et al. [20] introduced a multi-label classification method to forecast the entire network allocation. Compared with decision tree method and logistic regression method, the neural network method shows superior results and saves up to two-thirds of the running time of the algorithm. Carner et al. [13] compared the performance of traditional methods and neural network methods for network transmission delay prediction. By training a model, network delay is automatically predicted according to traffic load and overlapping routing strategy. The M/M/1 network model and NN model are introduced for network transmission delay prediction in their works. The experimental results show that the network transmission delay predicted based on neural network has better accuracy than the method based on M/M/1 model. In [32], the researchers used an 8-layer deep neural network to identify mobile applications. The quintuple included destination IP address and port number and so on is used to feature a flow, which is the training data of an 8-layer deep NN. The experiment show that the recognition accuracy of the trained model for 200 mobile applications reaches 93.5%. Abubakar et al. [1] introduced a SDN intrusion detection system using the neural network method, which achieves 97.3% high accuracy in NSL-KDD data sets.

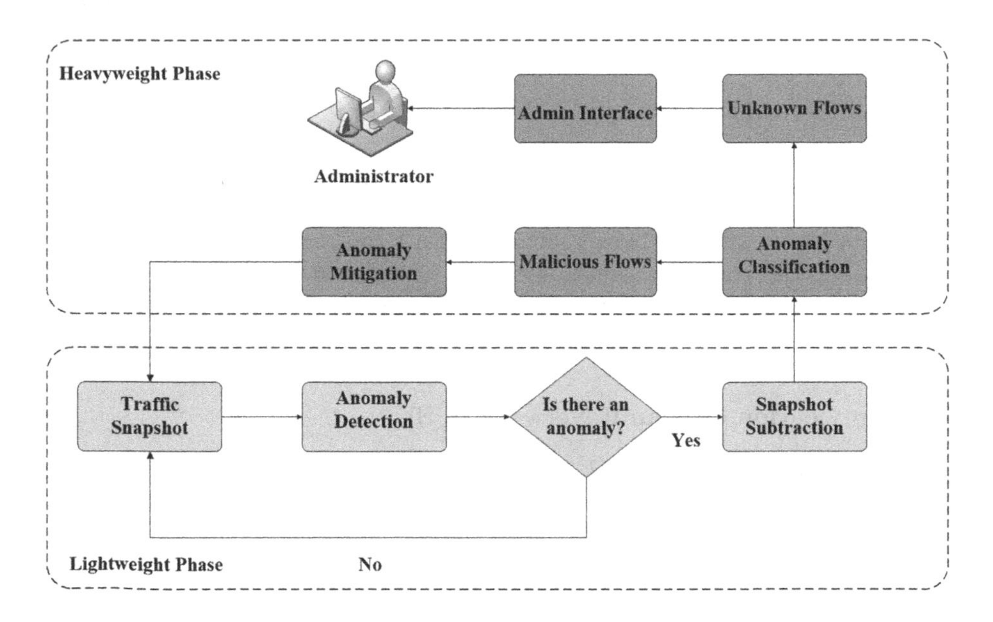

Fig. 2. Atlantic system workflow.

4.2 Application of Support Vector Machine in SDN

The support vector machine (SVM) has the advantage of processing high-dimensional data sets well, but it is difficult to train large data sets because of the large amount of training calculation. The support vector machine method is mainly used to intrusion detection and traffic classification in SDN paradigm [10,23,23,24,35,37,43,49].

Silva et al. [43] propose a prototype system called Atlantic for joint abnormal traffic detection, classification and mitigation in SDN. The Fig. 2 illustrates a work view of the Atlantic. The Atlantic framework implements anomaly detection and classification in two stages: lightweight stage and heavyweight stage. In the lightweight stage, The authors adopt some methods with low computational cost (such as information theory), The lightweight methods can be called more continually to quickly remark the potential malicious traffic. In the heavyweight stage, by using an SVM algorithm to leverage historical knowledge about past anomalies, the flows are analyzed and classified according to their abnormal behavior. Atlantic then takes appropriate mitigation measures to automatically handle malicious traffic, and human administrators manually analyze unknown traffic. Kokila et al. [23] introduced an approach to detect DDOS attacks on the SDN controllers. Compared with the traditional classifiers, their method based on SVM has higher accuracy and lower error rate. Boero et al. [10] used SVM to detect malicious software based on SDN, and the information gain (IG) measures were used to select the most dependent features. Their models achieve 80% and 95% malware and normal traffic detection rates. Furthermore, the false alarm rates of malware and normal traffic were 5.4% and 18.5% respectively. In [37], the authors implement an application aware traffic classification system using SVM. The system classifies UDP traffic according to NetFlow records (such as received packets and bytes). The experimental results show that the classification accuracy of the model is more than 90%.

4.3 Application of k-Means Clustering in SDN

The k-means clustering algorithm is easy to implement and explain clustering results, but the calculation cost is linear with the number of training data. k-means clustering method is mainly used to deploy intrusion detection system in SDN paradigm [5], routing decision [11], solve the placement problem of optimal controller [40], and analyze user traffic [9].

Bernaille et al. [9] introduced an approach based on a Simple K-Means algorithm that classified different types of TCP-based applications using a first few packets of the flows. Budhraja et al. [11] proposed a routing protocol in a strictly compliant environment. Firstly, the network traffic is divided into multiple risk ratio clusters by a k-means algorithm in an offline way. Then, the authors adopt an ant colony optimization (ACO) algorithm to select the path with the least risk of privacy exposure and compliance for a given data transmission session in an online way. Sahoo et al. [40] used k-means method to treat the placement of optimal controllers. They compared two kinds of clustering algorithms: k-Medoids and k-Center. The results of the comparative experiment show that k-Center algorithm has superior results than k-Medoid algorithm. Barki et al. [5] compared the performance of four ML methods (i.e., naive Bayesian, k-nearest neighbor, k-mean and k-center) in detecting DDoS attacks in SDN. The experiments show that naive Bayesian method achieved the highest detection rate. However, k-means clustering method achieves good results in processing time.

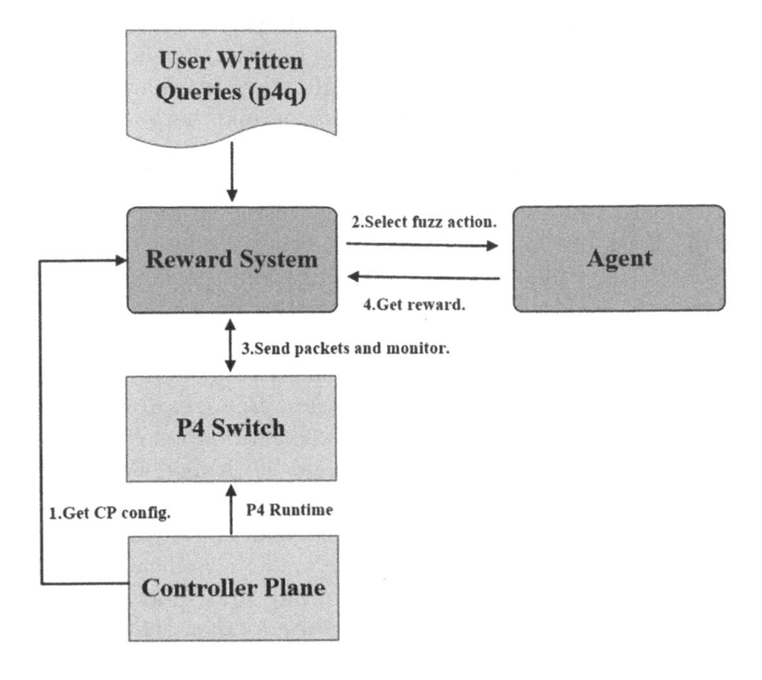

Fig. 3. p4rl system workflow.

4.4 Applications of Other Machine Learning Methods in SDN

In addition to some common ML approaches mentioned above, the combined use of other ML algorithms in SDN broadens our thinking for further expanding the performance of SDN. Apoorv Shukla et al. [42] presents a novel approach for P4 switch verification. They implement a prototype called p4rl which can use reinforcement learning model to guide the fuzz testing to verify the P4 switches automatically during execution time. The Fig. 3 illustrates a work view of the p4rl framework. First, the regulator sets the behavioral attributes of the network to be verified. It coming with configuration information about the network is used as the input of reward system which can provide the basis for verification. At the same time, the authors use an agent based on reinforcement learning to select variant network actions. These mutated actions are used to guide the generation of network packets as test cases. The information about how P4 switch processes the packets is used as the feedback of the reward system. The agent can updates its next action based on the feedback returned by the reward system. The experiment demonstrates that reinforcement learning approach can make the fuzzing test process focused and improve the test efficiency. In [29], the authors propose an architecture of Self-Driving Network. To enhance the QoE of bandwidth and latency sensitive applications, a ML-based classifier is introduced to class the current experience states of the applications, then the states are sent to a state-machine. If a critical event of the application behavior is detected by the state-machine (arising due to a transition among states), the network controller will execute the corresponding action in order to elevate the performance of the applications. Wang et al. [46] introduced an improved behavior-based SVM to classify network attacks. In order to improve the accuracy of intrusion detection and accelerate the learning rate in the normal mode and intrusion mode, the authors use a decision tree method to reduce features. Firstly, they sort the original feature set and select the features with the best representation. Then, they use these selected features as the input of the model to train a SVM classifier. The model also uses ID3 decision tree method for feature selection. The experimental record on KDD-CUP99 dataset show that the classification accuracy of the model is 97.60%.

4.5 Comparisons of Different Machine Learning Methods Enabling SDN

In order to carry out more in-depth research in ML enabled SDN, we further analyze the characteristics of various ML algorithms in SDN applications.

Overall, the supervised learning algorithm is the common algorithm in intrusion detection, because the core task of the intrusion detection system is often regarded as a classification work. In SDN, ML-based intrusion detection system has been studied extensively. QoS prediction is usually regarded as a regression work, while QoE prediction is regarded as a classification work. Therefore, supervised learning method can also be well handled the QoS or QoE prediction task. Nevertheless, the key to using supervised learning is whether it is convenient to obtain enough labeled training data sets. Compared with supervised learning, the semi-supervised learning approaches only require a small amount of labeled data. Thus, the semi-supervised learning approaches are more easily applied to QoS/QoE prediction. Compared with supervised learning and semi supervised learning, RL algorithm has obvious advantages and application potential in those applications where it is difficult to obtain a large number of training data. On the one hand, the RL algorithm does not require labeled training data sets. Moreover, the optimization objectives (such as network delay, bandwidth utilization, and energy utilization) can be flexibly set through various incentive functions. Specifically, we discuss the advantages and disadvantages of different ML algorithms in Table 1.

Table 1. Strengths and weaknesses of various ML models when applying to SDN

ML Algorithm	Strengths	Weaknesses
Neural Network	Once trained, execution speed is fast The ability to approximate arbitrary functions to predict complex network data Work well on high-dimensional network datasets	Expensive computing makes online training difficult Hard to guide researchers to set structure of NN
SVM	Work well on both linearly separable and non-linearly separable dataset	Hard to train large-scale network datasets Sensitivity to noise Data in SDN
K-Means	Simple to implement deployment in SDN	Sensitive to initial points and outliers Computing increases linearly with the size of network datasets
RL	Working well without prior knowledge can flexibly handle different optimization objectives	Hard to solve problems in high-dimensional space
Semi-supervised learning	Using labeled and unlabeled data to effectively deal with situations where subjective datasets, such as QoE, are difficult to obtain	Rely on assumptions

5 Conclusions

This paper summarizes the research work on the application of ML technology in SDN paradigm. The research shows that an increasing number of the ML technologies are used to solve a wide range of network problems. The ML technologies have been proved to be the valuable means in SDN. The advantages of ML technology in classification, prediction, and feature extraction can better solve the security protection, resource allocation, routing, load balancing, and other issues in SDN. Compared with traditional methods and other artificial intelligence technologies, ML technology shows a broader application. In addition, compared with traditional ML technology, deep learning, can provide better results. However, The ML also brings new challenges to SDN. ML models and related training data are faced with various security risks [27]. More attention should be paid to the robustness of ML in confrontational environments.

Acknowledgement. The work is supported by National Key Research and Development Program of China under Grant No. 2018YFB0204301, National Natural Science Foundation of China under Grant Nos. 61702539 and U1811462, Hunan Provincial Natural Science Foundation of China under Grant No. 2018JJ3611, and NUDT Research Project under Grant No. ZK-18-03-47.

References

1. Abubakar, A., Pranggono, B.: Machine learning based intrusion detection system for software defined networks. In: Proceedings of the 7th International Conference on Emerging Security Technologies, pp. 138–143 (2017)
2. Alvizu, R., Troia, S., Maier, G., Pattavina, A.: Matheuristic with machine learning based prediction for software defined mobile metro core networks. IEEE/OSA J. Opt. Commun. Network. **9**(9), D19–D30 (2017)
3. Arulkumaran, K., Deisenroth, M.P., Brundage, M., Bharath, A.A.: Deep reinforcement learning: a brief survey. IEEE Signal Process. Mag. **34**(6), 26–38 (2017)

4. Ashifuddin Mondal, M., Rehena, Z.: Intelligent traffic congestion classification system using artificial neural network. In: Companion Proceedings of The 2019 World Wide Web Conference, WWW 2019, pp. 110–116. , Association for Computing Machinery, New York (2019). https://doi.org/10.1145/3308560.3317053
5. Barki, L., Shidling, A., Meti, N., Narayan, D.G., Mulla, M.M.: Detection of distributed denial of service attacks in software defined networks. In: Proceedings of the IEEE International Conference on Advances in Computing, Communications and Informatics, pp. 2576–2581 (2017)
6. Bendriss, J., Yahia, I.G.B., Chemouil, P., Zeghlache, D.: AI for SLA management in programmable networks. In: Proceedings of the 13th International Conference on Design of Reliable Communication Networks, pp. 1–8 (2017)
7. Bendriss, J., Yahia, I.G.B., Zeghlache, D.: Forecasting and anticipating SLO breaches in programmable networks. In: Proceedings of the 20th Conference on Innovations in Clouds, Internet and Networks, pp. 127–134 (2017)
8. Bera, S., Misra, S., Vasilakos, A.V.: Software-defined networking for internet of things: a survey. IEEE Internet Things J. **4**(6), 1994–2008 (2017)
9. Bernaille, L., Teixeira, R., Akodkenou, I., Soule, A., Salamatian, K.: Traffic classification on the fly. SIGCOMM Comput. Commun. Rev. **36**(2), 23–26 (2006). https://doi.org/10.1145/1129582.1129589
10. Boero, L., Marchese, M., Zappatore, S.: Support vector machine meets software defined networking in IDS domain. In: Proceedings of the 29th IEEE International Teletraffic Congress, pp. 25–30 (2017)
11. Budhraja, K.K., Malvankar, A., Bahrami, M., Kundu, C., Kundu, A., Singhal, M.: Risk-based packet routing for privacy and compliance-preserving SDN. In: Proceedings of the 10th IEEE International Conference on Cloud Computing, pp. 761–765 (2017)
12. Cai, Z., Zhang, H., Liu, Q., Xiao, Q., Cheang, C.F.: A survey on security-aware network measurement in SDN. Security and Communication Networks **2018**, 14 (2018). Article ID 2459154
13. Carner, J., Mestres, A., Alarcon, E., Cabellos, A.: Machine learning-based network modeling: an artificial neural network model vs a theoretical inspired model. In: Proceedings of the 9th International Conference on Ubiquitous and Future Networks, pp. 2576–2581 (2017)
14. Chen, X.F., Yu, S.Z.: CIPA: a collaborative intrusion prevention architecture for programmable network and SDN. Comput. Secur. **58**, 1–19 (2016)
15. Chen-Xiao, C., Ya-Bin, X.: Research on load balance method in SDN. Int. J. Grid Distrib. Comput. **9**(1), 25–36 (2016)
16. Fan, X., Guo, Z.: A semi-supervised text classification method based on incremental EM algorithm. In: Proceedings of the WASE International Conference on Information Engineering, pp. 211–214 (2010)
17. Gabriel, M.I., Valeriu, V.P.: Achieving DDoS resiliency in a software defined network by intelligent risk assessment based on neural networks and danger theory. In: Proceedings of the 15th IEEE International Symposium on Computational Intelligence and Informatics, pp. 319–324 (2014)
18. Greenberg, A., et al.: A clean slate 4D approach to network control and management. ACM SIGCOMM Comput. Commun. Rev. **35**(5), 41–54 (2005)
19. Gude, N., et al.: NOX: towards an operating system for networks. ACM SIGCOMM Comput. Commun. Rev. **38**(2), 105–110 (2008)
20. He, M., Kalmbach, P., Blenk, A., Kellerer, W., Schmid, S.: Algorithm-data driven optimization of adaptive communication networks. In: Proceedings of the 25th IEEE International Conference on Network Protocols (ICNP), pp. 1–6 (2017)

21. Heorhiadi, V., Reiter, M.K., Sekar, V.: Simplifying software-defined network optimization using SOL. In: Proceedings of the 13th USENIX Symposium on Networked Systems Design and Implementation (NSDI 2016), pp. 223–237 (2016)
22. Jain, R.: Ten problems with current internet architecture and solutions for the next generation. In: Proceedings of the IEEE MILCOM 2006, pp. 1–9 (2006)
23. Kokila, R.T., Selvi, S.T., Govindarajan, K.: DDoS detection and analysis in SDN-based environment using support vector machine classifier. In: Proceedings of the 66h IEEE International Conference on Advanced Computing, pp. 205–210 (2014)
24. Latah, M., Toker, M.: A novel intelligent approach for detecting dos flooding attacks in software defined networks. Int. J. Adv. Intell. Inf. **4**(1), 11–20 (2018)
25. Latah, M., Toker, L.: Artificial intelligence enabled software defined networking: a comprehensive overview (2018)
26. Lin, P., Bi, J., Wolff, S., et al.: A west-east bridge based SDN inter-domain testbed. IEEE Commun. Mag. **53**(2), 190–197 (2015)
27. Liu, Q., Li, P., Zhao, W., Cai, W., Yu, S., Leung, V.C.M.: A survey on security threats and defensive techniques of machine learning: a data driven view. IEEE Access **6**, 12103–12117 (2018)
28. Lyu, Q., Lu, X.: Effective media traffic classification using deep learning. In: Proceedings of the 2019 3rd International Conference on Compute and Data Analysis, ICCDA 2019, pp. 139–146. Association for Computing Machinery, New York (2019). https://doi.org/10.1145/3314545.3316278
29. Madanapalli, S.C., Gharakheili, H.H., Sivaraman, V.: Assisting delay and bandwidth sensitive applications in a self-driving network. In: Proceedings of the 2019 Workshop on Network Meets AI & ML, NetAI 2019, pp. 64–69. Association for Computing Machinery, New York (2019). https://doi.org/10.1145/3341216.3342215
30. Mestres, A., et al.: Knowledge-defined networking. ACM SIGCOMM Comput. Commun. Rev. **47**(3), 2–10 (2017)
31. Mogul, J.C., Congdon, P.: Hey, you darned counters!: Get off my ASIC! In: Proceedings of the ACM SIGCOMM Workshop on HotSDN, pp. 25–30 (2012)
32. Nakao, A., Du, P.: Toward in-network deep machine learning for identifying mobile applications and enabling application specific network slicing. IEICE Trans. Commun. **E101.B**(7), 1536–1543 (2018). https://doi.org/10.1587/transcom.2017CQI0002
33. Negnevitsky, M.: Artificial Intelligence - A Guide to Intelligent Systems, 2nd edn. Addison-Wesley, Essex (2005)
34. Nguyen, T.T., Armitage, G.: A survey of techniques for internet traffic classification using machine learning. IEEE Commun. Surv. Tutor. **10**(4), 56–76 (2008)
35. Phan, T.V., Van Toan, T., Van Tuyen, D., Huong, T.T., Thanh, N.H.: OpenFlowSIA: an optimized protection scheme for software defined networks from flooding attacks. In: Proceedings of the 6th IEEE International Conference on Communications and Electronics, pp. 13–18 (2016)
36. Raza, S., Huang, G., Chuah, C.N., Seetharaman, S., Singh, J.P.: MeasuRouting: a framework for routing assisted traffic monitoring. IEEE/ACM Trans. Network. **20**(1), 45–56 (2012)
37. Rossi, D., Valenti, S.: Fine-grained traffic classification with Netflow data. In: Proceedings of the 6th International Wireless Communications and Mobile Computing Conference, pp. 479–483. Association for Computing Machinery, New York (2010). https://doi.org/10.1145/1815396.1815507
38. Russell, S., Norvig, P.: Artificial Intelligence (A Modern Approach), 3rd edn. Prentice Hall, New Jersey (1995)

39. Sabbeh, A., Al-Dunainawi, Y., Al-Raweshidy, H.S., Abbod, M.F.: Performance prediction of software defined network using an artificial neural network. In: Proceedings of the SAI Computing Conference (SAI), pp. 80–84 (2016)
40. Sahoo, K., Sahoo, S., Mishra, S., Mohanty, S., Sahoo, B.: Analyzing controller placement in software defined networks. In: Proceedings on National Conference on Next Generation Computing and Its Applications in Computer Science and Technology, pp. 12–16 (2016)
41. Sander, C., Rüth, J., Hohlfeld, O., Wehrle, K.: DeePCCI: deep learning-based passive congestion control identification. In: Proceedings of the 2019 Workshop on Network Meets AI & ML, NetAI 2019, pp. 37–43. Association for Computing Machinery, New York (2019). https://doi.org/10.1145/3341216.3342211
42. Shukla, A., Hudemann, K.N., Hecker, A., Schmid, S.: Runtime verification of P4 switches with reinforcement learning. In: Proceedings of the 2019 Workshop on Network Meets AI & ML, NetAI 2019, pp. 1–7. Association for Computing Machinery, New York (2019). https://doi.org/10.1145/3341216.3342206
43. da Silva, A.S., Wickboldt, J.A., Granville, L.Z., Schaeffer-Filho, A.: Atlantic: a framework for anomaly traffic detection, classification, and mitigation in SDN. In: Proceedings of the 2016 IEEE/IFIP Network Operations and Management Symposium, pp. 27–35 (2016)
44. Tennenhouse, D.L., Smith, J.M., Sincoskie, W.D., Wetherall, D., Minden, G.J.: A survey of active network research. IEEE Commun. Mag. **35**(1), 80–86 (1997)
45. Tennenhouse, D.L., Wetherall, D.J.: Towards an active network architecture. ACM SIGCOMM Comput. Commun. Rev. **37**(5), 81–94 (2007)
46. Wang, P., Chao, K.M., Lin, H.C., Lin, W.H., Lo, C.C.: An efficient flow control approach for SDN-based network threat detection and migration using support vector machine. In: Proceedings of the 13th IEEE International Conference on e-Business Engineering, pp. 56–63 (2016)
47. Xie, J., et al.: A survey of machine learning techniques applied to software defined networking (SDN). IEEE Commun. Surv. Tutor. **21**(1), 393–430 (2019)
48. Yan, H., Maltz, D.A., Gogineni, H., Cai, Z.: Tesseract: a 4D network control plane. In: Proceedings of the 4th USENIX Symposium on Networked Systems Design & Implementation (NSDI 2007), pp. 369–382 (2007)
49. Yuan, R., Li, Z., Guan, X., Xu, L.: An SVM-based machine learning method for accurate internet traffic classification. Inf. Syst. Front. **12**(2), 149–156 (2010)

Intelligent Applications

Mining Raw Trajectories for Network Optimization from Operating Vehicles

Lei Ning[1(✉)], Runzhou Zhang[1,2], Jing Pan[3], and Fajun Li[1]

[1] College of Big Data and Internet, Shenzhen Technology University, Shenzhen, China
ninglei@sztu.edu.cn

[2] College of Applied Technology, Shenzhen University, Shenzhen, China

[3] Chaincomp Technologies, Shenzhen, China

Abstract. Improving the user peak rate in hot-spots is one of the original intention of design for 5G networks. The cell radius shall be reduced to admit less users in a single cell with the given cell peak rate, namely Hyper-Dense Networks (HDN). Therefore, the feature extraction of the node trajectories will greatly facilitate the development of optimal algorithms for radio resource management in HDN. This paper presents a data mining of the raw GPS trajectories from the urban operating vehicles in the city of Shenzhen. As the widely recognized three features of human traces, the self-similarity, hot-spots and long-tails are evaluated. Mining results show that the vehicles to serve the daily trip of human in the city always take a short travel and activate in several hot-spots, but roaming randomly. However, the vehicles to serve the goods are showing the opposite characteristics.

Keywords: Trajectory mining · Vehicle mobility · Hyper-dense networks

1 Introduction

With the development of network and microelectronics technology in the past decades, the network world has gradually expanded its connection between people and things as well as between things from that between people. By 2025, the Internet of Things (IoT) will have more than 55 billion connections, the report said. The explosive growth of IoT expands the connectivity of the network and the way of data exchange, such as portable electronic equipment, household appliances, vehicles and manufacturing devices, and a series of integrated devices including electronics, software, sensors, drivers and networks. From consumer wearable devices to industrial production devices, these networked devices can sense environmental information, be controlled remotely, make decisions and take

Sponsored by the Young Innovative Project from Guangdong Province of China (No. 2018KQNCX403) and the Teaching Reform Project from Shenzhen Technology University (No. 2018105101002).

X. Wang et al. (Eds.): 6GN 2020, LNICST 337, pp. 207–217, 2020.
https://doi.org/10.1007/978-3-030-63941-9_15

actions by themselves; however, the number of users and their business requests are unevenly distributed in the network. Especially in the era of IoT, a large number of business requests are initiated in hot area collectively including indoor, business center, stadium, factory, farm and other node-intensive areas. Meanwhile, the business type of voice-intensive is changing to that of data-intensive and connection-intensive.

In order to meet the above requirements, in the fifth generation mobile communication technology (5G), its performance design index not only has high-speed rate, low delay and large connection, but also proposes the concept of hyper dense networking for hot areas, that is, improving the cell capacity per unit area under the condition of mixed multiple network systems. Due to the cellular characteristics of mobile networks, there are three technologies mainly to improve cell capacity as follows, increasing spectrum resources, improving wireless link performance and enhancing cell density respectively.

Hybrid Networking technology, as the multiple networks deployed in the same area, provides multi-mode terminals to select appropriate network access according to business needs and network status; Large Scale MIMO and New Multiple Access technology is to improve the wireless link performance in vertical space; while Small Cell technology improves the spectral efficiency in horizontal space by increasing the cell density in specific area [23]. In hot area, user's mobile behavior under the multiple and hyper dense deployment network greatly increases the difficulty for the network to guarantee the user service experience [22]. User's mobile behavior evolution under multiple networks cooperation is closely related to the development of hyper dense network technology, therefore, user's mobile behavior will significantly affect the data bearing distribution of different networks, as well as affect the overall performance and user experience of the network.

With the integrated development of IoT and wireless access technology, user's behavior in the network is complex and changeable, because its accessing users are no longer limited to human portable devices, but also include animals with sensors and communication devices, as well as machines with autonomous mobile functions. As the result, analyzing the characteristics of user's mobile trajectory under the hyper dense network, is helpful to optimize the network performance and promote the development of network technology for intelligent manufacturing and interconnection of all things [24].

Recently, Feng et al. [5] summarizes a survey on trajectory data mining, including main techniques and applications, which a wide spectrum of applications is driven by trajectory data mining, such as path discovery [3,4,10–12,18], movement behavior analysis [1,6,15,19], group behavior analysis [14], urban service [9,20,21] and so on.

Some of researches focus on human behavior analysis based on raw trajectory data, to understand people's real demand for transportation, driving preferences, extracting mobility behavioral patterns, which can be used to enhance utilization efficiency of public transportation [10], improve quality of user satisfaction [4], understand behavior of people moved in geographical context [15].

Moreover, another work [6] explores individual human mobility patterns by studying a large number of anonymous position data from mobile phone users and reveals a high degree of temporal and spatial regularity in human trajectories. While others focus on unhuman behavior analysis based on raw trajectory data, they believe that knowledge discovered with trajectory data mining techniques helps to improve quality of life in urban areas from several aspects [8,9,20,21]. Yuan et al. [21] address a problem of discovering regions of different functions in a city based on a large scale of trajectory data. Liu et al. [9] address a problem of map inference in a practical setting through inferring road maps from large-scale GPS traces. Especially, the vehicles behavior analysis based on trajectory data are of great importance in unhuman field, which abstract more attention. iPark in the literature of Yang et al.[20] aims to enable parking information, i.e., annotating an existing map with parking zones based on trajectory data of vehicles. Through analyzing a large scale of trajectory data collected from electronic vehicles, Li et al. [8] address that how to strategically deploy charging stations and charging points so that minimizing average time to the nearest charging station and average waiting time for an available charging point. Lee et al. [7] collected real users' data through Garmin-GPS-60CSx handheld terminal in the North American environment. Based on a large number of measured data collected above, a comprehensive moving model reflecting the characteristics of data samples is proposed. In such researches, real data plays more and more important role, and comparing with the researches on human behavior analysis. Thus, in this paper, the urban operating vehicles data will be used as the raw trajectories to mining and analyzing its characteristics and patterns and draw some conclusion, which will do some contribution for future modeling as well as network performance optimization. At the same time, user's mobile behavior tends to study the implicit constraints beyond the time and space information in user's mobile trajectory, in raw GPS trajectories from the urban operating vehicles data, there are many kinds of users in the network and their behaviors are complex and changeable. It is of great theoretical value and practical significance for the network performance optimization and intelligent deployment of IoT to study the mobile behavior characteristics of multiple types of users under the IoT and hyper dense network.

2 Trajectory Data Description

In this paper, the raw GPS data is published from Shenzhen Transportation Bureau. The trajectories contain 113,503 entries in a single day from the urban operating vehicles up to five types with the number of 29,218. This big data are gathered from 12:00 am, Oct. 08, 2018 to 11:59 pm Oct. 14, 2018 in the local time. The number of raw GPS coordinates in one typical trajectory is 295,966,347. For example, in Fig. 1, it shows the heap map of a random selected trajectory of the vehicle, which mainly take the activities around the central part ot the city.

In the following part, the basic definitions which are used throughout the paper are introduced for data descriptions.

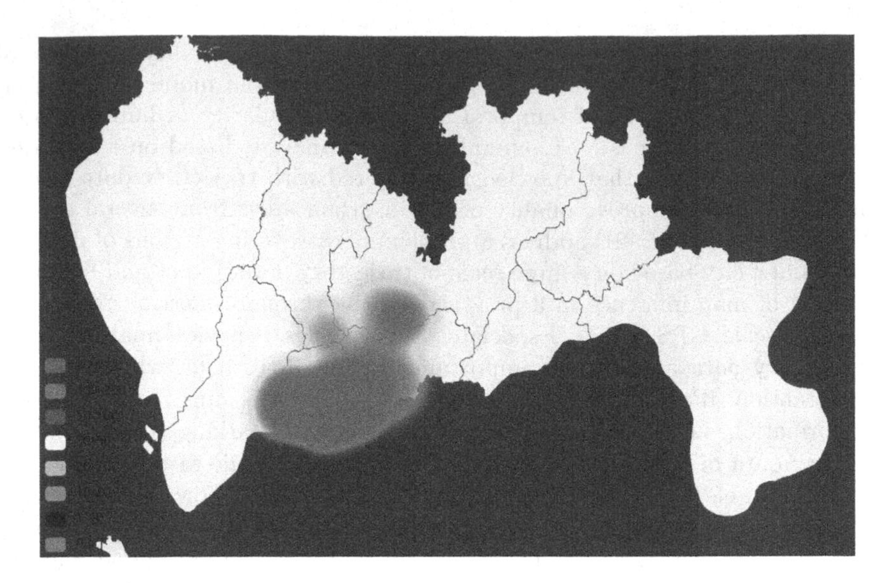

Fig. 1. The heap map of a random selected trajectory of the vehicle.

Definition 1 (Raw GPS Data): It is a 4-tuple of the form (*lat*, *lon*, *t*, *p*) where *lat* and *lon* is the vehicle's latitude and longitude respectively, *t* is the timestamp at which the record was tracked, *p* is the plate number of the vehicle that also indicates the type.

Definition 2 (Trajectory): It is a sequence of time-ordered raw GPS data for the specific vehicle in one day.

Definition 3 (Distance): It is a length of a line segment between two given coordinates. The earth radius shall be considered since the coordinates from the raw GPS data is presented by longitude and latitude respectively.

3 The Feature Extraction Method of the Trajectory

Previously some related work have been proposed that the human mobility has regularities, which are self-similar, hot-spots and heavy-tail [7,13]. Based on those discovered regularities and the raw GPS data of several vehicles, an extraction method of the trajectory is presented to evaluate the vehicle mobility features from the urban scale in this section.

3.1 Evaluation of Self-Similarity with Hurst Exponent

A system with Hurst statistical characteristics does not need the independent random event hypothesis of general probability statistics [2]. It reflects the result of a long series of interrelated events. What happens today will affect the future, and what happened in the past will affect the present. Accordingly, the human always selects a familiar path from the constant location to the temporary destination, which is called self-similarity. Therefore, the Hurst exponent is adopted to analyze the self-similarity of trajectories.

The aggregated variance and the R/S methods are full-blown implementations of Hurst exponent algorithm. However, the candidate raw GPS data are huge. So the algorithm proposed by Bill Davidson named BD procedure in what follows is to quantify the self-similarity of way-points, which is far faster than the conventional algorithm.

Algorithm 1. BD procedure

get the set of raw GPS data $\boldsymbol{L}$
while $\boldsymbol{L}_{length} \geq \boldsymbol{L}_{threshold}$ **do**
  $\boldsymbol{L}_y = std(\boldsymbol{L})$
  $\boldsymbol{L}_x = \boldsymbol{L}_x \times 2$
  $\boldsymbol{L}_{length} = fix(\boldsymbol{L}_{length}/2)$
  for all $index \in \boldsymbol{L}_{length}$ **do**
    $\boldsymbol{L}_{tmp} = (\boldsymbol{L}(2 \times index) + \boldsymbol{L}((2 \times index) - 1)) \times 0.5$
    get new $\boldsymbol{L}$ from $\boldsymbol{L}_{tmp}$
  end for
end while
make the linear fit for $\boldsymbol{L}_y$ and $\boldsymbol{L}_x$
get Hurst exponent from the slope of the linear fit of log-log plot

3.2 Evaluation of Hot-Spots with Density-Based Clustering

From a macro perspective, human always activates in a constant area, which can be called hot-spots [16]. However, the urban operating vehicles have public and specific attributes. It is necessary to cluster trajectory of operating vehicles to explore whether there are hot-spots. Density-Based Spatial Clustering of Applications with Noise (DBSCAN) can divide the area with enough high density into groups, and find clusters of arbitrary shape in noisy spatial database, which can be applied to the big raw GPS data of operating vehicles [17].

3.3 Evaluation of Long-Tails with Cumulative Distribution Function

The head and tail are two statistical terms, where the head is a protruding part in the middle of normal curve and the tail is a relatively flat part on both sides. From the perspective of human mobility, most of the daily trip will focus on the head, which can be called popular, while the demand distributed in the long-tail is personalized, scattered and small.

In order to evaluate the long-tail effects of vehicle trajectories, the cumulative distribution function is introduced as follows.

Algorithm 2. DBSCAN procedure

```
get the set of raw GPS data L
get the radius e and minimum points MinPts
for all L_i ∈ L do
  if L_i is the core point then
    find all the objects that can reach the density from this point and form a cluster
  end if
end for
obtain the clusters with each center Coordinate_{x,y}^{hot-spots}
```

Theorem 1. *For all the discrete distance from the generated set* $\boldsymbol{D}$*, the cumulative distribution function is defined as the sum of occurrence probability of all values less than or equal to the specific distance* d*.*

$$F_{\boldsymbol{D}}(d) = P(\boldsymbol{D} \leq d) \tag{1}$$

Theorem 2. *The generated set* $\boldsymbol{D}$ *is defined as the time series of travel distance from vehicles with the most hot-spots as the center. Therefore, the distance* d_i *in the sampling time* t_i *as an element of* $\boldsymbol{D}$ *is calculated as follows.*

$$d_i = \|Coordinate_{x,y}^{t_i} - Coordinate_{x,y}^{hot-spots}\| \tag{2}$$

4 Performance Evaluation

Framework of Trajectory Data Mining. The basic data mining can be divided into two parts, which are data storage and calculation. For the big data that cannot dealing with the conventional tools, the MongoDB database and the SPARK calculation are selected in this paper respectively. Those two open source tools shall be a close combination with the Python language. As it is shown in Fig. 2, the MongoDB takes responsible for data compression, index, and storage, while the SPARK is in charge of the data preprocessing and mining of the trajectory features.

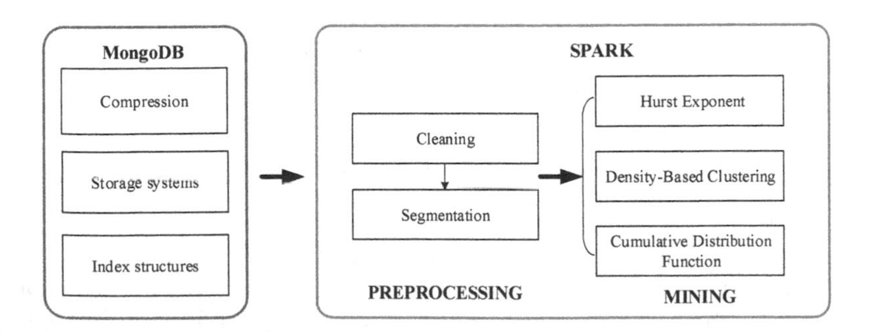

Fig. 2. The framework of trajectory data mining.

Self-Similarity. The BD procedure is adopted to calculate the Hurst value of the trajectories between the vehicles with blue and yellow license plate. In Fig. 3, it can be seen that the vehicle with the blue license plate has the lower value of Hurst, while the yellow vehicle has a litter higher one. From the definition of the Hurst parameter, the blue license plate that stands for the cabs is random roaming across the city, and the yellow license plate that stands for the trucks has a self-similarity path of moving.

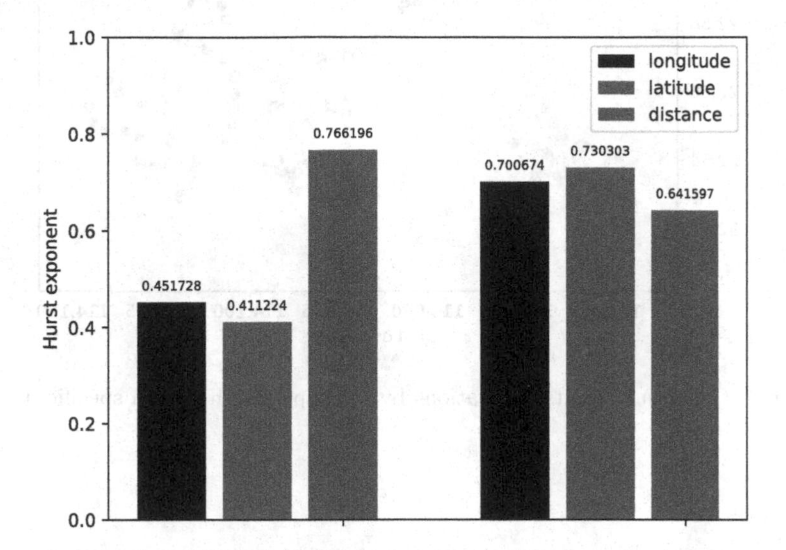

Fig. 3. The Hurst values between the vehicles with blue and yellow license plate respectively.

Hot-Spots. The DBSCAN procedure with the setup parameters listed in Table 1 is adopted to search for the clusters of the specific trajectory. Figure 4 that stands for the cabs and Fig. 5 that stands for the trucks has show the locations in one day for a specific type of the vehicle. After the DBSCAN procedure, the graph is colored based on the clustering results. Therefore, it is seen that the caps always roam in several certain hot-spots, while the trucks is always moving by a specific path.

Table 1. Data clustering for DBSCAN parameters.

Parameter	Value	Unit	Illustration
e	1000	m	Radius
$MinPts$	10	Null	Minimum points

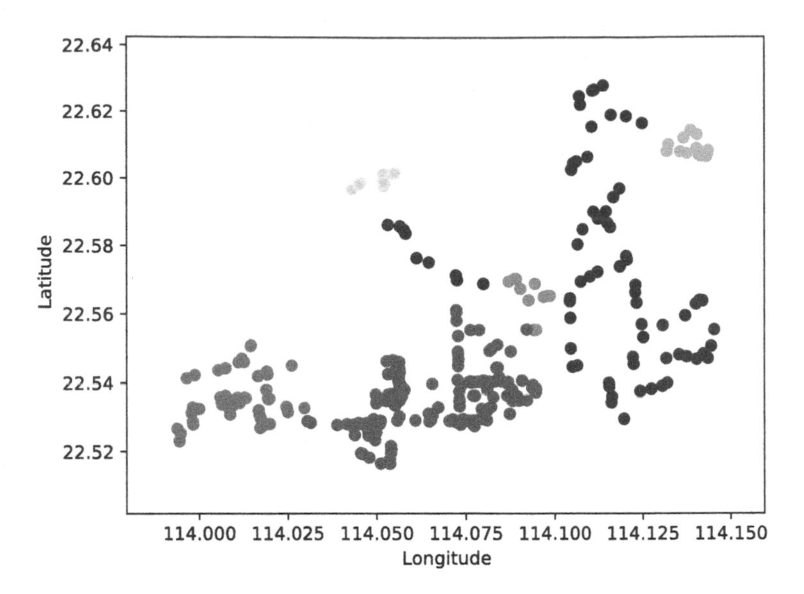

Fig. 4. Clustering results of locations by the caps activities in a specific day.

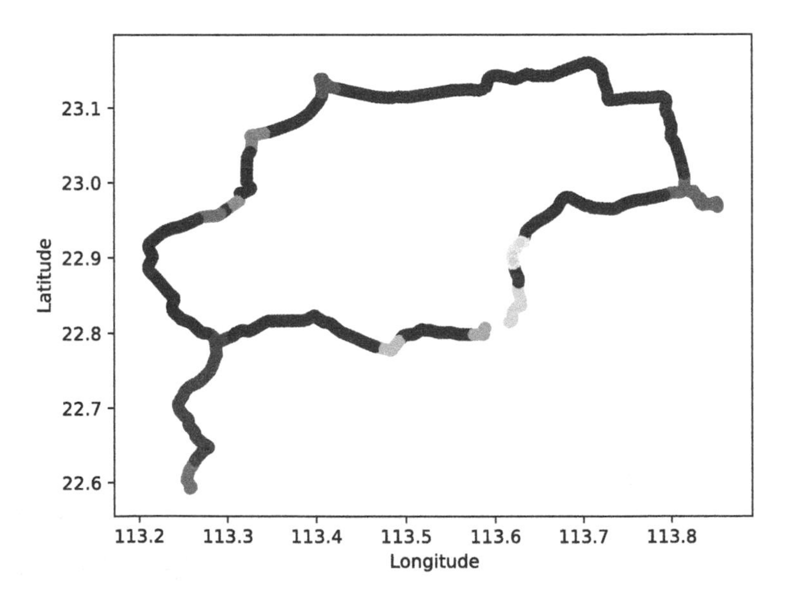

Fig. 5. Clustering results of locations by the trucks activities in a specific day.

Long-Tails. The cumulative distribution function is introduced in order to evaluate the long-tail effects of vehicle trajectories. Figure 6 shows that cabs in the city always take a short journey, while the distance that the trucks move is basically steady.

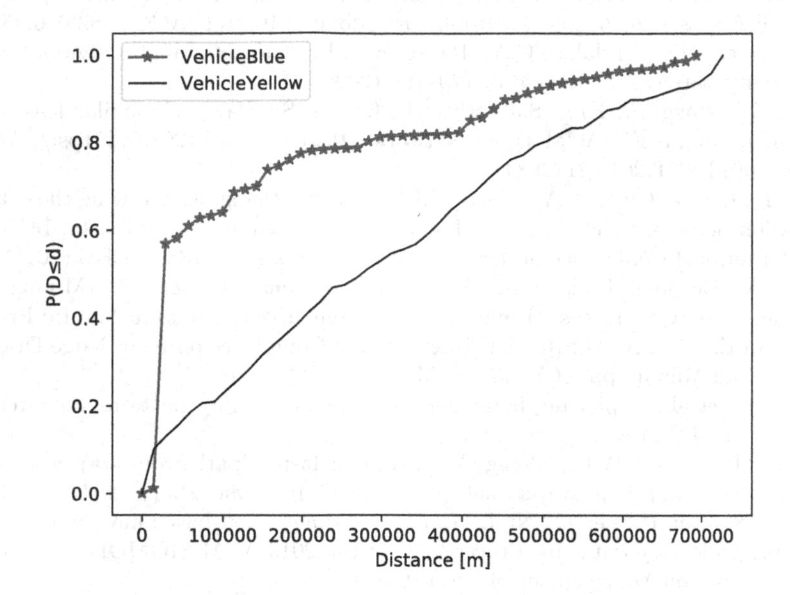

Fig. 6. The curve of the cumulative distribution function for the vehicles of cabs and trucks respectively.

5 Conclusion

In this paper, a data mining of the raw GPS trajectories is presented from the urban operating vehicles in the city of Shenzhen. The self-similarity, hot-spots and long-tails are evaluated as the widely recognized three features of human traces. Mining results show that the cabs which represent to serve the daily trip of human in the city always take a short travel and activate in several hot-spots, but roaming randomly. However, the trucks which represent to serve the goods are showing the opposite characteristics to cabs. Therefore, the vehicle types are the key feature of consideration to optimize algorithms for radio resource management in HDN.

References

1. Ando, S., Suzuki, E.: Role-behavior analysis from trajectory data by cross-domain learning (2012)
2. Chen, Y., Li, R., Zhao, Z., Zhang, H.: Fundamentals on base stations in urban cellular networks: from the perspective of algebraic topology. IEEE Wirel. Commun. Lett. **8**(2), 612–615 (2019). https://doi.org/10.1109/LWC.2018.2889041

3. Chen, Z., Shen, H.T., Zhou, X.: Discovering popular routes from trajectories. In: ICDE **6791**(9), 900–911 (2011)
4. Dai, J., Yang, B., Guo, C., Ding, Z.: Personalized route recommendation using big trajectory data (2015)
5. Feng, Z., Zhu, Y.: A survey on trajectory data mining: techniques and applications. IEEE Access **4**, 2056–2067 (2016). https://doi.org/10.1109/ACCESS.2016.2553681
6. Gonzalez, M.C., Hidalgo, C.A., Barabasi, A.L.: Understanding individual human mobility patterns. Nature **453**, 779–782 (2008)
7. Lee, K., Hong, S., Kim, S.J., Rhee, I., Chong, S.: Slaw: self-similar least-action human walk. IEEE/ACM Trans. Network. **20**(2), 515–529 (2012). https://doi.org/10.1109/TNET.2011.2172984
8. Li, Y., Luo, J., Chow, C.Y., Chan, K.L., Ding, Y., Zhang, F.: Growing the charging station network for electric vehicles with trajectory data analytics. In: IEEE 31st International Conference on Data Engineering, pp. 1376–1387. IEEE (2015)
9. Liu, X., Biagioni, J., Eriksson, J., Wang, Y., Forman, G., Zhu, Y.: Mining large-scale, sparse GPS traces for map inference: comparison of approaches. In: Proceedings of the 18th ACM SIGKDD International Conference on Knowledge Discovery and Data Mining, pp. 669–677. ACM (2012)
10. Liu, Y., et al.: Exploiting heterogeneous human mobility patterns for intelligent bus routing (2015)
11. Lu, E.H.C., Lee, W.C., Tseng, V.S.: Mining fastest path from trajectories with multiple destinations in road networks. Knowl. Inf. Syst. **29**(1), 25–53 (2011)
12. Luo, W., Tan, H., Lei, C., Ni, L.M.: Finding time period-based most frequent path in big trajectory data. In: Proceedings of the 2013 ACM SIGMOD International Conference on Management of Data (2013)
13. Lv, Q., Qiao, Y., Ansari, N., Liu, J., Yang, J.: Big data driven hidden Markov model based individual mobility prediction at points of interest. IEEE Trans. Veh. Tech. **66**(6), 5204–5216 (2017). https://doi.org/10.1109/TVT.2016.2611654
14. McGuire, M.P., Janeja, V.P., Gangopadhyay, A.: Mining trajectories of moving dynamic spatio-temporal regions in sensor datasets. Data Min. Knowl. Discov. **28**(4), 961–1003 (2014)
15. Renso, C., Baglioni, M., de Macedo, J.A.F., Trasarti, R., Wachowicz, M.: How you move reveals who you are: understanding human behavior by analyzing trajectory data. Knowl. Inf. Syst. **37**(2), 331–362 (2013)
16. Rhee, I., Shin, M., Hong, S., Lee, K., Kim, S.J., Chong, S.: On the levy-walk nature of human mobility. IEEE/ACM Trans. Network. **19**(3), 630–643 (2011). https://doi.org/10.1109/TNET.2011.2120618
17. Wang, X., Huang, D.: A novel density-based clustering framework by using level set method. IEEE Trans. Knowl. Data Eng. **21**(11), 1515–1531 (2009). https://doi.org/10.1109/TKDE.2009.21
18. Wei, L.Y., Zheng, Y., Peng, W.C.: Constructing popular routes from uncertain trajectories (2012)
19. Xuan, S., Zhang, Q., Sekimoto, Y., Shibasaki, R.: Prediction of human emergency behavior and their mobility following large-scale disaster. ACM (2014)
20. Yang, B., Fantini, N., Jensen, C.S.: iPark: identifying parking spaces from trajectories (2013)
21. Yuan, N.J., Zheng, Y., Xie, X., Wang, Y., Zheng, K., Xiong, H.: Discovering urban functional zones using latent activity trajectories. IEEE Trans. Knowl. Data Eng. **27**(3), 712–725 (2015). https://doi.org/10.1109/TKDE.2014.2345405

22. Zhang, H., Huang, S., Jiang, C., Long, K., Leung, V.C.M., Poor, H.V.: Energy efficient user association and power allocation in millimeter-wave-based ultra dense networks with energy harvesting base stations. IEEE J. Sel. Areas Commun. **35**(9), 1936–1947 (2017). https://doi.org/10.1109/JSAC.2017.2720898
23. Zhang, H., Liu, H., Cheng, J., Leung, V.C.M.: Downlink energy efficiency of power allocation and wireless backhaul bandwidth allocation in heterogeneous small cell networks. IEEE Trans. Commun. **66**(4), 1705–1716 (2018). https://doi.org/10.1109/TCOMM.2017.2763623
24. Zhang, H., Liu, N., Chu, X., Long, K., Aghvami, A., Leung, V.C.M.: Network slicing based 5G and future mobile networks: mobility, resource management, and challenges. IEEE Commun. Mag. **55**(8), 138–145 (2017). https://doi.org/10.1109/MCOM.2017.1600940

Group Paging Mechanism with Pre-backoff for Machine-Type Communication

Yong Liu(✉), Qinghua Zhu, Jingya Zhao, and Wei Han

School of Telecommunication Engineering, Beijing Polytechnic, Beijing 100176, China
loot.07@163.com

Abstract. In order to resolve congestion of the random access channel (RAC) caused by UE's concentrated access to the network in the process of group paging, this paper introduces the pre-backoff algorithm on the basis of studying MTC service characteristics to the group paging mechanism, and proposes the analysis model based on the group paging mechanism with pre-backoff. In the group paging mechanism with pre-backoff, when monitoring results indicate ongoing of group paging, all UE within the group, before access, will implement pre-backoff, and through pre-backoff, different pieces of UE are evenly distributed to a period of access time to alleviate collisions and conflicts resulted from concentrated access. Simulation results of three performance indexes, including the probability of conflict, probability of successful access and delay of average access, are used to analyze and verify the validity of the analysis model based on the group paging mechanism with pre-backoff.

Keywords: Machine-Type Communication (MTC) · Random Access Channel (RAC) · Group paging · Pre-backoff

1 Introduction

As the most important application form of the Internet of Things IOT) for the time being, Machine-type Communications or MTC for short has brilliant application prospects and tremendous market potential [1]. Also known as Machine-to-Machine M2M), MTC has been widely explored and implemented by three major telecommunications service providers in China, which has helped promote the MTC service development layout [2]. Nevertheless, the current Long-term Evolution LET) system, a primary supporting technology of MTC, for commercial use, which was initially designed to meet the needs of Human-type Communications HTC), has failed to take the mass equipment and emergencies of MTC into consideration. As a result, when a large number of MTC terminals are switched into the LTE system designed on the basis of Human-to-Human H2H) communications rules, access congestion would be inevitable. Therefore, it is of vital significance to improve the LTE mobile network to suit the application and development of MTC.

Though research into the MTC technology has been a general research interest, research findings are still limited. Apart from defining MTC service requirements and

X. Wang et al. (Eds.): 6GN 2020, LNICST 337, pp. 218–228, 2020.
https://doi.org/10.1007/978-3-030-63941-9_16

functional structure based on the 3GPP standards, most researchers concentrate on studying the realization of application technologies, system structures and communication plans but pay little attention to the M2M service modeling and analysis [4]. The traditional mobile cellular network was originated from H2H communications, and has kept on evolving and improving according to user demands. Different from H2H communications, MTC is characterized by a high synchronicity of data transmission, low mobility of terminals, high terminal distribution density, etc.

Literature [5] points out that, in receiving group paging information, all user equipment UE) immediately transmits its paging response through the random-access channel RAC). Simultaneous signal access of all UE might result in serious conflicts of the RACH within a short period of time. Therefore, this paper, through introduction of the pre-backoff algorithm, puts forward an analysis model based on the pre-backoff group paging mechanism under MTC, attempting to address competition conflicts caused by users' concentrated access to the network and meet users' demand for quality of service QoS).

2 Pre-backoff Group Paging Mechanism Analysis Model

This paper puts forward an analysis model based on the pre-backoff group paging mechanism, which can be used to analyze the performance of group paging strategies based on the pre-backoff algorithm. Three performance indexes, including probability of conflict P_C), probability of successful access P_S) and delay of average access D_A) are employed to verify the accuracy and effectiveness of the pre-backoff group paging mechanism. The method proposed by this paper is not a process of repeated paging but makes use of the pre-backoff mechanism to expand the transmission of the first preamble to increase the probability of success at a lower access delay.

In the pre-backoff group paging mechanism, every UE should implement pre-backoff of its first transmission, and adheres to and supports random access of the repeatedly-transmitted standard LTE. The pre-backoff timer will evenly choose within the scope from 0 to W_{PBO}.

First of all, the pre-backoff group paging is available at the group paging interval of the I_{max}-th RA (Random Access) timeslot. The group paging interval starts from and ends with the first RA timeslot. The pre-backoff algorithm can delay the first preamble transmission time to the pre-backoff window, W_{PBO}. The group paging is presented in Eq. (1). The definitions of various parameters are provided by Literature [6].

$$I_{max} = \left\lceil \frac{W_{PBO}}{T_{RA\_REP}} \right\rceil + 1 + (N_{PT\max} - 1) \left\lceil \frac{(T_{RAR} + W_{RAR} + W_{BO})}{T_{RA\_REP}} \right\rceil \tag{1}$$

Define $M_{i,S}[n]$ and $M_{i,F}[n]$ as the number of UE succeeding in transmitting and failing to send the n-th preamble at the i-th RA timeslot, respectively; $M_i[n]$ as the total number of UE transmitting the preamble at the i-th RA timeslot; P_P as the probability of the UE successfully receiving the paging information. After the paging information is received, "$P_P \times M$" UE will immediately implement even backoff before transmitting the first preamble. Export the total number of UE transmitting the first preamble at the i-th RA timeslot, M_i [1], from Eq. (2).

$$M_i[1] = \begin{cases} \dfrac{P_p M}{W_{PBO}} & if \quad i = 1 \\ \dfrac{T_{RA\_REP} P_p M}{W_{PBO}} & if \quad 2 \le i \le \left\lfloor \dfrac{W_{PBO}}{T_{RA\_REP}} \right\rfloor \\ \dfrac{(W_{PBO} \bmod T_{RA\_REP}) P_p M}{W_{PBO}} & if \quad i = \left\lfloor \dfrac{W_{PBO}}{T_{RA\_REP}} \right\rfloor + 1 \\ 0 & otherwise \end{cases} \tag{2}$$

There is no preamble retransmission in the first RA timeslot, so RA in the first RA timeslot tries to be investigated respectively with other RA timeslots. The number of successful UE, $M_{1,S}[n]$, UE in fault and $M_{1,F}[n]$ of the first RA timeslot are given by Eq. (3) and Eq. (4), respectively. Of special note is that, under the initial condition of $i = 1$, if $n \neq 1$, then $M_1[n] = 0$.

$$M_{1,S}[n] = \begin{cases} N_{UL} & if \quad n = 1 \quad and \quad M_i[1] e^{-\frac{M_i[1]}{R}} p_1 \ge N_{UL} \\ M_i[1] e^{-\frac{M_i[1]}{R}} p_1 & if \quad n = 1 \quad and \quad M_i[1] e^{-\frac{M_i[1]}{R}} p_1 < N_{UL} \\ 0 & if \quad n \ne 1 \end{cases} \tag{3}$$

$$M_{1,F}[n] = \begin{cases} M_i[1] - N_{UL} & if \quad n = 1 \quad and \quad M_i[1] e^{-\frac{M_i[1]}{R}} p_1 \ge N_{UL} \\ M_i[1](1 - e^{-\frac{M_i[1]}{R}} p_1) & if \quad n = 1 \quad and \quad M_i[1] e^{-\frac{M_i[1]}{R}} p_1 < N_{UL} \\ 0 & if \quad n \ne 1 \end{cases} \tag{4}$$

Where, N_{UL} denotes the maximum number of UE which can be confirmed in the response window; p_n denotes the detection probability of the n preamble transmission under the power slope effect [7], and $p_n = 1-(1/e^n)$; the UE failing at the first RA timeslot will implement backoff and be resent in the following RA timeslot. The total number of UE succeeding and failing at every RA timeslot, namely $M_{i,S}[n]$ and $M_{i,F}[n]$, can be respectively given by Eq. (5) and Eq. (6) recursively:

$$M_{1,S}[n] = \begin{cases} M_i[n] e^{-\frac{M_i}{R}} p_n & if \quad \sum_{n=1}^{N_{PT\max}} M_i[n] e^{-\frac{M_i}{R}} p_n \le N_{UL} \\ \dfrac{M_i[n] e^{-\frac{M_i}{R}} p_n}{\sum_{n=1}^{N_{PT\max}} M_i[n] e^{-\frac{M_i}{R}} p_n} & otherwise \end{cases} \tag{5}$$

$$M_{1,F}[n]=\begin{cases} M_i[n](1-e^{-\frac{M_i}{R}}p_n) & if \ \sum_{n=1}^{N_{PT\max}} M_i[n]e^{-\frac{M_i}{R}}p_n \le N_{UL} \\ M_i[n]\left(1-\frac{p_n}{\sum_{n=1}^{N_{PT\max}} M_i[n]p_n}\right)N_{UL} & otherwise \end{cases} \quad (6)$$

If n > 1, the total number at the n-th RA attempt, $M_i[n]$, of UE of the *i*-th RA timeslot can be approximately written as Eq. (7), where $M_{k,F}[n-1]$ denotes the $(n-1)$-th preamble transmitted by UE at the *k*-th RA timeslot but in vain. The $\alpha_{k,I}$ of UE suffering from the above failure is retransmitted in the *i*-th RA timeslot. K_{min} and K_{max} represent the minimum and maximum of k, respectively.

$$M_i[n]\approx\sum_{k=K\min}^{K\max}\alpha_{k,i}M_{K,F}[n-1] \quad if \ n>1 \quad (7)$$

In Eq. (7), the upper limit, K_{max}, and the lower limit, K_{min}, of the transmission probability, $\alpha_{k,I}$, can be derived by the sequence chart from Literature [8]. K_{max} and K_{min} are given by Eq. (8) and Eq. (9), respectively. If the backoff interval of the k-th RA is overlapped with the transmitting interval of the i-th RA timeslot, then the UE failing to transmit the preamble at the k-th RA timeslot will retransmit a new preamble at the i-th RA. Therefore, $\alpha_{k,I}$ denotes the backoff interval of the k-th RA interval, whose transmission interval is overlapped with that of the i-th RA timeslot (k < i). At $(k-1)$ T_{RA_REP}, the UE transmitting its preamble at the k RA timeslot will recognize the RA fault after the $(T_{RAR}+W_{RAR})$ subframe. Every UE failing to send the preamble will launch its backoff at the time of "$(k-1)\ T_{RA\_REP}+(T_{RAR}+W_{RAR})+1$". Therefore, the backoff interval at the k-th RA timeslot starts at "$k-1)\ T_{RA\_REP}+T_{RAR}+W_{RAR})+1$" and ends at "$(k-1)\ T_{RAREP}+(T_{RAR}+W_{RAR})+W_{BO}$". If their backoff fills in between the (i − 1)th RA timeslot and the i-th RA timeslot, then the UE can upload the preambles at the i-th RA timeslot. Therefore, the transmission interval of the i-th RA can be written as "$[(i-1)\ T_{RA\_REP}+1(i-1)\ T_{RA\_REP}]$".

When the right boundary of the backoff interval at the k-th RA timeslot arrives at the left boundary of the transmission interval at the i-th RA timeslot, namely "(Kmin − 1) TRA_REP + TRAR + WRAR + WBO ≥ 1 + (i − 2) TRA_REP", then K_{min} can be obtained, and written as below:

$$K_{\min}=\left\lceil (i-1)+\frac{1-(T_{RAR}+W_{RAR}+W_{BO})}{T_{RA\_REP}}\right\rceil \quad (8)$$

When the left boundary of the backoff interval at the k-th RA timeslot arrives at the right boundary of the transmission interval at i-th RA timeslot, namely "$(K_{max}-1)T_{RA\_REP}+T_{RAR}+W_{RAR}+1\le(i-2)\ T_{RA\_REP}$", then $k(K_{max})$ can be obtained, and written as below:

$$K_{\max}=\left\lfloor i-\frac{1+(T_{RAR}+W_{RAR})}{T_{RA\_REP}}\right\rfloor \quad (9)$$

3 Performance Indexes

As mentioned above, performance indexes of the analysis model based on the pre-backoff group paging mechanism include: (1) Probability of conflict (P_C); (2) Probability of successful access (P_S); and (3) Delay of average access (D_A).

Among them, probability of conflict (P_C) is defined as the ratio of the number of preambles with collision conflicts and the number of all preambles; probability of successful access (P_S) is defined as the ratio of the number of UE with success network access to the total number of UE accessing within the timeslot; delay of average access (D_A) is defined as the period of time from the start of access upon paging to the establishment of network connection after completion of four steps of random access. Delay normalization is conducted of all UE with successful access will undergo obtain the delay of average access, D_A [9, 10].

Literature [6] provides the definition for formulas and parameters, including P_C, P_S and D_A, and these three performance indexes can be given by Eq. (10), Eq. (11), and Eq. (12), respectively.

$$P_C = \frac{\sum_{i=1}^{I_{max}} \left(R - M_i e^{-\frac{M_i}{R}} - R e^{-\frac{M_i}{R}}\right)}{I_{max} R} \tag{10}$$

$$P_S = \frac{\sum_{i=1}^{I_{max}} \sum_{n=1}^{N_{PT\max}} M_{i,S}[n]}{M} \tag{11}$$

$$D_A = \frac{\sum_{i=1}^{I_{max}} \sum_{n=1}^{N_{PT\max}} M_{i,S}[n]\, T_i}{\sum_{i=1}^{I_{max}} \sum_{n=1}^{N_{PT\max}} M_{i,S}[n]} \tag{12}$$

Where, T_i denotes the access delay of UE with successful access via preamble transmission within the RA timeslot, which can be obtained through Eq. (13):

$$T_i = (i-1)T_{RA\_REP} + T_{RAR} + W_{RAR} \tag{13}$$

Of special note is that a high P_S, a low P_C, and a low D_A can be chosen as the desired results. The larger the backoff window, W_{PBO}, is, the lower the probability of conflict, P_C, is, and the higher the probability of successful access, P_S, is, namely the better the effects of users' access to the RCA. However, if the pre-backoff window changes, the delay of average access, D_A, of users will change as well. Therefore, it is necessary to find a suitable value for the backoff window, W_{PBO}.

4 Simulation Results and Discussions

By analyzing MATLAB simulation results of the probability of conflict (P_C), successful access probability (P_S) and delay of average access (D_A), this paper verifies the effectiveness of the pre-backoff group paging mechanism.

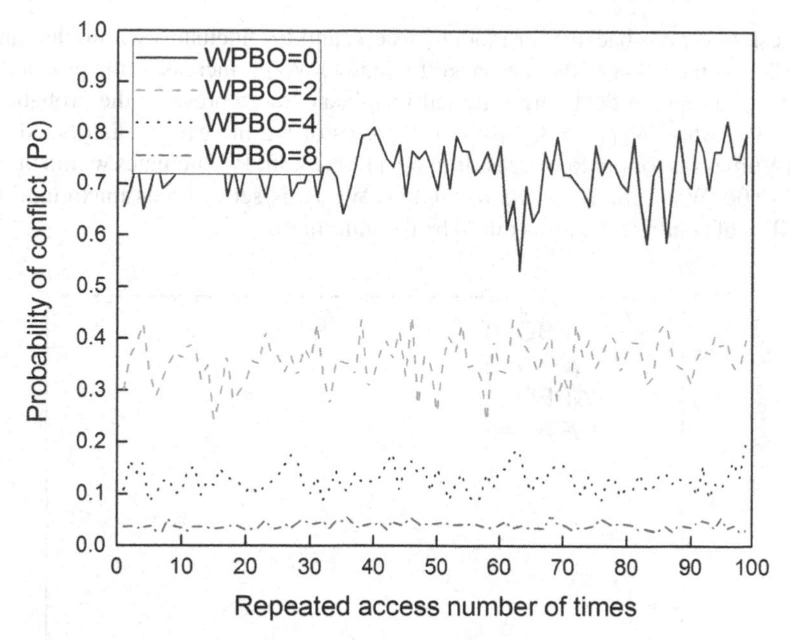

Fig. 1. Probability of conflict, P_C (Color figure online)

Currently, there have not yet been any suitable scenarios which can be used to compare the performance of the RAN (Radio Access Network) overload control plan based on the push and pull media. So, this paper compares the group paging strategies with other push plans. The simulation results of the pre-backoff group paging mechanism is shown in Fig. 1, Fig. 2, and Fig. 3. The numerical results are expressed by dots, and results of 100 repeated random access are adopted for analysis and discussion.

4.1 Charts of Simulation Results

Figure 1 presents the simulation results of probability of conflict (P_C), which shows the probability of conflict of every access of 160 pieces of UE in the group with repeated access to the RCH for 100 times, when the pre-backoff window, W_{PBO}, is set to different values. The blue full line in Fig. 1 denotes the probability of conflict when $W_{PBO} = 0$, namely the probability of conflict (P_C) of all UE receiving paging information from certain group and with direct access to the RCH within the RA timeslot. From Fig. 1, it can be seen that the probability of conflict, at $W_{PBO} = 0$ changes within the range of 0.5 to 0.85, so the ratio of the number of preambles with collision conflicts to the number of all preambles during the process of random access is very high, which might cause serious congestion. In Fig. 1, the green imaginary line represents the probability of conflict when $W_{PBO} = 2$. All the 160 users within the group are evenly distributed within two RA timeslots for random access. When P_C is fluctuating within the range of 0.2–0.45, the probability of conflict significantly declines compared with direct access without the introduction of pre-backoff. The black imaginary line represents the probability of conflict when $W_{PBO} = 4$. The 160 pieces of UE are evenly distributed within four

RA timeslots via pre-backoff for random access, and P_C fluctuates within the range of 0.08–0.20. As the value of the pre-backoff window, W_{PBO}, increases, the probability of conflict, P_C, keeps on declining. The red imaginary line represents the probability of conflict (P_C) when $W_{PBO} = 8$. All the 160 pieces of UE are evenly distributed within eight RA timeslots for random access through pre-backoff. P_C fluctuates within the range of 0.02–0.06. When the pre-backoff window, W_{PBO}, is set to be its maximum, 8, the probability of conflict, P_C, turns out to be the minimum.

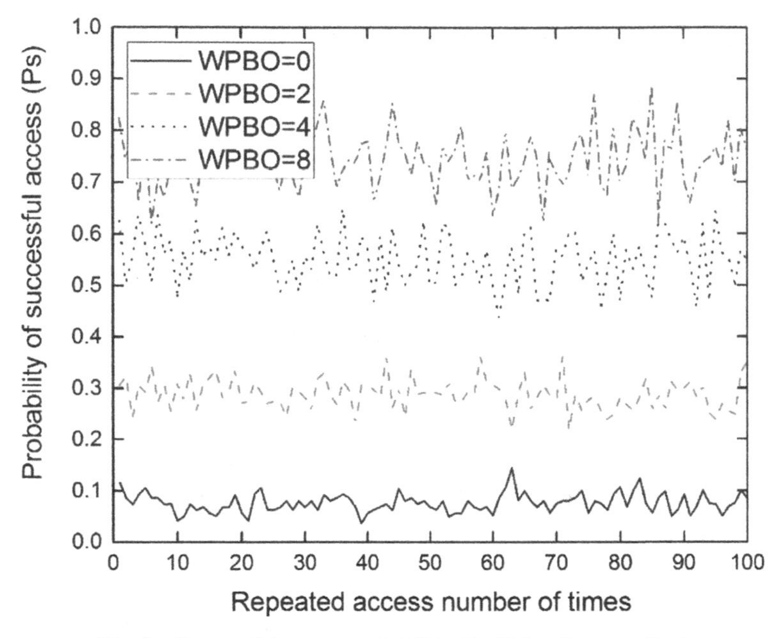

Fig. 2. Successful access probability, Ps (Color figure online)

Figure 2 shows the simulation results of probability of successful access (P_S), where the blue solid line represents the probability of successful access (P_S) when $W_{PBO} = 0$, namely the probability of successful access when all UE receiving paging messages from certain group has direct access to the RCH within the RA timeslot. From Fig. 2, one can observe that the probability of successful access (P_S) when $W_{PBO} = 0$ fluctuates within the range of 0.03–0.15. Therefore, the ratio of the number of UE with successful network access to the total number of UE accessing within the timeslot is very slow, which is obviously not the desired result. During the process of random access, as many pieces of UE receiving paging messages as possible are expected to realize successful network access. The green imaginary line represents the probability of success access (Ps). The 160 pieces of UE within the group are evenly distributed within two RA timeslots for random access via pre-backoff. When P_C is fluctuating within the range of 0.2–0.4, the probability of successful access (P_S), compared with direct access without introducing pre-backoff, obviously increases. The black imaginary line represents the probability of successful access (P_S) when $W_{PBO} = 4$, and P_S fluctuates within the range of 0.4–0.7. As the value of the pre-backoff window, W_{PBO}, increases, the successful access probability,

P_S, continues rising. The red solid line represents the probability of successful access (P_S) when $W_{PBO} = 8$, and P_S fluctuates within the range of 0.6–0.9. When the value of the pre-backoff window, W_{PBO}, is set to be the maximum, 8, the probability of successful access (P_S) turns out to be the maximum.

Figure 3 displays the simulation results of the delay of average access (D_A). Access delay refers to the period of time from the start of access upon paging to the establishment of network connection after completion of four steps of random access. Delay normalization is conducted of all UE with successful access to obtain the delay of average access, D_A. The blue solid line represents the delay of average access, D_A, when $W_{PBO} = 0$. D_A fluctuates around the 7.5 sub-frame. In random access, a low probability of conflict (P_C), a high probability of successful access (P_S), and a low delay of average access (D_A) are desired. The green imaginary line represents the delay of average access (D_A) when $W_{PBO} = 2$. All the 160 pieces of UE within the group are evenly distributed within two RA timeslots for random access via pre-backoff. D_A fluctuates around the 5.8 sub-frame. The delay of average access (D_A), compared with that of direct access without introducing pre-backoff, significantly declines. The black imaginary line represents the delay of average access (D_A) when $W_{PBO} = 4$, and D_A fluctuates around the 5.0 sub-frame. As the value of the pre-backoff window, W_{PBO}, increases, the delay of average access (D_A) decreases slightly. The red solid line represents the delay of average access (D_A) when $W_{PBO} = 8$. D_A fluctuates around the 6.3 sub-frame. As the value of the pre-backoff window (W_{PBO}) increases, the delay of average access (D_A) increases to the contrary.

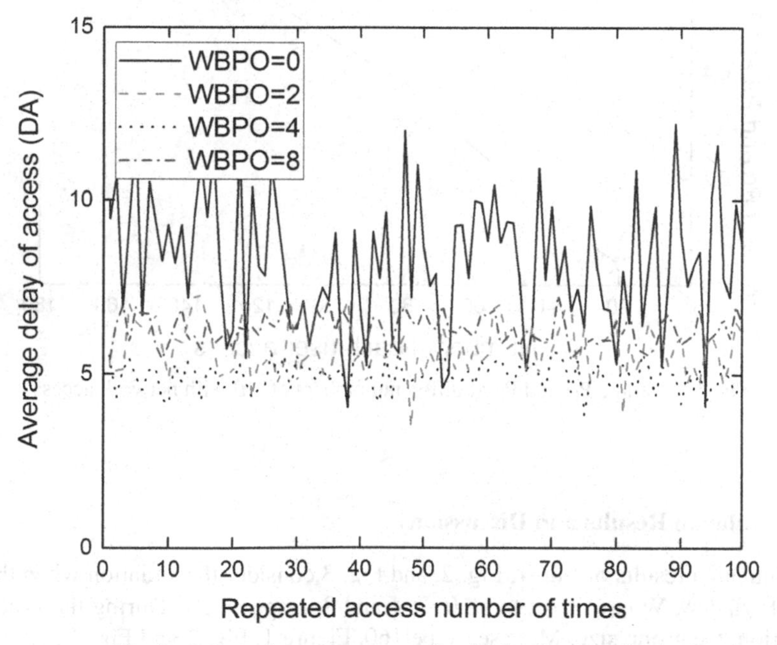

Fig. 3. Delay of average access, D_A (Color figure online)

Figure 4 presents the average probability of conflict (P_C) and probability of successful access (P_S) at different numbers of UE with network access. The asterisk indicates the average probability of conflict (P_C) and the circle indicates the average probability of successful access (P_S), both at different numbers of UE with network access. It can be clearly seen that, without pre-backoff, the number of UE is 160 and, at the moment, the probability of conflict (P_C) is the highest, while the probability of successful access (P_S) is the lowest, which suggests serious congestion. With the increasing number of pre-backoff windows (W_{PBO}), namely when the number of UE accessing at the same RA timeslot, the probability of conflict (P_C) keeps on declining, while the probability of successful access (P_S) keeps on increasing. This means that the competition conflict resulted from concentrated user access to the network during the process of group paging is satisfactorily improved.

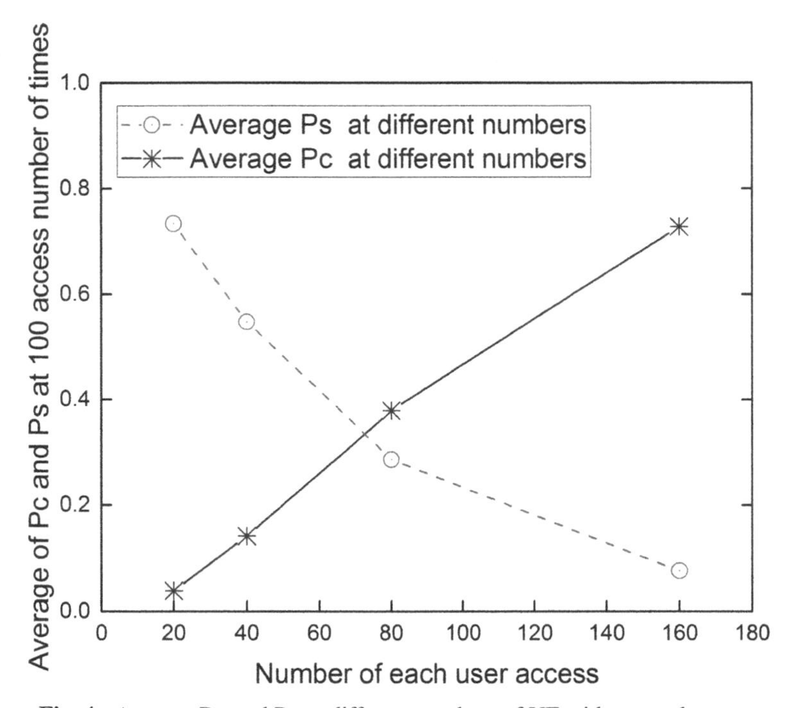

Fig. 4. Average P_C and P_S at different numbers of UE with network access

4.2 Simulation Results and Discussions

The simulation results of Fig. 1, Fig. 2, and Fig. 3 consider the situation when the pre-backoff window, W_{PBO}, is set to be 0, 2, 4 and 8, respectively. During the process of simulation, the group size, M, is set to be 160. Figure 1, Fig. 2 and Fig. 3 demonstrate the simulation results of the probability of conflict (P_C), probability of successful access (P_S), and delay of average access (D_A). The overall simulation results imply that the analysis model can relatively accurately estimate the performance of the pre-backoff group paging mechanism at different values of W_{PBO}. In Fig. 1, the probability of conflict (P_C) decreases as W_{PBO} increases. When the value of W_{PBO} reaches the maximum, the minimum of P_C is obtained. Simulation results also suggest that, as the value of W_{PBO} increases, the probability of conflict of the pre-backoff paging conflict keeps on decreasing. This is because the pre-backoff mechanism will expand the access of UE at the first RA timeslot to the even access of UE at multiple RA timeslots. In Fig. 2, the probability of successful access (P_S) increases as the value of W_{PBO} increases, and P_S reaches its maximum when the W_{PBO} is at its maximum. Simulation results mean that, as the value of W_{PBO} increases, the probability of successful access based on the pre-backoff group paging mechanism keeps on rising. In Fig. 3, the delay of average access, D_A, first increase and then decrease, as the value of W_{PBO} increase. Simulation results provide solid evidence for that the pre-backoff group paging mechanism can significantly increase the probability of successful access, and also maintain a low delay of average access for UE with successful access. In Fig. 4, the average probability of conflict (P_C) and the average probability of successful access (P_S), both at different numbers of UE, are compared. As the number of UE accessing within the same RA timeslot decreases, the probability of conflict (P_C) keeps on decreasing, while the probability of successful access (P_S) keeps on rising. To sum up, the above simulation results can verify the validity of the group paging mechanism analysis model with the pre-backoff algorithm introduced, in that, the analysis model can favorably improve the competition conflict resulted from users' concentrated access to the network, and also satisfy the requirement of overload control.

5 Conclusions

All in all, this paper proposes the pre-backoff group paging mechanism as a solution plan for congestion caused by MTC random access. Simulation results suggest that, when the group size is large, the pre-backoff method can significantly improve the performance of the group paging strategy, alleviate the competition conflict resulted from UE's concentrated access to the network, and satisfy users' requirement of access QoS.

Acknowledgment. This work was supported by the Beijing City Board of education project (NO. KM202010858005).

References

1. Xian-Feng, W., Yin-Feng, W.: Study on M2M based on PLMN. Commun. Tech. **40**(3), 66–68 (2012)
2. Dapeng, W., Hang, S., Honggang, W., et al.: A feature based learning system for internet of things applications. IEEE Internet Things J. **6**(2), 1928–1937 (2019)
3. Zhang, H., Huang, S., Jiang, C., et al.: Energy efficient user association and power allocation in millimeter wave based ultra dense networks with energy harvesting base stations. IEEE J. Sel. Areas Commun. **35**(9), 1936–1947 (2017)

4. Zhang, H., Dong, Y., Cheng, J., et al.: Fronthauling for 5G LTE-U ultra dense cloud small cell networks. IEEE Wirel. Commun. **23**(6), 48–53 (2017)
5. Dapeng, W., Qianru, L., Honggang, W., et al.: Socially aware energy-efficient mobile edge collaboration for video distribution. IEEE Trans. Multimedia **19**(10), 2197–2209 (2017)
6. Jiang, W., Wang, X.: Group paging based on pre-backoff access scheme in Machine-type Communications. Commun. Tech. **47**(2), 172–178 (2014)
7. Wei, C.-H., Cheng, R.-G., Tsao, S.-L.: Performance analysis of group paging for machine-type communications in LTE networks. IEEE Trans. Veh. Tech. **62**(7), 3371–3382 (2013)
8. Farhadi, G., Ito, A.: Group-based signaling and access control for cellular machine-to-machine communication. In: Vehicular Technology Conference, pp. 1–6. IEEE, Las Vegas (2013)
9. Chen, Yu., Guo, Z., Yang, X., Hu, Y., Zhu, Q.: Optimization of coverage in 5G self-organizing small cell networks. Mob. Netw. Appl. **23**(6), 1502–1512 (2017). https://doi.org/10.1007/s11036-017-0983-x
10. Dapeng, W., Zhihao, Z., Shaoen, W., et al.: Biologically inspired resource allocation for network slices in 5G-enabled internet of things. IEEE Internet Things J. 1, (2018). At present Online publishing

Research on Monitoring and Regularity of Coal Mine Methane Extraction State

Chao Wu[1], Yuliang Wu[2](✉), Changshegn Zhu[1], Chengyuan Zhang[1], and Qingjun Song[1]

[1] College of Intelligent Equipment, Shandong University of Science and Technology, Taian 271019, Shandong, China

[2] Sichuan Staff University of Science and Technology, Chengdu 610101, Sichuan, China

Luckfan2002@163.com

Abstract. The existing gas drainage monitoring system only can measure the gas concentration of the drainage, can't dynamically predict the residual gas and the drainage cycle of the coal seam according to the drainage parameters. The mining conditions of deep coal seams have become complicated by the increasing depth of mining. The measurement of the extraction status becomes more and more difficult. It is of great practical significance to effectively predict the drainage time and the gas content of coal seam. In order to increase the gas drainage concentration of coal seam under the coal mine, combined with the actual situation at the coal mine site, a gas drainage pipeline control system was designed and developed. It is mainly composed of gas sensors, microcontroller, miniature electric valves, power conversion circuits, and wireless communication modules. In view of the non-linear, multi-coupling and hysteresis characteristics of gas flow, a fuzzy control algorithm is adopted by the controller. By measuring the gas concentration and pressure of the field extraction pipeline, the microcontroller STM8 working as the control core controls opening of the pipeline valve according to the fuzzy control law. Finally the purpose of increasing the concentration of the extracted gas is achieved. Both the fuzzy control algorithm and the system circuit is designed in detail. The system has the characteristics of long-distance wireless communication, compact structure and low power consumption. It can increase the concentration of gas extraction, which is of great significance for the gas prevention and management and coal mine gas extraction.

Keywords: Gas extraction · Gas concentration · Wireless communication · Residual gas · Drainage time

1 Introduction

Coal mine gas (Coal Mine Methane, CMM, also known as gas) extraction technology which is the main technical measure to prevent gas disaster accidents has also been continuously improved and developed with the development of coal industry technology, but the overall level is still low [1]. The informatization, intelligence and standardized and refined means of gas prevention and control technology equipment have just

X. Wang et al. (Eds.): 6GN 2020, LNICST 337, pp. 229–239, 2020.
https://doi.org/10.1007/978-3-030-63941-9_17

started. The collection, transmission, and intelligent identification models of information closely related to the hidden dangers of gas accidents are not complete. The keys of coal mine gas extraction parameter monitoring, engineering construction, management, effect measurement and inspection have not yet been digitized and informatized. A lot of data is still in the stage of full manual determination [2].

According to the investigation on the coal mining and discussion with on-site technical staff, The reasons for low coal gas drainage rate and the existing problems of gas drainage concentration in coal seam drilling are as follows.

Single-hole metering devices is not equipped in some coal mine pre-drainage project, so the extraction effect cannot be investigated in time [3]. The gas measuring holes and control valves is supplied for main pipeline of each roadway but not for extraction branch, so the gas in branch pipeline is out of control and the leakage of branch will affect the entire extraction system.

Sometimes the extraction system has to be shut down in order to ensure the gas concentration throughout the area. In some coal mine gas drainage systems, a regulating valve and gas measuring hole works for two boreholes and valves is supplied for the sub-pipelines. In some coal mine gas extraction systems 10 drainage drilling are connected in parallel to a sub-pipelines. The drainage drilling is provided with regulating valve and gas measuring hole and the valve is stalled for sub-pipelines so each pipeline is under control. But the operation of valve is basically manual with wrench. An air leak in one borehole is sometimes handled by shutting off the whole system which is wasteful and inefficient. The greater the density of the arranged single-hole metering device, the more complete and the higher the reference value the data obtained. The more monitoring stations are arranged, the higher the mine construction cost.

There is no effective theoretical guidance but experience for the gas concentration of drainage pipeline. So the drainage control effect is poor, and there are a lot of drill hole closing which not only causes lot waste of resources but also leaves security risks for mine. The extraction concentration of different boreholes in the same area varies greatly in some mine. There were no online monitoring of pressure, flow and concentration for branch pipeline and even there were not a flowmeter or detection hole for main pipeline of pump station. The same situation happens in most gas drainage systems and it is common to be unable to detect gas drainage volume and analyze source of gas. When the concentration of most branch pipeline does not reach the usable concentration how to take advantage of gas is big problem.

Domestic and foreign researches on gas extraction systems have mostly focused on optimization of extraction technology and systems optimization. In recent years, there have also been applications of artificial intelligence control technology for gas extraction pipelines especially branch pipelines.

Zuo Jianghong [4] and Cai Feng [5] analyzed the underground gas drainage method from the aspects of adjusting negative pressure and sealing process, Jiang Zhigang [6] found that the optimal negative pressure for drilling varies with the coal seam conditions, the extraction period, and the sealing conditions. By adjusting the negative pressure, the optimal gas concentration can be obtained. Wang Zhenfeng [7] proposed that a multi-stage negative pressure adjustment method can be used during the extraction process to ensure that the gas extraction concentration is in a safe range for a long time. The disadvantage of that system is that the branch valve is still operated by manual operation. Jiang Yuanyuan [8] introduced an intelligent system which worked by AT89C52 single-chip microcomputer as the control core and fuzzy control as algorithms to control of the gas extraction process. Kaifeng Huang [9] studied a kind of intelligent control system for coal mine gas mobile extraction system. In order to meet the need of on-line control, the system uses PLC instead of single-chip microcomputer as the main controller. Ansai [10] introduced the KJ456 adaptive gas drainage pipeline valve automatic control system which automatically controls the opening of the pipeline valve through an adaptive control strategy to achieve the maximum gas drainage concentration or pure flow. Zhou Fubao [11] designed a gas intelligent extraction system, which uses the Arduino controller to control the speed of the gas extraction pump to improve the extraction efficiency ratio and extraction pure volume. This system is relatively complicated, which is not conducive to the installation of multiple pipelines on site. Wu Yuliang [12] studied the coal mine gas drainage system based on wireless sensor network, and applied conventional control and fuzzy control to the gas drainage system. Huang Ruifeng [13] shows that the extraction concentration can be kept in a safe and available range for a long time by means of multi-level and multiple regulation.

In short, due to the complexity of gas flow law, there are the following problems in coal seam gas drainage: accurate mathematical model can not be established, conventional control method is difficult to use, gas multi parameter measurement is needs for pipeline control, field construction of traditional gas monitoring equipment adopts wired communication is complex, and the occurrence characteristics and laws of residual gas need to be further studied.

To solve the above problems, it is necessary to design a gas extraction system to achieve accurate control of pipelines. Work of the system for improving gas drainage is to control the pumping pipeline through data acquisition and control algorithm efficiency and to estimate the residual gas parameters according to the measured data. Combined with control technology and wireless sensor network and model estimation for gas drainage, the monitoring system of gas drainage branch is designed. Data parameters including gas concentration are collected by sensors, the valve operation of pumping branch can be automatically regulated. The extraction period are estimated.

2 Design of Monitoring System for Coal Seam Gas Drainage Branch Pipe

2.1 Prototype of System Model Selection

The gas drainage cycle estimation and gas concentration control of gas pre drainage in Xuyong coal mine are studied as the object. Xuyong coal mine has the following characteristics. The relative gas emission of the mine is 20.1 m^3/t, and the absolute gas emission 34.29 m^3/min, C20 main seam is coal and gas outburst whose gas content is high and permeability is poor. C20 is chosen as the first coal seam and the protective layer at the same time. The technology of cutting through coal seam strike and intercepting longhole drilling is adopted. The drilling is type of ZDY-3200 m, with 73 mm drill pipe and 120 mm hole diameter. 50 mm PVC pipe with length of 12 m and sealing length of 8 m are reserved in the hole.

2.2 System Design

The structure diagram of coal seam gas drainage system is shown in Fig. 1. In order to achieve informatization of coal mine gas drainage, gas concentration and pressure, flow, temperature and other parameters of pipe network should be collected in time to gas drainage monitoring system. The system consists of a central node and several sensor acquisition nodes. Each node includes gas flow, concentration, negative pressure and other detection devices. Each gas drainage branch pipe is installed with a sensor acquisition node. Only one central node is needed for a gas drainage face not more than 500 m in length. The monitoring for drainage branch pipe is achieved by wireless sensor technology. Fuzzy control [14] is introduced into the control method. Each branch pipeline is equipped with a controller which collects data of gas concentration, flow rate and pressure periodically and operates valve for the negative pressure of pipe to gas concentration.

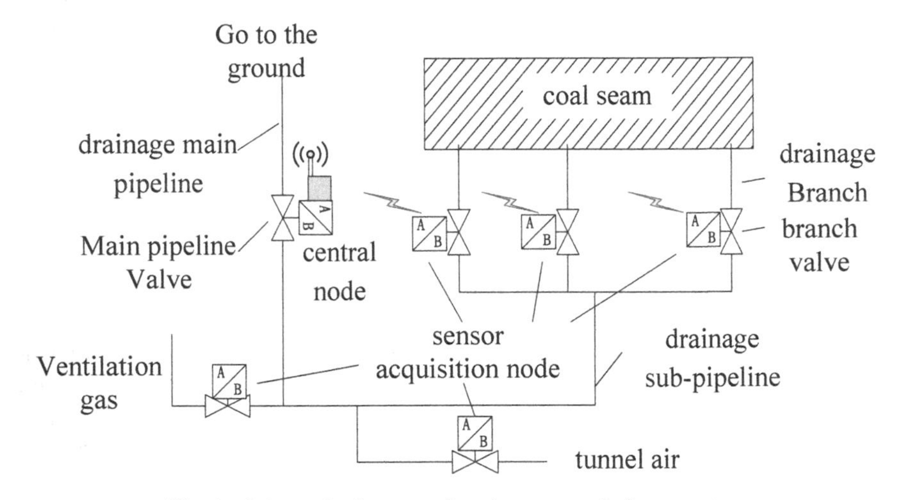

Fig. 1. Schematic diagram of coal seam gas drainage system

The function of sensor acquisition node is to collect gas drainage pipeline parameters, and transmit the data to the central node by wireless way. It is mainly composed of microcontroller, concentration sensor, flow sensor, pressure sensor, temperature sensor, micro electric valve, wireless communication circuit, etc.

Due to characteristics of the randomness, high nonlinearity and strong coupling between variables, it is unrealistic to establish an accurate mathematical model of the pumping system. It is difficult to achieve satisfactory control accuracy by traditional control algorithm. Therefore, the fuzzy control strategy is adopted for pipeline concentration control.

The node based on wireless sensor network can be combined with the inherent monitoring system. The node can be arranged according to the measuring position. The way of wireless technology networking and data transmission is fit for the complex and turbulent environment of coal mine. can avoid the influence of wired equipment wiring and pipeline movement, and facilitate the installation of equipment for central node and sensor acquisition node with the change of coal mining position. The adverse factors caused by wired equipment wiring and device movement with pipeline is avoided. It is convenient to install the node equipment as pipeline moves.

The sensor acquisition node is installed at the monitoring position of the drainage pipeline, and the gas concentration, flow rate, pressure and temperature are collected and converted by node. The intelligent control of gas concentration in the pipeline is coded by fuzzy control. The fuzzy control algorithm is designed according to staff experience so that the gas concentration is within the standard production range and the intelligent auto control of gas concentration is achieved. The acquisition node receives the command from the central node and sends the collected pipeline gas information back to the central node through wireless communication.

The central node transmits the data to the remote monitoring substation by bus communication to realize the monitoring of gas drainage pipeline parameters. The central node relays the command to the acquisition node through wireless communication, and the acquisition node transmits the collected information to the central node in order. The central node will analyze the information received, fit the change data of gas flow, predict the content and distribution of residual gas, and calculate the drainage cycle for engineering reference. The central node is mainly composed of microcontroller, wireless communication circuit, power conversion circuit and display.

When the level of gas in the main pipe is low, the main drainage pipe would be closed and the ventilation gas and roadway air valve is opened.

3 Design of Fuzzy Controller for Gas Concentration

The sensor acquisition node obtains the gas concentration of the branch pipe, and calculates the reasonable control signal through the fuzzy rule judgment. The signal is intended to opening of the drainage branch valve so as to adjust the gas drainage flow and to ensure the minimum gas concentration of the branch pipe.

The fuzzy controller is a kind of double input and single output controller. The deviation of gas concentration E and the change rate of gas concentration deviation E_c are selected as the input of fuzzy controller so called two-dimensional fuzzy controller. The regulating value U of micro electric valve is the output. Two inputs can reflect the dynamic characteristics of the process, and the control effect is much better than one-dimensional controller. The E_c is chosen to predict the trend of controlled variables which is equivalent to the derivative of PID control which is set as input in general large inertia system, such as boiler temperature control system.

The gas concentration value E_k in the branch pipe is the controlled quantity of system, E_0 is the minimum value, and E is the concentration deviation, $E = E_k - E_0$, where k for current time. $E_c = E_{k-1} - E_k$ is the change rate of concentration deviation, k − 1 for the previous moment. The principle is sectional multilevel negative pressure control method. Specifically, it is 4-section and 3-level. In the early stage of pumping, there is a reasonable negative pressure P_1kPa. As time goes on, the reasonable negative pressure will gradually decrease. The reasonable value of negative pressure is P_nkPa to later stage. That value is determined by on-site technicians staff. If the extraction concentration is too low, even less than 10%, the valve will be closed directly.

The pseudo code of concentration fuzzy control algorithm is as follows.

```
count_low=0
Do While time_delay=0
readSensordata()
LoraCommunication()
time_delay←5 //timer delay
if E_k < E_0 AND count_low=0
then P_1←P_1-15 count_low←count_low+1 time_delay←20
if E_k < E_0 AND count_low=1
then P_1←P_1-10 count_low←count_low+1 time_delay←15
if E_k < E_0 AND count_low=2
then P_1←P_1-5 count_low←count_low+1 time_delay←10
if E_k < E_0 AND count_low=3
then shut_vavle()
if E_c > E_cmax then alarm()
else close_alarm()
valve_adjust( P_1)
EndWhile
```

The drainage concentration of drilling branch can be expressed as follows (3 * 10 for example)

$$P_{3\times 10} = \begin{bmatrix} \rho_{1,1} & \rho_{1,2} & \cdots & \rho_{1,10} \\ \rho_{2,1} & \rho_{2,2} & \cdots & \rho_{2,10} \\ \rho_{3,1} & \rho_{m,n} & \cdots & \rho_{3,10} \end{bmatrix} \tag{1}$$

$P_{3\times 10}$ is the extraction volume concentration matrix, $\rho_{m,n}$ is the row of m and column of n extraction volume concentration.

The flow can be expressed as

$$F_{3\times 10} = \begin{bmatrix} f_{1,1} & f_{1,2} & \cdots & f_{1,10} \\ f_{2,1} & f_{2,2} & \cdots & f_{2,10} \\ f_{3,1} & f_{m,n} & \cdots & f_{3,10} \end{bmatrix} \tag{2}$$

$F_{3\times 10}$ is the pumping flow matrix, and $f_{m,n}$ is the row of m and column of n drainage flow.

The pure quantity of gas drainage is the evaluation basis to investigate whether the gas drainage and drainage pipe network are efficient. The larger the pure flow is, the more gas content in the target area can be reduced as much as possible. Therefore, the gas drainage volume is the most important evaluation standard for the drainage effect of drainage pipe network. Based on the measured gas flow f and gas drainage concentration

ρ, the pure gas drainage flow PF can be expressed as $PF = \rho \times f$. The pure flow of gas drainage system is

$$PF_{3\times10} = \begin{bmatrix} \rho_{1,1} \times f_{1,1} & \rho_{1,2} \times f_{1,2} & \cdots & \rho_{1,10} \times f_{1,10} \\ \rho_{2,1} \times f_{2,1} & \rho_{2,2} \times f_{2,2} & \cdots & \rho_{2,10} \times f_{2,10} \\ \rho_{3,1} \times f_{3,1} & \rho_{m,n} \times f_{m,n} & \cdots & \rho_{3,10} \times f_{3,10} \end{bmatrix} \quad (3)$$

where: $PF_{3\times10}$ is the pure flow matrix of gas drainage. for example, the branch pipe in the first row and the first column is $PF_{1,1} = \rho_{11} \times f_{1,1}$. With the matrix form storage mode it is convenient for the central node to carry out three-dimensional portrait of the whole coal seam gas content, gas concentration and change trend.

4 Estimation of Gas Pre Drainage Time

It shows that the quantity of gas pre drainage every day tended to retrogress, which indicates that pre drainage time can not be extended indefinitely, and there should be a reasonable time [15]. Reasonable time is not a constant but a variable affected by many factors. Therefore, reasonable pre drainage time is different for the same coal seam, while is same for different coal seam. At present, the conventional method to predict the reasonable pre drainage time is mainly through drilling in the underground to determine the initial gas drainage amount and attenuation coefficient of the gas drainage volume, which are determined as important parameters. The precondition of this method is that the pumping quantity is exponential attenuation.

Another attenuation characteristics different from exponential attenuation is exhibited in different coal mine drainage data. Therefore, combined with the actual data characteristics, we presented a new method to determine the reasonable pre drainage time of coal seam gas.

The electromagnetic environment of coal mine is complex so sensor will be disturbed or fault, its data will be missing or abnormal. Kalman filter is taken at the central point to refine the gas flow and concentration data of the main pipe.

The process of coal seam gas drainage, can be approximately equivalent to that of water tank drainage. The coal seam is equivalent to water tank, and gas to water in a tank, all branch pipes to a pump, the residual gas content to the water volume, the drainage flow of water pump to the flow of gas drainage. The relationship between water pump drainage flow and time is very close to that of gas drainage flow and time as shows in Fig. 2.

If it is the characteristic of curve I, the initial flow rate is small, and gradually increases to the maximum value with the extension of time, and then decays. With the continuous increase of drainage time, the gas flow in the branch pipeline decreases. The rising speed is different in many mine, which may be a characteristic of low permeability outburst coal seam [16]. The measured curves in references [17] and [18] also show the characteristics of step response of underdamped second-order systems. If it is the characteristic of curve II, the decreasing rate of gas flow with time decreases continuously, and the curve shows a hyperbolic trend,

If the flow curve is close to curve I, it can be expressed by power function formula. If curve II then the step response formula of the second-order system. If the dotted line

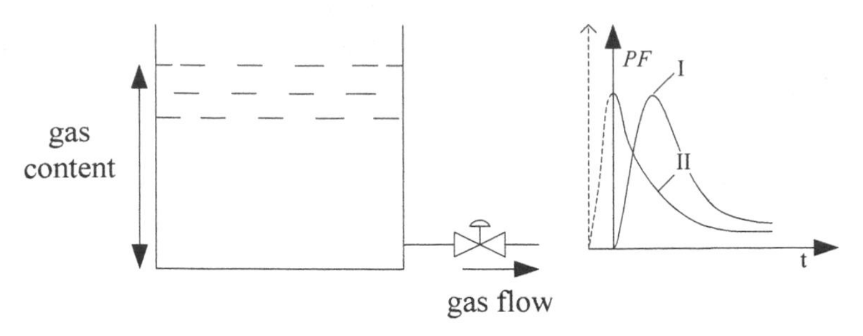

Fig. 2. Schematic diagram of the water tank pump drainage process

as shown in the Fig. 2 is added to curve I, a step response curve of the second-order system can still be obtained by translating the coordinate axis. Therefore, the model of coal seam gas drainage can be represented by a second-order system. The relationship between pumping flow and time can be approximately expressed by the step response curve of the second-order system.

The mathematical model of the second order system in Laplace transform is

$$G(s) = \frac{\omega_n^2}{s^2 + 2\zeta\omega_n s + \omega_n^2} \tag{4}$$

The model parameters ζ and ω_n are related to the gas flow of coal seam, namely both of them have a certain functional relationship with coal seam gas content. The input is step signal of $R(s) = R_0/s$, R_0 the signal amplitude. The step response expression of the second-order underdamped system is

$$c(t) = L^{-1}[R(s)G(s)] = 1 - e^{-\zeta\omega_n t}\sin(\omega_d t + \beta)/(\sqrt{1-\zeta^2}) \tag{5}$$

where L^{-1} for inverse Laplace transform.

The overshoot is

$$\sigma\% == e^{-\frac{\xi\pi}{\sqrt{1-\xi^2}}} \times 100\% \tag{6}$$

peak time is

$$t_p = \pi/\omega_d = \pi/(\omega_n\sqrt{1-\xi^2}) \tag{7}$$

The overshoot $\sigma\%$ is obtained by maximum value of flow $PF_{m\times n}$ which is easy to get, and the coefficient ξ can be obtained by reverse calculation. The coefficient ω_n can be calculated according to the corresponding peak time t_p. When the flow $PF_{m\times n}$ decreases to Δ, the settle time can be obtained according to the step response curve. That settle time is the estimated drainage time. For example, when the steady state value is 1, the steady-state error $\Delta = 0.05$, $t_s = 3/\xi\omega_n$, the unit is day. The estimated of drainage time is not fixed, but is obtained according to the calculation dynamic of maximum gas valuc and steady state value. It is easy to get the maximum value through the data, while the steady state value is get under condition of the variation of gas flow $PF_{m\times n}$

within a certain range. Only when the residual gas content and drainage rate meet the requirements can the estimated of drainage time be matter.

The concrete steps to determine the reasonable pre drainage time of coal seam gas are as follows:

1) Kalman filter to the data.
2) calculate the maximum gas flow value and corresponding peak time.
3) calculate the initial gas content of coal seam.
4) calculate the residual gas content and extraction rate.
5) according to the steady state flow value, calculate the drainage time.

The average gas data acquisition interval is 1 time/5 min. There are 25920 data in three months. The first 50 days are the initial training data, the next 20 days an average sliding window period, 10 days training, another 10 days forecast and adjustment.

Considering of the geological conditions and human factors, the actual reasonable pre drainage time should be slightly larger than the theoretical calculation results.

5 Calculation of Residual Gas Content

In order to obtain the residual gas content after a period of drainage, the original coal seam gas content is required first, and the drainage amount of roadway is also considered.

$$Q = Q_T + Q_c + Q_v \tag{8}$$

where Q for original gas content of coal, m^3/t, Q_T for total amount of gas drainage per ton of coal, m^3/t. Q_c for residual gas content of coal, m^3/t. Q_v ventilation air gas, m^3/t.

The residual gas content after gas drainage is calculated according to the following formula.

$$Q_c = Q - \frac{pf_t + pf_v}{kG} \tag{9}$$

where pf_t for the amount of gas drainage, $pf_t = \rho_{main} \cdot f_{main} \cdot D_{main}$, pf_v for the amount of ventilation gas $pf_v = \rho_V \cdot f_V \cdot D_V$, v for ventilation, k for test the influence coefficient of unit gas pre drainage, k = 1.2, G for the test unit to participate in the calculation of coal reserves, t, $G = L_1 L_2 m_0 \gamma$, L1 is the length of pre pumping drilling of inspection unit, m; L2 for the width of inspection unit, m; m0 for the average thickness of coal seam in the test unit, m; γ for the bulk density (pseudo density) of coal, t/m^3, ρ for the average gas concentration and the amount of gas discharged by air, f for the air volume, m^3/min, D for the running time, d.

The gas content at different levels of C19 is calculated by indirect method.

6 Conclusion and Prospect

With the increase of mining depth, the coal seam gas drainage technology will play a more important role in coal mine gas control. Combined the theoretical research of gas flow with the practical engineering application, this paper points out some problems existing in the gas drainage of coal seam, analyzes and summarizes the existing problems. The main conclusions are as follows:

A kind of monitoring system based on wireless communication for coal seam gas drainage pipeline is designed. Through sensors, microcontrollers, micro electric valves and wireless communication modules, the gas concentration in the branch pipe is under the control which is beneficial to ensure the concentration level of gas drainage.

In this paper, the fuzzy control is applied to the monitoring system of the gas drainage branch pipe, and the sectional multilevel method is presented which improves the automation level of the gas drainage system.

In this paper, the coal seam drainage model is regarded as a second-order system. By analyzing the characteristics of the gas drainage data curve, it is concluded that the second-order system step response curve is a good choice to fit the data, and a new method for predicting the drainage time is proposed.

Gas drainage involves interdisciplinary fields, and the research to be solved in the future includes the following aspects: (1) to predict model of coal seam gas drainage based on deep learning, which provides theoretical basis for intelligent drainage and prediction of residual gas in coal seam. (2) to develop advanced sensing technology for key parameters of pumping branch pipe and main pipe, to develop intelligent pumping equipment based on industrial IOT.

Acknowledgments. The authors gratefully acknowledge the Key Technologies Research and Development Program of Shandong China (2019GGX101011), Tai'an Science and Technology development program (2018GX0039).

References

1. Zhu, Z., Yin, Z.: Present situation and prospect of exploitation and utilization of coalbed methane (coalbed gas) in the Sichuan province. China Min. Mag. **10**, 89–91 (2007)
2. Sun, D.: Promote informatization and intelligence of the whole process of gas prevention. China Coal News **5**(09), 3 (2020)
3. Wu, K., Shi, S.: Discussion on problems and countermeasures of gas drainage in hunan coal mine. Min. Eng. Res. **33**(03), 31–34 (2018)
4. Jianghong, Z.: Deep-hole pre-splitting blasting technology for high gassy deep complex geology area and low permeability coal seam. Saf. Coal Mines **44**(10), 69–71 (2013)
5. Cai, F., Liu, Z., Zhang, C., et al.: Numerical simulation of improving permeability by deep-hole presplitting explosion in loose-soft and low permeability coal seam. J. China Coal Soc. **32**(5), 499–503 (2007)
6. Jiang, Z., Tao, Y., Wang, F.: Research and application on automatic adjusting control system of concentration of gas extraction. Coal Technol. **10**, 34–36 (2014)
7. Wang, Z.: Study on gas extraction joint tube system and method of concentration control. Henan Polytechnic University (2009)
8. Jiang, Y.: Intelligent control and optimization of mine gas mobile pump-exhaust system. Anhui University of Science and Technology (2006)
9. Huang, K.: Mobile coalmine methane pumping-exhaust system based on online interpolation fuzzy control. Anhui University of Science and Technology (2008)

10. An, S., Zhang, H., Li, H., et al.: Design and application of automatic adjusting control system for self-adaptive gas drainage pipeline valve. Coal Eng. **48**(3), 24–26 (2016)
11. Zhou, F., Liu, C., Xia, T., et al.: Intelligent gas extraction and control strategy in coalmine. J. China Coal Soc. **44**(8), 2377–2387 (2019)
12. Wu, Y., Wu, C.: Design of mining gas drainage system based on wireless sensor network. Coal Mine Electromech. **3**, 67–70 (2015)
13. Huang, R., Zhang, Z., Bo, C.: Study on control mechanism and method of gas concentration in gas drainage borehole of underground mine. Coal Sci. Technol. **45**(05), 128–135 (2017)
14. Liu, J.: Intelligent Control. Electronic Industry Press, Beijing (2005)
15. Li, G.: Discussion on determination method of coal seam gas rational pre-drainage period. Saf. Coal Mines **43**(04), 118–120 (2012)
16. Yu, B.: Coal mine gas disaster prevention and utilization technical manual. Coal Industry Publishing House (2005)
17. Wang, L.: Study on characteristics of gas occurrence and distribution of residual gas after drainage in Hudi mine. China University of Mining and Technology (2018)
18. Ming, D.: Study on prediction model of residual gas content in protected coal seam. Shandong University of Science and Technology (2017)

Improvement of Online Education Based on A3C Reinforcement Learning Edge Cache

Haichao Wang, Tingting Hou, and Jianji Ren(✉)

Henan Polytechnic University, Jiaozuo 454000, Henan, China
renjianji@hpu.edu.cn

Abstract. Online education is the complement and extension of campus education. Aiming at the situation that the mainstream network speed in online education scenarios is difficult to meet the requirements for smooth video playback, such as ultra-high-definition video resources, live broadcast, etc., this paper proposes an A3C-based online education resource caching mechanism. This mechanism uses A3C reinforcement learning-based edge caching to cache video content, which can meet the requirements of smooth video playback while reducing bandwidth consumption and improving network throughput. We use the Asynchronous Advantage Actor-Critic (A3C) technology of asynchronous advantage actors as a caching agent for network access. The agent can learn based on the content requested by the user and make a cache replacement decision. As the number of content requests increases, the hit rate of the cache agent gradually increases, and the training loss gradually decreases. The experimental comparison of LRU, LFU, and RND shows that this scheme can improve the cache hit rate.

Keywords: Edge cache · Reinforcement learning · Online education

1 Introduction

Online education, as a supplement and extension of traditional classroom education, especially during the COVID-19 epidemic, overcomes the limitations and disadvantages of traditional classroom education and can optimize the allocation of educational resources and improve the quality of teaching to a certain extent. However, with the popularity of online education and mobile Internet access devices, mobile devices (such as smartphones and tablets) are increasingly requesting multimedia services (such as audio, high-definition video, photos, etc.). The content provider's network traffic load has increased rapidly, and the quality of service (QoS) of users has decreased [1].

Online education has the following advantages: small environmental limitations, low teaching costs, and can alleviate the problem of unbalanced teaching resources. However, online education has high requirements for network infrastructure, so the biggest problem it faces is network latency. When users use

X. Wang et al. (Eds.): 6GN 2020, LNICST 337, pp. 240–249, 2020.
https://doi.org/10.1007/978-3-030-63941-9_18

online education and learning, if the network speed can not keep up, there will be a delay stuck; when transmitting data and lost packets, there will also be a picture and sound "stuck". Therefore, connectivity and networking must become the top priority for online education. However, this is a big challenge for schools with many distributed network resources.

In online education, some users will have more demand for online courses. These users may be located in more concentrated areas (urban primary and secondary schools) or multiple areas (university online education). For users in the distribution area, how to reduce the propagation flow through local short-distance communication and effectively reduce the duplicate backhaul traffic has become a hot research topic.

The emergence of 5G and edge cache can alleviate the dilemma faced by online education, which reduces latency by storing data locally, thereby obtaining a continuous and better user experience. Edge computing can disperse computing resources and move them closer to data sources. When schools use edge computing, they will prioritize connections and networks across multiple campuses to eliminate the slow speed, thereby significantly improving the online education experience for students and teachers. Therefore, it is the general trend to apply edge caching to online education.

Edge caching is a promising technology that can increase network throughput, improve energy efficiency, reduce service latency, and reduce network traffic congestion by caching cache nodes on the edge of a wireless network. Cache nodes can be placed on base stations and home gateways so that the content requested by the user is closer to the user, rather than being downloaded repeatedly from the content service provider over the network.

Therefore, it is very interesting and necessary for some effective prefetching strategies to place educational resources with high-frequency requests on edge cache nodes, thereby reducing resource access latency for end-users. We study the caching and replacement decision problems in BSs, model the content caching of edge nodes and user requests as Markov decision problems, and deploy a learning framework based on A3C for content caching of BSs. Based on the results of simulation experiments, compared with the existing LRU (least recently used), LFU (least frequently used), and RND (random number deletion) hit rates, there is a certain improvement.

The rest of this article is organized as follows: In the second part, we introduce related work; in the third part, we propose problems and algorithms; We conduct simulation experiments and give the experimental results in the fourth part; the last part is our research conclusions and future research challenges.

2 Related Work

In recent years, online education resources have grown rapidly, and the network traffic for educational applications has also increased tremendously, which has placed a heavy burden on the core network, which has ultimately led to long delays in resource access and user experience. Edge caching can help solve this problem by fetc.hing and caching new content.

In [6], a coordinated mobile edge network that supports caching was discussed. This network structure mainly reduces the load on the front-end link, while achieving low-latency and high-rate content delivery.

Recently, researchers have used artificial intelligence (AI) technology combined with edge caching to better understand user behavior and network characteristics. In particular, neural networks placed in edge nodes learn to predict popular content and make caching decisions. For example, in [2], collaborative filtering of stack-type automatic coding, recurrent neural network, deep neural network, and recurrent neural network [3] is proposed to predict the popularity of content.

The researches in [4] and [5] show that reinforcement learning RL has great application potential in the content caching scheme design. The author of [4] proposed a cache replacement strategy based on Q-learning to reduce the traffic load in the network and used a multi-armed robber MAB to place the cache.

Aiming at complex control problems, [7] proposed a collaborative deep cache framework for mobile networks based on dual deep q networks. The framework aims to minimize content acquisition delays and does not require any prior knowledge. [8] proposed a collaborative edge cache architecture based on 5G networks, which uses mobile edge computing resources to enhance edge cache capabilities, and proposes vehicle-assisted edge caches in combination with smart car cache resources. [9] use Q-learning to design the caching mechanism, reduce the backhaul traffic load and transmission delay from the remote cloud, and propose an action selection strategy for the caching problem. Find the right cache state through reinforcement learning.

Although the replay of the experience used by the DQN algorithm can be said to solve the problem of reinforcement learning satisfying the independent and identical distribution, it uses more resources and the calculation of each interaction process, and it needs an off-policy learning algorithm to update the data generated by the old strategy, [10] put forward the concept of "executively executing multiple agents in parallel on multiple environment instances". Its advantage is that unlike traditional methods using GPUs or large-scale distributed computing, It is running on a single multi-core CPU, which is the same task. The asynchronous A3C algorithm has achieved the same or even superior results, and the cost is a shorter training time. Inspired by this, we will use A3C technology in edge caching.

3 Edge Cache Model

In the future 5G network architecture, edge nodes (EN) will become an indispensable part. To meet the low-latency and high-traffic network services in online education scenarios, this study proposes an education resource pre-cache based on the combination of EN Caching and A3C technology.

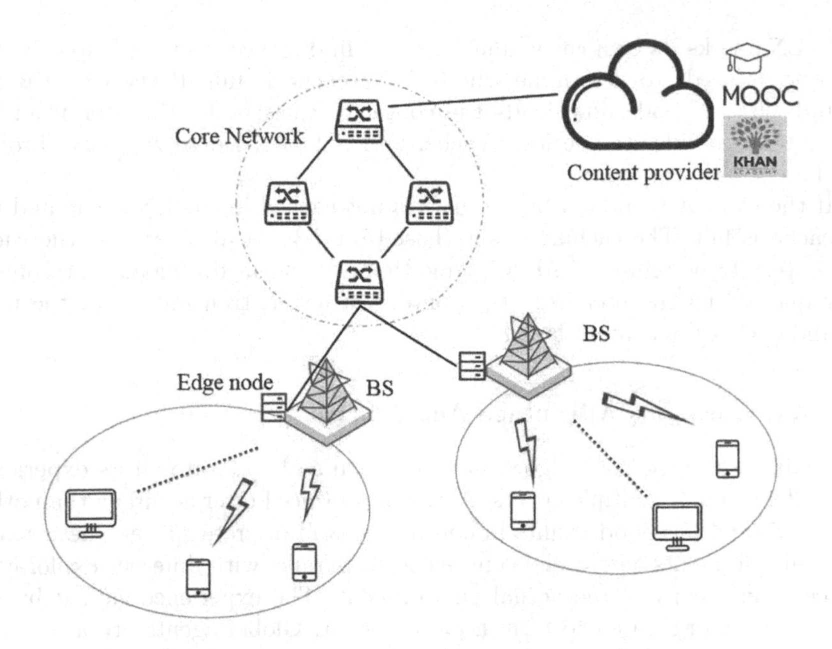

Fig. 1. Edge cache architecture.

3.1 Problem Formulation

The model building is illustrated in Fig. 1. In EN, there is an A3C-based caching agent. Because the agent involves a series of interactions between the agent and the environment during the edge caching process, the edge node caching issues can be modeled as a Markov decision process. The Markov process can be represented by five key elements:

A group of states S that the agent can really be in; a set of actions A performed by the agent when it moves from one state to another; transition probability, the execution of a certain behavior a, and transition from a state s, probability of going to another state s'; reward, the agent performs a certain behavior a, and transitions from one state s to another state s' to get reward probability; discount factor controls the importance of reward and future reward.

Edge caching is to strike a balance between broadband transmission costs and storage costs. Therefore, caching of popular content is very important. In general, the distribution of popular content is represented by the Zipf function distribution, that is, most users often request specific files that account for a small proportion of the requested content, while most files are rarely requested.

The following is the process by which the agent makes the cache decision:

(1) When a user's content request reaches EN, the EN first checks whether there is content requested by the user in the cache, and if so, returns the content directly to the user.

(2) If EN checks its own cache and does not find the content requested by the user, it needs to determine whether the cache is full. If the cache is not full, the EN node downloads the content requested by the user from the content provider to the local cache, and is forwarded to the user through the network.
(3) If the content requested by the user is not cached in the EN cache and the cache is full. The caching agent (based on A3C) will determine the cache content to be removed. At the same time, the agent downloads the content requested by the user from the content provider, transmits it to the user, and caches the content locally.

3.2 Asynchronous Advantage Actor-Critic

The main idea of the A3C algorithm is to learn and integrate all its experience in parallel through multiple agents. A3C can produce better accuracy than other algorithms and has good results in continuous and discrete space. The network uses multiple agents, and each agent learns in parallel with different exploration strategies in a copy of the actual environment. The experience gained by the agents is then integrated to form a global agent. Global agents are also called core networks or global networks, while other agents are called workers.

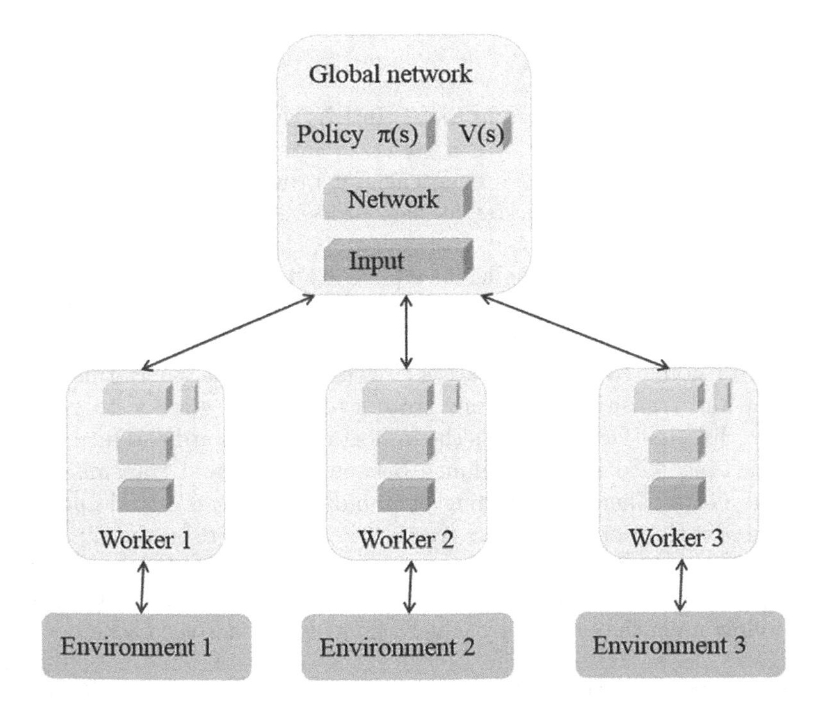

Fig. 2. A3C agent structure.

In A3C. The first 'A' is asynchronous. Multiple agents interact with the environment, a copy of the environment needs to be provided for each agent, and the states between the agents are asynchronous. The second 'A' is the advantage. The dominant function can be defined as the difference between the Q function and the value function. The Q function determines how well the behavior is in a certain state. The value function determines how good the state is the difference between Q function and value function. It means that the agent performs the behavior in the state s. The third 'A' is an actor-critic. There are two types of A3C neural network architecture, actors and critics. The role of the actor is to learn a strategy. The role of a critic is to evaluate the actor's learning strategy.

There are multiple worker agents in Fig. 2. Each agent interacts with its own copy of the environment. Then the worker agent learns the strategy and calculates the gradient of the strategy loss. And update the gradient in the global network. This global network is updated by each agent. One of the advantages of A3C is that it does not use experience playback memory. Because there are multiple agents interacting with the environment and integrating their respective information into the global network. Information between agents. There is little or no correlation. Since A3C does not require memory, it can greatly reduce storage space and calculation time.

3.3 Formula Derivation

The agent tries to maximize the cumulative rewards it receives from the environment. The total amount of rewards an agent receives from the environment is called the reward. The reward function R_t is calculated as follows:

$$R_t = t_{t+1} + r_{t+2} + \cdots + r_t = \sum_{k=0}^{\infty} \gamma^k r_{t+k+1} \tag{1}$$

The policy function maps state to behavior as π. This represents a mapping from state to behavior. In essence, the policy function indicates what behavior is performed in each state. The ultimate goal is to find the optimal strategy for each state to specify the correct behavior, thereby maximizing the reward. The strategy function π can be expressed as:

$$\pi(s) : S- > A \tag{2}$$

State value function referred to as the value function. Determine the optimal degree of an agent in a certain state under the strategy. It is usually recorded as $V(s)$, which represents the value of the state after the policy is executed. The value function $V(s)$ is defined as follows:

$$V^{\pi}(s) = E_{\pi}[R_t | s_t = s] \tag{3}$$

Substituting formula (1) into the above R_t value, we get

$$V^{\pi}(s) = E_{\pi}[\sum_{k=0}^{\infty} r_{t+k+1}|s_t = s] \tag{4}$$

State behavior value function. Also called the Q function. Used to indicate that the agent adheres to the policy π. The optimal degree of specific actions performed in a state. The Q function is denoted as $Q(s)$, which represents the value of the behavior taken in a certain state by following the policy π. The Q function is defined as follows:

$$Q^{\pi}(s,a) = E_{\pi}[s_t = s, a_t = a] \tag{5}$$

The difference between the value function and the Q function is that the value function is the best way to determine the state. The Q function determines the optimal degree of behavior in a certain state.

Algorithm 1. A3C-based edge cache algorithm

1: **Initialization parameters:** $T = 0, \theta, \theta', \theta_v, \theta'_v$;
2: Repeat if there is a user content request;
3: $d\theta = 0; d\theta_v = 0; \theta' = \theta; \theta'_v = \theta_v; t_{start} = t; s_t$;
4: **Repeat**;
5: Perform according to policy $\pi(a_t|s_t;\theta^t)$;
6: Receive reward r_t and new state s_{t+1} ;
7: $t = t + 1; T = T + 1$;
8: until s_t or $t - t_{start} == t_{max}$;
9: $R = \begin{cases} 0 \\ V(s_t, \theta'_v) \end{cases}$;
10: **For** $i \in \{t-1, \ldots, t_{start}\}$;
11: $R = r_n + \gamma r_{n-1} + \gamma^2 r_{n-2}$;
12: $\theta' : d\theta = d\theta + \nabla_{\theta'} log\pi(a_i|s_i;\theta')(R - V(s_i;\theta'_v))$;
13: $\theta'_v : d\theta_v = d\theta_v + \sigma(R - V(s_i;\theta'_v))^2/\sigma\theta'_v$;
14: **End for**;
15: update θ using $d\theta$ and θ_v using $d\theta_v$;
16: **Until** $T > T_{max}$.

The A3C workflow first resets the worker agent to the global network. Then start interacting with the environment. Each worker agent learns an optimal strategy according to different exploration strategies. Next, calculate the value and strategy loss. Then calculate the loss gradient. And update the gradient in the global network. The worker agent restarts resetting the global network and repeats the counting process. The dominant function $A(s,a)$ is the difference between the Q function and the value function:

$$A(s,a) = Q(s,a) - V(s) \tag{6}$$

According to the above formula, we can get the value loss, which is the mean square difference between the discount return and the state value:

$$L_v = \sum(R - V(s)^2) \tag{7}$$

According to the value loss, the agent can train the A3C neural network, As the training increases, the value loss will gradually decrease, and the agent's reward will continue to increase.

By deploying servers with caching and computing capabilities at the edge of the network, you can solve low latency problems and ease the burden of high traffic in the cloud. Specifically, with edge computing, the edge cache can pre-cache some popular content during off-peak hours. Therefore, during peak hours, edge caching can reduce transmission delay and front-end transmission burden, and can also reduce traffic, thereby achieving the goal of improving the network environment.

In this article, we only focus on the caching strategy, which is to determine the content and timing of caching or updating the cached content, which is critical to the overall performance of the cache-enabled communication system. In this article, we choose the A3C agent with the decision-making ability to cache. A3C agent continuously improves the performance of the network experience through continuous interaction and learning with the environment at the network.

4 Experiments and Results

With the rapid development of edge computing and communication technology, traditional education is changing. Through the application of new technologies, the utilization rate of existing campus resources can be improved, and the work efficiency of students and teachers can be improved. Popular teaching content is cached on edge servers, such as communities, teaching buildings, etc. This way, the network environment, and user quality are improved.

Simulation experiments verify the effectiveness of the A3C edge cache method. This article uses the simulation experiment method to simulate the edge cache node by instantiation. For A3C agents, the actor and critic learning rate is 0.001. The experiment uses CentOS7, the processor is E5-2650, and the hard disk size is 480G SSD + 4T. Enterprise hard disk, 32G memory, code interpreter is Python, version 3.6, code dependent libraries include Tensorflow, Numpy, etc. User requests in experimental data are randomly generated by calling the Zipf function in Numpy.

The loss curve of the agent is shown in Fig. 3. The y-axis is the loss value of the edge node agent, and the x-axis is the edge node agent training cycle. 80% smoothing is done in Fig. 3 for demonstration. Every 200 times as a training cycle and record the training loss. In the first 100 cycles, the loss remained at a high level. On the one hand, the amount of training data was small, and on the other hand, the worker passed the neural network parameters to the global network for the update. After several updates, the global network was updated.

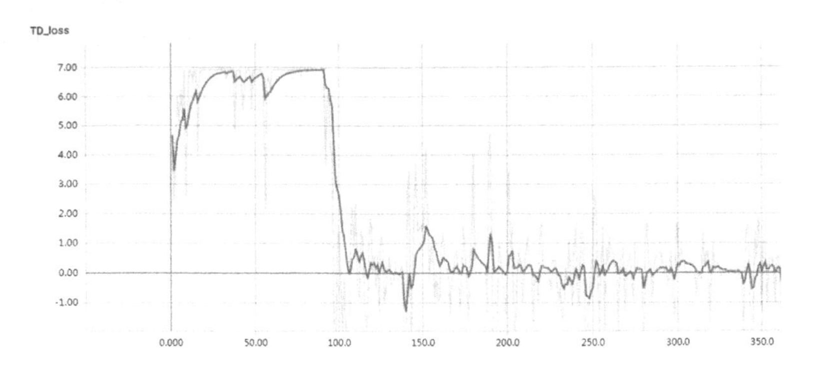

Fig. 3. A3C agent training loss.

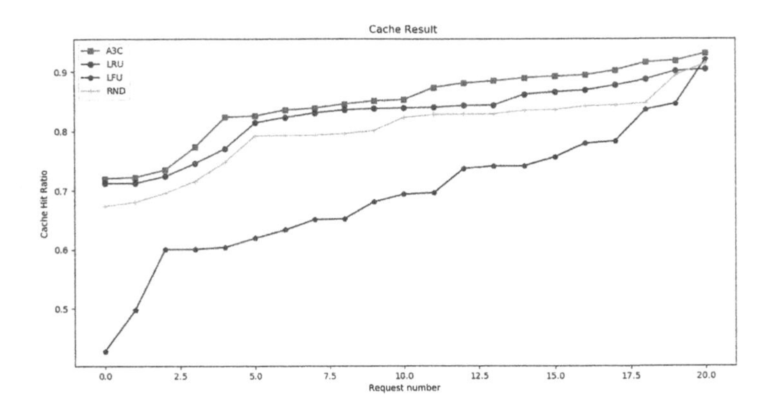

Fig. 4. A3C, LRU, LFU, RND hit ratio comparison.

Will reach a level of low loss. As the training progressed, the agent loss gradually decreased and reached around 0 after the training cycle reached 100, and then there was a large fluctuation. As the training continued, the fluctuation range became smaller and smaller, until the fluctuation range was less than plus or minus 1.

Figure 4 plots the comparative experimental results of A3C, RND, LRU, and LFU. The y-axis in the graph is the cache hit ratio, and the x-axis is the number of user requests (the unit is 1000). The cache hit rate we define is the number of cache hits divided by the sum of hits and misses. At first, due to the small number of user requests, the hit rates of the four caching strategies were very low. As user requests increase, training based on A3C caches also tends to stabilize. LFU, A3C performance is better. As the number of requests continues to increase, the cache hit rate of RND, LRU, and LFU also gradually increases, and the advantages of A3C also gradually weaken. Finally, when the number reaches the highest value, four cache methods reach similar levels. From the point of view of the caching process, a3c-based edge caching is stronger than LRU, LFU, and RND caching.

5 Conclusions and Future Work

In this paper, to alleviate the dilemma of online education facing high latency and stuttering, we propose an A3C-based edge cache online education architecture. By using resources placed in EN, content caching and hit ratio can be improved. Edge caching is an important way to improve the efficiency of 5G network content distribution. However, how to make more effective use of edge node caches in content storage and delivery still requires further exploration of computing and communication functions. Besides, the popularity of content will change with time, and how to design intelligent and adaptive caching mechanisms is a challenge in the future. We also need to further study how to motivate a large number of heterogeneous cache nodes to follow effective caching strategies, while ensuring content and network security.

References

1. Wang, X., Chen, M., Han, Z., et al.: TOSS: traffic offloading by social network service-based opportunistic sharing in mobile social networks. In: IEEE INFOCOM 2014-IEEE Conference on Computer Communications, pp. 2346–2354. IEEE (2014)
2. Chen, M., Saad, W., Yin, C., et al.: Echo state networks for proactive caching in cloud-based radio access networks with mobile users. IEEE Trans. Wirel. Commun. **16**(6), 3520–3535 (2017)
3. Devooght, R., Bersini, H.: Collaborative filtering with recurrent neural networks. arXiv preprint arXiv:1608.07400 (2016)
4. Gu, J., Wang, W., Huang, A., et al.: Distributed cache replacement for caching-enable base stations in cellular networks. In: IEEE International Conference on Communications (ICC), 2648–2653. IEEE (2014)
5. Blasco, P., Gündüz, D.: Learning-based optimization of cache content in a small cell base station. In: 2014 IEEE International Conference on Communications (ICC), pp. 1897–1903. IEEE (2014)
6. He, S., Huang, W., Wang, J., et al.: Cache-enabled coordinated mobile edge network: opportunities and challenges. arXiv preprint arXiv:1912.11626 (2019)
7. Li, D., Han, Y., Wang, C., et al.: Deep reinforcement learning for cooperative edge caching in future mobile networks. In: IEEE Wireless Communications and Networking Conference (WCNC), pp. 1-6. IEEE (2019)
8. Zhang, K., Leng, S., He, Y., et al.: Cooperative content caching in 5G networks with mobile edge computing. IEEE Wirel. Commun. **25**(3), 80–87 (2018)
9. Chien, W.C., Weng, H.Y., Lai, C.F.: Q-learning based collaborative cache allocation in mobile edge computing. Future Gener. Comput. Syst. **102**, 603–610 (2020)
10. Mnih, V., et al.: Asynchronous methods for deep reinforcement learning. In: Proceedings of The 33rd International Conference on Machine Learning, vol. 48, pp. 1928–1937, 19 June 2016

Wireless System and Platform

An Intelligent Elevator System Based on Low Power Wireless Networks

Congying Yu, Ruize Sun, Qijun Hong, Weiwen Chao, and Lei Ning(✉)

College of Big Data and Internet, Shenzhen Technology University, Shenzhen, China
ninglei@sztu.edu.cn

Abstract. The trend of Internet of Things (IoT) and wireless network techniques have resulted in a promising paradigm, and a more rational and intelligent elevator control system shall be considered. Many previous works are devoted to improving traffic congestion capability to increase time efficiency. However, these approaches confront the limitations of destination perception in advance or there is a steady stream of persons coming to wait the elevator that may increase the uncertainty of sensing the traffic load, since the user interface is still around elevator car. In this paper, an improved elevator system is proposed with remote calling and cloud scheduling based on low power wireless networks. It enables users to call the elevator remotely through portable devices, solves the problem of elevator invalid stop, reduces system energy consumption, and improves the service life of the elevator. It can match the running state of the elevator with the multi-user call request, shorten the time for users to take the elevator, and improve the comprehensive operation efficiency of the elevator.

Keywords: Elevator scheduling · Low-power wide-area network · Narrow Band Internet of Things · Elevator networking

1 Introduction

With the development of the information society, the connection objects of the network gradually expand from the connection between persons to the connection among people and things [1]. At the same time, the number of node connections in the network will also exceed 100 billion scale, and the information society has also moved from the Internet at the beginning of its birth to the era of IoT, which extends the way of network connection and data exchange, from consumer wearable devices to industrial production devices [3,5,14,17]. These terminals can sense environmental information, be controlled remotely, make decisions and take actions. Therefore, IoT technology plays an important role in the development of intelligent information society towards a higher level.

Sponsored by the Young Innovative Project from Guangdong Province of China (No. 2018KQNCX403) and the Teaching Reform Project from Shenzhen Technology University (No. 2018105101002).

X. Wang et al. (Eds.): 6GN 2020, LNICST 337, pp. 253–262, 2020.
https://doi.org/10.1007/978-3-030-63941-9_19

The expected explosive growth of IoT nodes depends on the evolving wireless communication technology, network infrastructure scale, terminal chip equipment and data and computing center. In recent years, in order to better support the network access of IoT nodes and meet the needs of IoT comprehensive application, IoT wireless access technology also presents the development trend of a hundred flowers and a hundred schools of thought. For example, in the field of authorized spectrum, based on the fourth generation cellular mobile communication (4G) technology, the enhanced Machine Type Communications (eMTC) supporting voice communication and Narrow Band Internet of things (NB-IoT) with low-power access technology and tiny packet data reporting service are developed [11,19]. In the field of unauthorized spectrum, the wireless access technologies such as WiFi, Bluetooth, ZigBee and Lora have also been widely deployed to meet the business requirements of different individuals and enterprises for node power consumption, coverage distance and transmission rate. However, because the Cellular Internet of things (CIoT) uses the authorized spectrum and is deployed by the operators on a large scale, it has better anti-interference and wide area coverage continuity, which can save the deployment cost of the user network, reduce the node access cost and guarantee the Quality of Service (QoS) [21].

Therefore, the trend of IoT and wireless network techniques have resulted in a promising paradigm, and a more rational and intelligent elevator control system could be considered [6,12,15]. By the end of 2016, there were more than 15 million elevators running in various buildings around the world, with billions of people taking elevators every day. Huawei expects 70% of the world's population to live in cities by 2050. In the next 20 years, there will be about 3 billion people entering the city [2,6,7,16]. After so many people enter the city, the limited space of the city will certainly promote the construction of high buildings, and the elevator will become more and more important. At present, China is probably the largest country in the world to carry out urbanization. At the beginning of 2015, a third-party report showed that in 2014, the proportion of new elevators in the global market in China was more than 68%, which means 68 of the new 100 elevators were installed in China. Obviously, it is necessary to deploy multi car elevator system with an intelligent scheduling to improve the transportation efficiency of large buildings with hyper dense people. However, due to the instantaneous increase of traffic load and the limitation of elevator car capacity, the elevator system is still faced with serious traffic congestion bottleneck in peak hours [5,14].

A lot of previous works are devoted to improve traffic congestion capability to increase time efficiency. In [4,9,20], it is an optimization of elevator scheduling that focuses on the stand-alone central controller for peak hours, while lacking the solution of sensing the external environment of elevator. In [10,18], multi-functional sensors were equipped at the floors for mainly detecting the passenger inside and outside the elevator car. However, these approaches confront the limitations of destination perception in advance. Therefore, proactively computing on the fine-grained traffic load information by the elevator system from each floor was proposed in [6,8]. This smart elevator system is to integrate traffic load by dynamically managing user interface and providing intelligent suggestions to guide passengers ride to other adjacent floors for time saving and physical health consideration. However, there is a steady stream of persons coming to wait the elevator that may increase the uncertainty of sensing the traffic load, since the user interface is still around elevator car.

In this paper, an improved elevator system is proposed with remote calling and cloud scheduling based on low power wireless networks. This work adopts remote call by the general portable equipment, which not only reduces the management cost, but also solves the problem of inaccurate identification of the number of incoming persons. The elevator dispatching center in the cloud receives the call data of users and the running state of the elevator at the same time, and judges the full load through multi factors. The design reduces the number of ineffective elevator door opening and closing, while improving the running efficiency of the elevator. The user call data and elevator data are presented on the web through the visual interface, which can give the maintenance personnel a more comprehensive and intuitive elevator operation health state, making the elevator maintenance become more active, rather than passive maintenance after failure or accident.

2 System Model

The novel elevator system is based on the layered concept of IoT including the perceptual recognition layer, network construction layer, cloud computing layer and application layer, which stands for smart elevators, the floor identification and related information transmission, the elevator scheduling center and the user interface, respectively.

In the layered concept of IoT, the perceptual recognition layer is a link between the physical world and the information world, so the smart elevators can report the running status and receive the instructions via wireless networks from the cloud. The network construction layer is a pipeline to exchange the information between the physical elevator and the remote control system. Meantime, with the support of high-performance computing and mass storage technology, the cloud computing layer organizes large-scale data efficiently and reliably, and provides intelligent scheduling for elevator applications.

The application layer is an integrated service provider and instructions collector. Thus, the unified interface of human-computer interaction is adopted in this paper, which is a cross platform architecture ranging from the web browser to the mobile mini program.

3 Design of Intelligent Elevator System

As it is shown in Fig. 2, users utilize the portable devices to detect the corresponding wireless identification signals of their particular floors, and send the messages including the present floor information, call requests and target

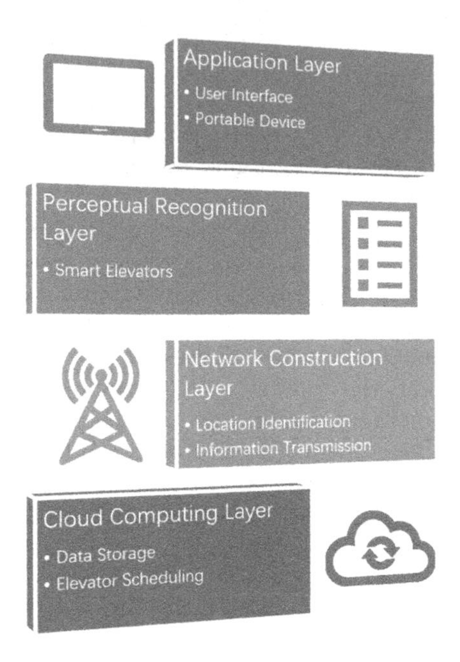

Fig. 1. The novel elevator system architecture based on the layered concept of IoT.

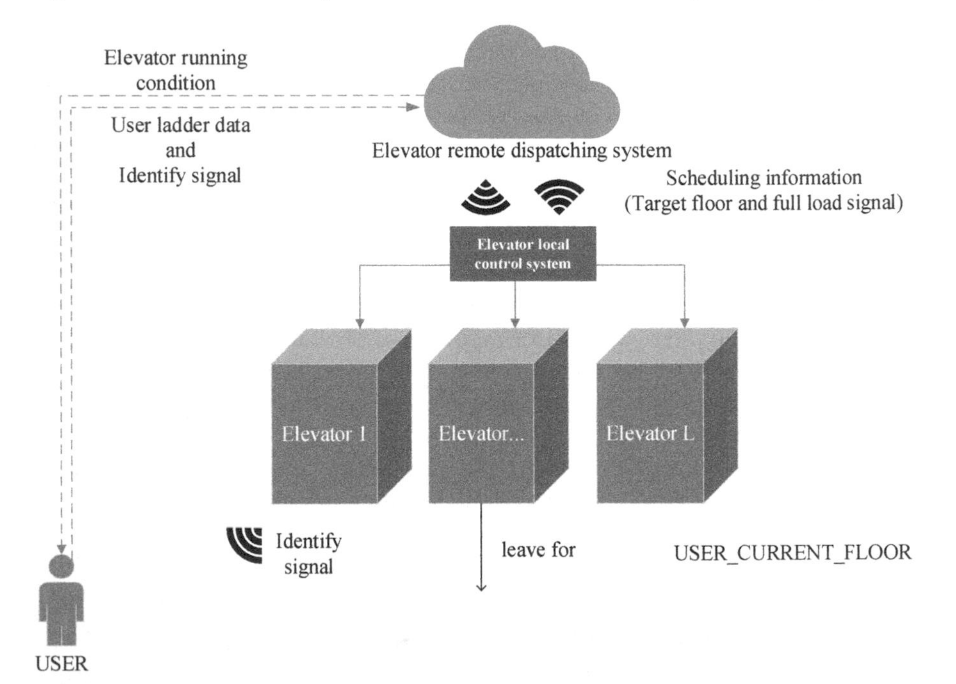

Fig. 2. The whole work flow of the intelligent elevator system.

floor information to the elevator remote dispatching system. Then the elevator remote dispatching system obtains the elevator stop oor information through the load evaluation and similarity matching method. In the meanwhile, the stop oor instruction is sent to the elevator local control system. The local control system controls the elevator to the corresponding oor according to the instructions (Fig. 1).

3.1 Application Layer

As it is shown in Fig. 3, users utilize a WeChat applet to realize the remote call to the elevator. After the WeChat applet, via Bluetooth, obtains the user' floor information and get connected with the server, the user could select the target floor on it. Then the WeChat applet sends the obtained data, namely the user's present and target floor information, to the cloud platform, where the user's request data and the running status of the elevator are monitored in the real time.

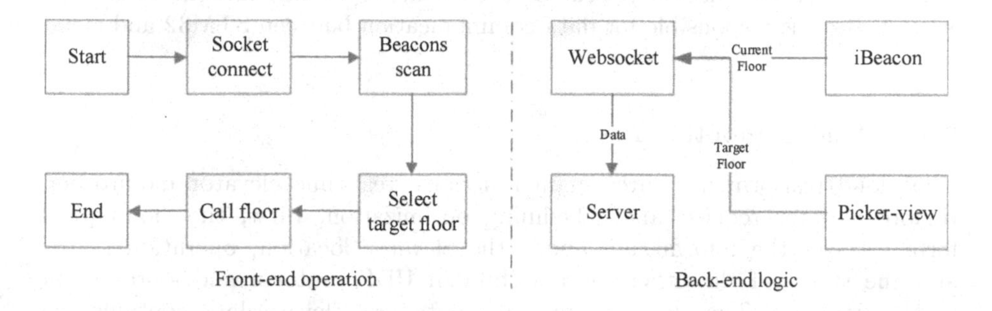

Fig. 3. The user interface design of the mobile mini program.

3.2 Perceptual Recognition Layer

The serial port instruction from the microchip unit is used to control the NB module to receive data. The operation process of the elevator is simulated on the single chip microcomputer, and the corresponding data is displayed on the Organic Light-Emitting Diode (OLED) screen. Meanwhile, the real-time operation data of the elevator is uploaded, by the NB-IoT module, to the cloud platform, where the latest call data is downloaded. The specific process of simulated elevator operation is as follows. When a user has hall request, judge the position of the elevator relative to the user, thus it is concluded that the direction of the elevator. The elevator operates to the user's floor. After opening and closing the door, the user is successfully received. Then, judge the current position of the elevator relative to the user's target floor, thus it is concluded that the direction of the elevator. The elevator operates to the user's target floor, that is, the user successfully reaches the destination.

3.3 Network Construction Layer

iBeacon is a low-power Bluetooth technology. The iBeacon device transmits signals, and the portable device of the user receive and freeback signals. There are four main data of iBeacon: Universal Unique Identifier (UUID), Major, Minor and Measured Power. UUID can create a new identifier for the new service. If the service provider can match the available service with this UUID, it will return a response. NB-IoT is a new technology in the field of IoT, which supports cellular data connection of low-power devices in Wide Area Network (WAN) so that NB-IoT is also known as low-power WAN. Compared with wired or cellular data transmission, NB-IoT has the advantages of low power consumption and easy installation, which is more suitable for the real-time monitoring of elevators. Compared with 4G, NB-IoT can also satisfy a large number of connections. When the iBeacon device transmits signals, the software of the client confirms the UUID of the iBeacon device at first. After the confirmation, the floor information is identified by the major value in the signal and the current floor is indirectly judged by the Received Signal Strength Indication (RSSI) value. And then the floor information is sent to the cloud server through the websocket API. NB-IoT is responsible for data communication between STM32 and cloud server.

3.4 Cloud Computing Layer

The cloud platform has three main functions: real-time elevator monitoring, remote elevator request and scheduling optimization. First, the cloud platform receives the information about the elevator location, operation status and the status of the other sensors through UDP protocol and stores them in the database. The elevator operation database Elevatordata contains the fields such as ElevatorFloorElevatorStateElevatorSpeedNumber_of_People, etc. The web interface reads the data in the database and presents the corresponding data of the elevator in a visual form. Second, the cloud platform obtains the users' call data including the users' present and target floor through websocket protocol, receives the elevator data through UDP protocol and stores the user data and the elevator data into the database. The user call database contains the fields UserFloorTargetFloorProcessedinElevator. Then, the following full load algorithm is used to judge whether the elevator is full. The fuzzy clustering algorithm is used to calculate the scheduling scheme. And finally the dispatching result and full load signal are sent to the elevator.

Scheduling of Elevators. The elevators scheduling procedure is shown as below in Algorithm 1. $E_{FULL\_LOAD\_CONDITION}$ stands for the full load condition of the elevator, n stands for the current number of people in the elevator, $E_{USER\_NUM\_THD}$ represents the number of people judging the elevator to be full, X stands for the number of invalid stop of the elevator, and m is the number of people in the elevator after it was last opened or closed. This algorithm can solve the problem of invalid stop of elevator floor. When the elevator is full

Algorithm 1. Elevators scheduling procedure

```
while TRUE do
  if E_n < E_USER_NUM_THD then
    if E_m is equal to E_n then
      E_X=E_X+1
      if E_X > E_INVALID_STOP_THD then
        E_X=0,E_FULL_LOAD_CONDITION = 1
      else
        E_FULL_LOAD_CONDITION = 0
      end if
    else
      E_X=0,E_m=E_n,E_FULL_LOAD_CONDITION = 0
    end if
  else
    E_FULL_LOAD_CONDITION = 1
  end if
end while
```

but not exceeding the predetermined threshold, it will not stop the elevator floor by floor but not able to carry more people, resulting in the waste of power and time.

Fuzzy Clustering of the Elevator Callings. In the scene of the elevator calling by various users, it is necessary for the central control to coordinate the concurrent user callings in order to provide more services. In this paper, a fuzzy clustering based method is adopted to achieve it. As illustrated in [13], clustering is the classification of similar objects into different groups, or more precisely, the partitioning of a data set into clusters, so that the data in each subset share some common trait, often proximity according to some defined distance measure.

In this paper, the input matrix to be fuzzy clustering is $Elevator_i$ $(i = 1, 2, \ldots, n)$ which represents for attribute aggregation about all users requesting for elevators E_i. It is defined as

$$Elevator_i = \begin{pmatrix} U_1 \\ U_2 \\ \cdots \\ U_n \\ E_1 \\ \cdots \\ E_l \end{pmatrix} = \begin{pmatrix} C_1 & D_1 & T_1 \\ C_2 & D_2 & T_3 \\ \cdots & \cdots & \cdots \\ C_n & D_n & T_n \\ E_1 & E_1 & E_1 \\ \cdots & \cdots & \cdots \\ E_l & E_l & E_l \end{pmatrix} \tag{1}$$

where U represents for the elevator calling by users, C represents for the floor that the user locates, D represents for the relative location with the elevator, T is the user target floor, n represents all users waiting for the elevator, and l stands for the number of the elevators.

After the data source is formed, it comes to the core of the decision-making algorithm which is fuzzy clustering. The matrix **Elevator** is fuzzy similar but it maybe not fuzzy equivalent. For the classification of similar objects into different groups, it is necessary to make **Elevator** convert to $\mathbf{Elevator}^*$. As **Elevator** is fuzzy similar, it exists minimum $k\,(k \leq n, k \in N)$ to make $t\,(\mathbf{Elevator}) = \mathbf{Elevator}^k$. Meantime, for all $l\,(l \leq l, l \in N)$, there definitely takes the equation form $\mathbf{Elevator}^l = \mathbf{Elevator}^k$ so that $t\,(\mathbf{Elevator})$ is fuzzy equivalent matrix $\mathbf{Elevator}^*$.

4 System Evaluation

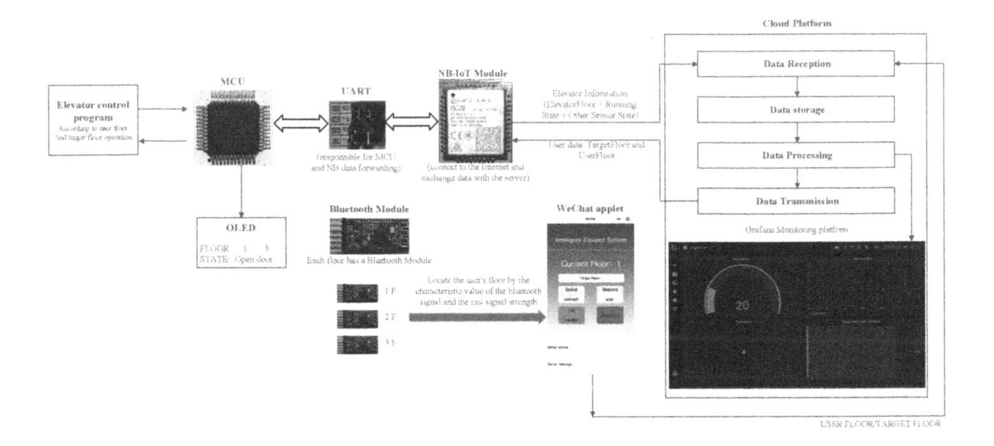

Fig. 4. The real intelligent elevator system.

At present, the elevator protocol is not open to the public, so the operation process of the elevator is simulated on the single chip microcomputer. STM32 is adopted not only because of its high performance and low power consumption but also because of its fast running speed, rich interface and rich communication modules (Fig. 4).

The specific operation process is as follows. Users select the target floor by utilizing the Wechat applet in the portable device, mobile phones, for instance. Then the background process of the applet scans the Bluetooth broadcast signal at the elevator waiting area. The users' floor location are identified by the characteristic value and RSSI signal strength, and the users' present and target floor information is sent to the elevator remote dispatching system. Then the call elevator data is stored in the database. The server sends the new user request, including the target floor and the user's floor data to the simulated elevator control terminal STM32 and sends the elevator floor and the running status of the elevator to the server. The Grafana monitoring platform displays the running status of the elevator and the user's call request in the real time.

5 Conclusion

Through the low power wireless networks including the NB-IoT and Bluetooth iBeacon technology, an intelligent elevator system is proposed with remote calling and cloud scheduling in this paper. The remote elevator calling is presented for the general portable equipment, which not only reduces the management cost, but also solves the problem of inaccurate identification of the number of incoming persons. The elevator dispatching center in the cloud receives the call data of users and the running state of the elevator at the same time, and judges the full load through multi factors. The design reduces the number of ineffective elevator door opening and closing, while improving the running efficiency of the elevator. The user call data and elevator data are presented on the web through the visual interface, which can give the maintenance personnel a more comprehensive and intuitive elevator operation health state, making the elevator maintenance become more active, rather than passive maintenance after failure or accident. The elevator entity that installed the proposed system will be tested for future work.

References

1. Al-Fuqaha, A., Guizani, M., Mohammadi, M., Aledhari, M., Ayyash, M.: Internet of Things: a survey on enabling technologies, protocols, and applications. IEEE Commun. Surv. Tutor. **17**(4), 2347–2376 (2015). https://doi.org/10.1109/COMST.2015.2444095
2. Chang, B., Catpinar, S.F., Jayasuriya, N., Kwatny, H.: Control of impaired aircraft with unanticipated elevator jam to a stable level flight. In: 2019 IEEE 15th International Conference on Control and Automation (ICCA), pp. 543–548, July 2019. https://doi.org/10.1109/ICCA.2019.8899603
3. Farooq, M.O., Wheelock, I., Pesch, D.: IoT-connect: an interoperability framework for smart home communication protocols. IEEE Consum. Electron. Mag. **9**(1), 22–29 (2020). https://doi.org/10.1109/MCE.2019.2941393
4. Fernández, J., Cortés, P., Muñuzuri, J., Guadix, J.: Dynamic fuzzy logic elevator group control system with relative waiting time consideration. IEEE Trans. Ind. Electron. **61**(9), 4912–4919 (2014). https://doi.org/10.1109/TIE.2013.2289867
5. Gao, Y., Xu, X., Lu, J., Sun, Z., Chen, S., Liu, Z.: Energy consumption braking characteristics analysis for multi-car elevator system. In: 2019 22nd International Conference on Electrical Machines and Systems (ICEMS), pp. 1–6, August 2019. https://doi.org/10.1109/ICEMS.2019.8921491
6. Ge, H., Hamada, T., Sumitomo, T., Koshizuka, N.: Intellevator: enhancing elevator system efficiency by proactive computing on the traffic flow. In: 2019 IEEE 1st Global Conference on Life Sciences and Technologies (LifeTech), pp. 80–84, March 2019. https://doi.org/10.1109/LifeTech.2019.8884070
7. Hacks, M.: Huawei elevator networking: connecting millions of elevators. J. Big Data Era **11**, 12–19 (2018)
8. Hangli, G., Hamada, T., Sumitomo, T., Koshizuka, N.: Precaelevator: Towards zero-waiting time on calling elevator by utilizing context aware platform in smart building. In: 2018 IEEE 7th Global Conference on Consumer Electronics (GCCE), pp. 566–570, October 2018. https://doi.org/10.1109/GCCE.2018.8574706

9. Ikuta, M., Takahashi, K., Inaba, M.: Strategy selection by reinforcement learning for multi-car elevator systems. In: 2013 IEEE International Conference on Systems, Man, and Cybernetics, pp. 2479–2484, October 2013. https://doi.org/10.1109/SMC.2013.423
10. Kwon, O., Lee, E., Bahn, H.: Sensor-aware elevator scheduling for smart building environments. Build. Environ. **72**, 332–342 (2018)
11. Li, J., Siddula, M., Cheng, X., Cheng, W., Tian, Z., Li, Y.: Approximate data aggregation in sensor equipped IoT networks. Tsinghua Sci. Technol. **25**(1), 44–55 (2020). https://doi.org/10.26599/TST.2019.9010023
12. Lin, S., Luo, F., Zhang, Z., Wang, X., Chen, Z.: Elevator scheduling based on virtual energy level transition of floors. In: 2019 Chinese Control Conference (CCC), pp. 2274–2278, July 2019. https://doi.org/10.23919/ChiCC.2019.8865576
13. Macario, V., de Carvalho, F.d.A.: An adaptive semi-supervised fuzzy clustering algorithm based on objective function optimization. In: 2012 IEEE International Conference on Fuzzy Systems (FUZZ-IEEE), pp. 1–8, June 2012. https://doi.org/10.1109/FUZZ-IEEE.2012.6251345
14. Mangera, M., Panday, A., Pedro, J.O.: Ga-based nonlinear pseudo-derivative feedback control of a high-speed, supertall building elevator. In: 2019 IEEE Conference on Control Technology and Applications (CCTA), pp. 982–987, August 2019. https://doi.org/10.1109/CCTA.2019.8920625
15. Mishra, K.M., Krogerus, T.R., Huhtala, K.J.: Fault detection of elevator systems using deep autoencoder feature extraction. In: 2019 13th International Conference on Research Challenges in Information Science (RCIS), pp. 1–6, May 2019. https://doi.org/10.1109/RCIS.2019.8876984
16. Nazarova, O., Osadchyy, V., Shulzhenko, S.: Accuracy improving of the two-speed elevator positioning by the identification of loading degree. In: 2019 IEEE International Conference on Modern Electrical and Energy Systems (MEES), pp. 50–53, September 2019. https://doi.org/10.1109/MEES.2019.8896414
17. Rodrigues, D.V.Q., Rodriguez, D., Wang, J., Li, C.: Smaller and with more bars: a relay transceiver for IoT/5G applications. IEEE Microw. Mag. **21**(1), 96–100 (2020). https://doi.org/10.1109/MMM.2019.2945151
18. Strang, T., Bauer, C.: Context-aware elevator scheduling. In: 21st International Conference on Advanced Information Networking and Applications Workshops (AINAW 2007), vol. 2, pp. 276–281, May 2007. https://doi.org/10.1109/AINAW.2007.131
19. Sun, M., Tay, W.P.: On the relationship between inference and data privacy in decentralized IoT networks. IEEE Trans. Inf. Forensics Secur. **15**, 852–866 (2020). https://doi.org/10.1109/TIFS.2019.2929446
20. Tartan, E.O., Erdem, H., Berkol, A.: Optimization of waiting and journey time in group elevator system using genetic algorithm. In: 2014 IEEE International Symposium on Innovations in Intelligent Systems and Applications (INISTA) Proceedings, pp. 361–367, June 2014. https://doi.org/10.1109/INISTA.2014.6873645
21. Wang, H., Fapojuwo, A.O.: A survey of enabling technologies of low power and long range machine-to-machine communications. IEEE Commun. Surv. Tutor. **19**(4), 2621–2639 (2017). https://doi.org/10.1109/COMST.2017.2721379

Construction of Smart Carbon Monitoring Platform for Small Cities in China Based on Internet of Things

He Zhang[1], Jianxun Zhang[1], Rui Wang[1](✉), Qianrui Peng[1], Xuefeng Shang[1], and Chang Gao[2]

[1] School of Architecture, Tianjin University, Tianjin, China
326047221@qq.com

[2] Urban Planning Institute of Tianjin University, Tianjin, China

Abstract. The rapid development of the Internet of Things has promoted the construction of smart cities around the world. Research on carbon reduction path based on Internet of Things technology is an important direction for global low carbon city research. Carbon dioxide emissions in small cities are usually higher than in large and medium cities. However, due to the large difference of data environment between small cities and large and medium-sized cities, the weak hardware foundation of the Internet of Things and the high input cost, the construction of a small city smart carbon monitoring platform has not yet been carried out. This paper proposes a smart carbon monitoring platform that combines traditional carbon control methods with IoT technology. It can correct existing long-term data by using real-time data acquired by the sensing device. Therefore, the dynamic monitoring and management of low-carbon development in small cities can be realized. The conclusion are summarized as follows: (1) Intelligent thermoelectric systems, industrial energy monitoring systems, and intelligent transportation systems are the three core systems of the monitoring platform. (2) The initial economic input of the monitoring platform can be reduced by setting up IoT identification devices in departments and enterprises with data foundations and selecting samples by using classification and stratified sampling.

Keywords: Internet of Things · Smart city · Low carbon city · Intelligent control of carbon emission · Real-time monitoring · Small cities in China

1 Introduction

The Internet of Things (IoT) is an information network, which connects information from any object in the world to the Internet through information sensing devices such as radio frequency identification (RFID), function sensors, global positioning systems, laser scanners, etc., for intelligent identification, monitoring and management of objects [1]. It is estimated that in 2020, 30 billion information sensing devices will be put into use, including some information-sensing devices such as vehicles embedded in electronic

X. Wang et al. (Eds.): 6GN 2020, LNICST 337, pp. 263–277, 2020.
https://doi.org/10.1007/978-3-030-63941-9_20

software and household appliances [2]. The real-time monitoring platform based on Internet of Things technology realizes early warning and decision making by using identifiable, capturable and sharable data, which is able to achieve the storage, query, analysis, mining and understanding of massive sensing data through Internet technology.

Intelligent low-carbon technology based on the Internet of Things is an important direction for the research of global low-carbon cities. Exploring and advancing the way to achieve smart carbon control has become the epoch topic of for government and urban designers [3]. For example, through RFID (Radio Frequency Identification) technology and the development of an electronic environmental protection sign system, cars can be classified and controlled according to different emission standards, on the basis of which, a low-carbon traffic zone is established to gain an aim of low-carbon traffic, energy conservation and emission reduction [4]. In addition, some scholars have proposed a future urban low-carbon community model supported by digital infrastructure and data management systems, which constitutes a smart, sustainable and inclusive growth strategy through cloud computing and IoT [5].

In China, a county or county-level city with an urban population of less than 500,000 is called a small city. According to the study, carbon emissions in small urban areas with low levels of urbanization are usually three times higher than those in big cities [6]. However, the low-carbon technology based on Internet of Things technology has not yet been launched in small cities in China. On the one hand, the technical system needs to be further improved. On the other hand, it is also affected by the differences in carbon emissions pathways in small cities and the limitations of the economic foundation. Firstly, the main influencing factors of carbon emissions in large cities such as traffic congestion and high population density are not applicable to small cities. However, the carbon emissions generated by industrial enterprises account for 50% of carbon emissions in small cities. Therefore, in the small-city smart city-controlled carbon monitoring technology, the objects and data collected by the Internet of Things perception layer are quite different from those of large cities. Secondly, as the economic development level of small cities in China still has a large gap with big cities, the new generation of mobile broadband network technology, internet technology and digital management technology have not been fully popularized in households and businesses. It leads to the results that the perception of real-time data on carbon emissions still requires a large number of hardware devices such as sensor technology, micro-electro-mechanical systems and so on, while the investment of this equipment will cause the economic cost burden of small cities. Therefore, the smart city carbon control method based on the Internet of Things in small cities still needs to consider both the precision monitoring and the economic cost.

The purpose of this paper is to build a smart low-carbon monitoring platform applied to small cities in China based on Internet of Things technology. It mainly solves two major problems: 1) constructing a real-time monitoring system to monitor carbon emission data from carbon sources in Chinese small cities; 2) initially implementing a smart carbon monitoring platform for small cities with lower cost hardware investment.

The rest of the paper is organized as follows. The second part briefly reviews the existing related research. The third part puts forward the construction of the smart station monitoring platform based on IoT. The fourth part takes Changxing as an example to explore the actual construction steps of the monitoring platform. The fifth part discusses how to implement a low-cost, high-precision, intelligent carbon-based system based on a limited source of perceptual data.

2 Literature Survey

The carbon control monitoring in small cities mainly relies on the continuous development and renewal of carbon emission estimation and measurement technology. It is mainly divided into three research stages: 1. the first stage, the calculation method of energy consumption carbon emissions based on IPCC calculation method; 2. the second stage, the carbon emission coefficient estimation based on big data; 3. the third stage, the real-time carbon emission measurement method with IoT as the core.

Since direct data on CO_2 emissions is difficult to obtain, most scholars often use existing energy consumption data and estimate it according to the calculation method provided by IPCC [7]. This is also one of the most common method. The 2006 IPCC guidelines for national greenhouse gas inventories provide two methods for estimating carbon dioxide emissions based on energy consumption. Firstly, the departmental analysis method, also known as top-down, uses the energy consumption data of transportation, industry and so on with their related emission coefficients for conversion. For example: (Yanchun Yi) obtained data on energy consumption in 108 prefecture-level cities in China, such as: raw coal, fuel coal, and fuel oil data, and introduced the carbon emissions measurement model of the 2007 IPCC (Formula 1) to calculate urban carbon emissions and analyze urban carbon emission levels and their influencing factors [8]. Secondly, the per capita consumption law, also known as bottom-up, uses data of natural gas, gas and others from urban residents to estimate carbon dioxide emissions. (Chou-Tsang Chang) Some scholars obtained data on gas consumption and electricity consumption of households in eight administrative districts of the Taichung metropolitan area in Taiwan, and converted gas and electricity consumption data based on carbon dioxide conversion coefficient to estimate the carbon dioxide emissions of city buildings [9]. In China, the energy consumption data from relevant departments can only be obtained in large and medium-sized cities, which are constituted with administrative units of prefecture-level cities, while the data of small cities including counties as administrative units are usually unavailable. Therefore, the carbon emission estimation of small cities adopts a top-down energy consumption decomposition method, which is based on the departmental analysis method to estimate the total carbon emissions of the production, living, and transportation sectors of the prefecture-level cities, to estimate the carbon emission of different parts, including economic output value, total population, and traffic road mileage. Etc.

With the development of network big data technology and the popularity of remote sensing technology, carbon emission estimation methods are experiencing a revolution. Many scholars have explored the estimation methods of total carbon emissions based on real-time population, occupational and commuting data, urban nighttime lights data, and car ownership number. Among them, the research on the estimation of carbon emissions from transportation and residential carbon emissions are the most common direction. In the carbon emission estimation of transportation, based on the interaction mechanism between carbon emissions and different elements, including residents' transportation [6, 10, 11], urban occupational residence balance and urban built environment factors, to determine the carbon emission impact coefficient under different factors, and then estimate the carbon emissions. Among the methods for estimating carbon emissions in residential life, the most commonly used method is according to night lighting data. For example, a recent study validated and simulated urban carbon dioxide emissions based on nighttime light image data from 327 prefecture-level cities in China [12]. The above method provides a new idea for carbon dioxide measurement and control methods in small cities in China.

In recent years, with the rapid development of Internet of Things technology, smart low-carbon technology has formed a cross-disciplinary research field with urban intelligent infrastructure research. By implanting inductive equipment, including urban infrastructure of electricity, energy and transportation, and mobile communication facilities, it is possible to monitor, analyze, and intelligently guide urban low-carbon operations and form an early warning mechanism. Among them, the research on the Internet of Things technology for car carbon emissions is of a great number. Somne studies suggest that the Bus Stop Interface (BSI) can be installed at a specific bus stop, and the optimal driving route can be generated according to the driver's destination selection, thereby reducing traffic time [13, 14] and road congestion time to reduce traffic carbon emissions. At the same time, the application of smart grid technology also provides a real-time monitoring method for the fine management of carbon emissions [15]. Thus, it can be further applied to low-carbon logistics [16], low-carbon energy conservation [17] and so on.

The continuous updating of carbon emission measurement technology provides many methods to support the control and monitoring of carbon emissions in small cities. However, each method has certain limitations in carbon monitoring in small cities: 1) Estimation techniques based on IPCC and big data methods can cover major consumer sectors of carbon emissions. However, due to the characteristic of static data, its timeliness is low, and it is difficult to support the application field of real-time monitoring; 2) Based on the carbon emission monitoring technology of Internet of Things technology, Real-time monitoring and refined management of carbon emissions can be achieved. However, it has high requirements for mobile internet and smart grid technology and hardware, and it still cannot achieve comprehensive coverage in the carbon accounting of the living and production sectors. Combining the advantages and disadvantages of the above methods, taking into account the practical basis and economic cost of the application of IoT technology in small cities, this paper introduces a real-time data monitoring and correction method for the IoT, based on traditional carbon emission measurement method, to initially construct a smart city carbon monitoring system and application platform for small cities.

3 Proposed Framework

3.1 System Framework with the Linkage of "Long Term Data – Real Time Data"

Based on the IPCC-based carbon inventory method, the carbon emission consumer departments of small cities mainly include public power and heat, industrial production processes, transportation, service industries, and residential lives. The power and heat energy consumption in the service industry and the residential life department has been calculated in the public power heat department, and other energy consumption such as gas is negligible. Therefore, the two departments are eliminated. The main departments that jointly monitor "long-term data-real-time data" in small cities are identified, which are the thermal power sector, the industrial sector, and the transportation sector. On this basis, a real-time data acquisition system is constructed based on the revised requirements of long-term data. They also constitute the three major systems of the framework of the intelligent carbon monitoring platform, which are thermal-electricity monitoring system, energy monitoring system of industrial process and transportation monitoring system (Fig. 1).

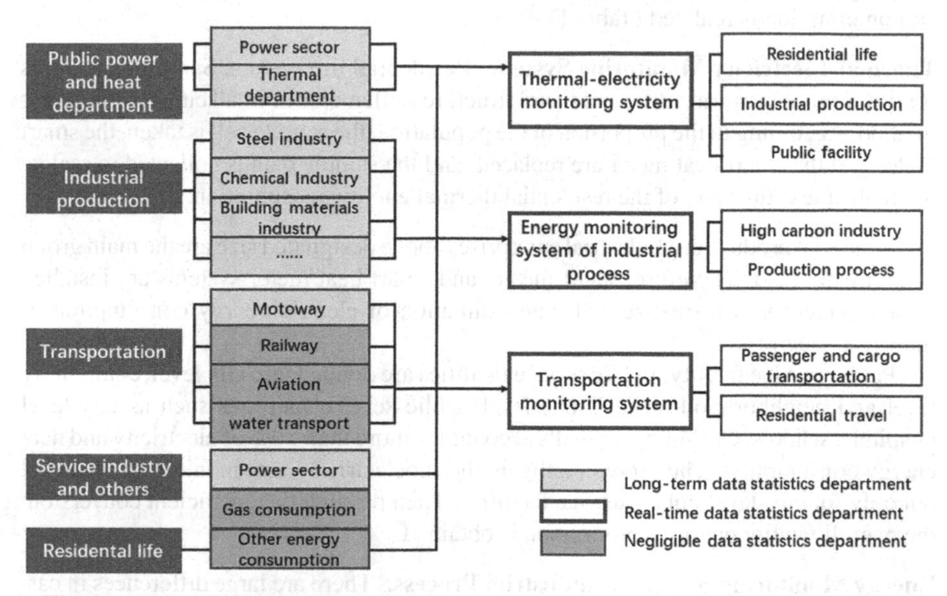

Fig. 1. "Long-term data - real-time data" linkage of carbon monitoring system framework

Thermal-electricity monitoring system is a real-time monitoring system based on sensing devices such as smart power meter and intelligent heat meter. It includes three major systems: residential life, industrial production, and public facilities. The carbon emission estimation results of long-term data are corrected in real time through real-time data of three systems.

Energy monitoring system of industrial process is to realize the correction of industrial process carbon emission measurement results through variable calculation, based on real-time data monitoring of material consumption and energy consumption data in industrial technology processes.

Transportation monitoring system includes passenger and cargo transportation systems and resident travel systems. The passenger and cargo carbon emission monitoring system is a real-time monitoring of the emission with magnetic the induction equipment and cameras monitoring the emission standards of passenger and freight vehicles, vehicle load and transit time in real time. Through the monitoring of fueling in town gas stations, the carbon emissions of residents in small cities are estimated.

3.2 Collection Method of Basic Environmental Data

It takes a long time for smart sensing devices based on individuals, families or vehicles to be popularized in small cities. Considering the feasibility of the implementation of the Internet of Things in small cities, data collection should be carried out in management departments and enterprises with certain data foundations. Based on the sociological survey method of sample sampling, the real-time fluctuation estimation technology of carbon emission is realized (Table 1).

Thermal-Electricity Monitoring System. Residential life system: Sampling methods for different capacity and different social structure settlements in small cities by sampling method. According to the proportion of the population, the sample cell is taken, the smart meter and the smart heat meter are replaced, and the sample data is collected to realize the real-time estimation of the residential thermal energy consumption.

Industrial production: Industrial enterprises above designated size are the main group of industrial land. Therefore, smart meters and smart heat meter systems are installed in such enterprises to realize real-time estimation of electric energy consumption of industrial enterprises.

Public service facility: Public service facilities are divided into city level, community level and neighborhood level. Urban-level public service facilities such as city-level hospitals, schools, and shopping malls account for more than 70% of electricity and heat energy consumption. Therefore, intelligent thermoelectric system monitoring is carried out only for city-level public service facilities. Then through the coefficient conversion, the overall facility energy consumption is obtained.

Energy Monitoring System of Industrial Process. There are large differences in carbon emissions from production processes in different industries. Therefore, energy consumption monitoring is selected for industrial enterprises whose energy consumption accounts for more than 20% of all industrial processes in small cities. Energy consumption monitoring devices and material total monitoring device are arranged in the production process that generates carbon emissions for real-time monitoring purposes.

Transportation Monitoring System. Passenger and cargo transportation: With the highway entrance and exit and section speed monitoring equipment, the speed, fuel consumption standard, driving history of passenger and freight cars are collected. And through the highway traffic flow ratio coefficient, real-time carbon emissions data of transit passenger and freight systems in small cities are estimated.

Table 1. Data attributes and acquisition methods of long-term data and real-time

Carbon monitoring system	Long-term data		Real-time data		
	Data attribute	Data Sources	Subsystem	Acquisition method	Data attributes and devices
Thermal-electricity monitoring system	Power energy consumption	Statistical data	Residential living	Stratified sampling	**Electricity and heat energy consumption:** Intelligent power meter Intelligent heat meter
			Industrial production	Stratified sampling	
	Thermal energy consumption	Statistical data	Public service facility	Classification sampling	
Energy monitoring system of industrial process	Energy consumption of raw materials and process streams	Statistical data	Subsystem by industry type (energy consumption accounting for more than 20% of all industries)	Classification sampling	**Industrial consumption energy consumption:** Energy consumption monitoring device Material total monitoring device
Transportation monitoring system	Passenger and cargo transportation	Data decomposition estimation	Passenger and cargo transportation	High speed station data acquisition	**Speed, fuel consumption standard, driving history:** Geomagnetic sensor, camera device
	Residential travel	Sample survey	Residential travel	Gas station data collection	**Amount of gasoline:** Oil quantity monitor

Resident travel system: Real-time monitoring and daily statistics on the amount of gasoline in small city gas stations. Estimate the dynamic trend of carbon emissions from residents' travel through the amount of gasoline. The amount of gasoline in small city gas stations is monitored in real time and daily statistics. And we use the trend of gasoline volume to replace the trend of carbon emissions in car travel.

3.3 Carbon Emission Measurement Correction Method Based on Real Time Data

Typical sample collection and departmental integration methods are selected for real-time data. The carbon emission method is different from the traditional static data calculation method, and it is difficult to perform simple calculation using superposition. Therefore, this paper uses the coefficient of variation method to perform real-time dynamic data monitoring based on traditional carbon emission measurement results.

Public Power and Thermal Department. Estimate the total carbon emissions of energy consumption in public utilities and thermal systems respectively. The calculation method is as follows:

$$CE_{Ph} = \sum_{j=1}^{2} CE_{rj}(1 + R_{rj}) + \sum_{j=1}^{2} CE_{ij}(1 + R_{ij}) + \sum_{j=1}^{2} CE_{pj}(1 + R_{pj})$$

Among them, CE_{ph} is the revised real-time total carbon emissions for the public electric power department. CE_{rj}, CE_{ij}, and CE_{pj} respectively represent the basic carbon emission values of residential, industrial, and public services, and they all use the calculation method provided by IPCC to estimate the static energy consumption data of each department. R_{rj}, R_{ij}, and R_{pj} are respectively calculated coefficients of real-time monitoring data of the three departments, and the calculation method is as follows:

$$R_r = \sum_{1}^{n} \frac{\Delta CER \cdot r_n}{P_n} / \sum_{1}^{n} \frac{CER \cdot r_n}{P_n};$$

$$R_i = \sum_{1}^{n} \Delta CEI / \sum_{1}^{n} CEI;$$

$$R_p = \alpha\left(\sum_{1}^{n} \Delta CEP / \sum_{1}^{n} CEP\right);$$

Among them, ΔCER, ΔCEI, ΔCEP represent real-time energy consumption changes of the three sectors, n is the number of samples, r_n is the proportion of population in different types of residential areas, and α is the energy conversion coefficient of total public service facilities and urban public service facilities.

Industrial Production Department. Based on the energy consumption monitoring of the production links of typical enterprises, the real-time data of carbon emissions are obtained to correct the total carbon emissions of existing industrial production processes. The calculation method is as follows:

$$CE_{PP} = CE'_{pp}(1 + R_{pp}),$$

Among them, CE_{pp} is the real-time total carbon emissions corrected by the industrial sector. CE'_{pp} is the basic carbon emission value of industrial production processes calculated by static data. R_{pp} is the coefficient of variation calculated for real-time monitoring coefficient. The calculation method is as follows:

$$R_{pp} = \sum_{1}^{n} (\Delta CEPP \cdot r_n) / \sum_{1}^{n} (CEPP \cdot r_n)$$

Among them, $\Delta CEPP$ is the real-time change value of energy consumption in industrial production, and r_n is the proportion of energy consumption of enterprises in different scales.

Transportation Department. The transportation sector uses top-down traditional estimation methods for road carbon emissions. It uses highway mileage to decompose provincial and municipal transportation carbon emissions data. In addition, the carbon emissions of residents' travel are converted by the coefficient of resident car ownership. However, the above methods all result in great errors. Therefore, this paper uses the real-time energy consumption carbon emission data of passengers and freight vehicles on transit roads to replace the original road carbon emission estimation method. Meanwhile, the estimation of carbon emissions in the real-time gasoline fueling capacity of the gas station is to replace the resident car ownership.

3.4 Smart Control Carbon Monitoring Platform System

The Internet of Things usually has three characteristics. One is comprehensive sensing, that is, using sensors and other devices to acquire object information at any time. The second is reliable transmission. The sensor network is combined with the Internet to transmit object information to the Internet. The third is intelligent processing, using intelligent algorithms such as cloud computing and fuzzy recognition to analyze and process the data, and then realize intelligent control of objects [18, 19]. Based on this, this paper proposes a framework for Intelligent Control Carbon Monitoring Platform for small cities in China (Fig. 2), which mainly includes the basic environment data layer, database and information management layer, computing layer and management control layer.

Basic Environment Data Layer. The base environment data layer includes a perceptual environment, sensing devices, and long-term data ends collected by traditional methods. The sensing environment is the industry, transportation, and residential area; the sensing equipment includes various sensors, chips, cameras, etc.; the long-term data end collected by the traditional method refers to the energy consumption data collection of different departments in the unit by the carbon emission inventory method.

Database and Information Management. The database and information management are divided into two parts: database and information management. The database includes energy consumption data of the public power and heat departments, industrial production departments and transportation departments. The information management layer mainly loads, optimizes, transforms and integrates the data.

Calculation Layer. Computational layer refers to the use of cloud computing, programming models, fuzzy computing and other intelligent calculation methods to calculate and analyze the carbon emissions of energy consumption data in the database, and build a real-time measurement model of carbon emissions.

Management Control Layer. The management control layer refers to the relevant management decisions realized after data analysis. Among them, for the industrial sector it can achieve over-standard warning, control and monitoring decisions; for the transportation sector it can implement road restrictions, vehicle limit and other decisions; for the family it can achieve building energy warning, energy supply adjustment and other decisions.

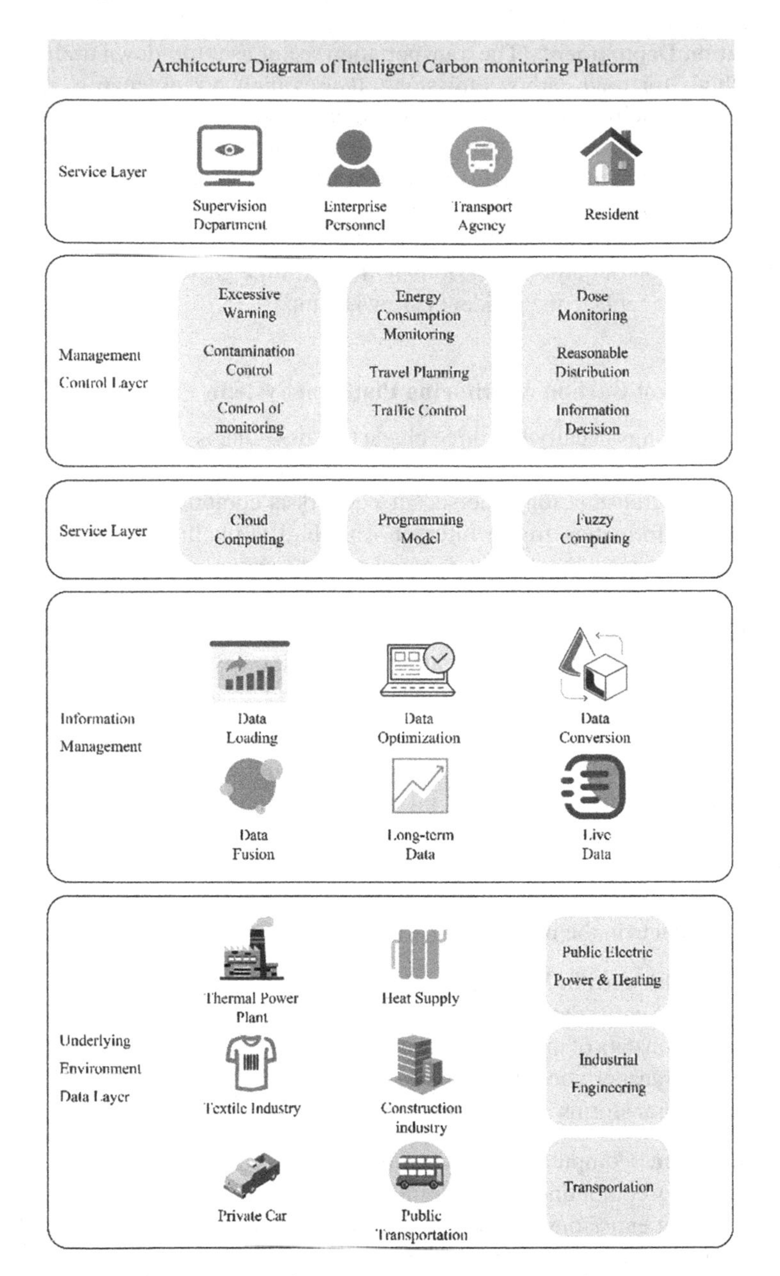

Fig. 2. The Layer of smart carbon monitoring platform system

4 Empirical Research

4.1 Empirical Introduction

The county of Changxing is affiliated to Huzhou City, Zhejiang Province. It is located on the southwestern shore of Taihu Lake between Suzhou and Hangzhou. The county governs 3 streets, 9 towns and 2 townships. The population of Changxing County is 233,200. Through the accounting of carbon emissions in Changxing County for many years, its carbon emissions are mainly concentrated in the energy industry, which is mainly based on electricity production, and the production process of building materials industry, which is mainly based on cement production (Fig. 3). And the carbon emissions of transportation and residential life are relatively low. In terms of carbon emission characteristics, Changxing has the representativeness of small cities in China.

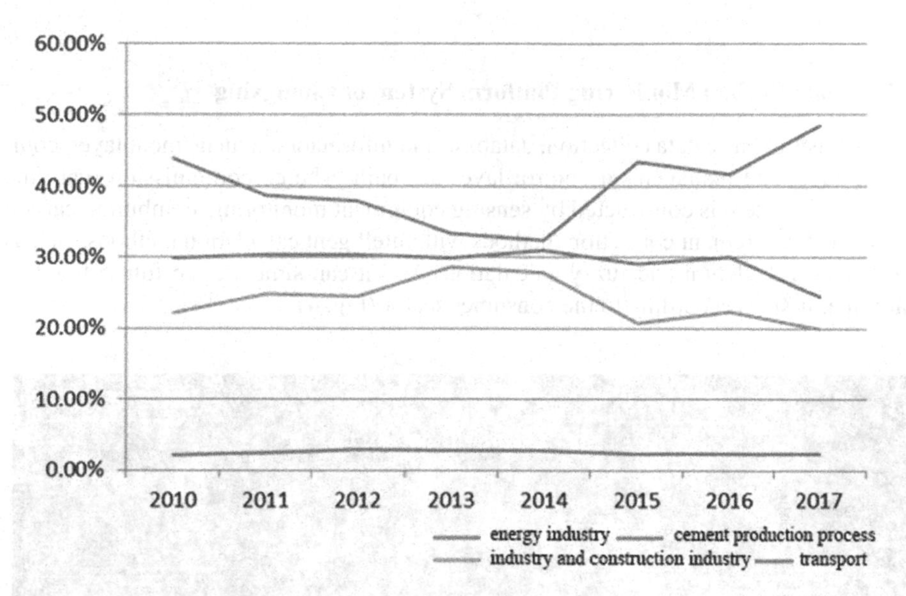

Fig. 3. Trends in carbon emission ratio of sub-sectors in Changxing (2010–2017)

4.2 Basic Environmental Data Acquisition Method of Changxing

Public Power and Heat Department. Residential life system: According to the classification method of low floor area ratio (0.8–1.1), medium floor area ratio (1.2–1.4), and high floor area ratio (>1.4), the settlement was sampled according to the proportion of the population, and finally 7 sample samples were obtained. It is proposed to install smart thermoelectric devices for the residents of the seven communities.

Industrial production system: Select 20 or more companies in different industries to install smart thermoelectric devices.

Public facility: Intelligent hotspot installations were carried out for 9 educational facilities (technical schools, secondary schools, primary schools), 8 integrated and specialist hospitals, and 7 cultural and sports facilities.

Industrial Process Department. According to the carbon emission list of Changxing County, the energy consumption of cement rotary kiln in the building materials industry accounts for 90% of the energy consumption of all industrial production. Therefore, only 9 samples of cement production enterprises are monitored for industrial process energy consumption.

Transportation Department. The intelligent monitoring equipments are installed at the super-superior station, interval speed, and camera of 18 highway entrances and exits. And oil monitoring and sensing equipment was installed at 29 gas stations in the county and townships.

4.3 Smart Carbon Monitoring Platform System of Changxing

On the basis of basic data collection, database and information management layer, computing layer and management control layer are built. The carbon emission real-time monitoring system is constructed by sensing equipment monitoring, combining carbon emission measurement correction methods with intelligent calculation methods such as data cloud calculation and fuzzy calculation. And it can simulate the future trend of carbon emissions according to the consumer sector (Fig. 4).

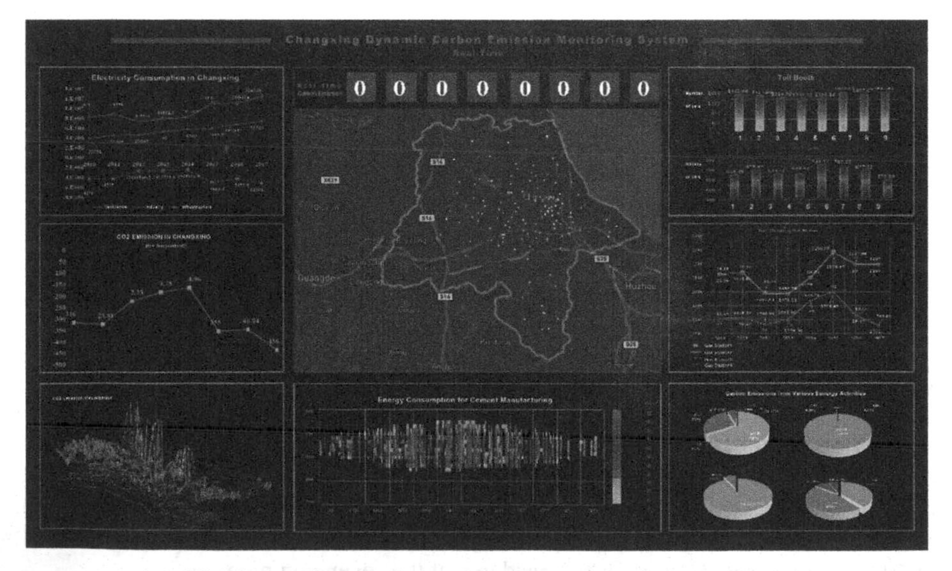

Fig. 4. Intelligent carbon monitoring platform in Changxing

5 Discussion

Based on the construction of "long-term data-real-time data" linkage control carbon monitoring system, this paper explores the real-time correction technology of traditional carbon emission measurement methods. And the intelligent control carbon monitoring platform system based on the Internet of Things was initially constructed. In this process, we mainly discuss the following issues, focusing on the characteristics and limitations of carbon control in small cities:

The small city carbon control system based on the Internet of Things technology mainly focuses on industrial thermoelectric systems, industrial production processes and transportation systems, and is quite different from large and medium cities. The focus of carbon control in large and medium-sized cities is mainly concentrated in two aspects: (1) carbon emission control brought by residential energy consumption under a large population concentration [20] and (2) carbon emission control of residential travel caused by traffic congestion [21]. The focus of carbon control in small cities focuses on (1) industrial energy consumption and carbon emissions from industrial production processes and (2) carbon emissions control for passenger and freight transportation. Therefore, low-carbon transportation and low-carbon residential Internet of Things technologies with application prospects in large cities are not very effective in controlling carbon in small cities. And increasing the IoT management and control methods for industrial processes and transit traffic is more targeted.

Selecting departments and enterprises with data foundations to collect data can reduce the input cost of the carbon monitoring platform and increase the possibility of application of IoT technology in small cities. The application of Internet of Things technology in large and medium-sized cities is often based on massive data analysis obtained from the extensive use of mobile communication devices and sensing devices [22]. However, for small cities, the data environment of their individual level is poor, and it is difficult to realize a large investment in IoT hardware devices in a short period of time. Therefore, the sensing equipment and the estimation model are deployed at the urban management department and enterprise level with a certain data foundation to avoid the dilemma of the implementation of the Internet of Things technology at the individual level in small cities.

The combination of Internet of Things technology and traditional data statistics methods is an important means of low-carbon management in small cities. Compared with small cities, the spatial structure, social structure and built environment of large and medium-sized cities show strong complex features. The perception, learning and simulation of the individual behavioral characteristics and life patterns of the Internet of Things technology make up for the gap in the management level of big cities. But for small cities, their demographic, spatial and economic structures are relatively stable and simple. And Individual IoT data environment is still immature. Therefore, the sampling method of sociology can still be used to solve the problem of unsatisfactory data environment through reasonable stratification and classification sampling. And through the combination of Internet of Things technology, real-time statistical results can be achieved.

6 Conclusion

The application of existing Internet of Things technology in low-carbon fields is often based on the superior data base, rapid economic development and low-carbon management system in large cities. However, it is not fully applicable to small cities with poor economic levels and weak data environment. In view of the above problems, this paper constructs a carbon-controlled monitoring system linked to "long-term data-real-time data", which is an exploration of the correcting method of traditional carbon emission measurement. Based on this, the intelligent carbon monitoring platform is initially constructed. This study provides a preliminary framework for the smart city carbon monitoring platform in small cities, but it still needs empirical support to explore the experience in data collection, processing and analysis.

Acknowledgments. Supported by National Key Research and Development Project (Grant No. 2018YFC0704701).

References

1. Rajanpreet, K.C., Neeraj, K., Shalini, B.: Trust management in social Internet of Things: a taxonomy, open issues, and challenges. Comput. Commun. **150**(15), 13–46 (2020)
2. Sofana, R.S., Tomislav, D.: Future effectual role of energy delivery: a comprehensive review of Internet of Things and smart grid. Renew. Sustain. Energy Rev. **91**(08), 90–108 (2018)
3. Ruike, Y., Yiwei, L., Zhuangfei, G., Wang, L.: The interactive development of low-carbon city and smart city. Sci. Technol. Econ. **30**(04), 12–85 (2017)
4. Kovavisaruch, L., Suntharasaj, P.: Converging technology in society: opportunity for radio frequency identification (RFID) in Thailand's transportation system. In: Conference of the Portland-International-Center-for-Management-of-Engineering-and-Technology 2007, Portland, vol. 1–6, pp. 300–304. Springer, Heidelberg (2006)
5. Deakin, M., Reid, A.: Smart cities: under-gridding the sustainability of city-districts as energy efficient-low carbon zones. J. Clean. Prod. **173**(2), 39–48 (2016)
6. Waygood, E., Yilin, S., Yusak, O.S.: Transportation carbon dioxide emissions by built environment and family lifecycle: case study of the Osaka metropolitan area. Transp. Res. Part D Transp. Environ. **31**(08), 176–188 (2014)
7. Eggleston, H., Buendia, K., Miwa, T., Ngara, K.: Tanabe, IPCC guidelines for national greenhouse gas inventories. Institute for Global Environmental Strategies (IGES), Hayama (2006)
8. Yanchun, Y., Sisi, M., Weijun, G., Ke, L.: An empirical study on the relationship between urban spatial form and CO2 in Chinese cities. Sustainability **672**(9), 1–12 (2017)
9. Chou, T.C., Chih, H.Y., Tzu, P.L.: Carbon dioxide emissions evaluations and mitigations in the building and traffic sectors in Taichung metropolitan area, Taiwan. J. Clean. Prod. **230**(09), 1241–1255 (2019)
10. Kitamura, R., Sakamoto, K., Waygood, O.: Declining sustainability: the case of shopping trip energy consumption. Int. J. Sustain. Transp. **2**(3), 158 (2008)
11. Anne, A., Marion, V.: Urban form, commuting patterns and CO2 emissions: what differences between the municipality's residents and its jobs? Transp. Res. Part **69**, 243–251 (2014)
12. Shaojian, W., Chenyi, S., Chuanglin, F., Kuishuang, F.: Examining the spatial variations of determinants of energy-related CO2 emissions in China at the city level using Geographically Weighted Regression Model. Appl. Energy **235**(02), 95–105 (2019)

13. NodeMCU, Team: Nodemcu-an open-source firmware based on esp8266 wifi-soc (2014). http://nodemcu.com/indexen.html
14. Lakshmi, S., Nithin, S.: A Smart transportation system facilitating on demand bus and route allocation. In: International Conference on Advances in Computing, Communications and Informatics (ICACCI), Manipal, INDIA, SEP 13–16, pp. 1000–1003. IEEE (2017)
15. Dileep, G.: A survey on smart grid technologies and applications. Renew. Energy **146**(02), 2589–2625 (2020)
16. Zhiyao, L., Zilong, Z., Zizhao, H., Qi, W.: A framework of multi-agent based intelligent production logistics system. Procedia CIRP **83**, 557–562 (2019)
17. Samuels, J.A., Booysen, M.J.: Chalk, talk, and energy efficiency: saving electricity at South African schools through staff training and smart meter data visualization. Energy Res. Soc. Sci. **56**(10), 101212 (2019)
18. Jia, B., Hao, L., Zhang, C., Zhao, H., Khan, M.: An IoT service aggregation method based on dynamic planning for QoE restraints. Mob. Netw. Appl. **24**(1), 25–33 (2018). https://doi.org/10.1007/s11036-018-1135-7
19. Chen, H.M., Cui, L., Xie, K.B.: A comparative study on architectures and implementation methodologies of Internet of Things. Chin. J. Comput. **36**(1), 168–188 (2013)
20. Shanshan, S., et al.: Research on low-carbon energy management system of intelligent community. In: ICAEMAS Conference, pp. 488–489 (2014)
21. Gargiulo, C., Russo, L.: Cities and energy consumption: a critical review. J. Land Mobil. Environ. **10**(3), 259–278 (2017)
22. Mehmood, Y., Ahmad, F., Yaqoob, I., Adnane, A., Imran, M., Guizani, S.: Internet of Things based smart cities: recent advances and challenges. IEEE Commun. Mag. **55**(9), 16–24 (2017)

Research on Simulation Method of 5G Location Based on Channel Modeling

Chi Guo[2], Jiang Yu[1], Wen-fei Guo[2](✉), Yue Deng[1], and Jing-nan Liu[1,2]

[1] School of Geodesy and Geomatics, Wuhan University, Wuhan 430079, China
[2] GNSS Research Center, Wuhan University, Wuhan 430079, China
wf.guo@whu.edu.cn

Abstract. It is of great significance to study 5G positioning as the series of new technologies introduced by 5G not only meeting the needs of communication but also improve the positioning accuracy and help achieving indoor and outdoor seamless navigation and positioning. However, 5G base stations and devices are far from popular, and the standards of 5G positioning has not fully formed. Thus, simulation is the main research method for studying 5G positioning. Accordingly, we propose a 5G positioning simulation experiment scheme and give its flow chart. The implementation ideas and the specific processes of the three consisted parts of the simulation experiment including scene generation, signal propagation simulation and position estimation are introduced. Furthermore, we carry out some experiments to verify the 5G simulation environment established by the scheme and make a simple discussion about the results.

Keywords: 5G · Location · Simulation · Channel modeling

1 Introduction

In recent years, the development of the fifth-generation (5G) mobile communication system is in full swing. Enhanced mobile broadband (eMBB), massive machine type communication (MMTC), and ultra reliable & low latency communication (URLLC) are the three main application scenarios of 5G [1]. In order to meet the needs of the above scenarios, many new technologies and features have been introduced by 5G, including high-frequency carrier and large bandwidth, massive multiple-input multiple-output (MIMO) and beamforming, ultra-dense network (UDN), device-to-device (D2D) [2] and others. These technologies can solve communication problems and also provide a lot of key features required for positioning, making 5G positioning be of great research value. The traditional global navigation satellite system (GNSS) is hard to apply to indoor environment. The databases of fingerprints used in Wi-Fi and Bluetooth positioning are difficult to build. By comparison, 5G positioning is expected to achieve both accuracy and convenience, and realize indoor and outdoor seamless navigation and positioning.

Currently, there are two ideas for 5G positioning. One is to establish a 5G-based assisted-GNSS (A-GNSS) system [3] based on the evolution of the existing LTE network-based A-GNSS. The other is to regard the 5G network as a satellite positioning network

X. Wang et al. (Eds.): 6GN 2020, LNICST 337, pp. 278–297, 2020.
https://doi.org/10.1007/978-3-030-63941-9_21

on the ground [4]. Each 5G base station is similar to a pseudo-satellite, the base station receives the GNSS signal for its own positioning, and uses the time service to obtain a high precision time reference and maintain synchronization in the network. Its synchronization accuracy can reach nanoseconds [5]. And thus, time synchronization is a precondition for positioning in this paper. Base station can simultaneously broadcast the communication information and the positioning information, as shown in Fig. 1. According to the known base station location, base station can locate the user by receiving and processing the positioning information.

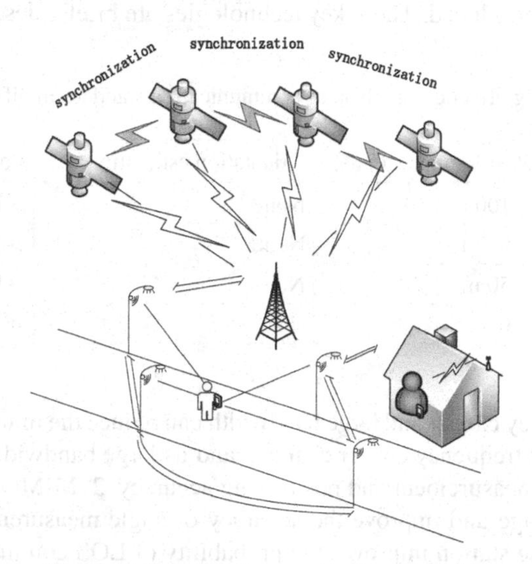

Fig. 1. Schematic diagram of 5G positioning

This paper adopts the second idea and proposes a 5G positioning simulation scheme with reference to the GNSS system. The scheme has the following advantages:

(1) The scheme avoids the complex signal processing process. It has improved a mature channel simulator from the perspective of positioning, which makes the acquisition of observations easier and reduces the difficulty of simulation.
(2) The scheme has rich parameter/scene settings, making it convenient to study the impact of different factors on the positioning which mainly including the number of base stations using for positioning, the distribution of base station, line of sight(LOS), non-line-of-sight (NLOS), multipath, and carrier bandwidth.
(3) The scheme introduces the map of real world and will play a guiding role when deploying base stations.
(4) The scheme is able to learn from the rich research results in GNSS field to solve 5G positioning problems.

2 Related Work

2.1 Development of Mobile Location and New Features of 5G

Researches on the use of mobile communication systems for positioning have increased since the enactment of the E911 Act in the United States, generating a variety of positioning technologies such as the cell ID (CID) positioning in the 2G era, and the observed time difference of arrival (OTDOA) positioning in the 3G and 4G eras [6, 7]. The positioning effect of 5G is expected to make a breakthrough [6, 7] (see Table 1) cause the new technologies introduced. These key technologies are briefly described below.

Table 1. Positioning effect comparison of communication standard in different generations.

Generations	Positioning error	Orientation estimation	Positioning delay
2G	>100 m	None	>1 s
3G	>50 m	None	>1 s
4G	<50 m	None	>1 s
5G (expected)	<0.1 m	<1°	≪1 s

1. High frequency carrier and large bandwidth can reduce the multipath effect due to the sparsity of high frequency carrier channel, and its large bandwidth can improve the resolution of time measurement and positioning accuracy. 2. MIMO and beamforming can enhance coverage and improve the accuracy of angle measurement. 3. The UDN structure of the base station improves the probability of LOS communication between the user and base stations, helping to achieve high precision positioning and indoor positioning. In addition, the newly-introduced device-to-device (D2D) communication through 5G standard makes it possible to achieve cooperative locating of equipment, which is helpful to solve the problem when it lacks reference station [8].

Based on these new features, the researchers conducted a series of studies on 5G positioning.

2.2 Research Status and Relevant Standards

Positioning is quite complex because there are a lot of problems to be solved such as frequency selection, waveform designing, channel modeling, the acquisition of observations and the derivation of localization algorithms [9, 10]. The research of 5G positioning also has many categories.

There are two main categories in terms of the overall structure of positioning. One is to establish a 5G-based A-GNSS system or similar system [3], and the other is to extend the 5G network into a pseudo-satellite system [4], using pseudo-range for positioning. As far as the current researches go, more studies have conducted based on the second structure (see Table 2).

In terms of frequency band for positioning, these two carrier frequencies (sub6 GHz and millimeter wave band) will coexist for a long time in 5G network due to the need of

Table 2. Major research directions of 5G positioning

Directions	Content
Frequency band	Sub6 GHz, Millimeter wave
Signal waveform	Common waveform/self-designed
Positioning strategy	Uplink signal of network side/downlink signal of user side
Time synchronization	Requirement, error source, solution
Channel Modeling	METIS, WINNER II, 3GPP, NYU

both coverage and rate. The two kinds of signals show different features in many aspects due to the differences in frequency, and thus different researches on the two frequency bands have been conducted. Koivisto M et al. made a detailed study on the architecture and positioning method of 5G positioning system in sub6 GHz frequency band. They deduced the theoretical performance of 5G positioning [11], and proposed the method of multi-layer filtering to gradually extract the observations and position [12]. They also considered the effect of synchronization. In the research of millimeter wave positioning, Wymeersch H et al. [6] focused on vehicle positioning. They comprehensively analyzed the impact of 5G new features on positioning, summarized the relevant research directions of 5G positioning, and gave a theoretical positioning accuracy of the millimeter wave band. Shahmansoori A et al. [13, 14] had a more detailed study. Under the carrier of 30 GHz, they derived the Cramer-Rao Lower Bound in 5G positioning under LOS/NLOS with TOA and DOA as observations.

In addition, some scholars have also done research on waveform, positioning strategy, time synchronization and other aspects. Cui et al. [8] and Dammann A et al. [15] made a detailed comparative study separately to discuss the influence of different signal waveforms on the positioning effect. Abu-Shaban Z et al. explored the impact of using the uplink/downlink on the positioning effect [16]. Li H et al. studied the requirements of time synchronization for 5G positioning [17], and explored the factors affecting synchronization and the degree of the effects.

The channel model plays an important role in 5G positioning. Firstly, the channel model contains the geometric relationship and propagation information between the observed value and the position to be estimated, and the accuracy of the position estimation is highly dependent on this information [9]. Secondly, the channel model can be used to derive positioning algorithms, and analyze the positioning performance [18, 19]. Finally, constructing a 5G positioning simulation environment also requires a channel model to provide observations. Therefore, articles or standards on 5G channel modeling are also introduced. At present, there is no unified standard for channel modeling of 5G. The commonly-used channel models include METIS [20], WINNER II [21], and related models of 3GPP [22]. It is worth mentioning that Rappaport T S et al. proposed a new channel models of 5G based on a large number of actual measurements and had compared performance with the 3GPP model [23, 24]. At the same time, they developed a channel simulator which was open soured [25].

Not only does the academic community care about 5G positioning, the industry also regards positioning as an important part of 5G in the future. The 3GPP has established a series of standards on positioning [26, 27]. The concept of 5G positioning has been extended in the standards, which is not a single method, but comprehensively adopts various current mainstream positioning methods, including A-GNSS, OTDOA, enhanced CID, Wi-Fi, Bluetooth, and air pressure positioning. It coincides with the current idea of multi-source integration, and it is also a major direction for future research.

In summary, there are three aspects can be studied in 5G positioning: (1) design aspects, including frequency selection, signal coding, and waveform designing; (2) interference, including multipath effect, LOS, NLOS and so on; (3) system requirements, including base station synchronization, the positioning strategy, and the solution of different requirements of locating and communicating such as base station distribution and base station-user interaction modes.

It can be noted that most of the articles are verified by simulation. Simulation is an important research method when 5G base stations and devices are not popular, and the standards of 5G positioning have not yet fully formed. However, only few article focus on the construction of 5G simulation environment. Therefore, how to establish an effective simulation environment may be a major obstacle. Based on this, this paper studies the system architecture of 5G positioning and proposes a simulation experiment scheme, and perform the implementation of simulation as verification, which will be significant references for researchers.

3 Scheme of the Simulation Experiment

The flowchart of 5G positioning proposed in this paper is shown in Fig. 2. The thought was inspired by works [11, 12] of Koivisto M et al. The user moves in uniform motion along a certain track in a certain scene, and makes a positioning request at regular intervals. At this time, the channel simulator simulates the signal propagation process, and gives the observations, that is, the time of arrival (TOA) and the angle of arrival (AOA). After the observations have been imported, the positioning program estimates the user's position and sends it to the user, completing a circle of positioning.

According to the above process, the simulation experiment of 5G positioning can be divided into three parts: scene generation, signal propagation simulation, and position estimation. The implementation steps are discussed below.

3.1 Scene Generation

The scene for positioning is the basis of simulation experiment. Both authenticity and complexity need to be considered in scene generation. Referring to the scenes in the simulation guide provided by authoritative organizations not only can reduce the workload, but also improve the reliability of the simulation. The scene used in this paper refers to the map of Madrid in the simulation guide [20] provided by Metis.

Two methods are taken to generate scene in this paper. One is to generate scene randomly by the program, and the other is to import data from open source maps (such as Open Street Map) with appropriate modification. These two methods have the same

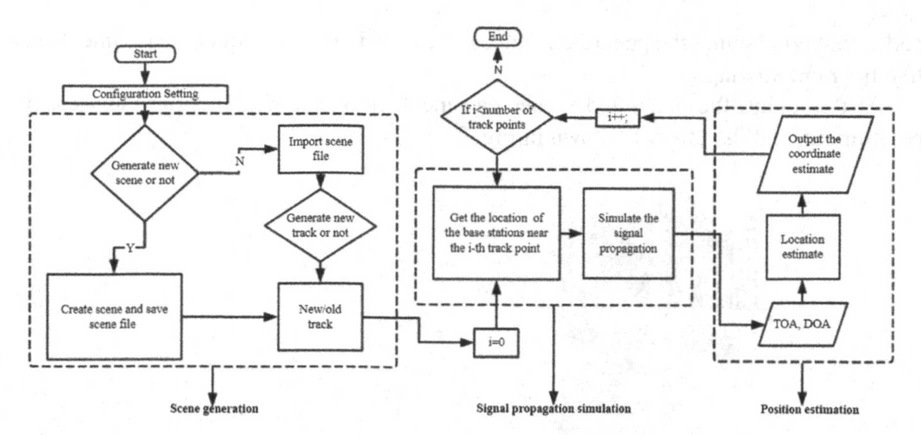

Fig. 2. Flowchart of 5G simulation experiment

positioning principles and processes. The introduction of the real map can increase the practical value of the simulation, for example, guiding the layout of the base station to obtain a better positioning effect, so we adopt these two methods.

Firstly, random scene generation is described. Four scene elements including buildings, roads, base stations and trajectories are finally selected after investigation [11, 12]. Random scene generation is based on buildings. And buildings with regular form are considered in this paper, which means the buildings are arranged in a square matrix, and the intervals between the buildings are equal and regarded as visual roads. The properties of the building include length, width, height and location information. The simulation guide [20] is the main reference for setting length, width and height, while the location information can be set randomly, just make sure the calculation is convenient. Then, base stations need to be generated. Combining the reality and simplifying it, we assume that base stations are randomly distributed on edge of the buildings [20]. So the base stations can be generated after obtaining data of the buildings. Another issue is how to confirm the number and distribution of the base stations. According to the conclusions in [12, 22, 28], the user can perform line-of-sight communication with the base station in 80% of cases when the base station interval is not more than 50 meters. We use this as a basis to deploy the base station. After that, the data of roads need to be generated. This paper specifically designs the topology structure for the road network. The roads and the directions are represented by the connection relationship of the intersections, and the connection relationship between the roads is also generated. The coordinates of the intersection are generated based on the building coordinates.

Next, scene generation by importing an open source map is briefly described. The open source map that contains the buildings information and the topological information of the road is stored in Extensive Markup Language (XML) format. It is parsed by python script and saved separately as a text file of a specific format. By reading the text file can reproduce the buildings and the roads, then the base station is generated based on the similar principle as described above. The topological relationships between the roads are automatically generated based on the information contained in the XML file. Without additional processing, the generated map can be used in the subsequent experiments just like the random map.

At this point, the generated scenes can meet the needs of the experiment for 5G positioning, and the effect is shown in Fig. 3.

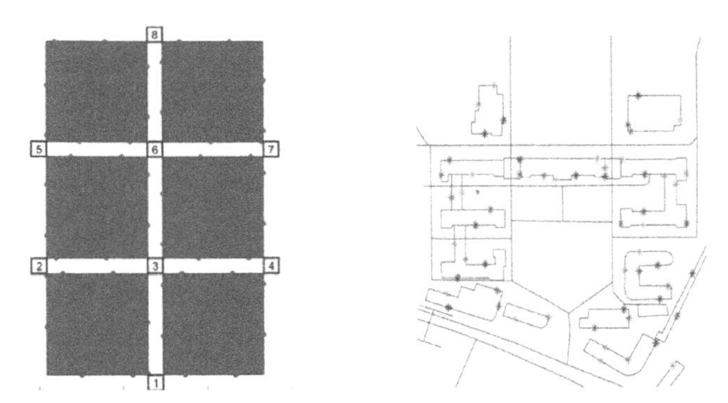

Fig. 3. Scenes of positioning. The left is the scene generating randomly and the right is the scene generated according to Open Street Map (OSM).

According to the positioning process, a track randomly generated on the simulation scene after the scene is generated. The trajectory generated in this paper is based on the road network. It includes two generation rules [11]: 1. when comes across the intersection, it will randomly turn to the other three directions but cannot return the way it came from; 2. the trajectory is considered to be over after six turns or when it go out of the map. It is easy to obtain the trajectory required for the experiment according to the above two generation rules and the road network data. The trajectory is represented by a series of road numbers, similar to the polyline. To achieve periodic positioning, the trajectory is split into a series of points based on a certain distance, and the coordinates of the points are recorded. The splitting can be realized from the point of vector or linear equation according to the line string of track and the known coordinates. The track point is the minimum unit of positioning in the simulation experiment.

3.2 Signal Propagation Simulation

The track points are taken out in turn from the obtained aggregate of points. Several base stations closest the track points are selected as the reference station for positioning. The standard of selection is the Euclidean distance, and the number of selected base stations can be set as needed. The signal propagation simulation can be conducted after the base stations being selected. The simulation experiment uses the channel simulator of New York University (NYU), an open source channel simulator of 5G developed by Rappaport et al. based on a large amount of measured data [25]. The simulator is simple and easy-to-use, and can provide plentiful parameters. And its simulation results improve compared with the commonly-used models such as WINNER and 3GPP [29]. Thanks to the open source of the software, the simulator after proper modification can be seamlessly integrated with the scene of positioning. Then the operation and principle of the simulator will be briefly introduced, and how to use the simulator to conduct signal propagation will be discussed.

The simulator is easy to operate. The simulation process is automatically completed by the program after the parameters are manually set. The input parameters are mainly used for channel and antenna setting [25]. As far as the impact on the simulation process is concerned, the most critical parameters are the scene parameters (sceType) and the environment parameters (envType). The simulator selects different built-in parameters for the subsequent simulation operations based on the combination of different scenarios and environments. These built-in parameters are from the statistical data obtained in field experiments under similar combinations by Rappaport team, which can better match the real environment. The simulator conducts signal propagation simulation in 12 steps as shown in Fig. 3. It finally gives the key values required for channel modeling, mainly including power distribution of the angle of departure, power distribution of the angle of arrival, omnidirectional power delay profiles, directional power delay profiles and small-scale power delay profiles. Several key steps are described below (Fig. 4):

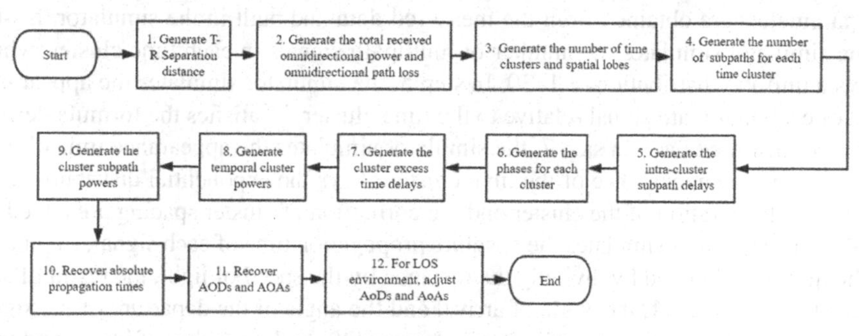

Fig. 4. The process of signal propagation simulation

(1) Path loss model [25]. The path loss model adopted by NYU satisfies the following formula:

$$P_L = 20\lg\left(4\pi\frac{d_0}{\lambda}\right) + n10\lg\left(\frac{d_{ist}}{d_0}\right) + d_{ist} \times aF_A + \Delta \tag{1}$$

P_L represents the path loss; $d_0 = 1$ m is the reference distance of free space; λ is the carrier wavelength; $n = 2$ is the path loss index; d_{ist} is the distance between the base station and the user; aF_A is the atmospheric attenuation factor and Δ is the shadow fading, which satisfies the normal distribution with a mean value of 0 and standard deviation of 4 dB. It is worth noting that the values of n and the distribution of Δ are different as parameters are different in different scene and environment.

(2) Simulation of multipath delay and angle

Two concepts time cluster and lobe are introduced to simulate the distribution of multipath signals in time and space dimensions, respectively in order to simulate the multipath in a more real way [30]. The time cluster refers to the multipath signal reaching the receiving end intensively in a certain period of time, and the lobe describes the spatial aggregation of the multipath signal. These two concepts are not firstly proposed in this simulator, but in previous models or literature the multipath signals of the same time cluster belong to the same lobe by default. However, Rappaport et al. had different research results [30]. This is also reflected in the simulator. Time clusters and lobes are generated separately with no binding between them.

According to the source code, the simulator completed the simulation of multipath in steps 3, 4, 5, 7, 10, and 11 step by step. In step 3, the number of time clusters and lobes that occur during a signal propagation process is simulated. According to the measured data, the number of time clusters obeys a uniform distribution of 1–6, and the number of lobes of the transmitting and the receiving end satisfies the Poisson distribution. And the parameters are obtained from the measured data and built in the simulator. In step 4, the simulator simulates the number of multipath signals in each time cluster, which obeys a uniform distribution of 1–30. In step 5, the simulator simulates the appearance time of each multipath signal relative to the time cluster, it satisfies the formula derived from the measured data. In step 7, the simulator simulates the appearance time of each time cluster. The appearance of the time cluster obeys the exponential distribution. On this basis, the duration of the cluster and the correction of cluster spacing are added. In step 10, the simulator simulates the absolute propagation time of each signal, by sum up of the quotient obtained by dividing the distance by the speed of light, the result of step 5 and step 7. In step 11, the angle of arrival and the angle of the departure of the signal are simulated. The space is equally divided according to the number of lobes, and then an angle is randomly selected as the central angle in each aliquot. For each signal, first randomly select the interval in which it falls, then it will fall near the central angle of the interval, and the distribution of obedience is Gaussian or Laplacian, and the parameter is angle. The angle value and the specific distribution are obtained from the measured data and are built in the simulator.

The propagation delay and the angle of departure and arrival of 5G signals are gradually generated during the simulation process of the channel simulator, which are the observations required for positioning. But there is a contradiction between the original simulator and the experiment of positioning that is the distance and angle of signal propagation are randomly generated by the simulator but the positional relation between the user and the base station in the positioning scenario is fixed. The contradiction is a key issue to be solved in this paper. It is solved by modifying the simulator based on the in-depth study on the simulator code. The specific idea is to calculate the geometric position relationship between the user and each base station at each track point, including the distance and angle, and then input it as a parameter into the channel simulator. After that, the random generation of the channel simulator is changed to a semi-random generation based on the input parameters. The output is the delay and angle from the user to each base station. Therefore, the simulation of the signal propagation process and the acquisition of observations are completed.

3.3 Position Estimation

Position estimation is the final process after completing the simulation of signal propagation and obtaining the TOA and DOA data. It is not difficult to make an observation equation based on the known geometric relationships. The result of position estimation can be obtained after the observation equation is linearized and solved iteratively by the least squares.

The observation equation is as follows:

$$\boldsymbol{Z}_i(\boldsymbol{k}) = \begin{bmatrix} AOD_{Azi} \\ AOD_{ele} \\ Ct \end{bmatrix} = \begin{bmatrix} \arctan\frac{y-y_s}{x-x_s} \\ \arctan\frac{z-z_s}{\sqrt{(x_s-x)^2+(y_s-y)^2}} \\ \sqrt{(x_s-x)^2+(y_s-y)^2+(z_s-z)^2} \end{bmatrix} \tag{2}$$

$\boldsymbol{Z}_i(\boldsymbol{k})$ is the observation value of the i-th base station at time k; AOD_{Azi} is the angle of arrival in the horizontal direction; AOD_{ele} is the angle of arrival in the vertical direction; t is the propagation delay and $C = 3 \times 10^8$ m/s. (x_s, y_s, z_s) is the coordinates of the i-th base station closest to the user at time k; (x, y, z) is the coordinates of the user at time k.

Linearize the Eq. (2) at the approximate coordinate $\boldsymbol{X}^0(\boldsymbol{k})$ and gain:

$$\boldsymbol{Z}_i(\boldsymbol{k}) = \boldsymbol{H}_i(\boldsymbol{k})\boldsymbol{x}(\boldsymbol{k}) + \boldsymbol{Z}_i^0(\boldsymbol{k}) \tag{3}$$

Where $\boldsymbol{H}_i(\boldsymbol{k})$ is the linearized Jacobian matrix; $\boldsymbol{Z}_i^0(\boldsymbol{k})$ is the approximate value of the observation at $\boldsymbol{X}^0(\boldsymbol{k})$ and $\boldsymbol{x}(\boldsymbol{k})$ is the difference between $\boldsymbol{X}^0(\boldsymbol{k})$ and the true position (x, y, z). The specific expressions are listed below:

$$\boldsymbol{x}(\boldsymbol{k}) = \begin{bmatrix} \Delta x \\ \Delta y \\ \Delta z \end{bmatrix} \quad \boldsymbol{Z}_i^0(\boldsymbol{k}) = \begin{bmatrix} arctan\frac{y_0-y_s}{x_0-x_s} \\ arctan\frac{z_0-z_s}{\sqrt{d_3}} \\ \sqrt{d_2} \end{bmatrix}$$

d_2 represents the geometric distance from the base station to $\boldsymbol{X}^0(\boldsymbol{k})$ and d_3 represents the length of projection on the plane where $\boldsymbol{X}^0(\boldsymbol{k})$ lies.

$$d_2 = (x_s - x_0)^2 + (y_s - y_0)^2$$

$$d_3 = (x_s - x_0)^2 + (y_s - y_0)^2 + (z_s - z_0)^2$$

$$\boldsymbol{H}_i(\boldsymbol{k}) = \begin{bmatrix} \frac{y_s-y_0}{d_2} & \frac{-(x_s-x_0)}{d_2} & 0 \\ \frac{(z_0-z_s)(x_s-x_0)}{d_2\sqrt{d_3}} & \frac{(z_0-z_s)(y_s-y_0)}{d_2\sqrt{d_3}} & \frac{\sqrt{d_3}}{d_2} \\ \frac{-(x_s-x_0)}{\sqrt{d_2}} & \frac{-(y_s-y_0)}{\sqrt{d_2}} & \frac{-(z_s-z_0)}{\sqrt{d_2}} \end{bmatrix}$$

In formula (3), let:

$$z(k) = \begin{bmatrix} Z_1(k) - Z_1^0(k) \\ \vdots \\ Z_i(k) - Z_i^0(k) \\ \vdots \end{bmatrix} H(k) = \begin{bmatrix} H_1(k) \\ \vdots \\ H_i(k) \\ \vdots \end{bmatrix}$$

Then:

$$z(k) = H(\mathrm{k})x(k) \tag{4}$$

For Eq. (4), the least square solution is

$$x(k) = \left(H^T(\mathrm{k})H(\mathrm{k})\right)^{-1} H^T(\mathrm{k})z(k) \tag{5}$$

Then, the estimation of the user's location is:

$$\hat{X}(k) = X^0(k) + x(k) \tag{6}$$

The above is a complete process of position estimation. Usually, there are some requirements for the positioning accuracy. Thus, it is often difficult to meet this requirement by performing estimation only once. It is generally necessary to improve the accuracy by an iterative method.

The algorithm is as follows:

Algorithm 1: Least Square
Input: $X^0(k), Z_i(k)$
Output: $X^0(k)$
1.While $\mathrm{norm}(x(k)) > 1e-8$
2. Calculate $H(k)$、$z(k)$;
3. $x(k) = (H^T(\mathrm{k})H(\mathrm{k}))^{-1}H^T(\mathrm{k})z(k)$
4. $X^0(k) = X^0(k) + x(k)$;
5.end

$X^0(k) = [x_0, y_0, z_0]^T$ can be initially given by the user through GNSS, or can be assumed to be the center position of several nearby base stations. It is then iterated continuously according to the above algorithm until it meets the given accuracy requirements.

4 Experiment and Discussion

The 5G simulation environment can be built up based on the previous section, however, it needs to be verified whether it is effective or it can support further research. For this purpose, some important factors that will influence the positioning effect are found out.

These factors are used in experiments as variable, and we can test the environment by observing whether the results change as expected or not. At the same time, there are many meaningful results in these experiments to be described. However, the reason is not the theme of this paper.

We get the positioning results from the observation Eqs. (2). These important factors can be found by analyzing the observation Eqs. (2). TOA and AOD finally decide the results, so each factor that affects TOA and AOD will influence the positioning effect. Considering the whole propagation process, such factors mainly include the number of base station, LOS/NLOS, multipath effect, and carrier frequency & bandwidth. Besides, in GNSS positioning, the distribution of base station will affect the Geometry dilution of position which has great influence in positioning result. The distribution may also have influences as our scheme is based on GNSS. In here, factors and experiments settings are listed in Table 3.

Table 3. Settings of simulation experiments

Scene and estimation method	Carrier frequency and bandwidth	Number of base stations	Base station distribution	LOS/NLOS	Multipath effect
Random map + least squares	3.5–100 28–400 77–800	1/2/3/4	None	None	None
Random map + least squares	3.5–100 28–400 77–800	1/2/3/4	Random/Adjustment	None	None
OSM map + EKF	3.5–100 28–400 77–800	1/2/3/4	Random/Adjustment	LOS/NLOS	Single/multi

Experiments were conducted in the simulation environment according to the experimental settings described above. Only a few representative results were selected to display and discuss because the results of the experiments were numerous and somewhat repetitive.

4.1 The Influence of the Number of Base Stations for Positioning

The number of base station decides the number of observation. Each base station will offer three observations. As there are three unknowns to be solved, 1 base station is needed at least. Adding base stations can make positioning result get better. In this experiment, we test the simulation environment's ability of simulating the specific impact of the number of base stations on the positioning effect.

Positioning was performed using a 3.5 GHz–100 MHz signal on the random map. Under the condition the base stations were randomly set, the number of base stations used for positioning increased from one to four. The location estimation was performed by the least squares, and the positioning effects of different number of base stations are shown in Fig. 5.

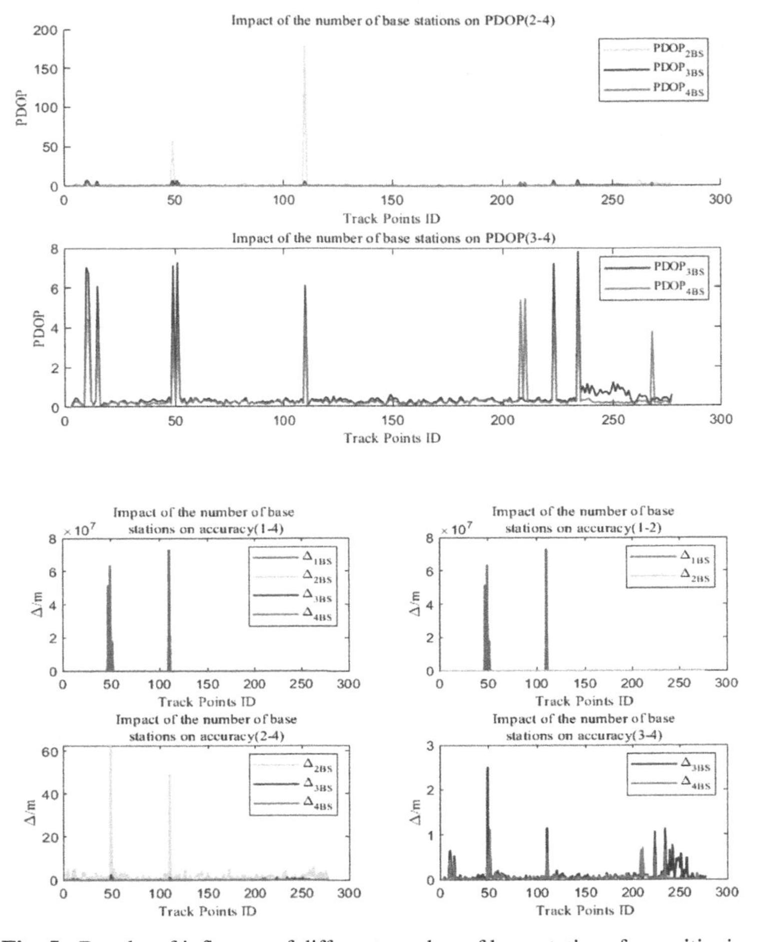

Fig. 5. Results of influence of different number of base stations for positioning

The results include the upper part and the lower part. The upper part shows the effect of the number of base stations on the Position Dilution of Precision (PDOP). PDOP characterizes the influence of measurement error on the final result of positioning. Generally speaking, the smaller the PDOP is, the better the observation condition is. The horizontal axis is the Track Point ID, represents different track points, and the vertical axis is the PDOP. The upper images show the PDOP value at each track point. The lower part shows the influence of the number of base stations on the positioning accuracy. Here, the horizontal axis is still the Track point ID, but the vertical axis is Δ, which represents the geometric distance in meters between the estimation result and the true value. The lower images show the positioning result on each track point. The later experimental results also conform to the above description.

There is no PDOP value when there is only one base station in Fig. 5 because there is no redundant observation with only one base station. When further analysis is carried out, it can be found that some outliers greatly affect the performance of the graphics as shown in Fig. 5. These outliers are actually the points with large positioning error, for example, the iteration does not converge. By adding constraints to the program of position estimation, these points can be found and eliminated, and the success rate of positioning can be used to characterize this situation. The success rate of positioning obtained by dividing the total number of experiments by the number of successful experiments under a certain condition of experiment. The success rate of positioning with different number of base stations is shown in the table below (Table 4).

Table 4. Success rate and number of base stations

The number of base stations	1	2	3	4
The success rate of positioning	98.56%	98.56%	99.64%	100%

On the basis of the above, the following results are obtained by removing the outliers (Fig. 6).

From the perspective of PDOP, the PDOP values are different with different number of base stations. As a whole, the more the base stations there are, the smaller the PDOP is. Specifically, the PDOP becomes smaller in most points when the number of base stations increases from 2 to 3, and when the number increases from 3 to 4, only points with larger PDOP can become smaller. In addition, the PDOPs at different track points are different when the number of base stations is the same. It may be caused by different distribution of base stations. From the perspective of positioning accuracy, the results are similar to that of PDOP, that is the more the number of base stations there is, the higher the accuracy of positioning is. The positioning accuracy is about 2 m when the number of base stations increases from 1 to 2, and it is about 0.1 m when the number of base stations increases from 3 to 4, neither of them have obvious improvement. However, the positioning accuracy changes significantly when the number of base stations increases from 2 to 3. Additionally, the positioning accuracy is different on different track points. In general, the positioning accuracy is agreed with the curve of the PDOP. The positioning accuracy is low when the PDOP is large, and the positioning accuracy is high when the PDOP is small, which is similar with GNSS.

4.2 Distribution Influence of the Base Station

In GNSS system, the geometrical distribution of satellites has a large impact on the results of positioning. It may also have influence in the proposed scheme. This experiment is used to test whether the environment can explore the influence of the geometrical distribution of base stations.

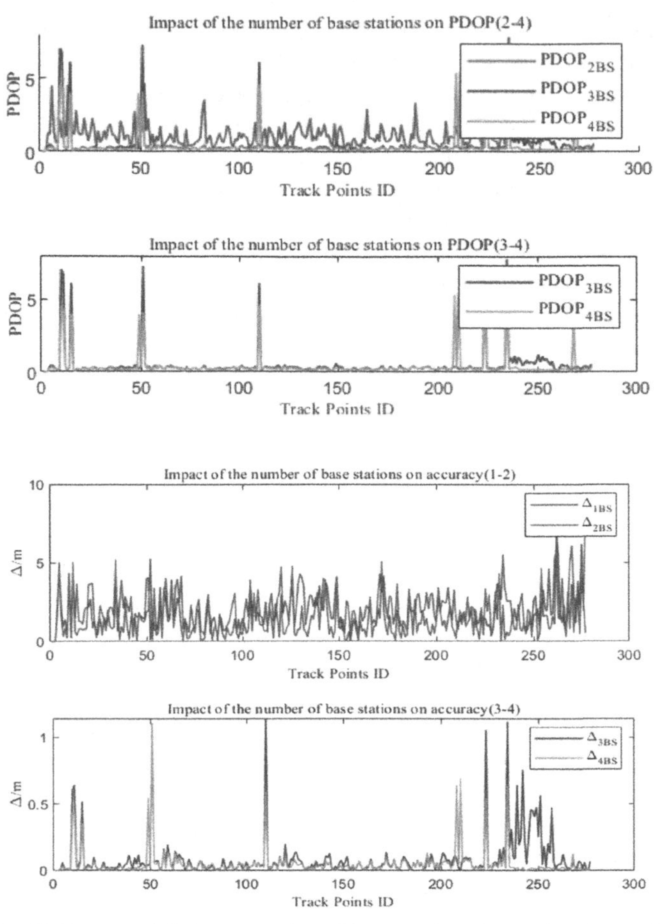

Fig. 6. Result of influence of different number of base stations for positioning (without outliers)

In the OSM map, the frequency of the signal used for positioning is set to 3.5 GHz–100 MHz; the number of base stations used for positioning is 4, and the estimation method of positioning is least squares. The geometric distribution of the base station and the final effect of positioning are shown below (Figs. 7, 8 and 9).

Comparing the scene graphs before and after the base station supplementation (Fig. 7, Fig. 8), it can be seen that two base stations are added on both sides of the scene, and a total of four base stations are added. It can be seen from the result graph (Fig. 9) that before the base station is supplemented, there are three regions in the trajectory with large PDOP, correspondingly, resulting in three regions with large errors of positioning. By supplementing the base station, it can be seen that the PDOP of the original three regions is greatly reduced, and the accuracy of positioning is improved. It can be seen that the arrangement of the base station has a great influence on the positioning accuracy.

The simulation program in this paper can represent the real-time positioning effect of the user in the scene in the form of error ellipse. It is also the basis for supplementing the

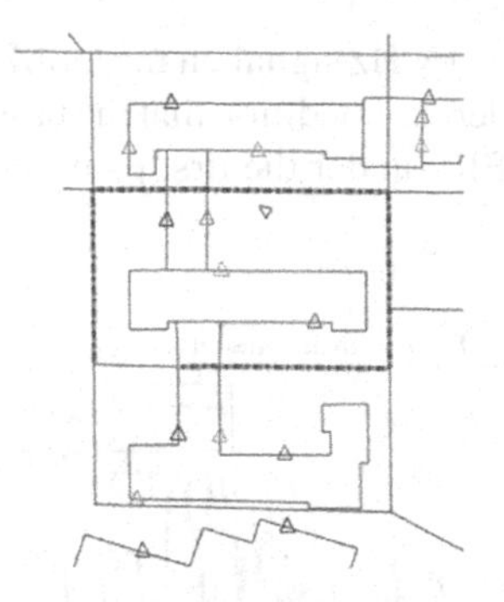

Fig. 7. Base station in random

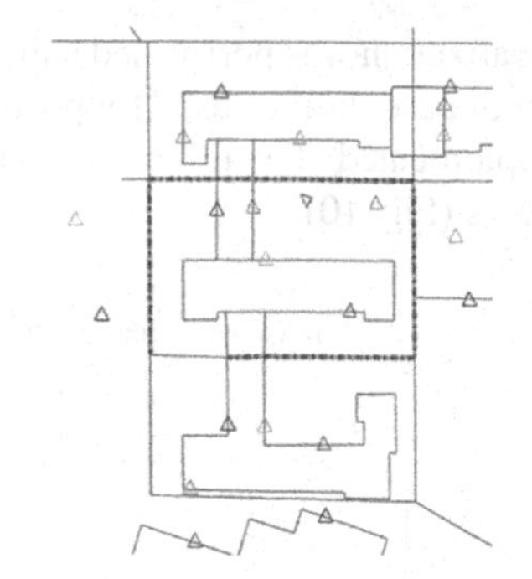

Fig. 8. Base station with judgement

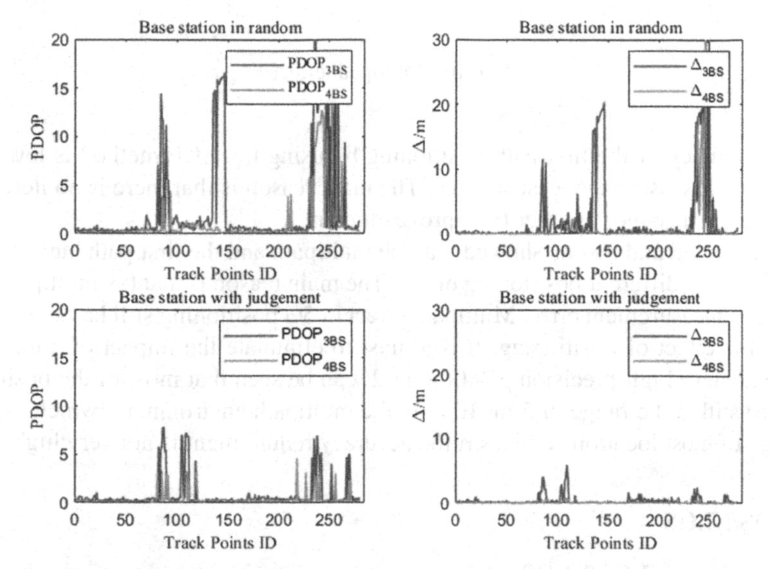

Fig. 9. The influence of the arrangement of the base stations

base station, that is, supplementing the base stations in the area where the error ellipse is large. The experimental results show that supplementing a small number of base stations in key areas can greatly improve the effect of positioning. The simulation program in this paper helps to find these key areas.

4.3 Multipath Effect

The multipath effect still exists in the 5G positioning, and this experiments is to test can it be studied through the simulation environment established in this paper. All path signals are weighted to calculate their observations according to the received intensity of signal, then the positioning under the multipath effect is simulated, and the positioning result of the first-path used as reference.

Localization was performed using a 3.5 GHz–100 MHz signal on the OSM map; the number of base stations used for positioning is 4 under the condition that the base stations are supplemented. The effect of positioning with EKF under the first path/multipath is as follows (Fig. 10):

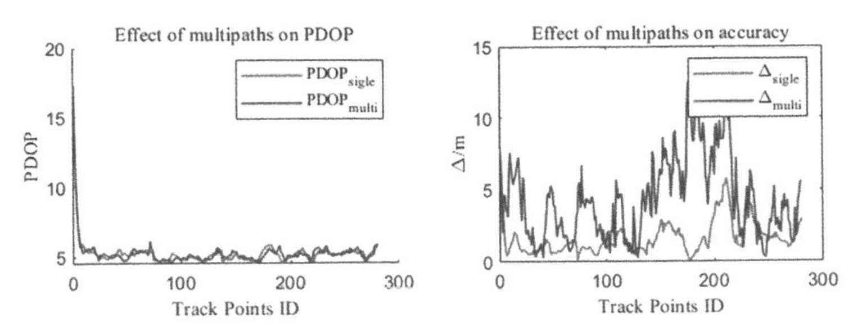

Fig. 10. Multipath effect

The accuracy of the first-path positioning by using the EKF method is lower than that obtained by using the least squares. The main reason is that there is no iteration in EKF code in this paper in order to improve efficiency.

The experimental results showed that the multipath and the first path have the same PDOP but quite different positioning effect. The main reason is that the multipath introduces large measurement error. Multipath effect in 5G positioning still has a great influence on the effect of positioning. It is a must to eliminate the impact of multipath in order to achieve high-precision positioning. It can be seen that most of the positioning errors are still in the range of 5 m–10 m in the multipath environment, which can meet the needs of most location services if the accuracy requirement is not very high.

4.4 LOS/NLOS

Choosing different LOS/NLOS parameters can simulate signal propagation in both LOS and NLOS environments. With this, the simulation environment of this paper may have ability in studying the influence of line-of-sight on the effect of positioning, and it is tested in this experiment.

In the channel simulator, the observations are different under LOS and NLOS conditions. While in the LOS condition the arrival angle of the first-path signal is corrected so that the difference between the arrival angle and the departure angle is maintained at 180°, then all the multipath are adjusted according to the corrected results, so that their characteristics satisfy the LOS condition better. And this correction is not conducted in the NLOS condition. In order to fully demonstrate the difference between LOS and NLOS, multipath position is used to compare the influence of LOS and NLOS on positioning.

Localization was performed using a 3.5 GHz–100 MHz signal on the OSM map; the number of base stations used for positioning is 4 under the condition that the base station is supplemented. The effect of positioning using EKF under LOS/NLOS is as follows (Fig. 11):

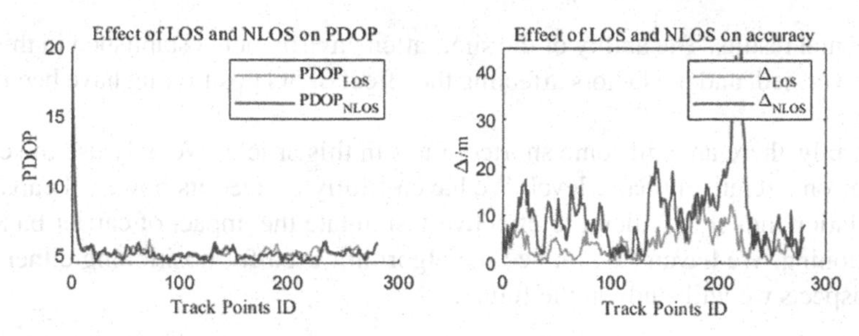

Fig. 11. LOS/NLOS effect

The results show that the positioning error in the NLOS condition is larger than that in the multipath effect, and the positioning error of about 5 m is expanded to nearly 20 m. Therefore, it is essential to ensure the LOS in order to achieve high-precision positioning. That is, a high-density base station network is the basis for high-precision 5G positioning.

4.5 Other Findings

There are some other findings in this article except for the above representative experiments.

In this paper, the positioning effects of different carrier bandwidth signals are tested. In various combination scenes. The positioning effect is somewhat different but not significant. In short, carrier bandwidth signals with frequency from 3.5 GHz to100 Mhz work best in the simulation program.

Comparing the positioning effects of EKF and least squares, the least squares requires more positioning base stations and better geometric distribution, and its positioning effect is slightly better than EKF. Two methods are used depending on the actual positioning condition. EKF is preferred in poor positioning condition while the least squares are used when the positioning condition is superior.

5 Conclusion

As one of the most popular technologies, 5G can be used not only in the field of communication, but also in the field of positioning. This paper first discusses the general idea of 5G positioning, namely 5G A-GNSS and pseudo-satellite. Next, new 5G technologies can help locating are introduced, including millimeter wave, MIMO and beamforming, and UDN. At the same time, the current research and standards related to 5G positioning are classified and explained, mainly comprising frequency, waveform, positioning strategy, time synchronization, channel model and so on.

We have proposed a 5G positioning simulation experiment scheme and presented its flow chart. Then the implementation ideas and the specific processes of the three main parts of the simulation experiment including scene generation, signal propagation simulation and position estimation are introduced. On this basis, we have carried out a large number of experiments by changing the system settings, and obtained corresponding experimental results. The ability of the simulation environment established in this paper has been verified, and the factors affecting the effect of 5G positioning have been briefly analyzed.

Certainly, there are still some shortcomings in this article. We only use the channel simulator on a relatively basic level. We haven't fully utilized its powerful capabilities on simulation thus it is difficult to effectively simulate the impact of carrier bandwidth on positioning. We haven't optimized the algorithm used for positioning either. These are the aspects we will study in the future.

Acknowledgments. This research was supported by the National Key Research and Development Program of China (2018YFC0809804).

References

1. Zhang, P., Tao, Y.Z., Zhang, Z.: Survey of several key technologies for 5G. J. Commun. (2016)
2. Qian, Z., Wang, X.: Reviews of D2D technology for 5G communication net-works. J. Commun. **7**, 1–14 (2016)
3. Zhao, Y.D., Wei, Z.Q., Feng, Z.Y., et al.: Fusion architecture and key technologies of satellite navigation and 5G mobile communication. Telecom Eng. Tech. Stand. **30**(1), 48–53 (2017)
4. Han, S., Gong, Z., Meng, W., et al.: Future alternative positioning, navigation, and timing techniques: a survey. IEEE Wirel. Commun., 2–9 (2016)
5. Guo, W., Song, W., Niu, X., et al.: Foundation and performance evaluation of real-time GNSS high-precision one-way timing system. GPS Solut. **23**(1) (2019)
6. Wymeersch, H., Seco-Granados, G., Destino, G., et al.: 5G mmWave positioning for vehicular networks. IEEE Wirel. Commun. **24**(6), 80–86 (2017)
7. Del Peral-Rosado, J.A., Raulefs, R., Lopez-Salcedo, J.A., et al.: Survey of cellular mobile radio localization methods: from 1G to 5G. IEEE Commun. Surv. Tutor., 1 (2017)
8. Cui, X., Gulliver, T.A., Song, H., et al.: Real-time positioning based on millimeter wave device to device communications. IEEE Access (2016)
9. del Peral-Rosado, J.A., et al.: Whitepaper on new localization methods for 5G wireless systems and the Internet-of-Things, 1–27 (2018)
10. Shahmansoori, A., Seco-Granados, G., Wymeersch, H.: Survey on 5G positioning. In: Nurmi, J., Lohan, E.-S., Wymeersch, H., Seco-Granados, G., Nykänen, O. (eds.) Multi-Technology Positioning, pp. 165–196. Springer, Cham (2017). https://doi.org/10.1007/978-3-319-50427-8_9
11. Werner, J.: Directional antenna system-based DoA/RSS estimation, localization and tracking in future wireless networks: algorithms and performance analysis. Tampere University of Technology. Publication, vol. 1350. Tampere University of Technology (2015)
12. Koivisto, M., Costa, M., Werner, J., et al.: Joint device positioning and clock synchronization in 5G Ultra-dense networks. IEEE Trans. Wirel. Commun. **PP**(99), 2866–2881 (2016)
13. Shahmansoori, A., Garcia, G.E., Destino, G., et al.: Position and orientation estimation through millimeter wave MIMO in 5G systems. IEEE Trans. Wirel. Commun. **17**(3), 1822–1835 (2017)
14. Shahmansoori, A., Garcia, G.E., Destino, G., et al.: 5G position and orientation estimation through millimeter wave MIMO (2017)
15. Dammann, A., Jost, T., Raulefs, R., et al.: Optimizing waveforms for positioning in 5G. In: IEEE International workshop on Signal Processing advances in Wireless Communications. IEEE (2016)

16. Abu-Shaban, Z., Zhou, X., Abhayapala, T., et al.: Error bounds for uplink and downlink 3D localization in 5G mmWave systems. IEEE Trans. Wirel. Commun., 1 (2018)
17. Li, H., Han, L., Duan, R., et al.: Analysis of the synchronization requirements of 5G and corresponding solutions. IEEE Commun. Stand. Mag. **1**(1), 52–58 (2017)
18. Begusic, S., et al.: Wireless indoor positioning relying on observations of received power and mean delay. In: 2013 IEEE International Conference on Communications Workshops (ICC), pp. 74–78, June 2013
19. Witrisal, K., Leitinger, E., Hinteregger, S., Meissner, P.: Bandwidth scaling and diversity gain for ranging and positioning in dense multipath channels. IEEE Wirel. Commun. Lett. **5**(4), 396–399 (2016)
20. METIS: D6.1 Simulation guidelines, October 2013. https://www.metis2020.com/wp-content/uploads/deliverables/METISD6.1v1.pdf
21. Meinilä, J., Kyösti, P., Jämsä, T., et al.: WINNER II channel models. Radio Technologies and Concepts for IMT-Advanced (2008)
22. GPP TR 36.873: Study on 3D channel model for LTE (release 12) (2015). http://www.3gpp.org/dynareport/36873.htm
23. Rappaport, T.S., Sun, S., Shafi, M.: Investigation and comparison of 3GPP and NYUSIM channel models for 5G wireless communications (2017)
24. Shu, S., Rappaport, T., Shafi, M., et al.: Propagation models and performance evaluation for 5G millimeter-wave bands. IEEE Trans. Veh. Technol. **PP**(99), 1 (2018)
25. Shu, S., Maccartney, G.R., Rappaport, T.S.: A novel millimeter-wave channel simulator and applications for 5G wireless communications (2017)
26. GPP TS 38.455: NG-RAN NR Positioning Protocol A (NRPPa)
27. GPP TS 38.305: NG Radio Access Network (NG-RAN); Stage 2 functional specification of User Equipment (UE) positioning in NG-RAN
28. METIS: D1.4 Channel models February 2015. https://www.metis2020.com/wp-content/uploads/METISD1.4v3.pdf
29. Rappaport, T.S., Sun, S., Shafi, M.: Investigation and comparison of 3GPP and NYUSIM channel models for 5G wireless communications. In: 2017 IEEE 86th Vehicular Technology Conference (VTC-Fall), pp. 1–5. IEEE (2017)
30. Samimi, M.K., Rappaport, T.S.: 3-D millimeter-wave statistical channel model for 5G wireless system design. IEEE Trans. Microw. Theory Tech. **64**(7), 2207–2225 (2016). http://ieeexplore.ieee.org/document/7501500/

UAV Powered Cooperative Anti-interference MEC Network for Intelligent Agriculture

Chenkai Li, Weidang Lu(✉), Xiaohan Xu, Hong Peng, and Guoxing Huang

College of Information Engineering, Zhejiang University of Technology, Hangzhou 310023, China
luweid@zjut.edu.cn

Abstract. Mobile edge computing (MEC) provides computing resources for lots of Internet-of-things (IoT) devices in intelligent agriculture, improving the efficiency of agriculture. The energy harvesting efficiency of wireless power MEC system can be improved through combining unmanned aerial vehicle (UAV) technique. In this paper, we propose an UAV powered cooperative anti-interference mobile edge computing strategy for intelligent agriculture, in which IoT devices transmit their information through distinct subcarriers. Required energy for information transmission of UAV is minimized through optimizing power allocation. Simulation results demonstrate the performance of our proposed strategy.

Keywords: Wireless power transfer (WPT) · Mobile edge computing · Cooperation · UAV

1 Introduction

With the maturity and commercial deployment of Internet-of-Things (IoT), smart applications have been gradually applied to intelligent agriculture. Due to the explosive growth of IoT devices in intelligent agriculture systems, a variety of real-time environmental information needs to be collected for ensuring yield and quality, insufficient computing resources have become a problem [1, 2].

MEC technology is able to provide cloud computing services for IoT devices, and assist IoT devices in completing computation-intensive and time-critically tasks through wireless access networks [3]. However, inadequate battery supply for IoT devices can impact MEC performance severely. WPT is able to provide sustainable and low-cost energy supply for low-power mobile devices from radio frequency (RF) signals, greatly extending the standby time of mobile devices [4]. [5] studied a WPT collaborative MEC system, aim at minimizing the emission energy under the energy consumption and delay constraints. A binary computation offloading strategy for multi-user WPT based MEC networks was studied in [6], which maximized the total computing rate of the mobile device by optimizing transmission rate and computation selection mode.

IoT devices are usually located in remote areas in intelligent agriculture. If the IoT device exceeds the range of wireless communication during the communication process,

X. Wang et al. (Eds.): 6GN 2020, LNICST 337, pp. 298–307, 2020.
https://doi.org/10.1007/978-3-030-63941-9_22

it will cause the communication with MEC to fail [7]. Because of its highly flexible characteristics, UAVs have been widely used in a variety of scenarios [8]. Moreover, the energy transfer efficiency is improved through the UAV-assisted wireless power transmission structure to create a short-distance power transmission link in [9]. In [10], optimizing the power allocation and trajectory of UAVs which are equipped with MEC servers and provide services for multiple mobile devices within a limited range, minimizes the total consumed energy.

However, WPT enabled wireless communication networks are vulnerable to the double "near-far" effect. IoT device located far away from ET acquires less energy but needs to communicate for a long distance. Through cooperation between closer-device and farther-devices, communication performance can be improved. The cooperation of the devices in wireless powered communication was studied in [11] to surmount the double "near-far" effect, in which the closer-device uses the same channel to assist the farther-devices, causing interference.

In order to surmount the interference in the process of IoT devices' cooperation, we propose an UAV powered cooperative anti-interference mobile edge computing strategy for intelligent agriculture, where the IoT device which is closer to the UAV transmits the information of the farther IoT devices and its own information by utilizing distinct subcarriers.

2 System Model and Problem Formulation

2.1 System Model

The UAV powered MEC network consists of one UAV and two IoT devices IoD_1 and IoD_2. We assume that IoD_1 and IoD_2 need to accomplish computationally intensive latency critical tasks. IoD_1 and IoD_2 obtain energy from UAV through WPT method and use the harvested energy to accomplish their computation tasks with partial offloading. IoD_1 and IoD_2 transmit their tasks with half-duplex mode. We adopt a harvest-then-offload protocol for wireless powered computation offloading, which uses a block-based TDMA structure with the duration of each block being T seconds.

Let I_i and O_i denote the size of the input computation data and output data, respectively. Assuming the maximum tolerable delay time $T_i = T(i \in \{1, 2\})$ of IoD_1 and IoD_2 are the same. Since input data is much larger than output data, i.e., $O_i \ll I_i$. We just consider the time of the WPT and uplink offloading as the total latency of the proposed UAV powered MEC system.

The signal is OFDM modulated on K subcarriers, and the subcarrier set is denoted as $N = \{1, 2, \cdots, K\}$. In order to surmount the interference, IoD_2 utilizes first half of subcarriers $k \in G_I$ to relay the tasks of IoD_1 and the other half subcarriers $k \in \overline{G_I}$ to transfer its own offloading tasks to UAV, where $G_I \in N$ and $\overline{G_I} \in N$.

In the first period $\frac{T}{3}$, UAV transmits signal to IoD_1 and IoD_2 over subcarrier k with power p_k at the same time, which is equally allocated. The harvest energy of IoD_i is given by

$$E_h^i = \frac{T}{3} \sum_{n=1}^{K} v_i \gamma_{i,k} p_k, i \in \{1, 2\} \tag{1}$$

where v_i denotes the energy conversion efficiency at IoD_i, $\gamma_{i,k}$ denotes the channel power gain on subcarrier k from UAV to IoD_i.

In the second period $\frac{T}{3}$, IoD_1 uses harvest energy to send offloading task to IoD_2 with power $p_{1,k}$. The offloading task $L_{1,2}$ is given by

$$L_{1,2} = \frac{T}{3} B \sum_{k=1}^{K} \log_2\left(1 + \frac{p_{1,k}\gamma_{3,k}}{N_2}\right) \tag{2}$$

where B denotes the channel bandwidth of subcarrier, $\gamma_{3,k}$ denotes the channel power gain on subcarrier k from IoD_1 to IoD_2, and N_2 denotes the receiver noise power at the IoD_2.

The energy consumed by offloading tasks from IoD_1 to IoD_2 is given by

$$E_{\text{off}}^{1} = \frac{T}{3} \sum_{k=1}^{K} p_{1,k} \tag{3}$$

In the third period $\frac{T}{3}$, IoD_2 uses harvest energy to forward the IoD_1's offloading task to UAV over subcarrier $k \in G_I$ with power $p_{21,k}$. The offloading task $L_{2,1}$ is given by

$$L_{2,1} = \frac{T}{3} B \sum_{k \in G_I} \log_2\left(1 + \frac{p_{21,k}\gamma_{4,k}}{N_0}\right) \tag{4}$$

where $\gamma_{4,k}$ denotes the channel power gain over subcarrier k from IoD_2 to UAV, and N_0 denotes the receiver noise power at the IoD_2.

Meanwhile, IoD_2 sends its own offloading task to UAV over subcarrier $k \in \overline{G_I}$ with power $p_{22,k}$. The offloading task L_2 is given by

$$L_2 = \frac{T}{3} B \sum_{k \in \overline{G_I}} \log_2\left(1 + \frac{p_{22,k}\gamma_{4,k}}{N_0}\right) \tag{5}$$

The energy consumed by offloading tasks from IoD_2 to UAV is given by

$$E_{off}^{2} = \frac{T}{3}\left(\sum_{k \in G_I} p_{21,k} + \sum_{k \in \overline{G_I}} p_{22,k}\right) \tag{6}$$

Thus, the offloading tasks size of IoD_1 with the relaying of the IoD_2 is given by

$$L_1 = \min\{L_{1,2}, L_{2,1}\} = L_{2,1} \tag{7}$$

The rest of data $I_i - L_i$ need to be computed locally at IoD_i, which should satisfy

$$(I_i - L_i(\mathbf{p}))C_i/f_i \leq T \tag{8}$$

where C_i denotes the amount of computing resources required to calculate 1-bit input data, f_i denotes the CPU frequency.

The minimum offloading data of IoD_i is denoted as

$$L_i(\mathbf{p}) \geq M_i^+ \tag{9}$$

where $M_i = I_i - f_i T / C_i$, $(x)^+ = \max\{x, 0\}$.

The energy consumption of IoD_i for local computation is given by

$$E_{loc}^i = (I_i - L_i) C_i Q_i \tag{10}$$

where $Q_i = k_i f_i^2$, k_i is the effective capacitance coefficient, depending on the chip architecture.

Therefore, IoD_i's saving energy is denoted as

$$E_s^i(P_0, \mathbf{p}) = E_h^i - E_{off}^i - E_{loc}^i \tag{11}$$

where $\mathbf{p} = [p_{1,k}, p_{21,k}, p_{22,k}]$.

2.2 Problem Formulation

In order to minimize the transmission energy of the UAV, the power allocation is optimized under the time delay constraint and the task size constraint. The optimization problem is given by

$$(\text{P1}): \min_{\mathbf{p}} P_0 \tag{12}$$

subject to

$$E_s^1(P_0, \mathbf{p}) \geq 0 \tag{13a}$$

$$E_s^2(P_0, \mathbf{p}) \geq 0 \tag{13b}$$

$$M_1^+ \leq L_1(\mathbf{p}) \leq I_1 \tag{13c}$$

$$M_2^+ \leq L_2(\mathbf{p}) \leq I_2 \tag{13d}$$

3 Optimal Solution

P1 can be equivalently converted into P2, which is given by

$$(\text{P2}): \min_{\mathbf{p}} P_0 \tag{14}$$

subject to

$$P_0 \geq \frac{\sum_{k=1}^{K} p_{1,k} + \frac{3C_1Q_1}{T}(I_1 - L_{2,1})}{v_1 \sum_{k=1}^{K} \gamma_{1,k}} \tag{15a}$$

$$P_0 \geq \frac{\sum_{k \in G_I} p_{21,k} + \sum_{k \in \overline{G_I}} p_{22,k} + \frac{3C_2Q_2}{T}(I_2 - L_2)}{v_2 \sum_{k=1}^{K} \gamma_{2,k}} \tag{15b}$$

$$m_1 \leq p_{21,k} \leq m_2 \tag{15c}$$

$$n_1 \leq p_{22,k} \leq n_2 \tag{15d}$$

where the value of m_1, m_2, n_1 and n_2 come from the equations $L_{2,1} = M_1^+$, $L_{2,1} = I_1$, $L_2 = M_2^+$, $L_2 = I_2$, respectively. Considering $M_1^+ \leq L_{1,2} \leq I_1$, the range of $p_{1,k}$ is $m_3 \leq p_{1,k} \leq m_4$.

Let

$$f = \sum_{k=1}^{K} p_{1,k} - C_1Q_1B \sum_{k \in G_I} \log_2\left(1 + \frac{p_{21,k}\gamma_{4,k}}{N_0}\right) \tag{16}$$

From (15a), we can see that when f reaches its maximum value, P_0 reaches its minimum value. Calculating the first derivation of f with $p_{1,k}$ and $p_{21,k}$, we can get

$$\frac{\partial f}{\partial p_{1,k}} = 1 \tag{17a}$$

$$\frac{\partial f}{\partial p_{21,k}} = -\frac{C_1Q_1B\gamma_{4,k}}{\ln 2(N_0 + p_{21,k}\gamma_{4,k})} \tag{17b}$$

Therefore, we can draw the conclusion that f increase monotonically with $p_{1,k}$ and decrease monotonically with $p_{21,k}$.

Let

$$y = \sum_{k \in G_I} p_{21,k} + \sum_{k \in \overline{G_I}} p_{22,k} - C_2Q_2B \sum_{k \in \overline{G_I}} \log_2\left(1 + \frac{p_{22,k}\gamma_{4,k}}{N_0}\right) \tag{18}$$

From (15b), we can get that P_0 reaches its minimum value when y reaches its maximum value. Calculating the first derivation of y with $p_{21,k}$ and $p_{22,k}$, we can get

$$\frac{\partial y}{\partial p_{21,k}} = 1 \tag{19a}$$

$$\frac{\partial y}{\partial p_{22,k}} = 1 - \frac{C_2Q_2B\gamma_{4,k}}{\ln 2(N_0 + p_{22,k}\gamma_{4,k})} \tag{19b}$$

We can find that y increase monotonically with $p_{21,k}$. When $p_{22,k} = p_{22,k}^* = \frac{C_2 Q_2 B}{\ln 2} - \frac{N_0}{\gamma_{4,k}}$, $\frac{dy}{dp_{22,k}} = 0$. Thus, we can conclude that y decreases with $p_{22,k}$ when $p_{22,k} < p_{22,k}^*$ and increases with $p_{22,k}$ when $p_{22,k} > p_{22,k}^*$.

The minimum value of P_0 can be given by

$$P_0 = \max\{P_{01}, P_{02}\} \tag{20}$$

where P_{01} and P_{02} are the minimum value of P_0 obtained from (15a) and (15b), respectively. By analyzing the relative value of $p_{21,k}$ and $p_{22,k}$ as follows, we can obtain both values of P_{01} and P_{02}.

Case 1. When $p_{21,k} = m_1$, the minimum value of P_{01} is given by

$$P_{01} = \frac{1}{v_1 \sum_{k=1}^{N} \gamma_{1,k}} \left[\sum_{k=1}^{N} m_4 + \frac{3C_1 Q_1}{T} (I_1 - M_1^+) \right] \tag{21}$$

(**a**) When $n_1 \le p_{22,k}^* \le n_2$, the minimum value of P_{02} is given by

$$P_{02} = \max \begin{cases} \dfrac{\sum_{k \in G_I} m_1 + \sum_{k \in \overline{G_I}} n_1 + \frac{3C_2 Q_2}{T} (I_2 - M_2^+)}{v_2 \sum_{k=1}^{N} \gamma_{2,k}} \\ \dfrac{\sum_{k \in G_I} m_1 + \sum_{k \in \overline{G_I}} n_2}{v_2 \sum_{k=1}^{N} \gamma_{2,k}} \end{cases} \tag{22}$$

(**b**) When $p_{22,k}^* < n_1$, the minimum value of P_{02} is given by

$$P_{02} = \frac{\sum_{k \in G_I} m_1 + \sum_{k \in \overline{G_I}} n_2}{v_2 \sum_{k=1}^{N} \gamma_{2,k}} \tag{23}$$

(**c**) When $p_{22,k}^* > n_2$, the minimum value of P_{02} is given by

$$P_{02} = \frac{\sum_{k \in G_I} m_1 + \sum_{k \in \overline{G_I}} n_1 + \frac{3C_2 Q_2}{T} (I_2 - M_2^+)}{v_2 \sum_{k=1}^{N} \gamma_{2,k}} \tag{24}$$

Case 2. When $p_{21,k} = m_2$, the minimum value of P_{01} is given by

$$P_{01} = \frac{\sum_{k=1}^{N} m_4}{v_1 \sum_{k=1}^{N} \gamma_{1,k}} \tag{25}$$

(**a**) When $n_1 \le p_{22,k}^* \le n_2$, the minimum value of P_{02} is given by

$$P_{02} = \max \begin{cases} \frac{\sum_{k \in G_I} m_2 + \sum_{k \in \overline{G_I}} n_1 + \frac{3C_2Q_2}{T}(I_2 - M_2^+)}{v_2 \sum_{k=1}^{N} \gamma_{2,k}} \\ \frac{\sum_{k \in G_I} m_2 + \sum_{k \in \overline{G_I}} n_2}{v_2 \sum_{k=1}^{N} \gamma_{2,k}} \end{cases} \tag{26}$$

(**b**) When $p_{22,k}^* < n_1$, the minimum value of P_{02} is given by

$$P_{02} = \frac{\sum_{k \in G_I} m_2 + \sum_{k \in \overline{G_I}} n_2}{v_2 \sum_{k=1}^{N} \gamma_{2,k}} \tag{27}$$

(**c**) When $p_{22,k}^* > n_2$, the minimum value of P_{02} is given by

$$P_{02} = \frac{\sum_{k \in G_I} m_2 + \sum_{k \in \overline{G_I}} n_1 + \frac{3C_2Q_2}{T}(I_2 - M_2^+)}{v_2 \sum_{k=1}^{N} \gamma_{2,k}} \tag{28}$$

4 Simulation Results

The channel is modeled as Rician fading, since the signal strength of the line-of-sight signal is stronger than that of the indirect signal, which is modeled as $\gamma_k = \sqrt{\frac{M}{M+1}}\tilde{f} + \sqrt{\frac{1}{M+1}}\hat{f}(k)$, where $M = 3$, $\tilde{f}$ denotes the line-of-sight deterministic factor, $\hat{f}(k)$ denotes the Rayleigh fading. The number of subcarriers is $K = 32$ and the total bandwidth is \$32\$ MHz. For simplicity, we set $v_i = 1.0$, $k_i = 10^{-28}$ and the receiver noise power $N_0 = N_2 = N$. The block time $T = 0.3$–0.5 s. The CPU frequency $f_i = 0.50$ GHz. The number of CPU cycles required per bit of data is $C_i = 1000$ cycle/bit.

Figure 1 shows the relationship between UAV's minimum transmit power and the blocking time T with different values of input data I_i when $N = 10^{-6}$ W. In this condition, the distance between UAV and IoD_1, UAV and IoD_2, and IoD_1 and IoD_2 are set to be $d_1 = 5$ m, $d_2 = 3$ m and $d_{12} = 3$ m, respectively. From Fig. 1, we can discover that as the size of the input calculation data increases, the minimum transmit power of the UAV increases as well. This is because the IoT devices consume more power to transmit data to UAV and compute locally as the input data size increases. We can also find in Fig. 1 that the minimum transmit power of UAV increases as T decreases. This is because in the case of shorter block time, in order to finish the transmission, the IoT devices require a higher transmission rate, which will lead to the UAV transmitting more power to provide energy for the IoT devices.

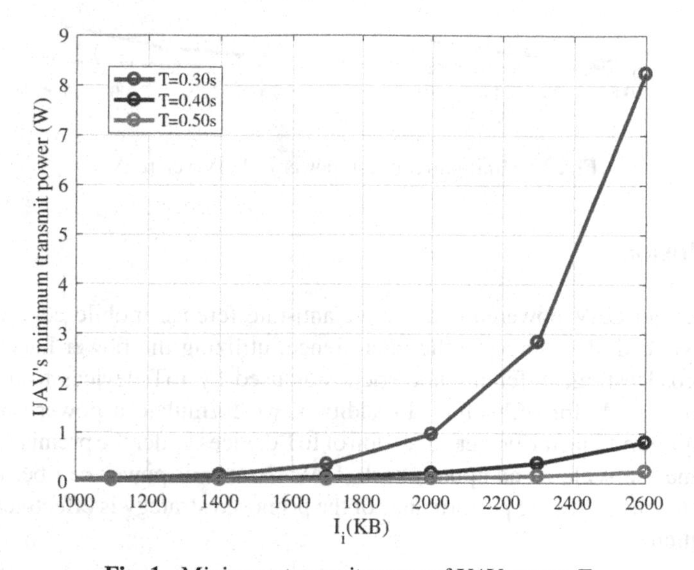

Fig. 1. Minimum transmit power of UAV versus T

Figure 2 shows the minimum transmit power of UAV versus noise power with different values of input data I_i when $T = 0.4$ s. In this condition, the distance parameters are set as same as Fig. 1. In Fig. 2, we can find that the minimum transmit power of UAV becomes smaller when the receiver noise power becomes smaller. That is because with a certain amount of offloading data, as shown in (2), the lower noise power, the lower power required.

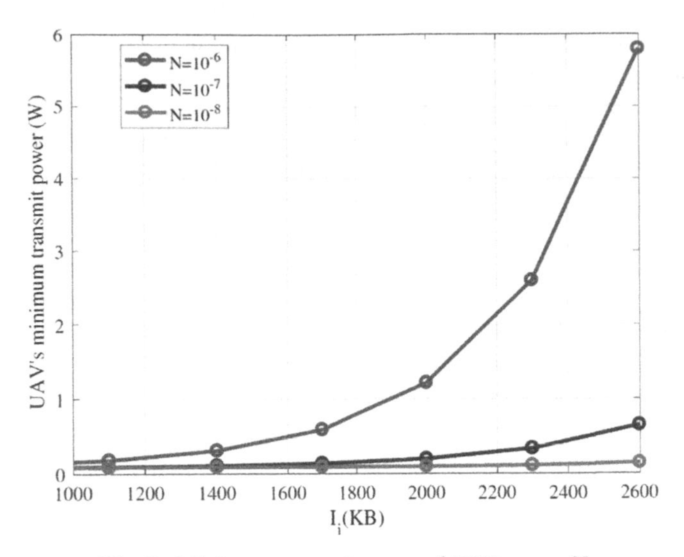

Fig. 2. Minimum transmit power of UAV versus N

5 Conclusion

In this paper, an UAV powered cooperative anti-interference mobile edge computing strategy is studied. To overcome the interference, utilizing the power harvested from the UAV broadcasting, different subcarriers was used by IoT devices to transmit the information to UAV for offloading. In addition, we formulate a power optimization problem. By optimizing the power allocation of IoT devices under the premise of meeting the delay and the scale of computing task, UAV's transmit power can be minimized. Simulation results prove the performance of the proposed strategy is effective and meets the expectation.

References

1. Lee, S., Oumaima, M., Ryu, S., Park, J.: A realtime spatiotemporal data acquisition system for precision agriculture. In: 2017 International Conference on Intelligent Environments (IE), pp. 149–152 (2017)
2. Bayrakdar, M.E.: A smart insect pest detection technique with qualified underground wireless sensor nodes for precision agriculture. IEEE Sens. J. **19**, 10892–10897 (2019)
3. Hua, M., Wang, Y., Li, C., Huang, Y., Yang, L.: UAV-aided mobile edge computing systems with one by one access scheme. IEEE Trans. Green Commun. Network. **3**, 664–678 (2019)
4. Wang, K., Yang, K., Chen, H., Zhang, L.: Computation diversity in emerging networking paradigms. IEEE Wirel. Commun. **24**, 88–94 (2017)
5. Zhang, K., et al.: Energy-efficient offloading for mobile edge computing in 5G heterogeneous networks. IEEE Access **4**, 5896–5907 (2016)
6. Bi, S., Zhang, Y.J.: Computation rate maximization for wireless powered mobile-edge computing with binary computation offloading. IEEE Trans. Wirel. Commun. **17**, 4177–4190 (2018)

7. Wang, S., Urgaonkar, R., He, T., Chan, K., Zafer, M., Leung, K.K.: Dynamic service placement for mobile micro-clouds with predicted future costs. IEEE Trans. Parallel Distrib. Syst. **28**, 1002–1016 (2017)
8. Hua, M., Yang, L., Li, C., Wu, Q., Swindlehurst, A.L.: Throughput maximization for UAV-aided backscatter communication networks. IEEE Trans. Commun. **68**, 1254–1270 (2020)
9. Zeng, Y., Zhang, R., Lim, T.J.: Wireless communications with unmanned aerial vehicles: opportunities and challenges. IEEE Commun. Mag. **54**, 36–42 (2016)
10. Zhang, T., Xu, Y., Loo, J., Yang, D., Xiao, L.: Joint computation and communication design for UAV-assisted mobile edge computing in IoT. IEEE Trans. Industr. Inform. **16**, 5505–5516 (2019)
11. Liang, H., Zhong, C., Suraweera, H.A., Zheng, G., Zhang, Z.: Optimization and analysis of wireless powered multi-antenna cooperative systems. IEEE Trans. Wirel. Commun. **16**, 3267–3281 (2017)

Network Performance Evaluation

System-Level Performance Analysis in 3D Drone Mobile Networks

Jiayi Huang[1(✉)], Arman Shojaeifard[2], Jie Tang[1], and Kai-Kit Wong[3]

[1] South China University of Technology, Guangzhou, China
eejiayihuang@mail.scut.edu.cn
[2] BT Labs, Adastral Park, Ipswich IP5 3RE, UK
arman.shojaeifard@bt.com
[3] University College London, London WC1E 7JE, UK
kai-kit.wong@ucl.ac.uk

Abstract. We present a system-level analysis for drone mobile networks on a finite three-dimensional (3D) space. A performance boundary derived by deterministic random (Brownian) motion model over Nakagami-m fading interfering channels is developed. This method allows us to circumvent the extremely complex reality model and obtain the upper and lower performance bounds of actual drone mobile networks. The validity and advantages of the proposed framework are confirmed via extensive Monte-Carlo (MC) simulations. The results reveal several important trends and design guidelines for the practical deployment of drone mobile networks.

Keywords: Drone cellular networks · System-level analysis · Stochastic geometry theory · Performance boundary · Monter-Carlo (MC) simulations

1 Introduction

The exponential growth of wireless data services driven by mobile smart devices (e.g., smartphone, pad) has triggered the investigation of assisted terrestrial networks in the era of the Internet of Things (IoT) [1]. However, it's a crucial task under the circumstances of large-gatherings (e.g., sports games, concerts) and natural or man-made disasters (e.g., floods, earthquake). Due to the advantages of Unmanned aerial vehicles (UAVs), UAVs assisted cellular network is considered a prominent solution for enhancing or recovering terrestrial cellular networks, which has attracted great attention in both academia and industry recently.Therefore, the fifth-generation (5G) communication system also considers the application of low-altitude drones in the system [2].

Among current wireless network researches, there are many published works related to the UAVs assisted networks, many works in the literature consider simplified movement models (e.g., fixed height around a circle). However, the reality is more complicated than these models and we can only obtain drone

X. Wang et al. (Eds.): 6GN 2020, LNICST 337, pp. 311–322, 2020.
https://doi.org/10.1007/978-3-030-63941-9_23

system performance in simple cases using these models. In [3], Poisson cluster process was applied to distribute user and drone hover in a certain height above cluster center in order to compare the performance of millimeter wave (mmWave) and sub-6 GHz. The work in [4] converted the problem of three-dimensional space into two-dimensional plane by distributing drones in a fixed altitude. In [5] the authors concluded that a cellular network with an omnidirectional antenna can support drone base stations downlinks and control channels in a low altitude, but a high altitude is still struggle. The authors in [6] investigated the coverage probability and average achievable rate in the post-disaster area by using two cooperative drones in a fixed height. This paper provides a unified model for performance analysis of drone mobile networks and obtain the lower and upper bounds of actual drone mobile networks by introduced deterministic motion model and 3D Brownian motion model. We explicitly account for certain constraints, such as small-scale and large-scale fading characteristics depending on line-of-sight (LOS) and non line-of-sight (NLOS) propagation, and the impact of drone mobility based on 3D deterministic random (Brownian) motion. The analytical formulations are validated via Monter-Carlo (MC) simulations.

Notation

$\boldsymbol{X}$ is a matrix; $\boldsymbol{x}$ is a vector; T, $\dagger$, and $+$ are the transpose, Hermitian, and pseudo-inverse operations; $\mathbb{E}_x[.]$ is the expectation; $\Pr[.]$ is the probability; $\mathcal{F}_x[.]$ is the cumulative distribution function (CDF); $\mathcal{P}_x[.]$ is the probability density function (PDF); $\mathcal{L}_x[.]$ is the Laplace transform (LT) function; $|x|$ is the modulus; $\|\boldsymbol{x}\|$ is the Euclidean norm; $\mathbf{I}_{(.)}$ is the identity matrix; $\mathcal{H}(.)$ is the Heaviside step function; $\delta(.)$ is the Delta function; $\mathcal{CN}(\mu, \nu^2)$ is the circularly-symmetric complex Gaussian distribution with mean μ and variance ν^2; $\Gamma(.)$ and $\Gamma(.,.)$ are the Gamma and incomplete (upper) Gamma functions; $\mathcal{G}(\kappa, \theta)$ is the Gamma distribution with shape parameter κ and scale parameter θ, respectively.

2 Preliminaries

In this word, we consider a large-scale UAV network in which K BSs (i.e., drones) are deployed on the finite 3D ball of radius R according to a homogeneous PPP Φ with spatial density λ at time $t = 0$. Let $\mathsf{K}_{\mathrm{LOS}}$ and $\mathsf{K}_{\mathrm{NLOS}}$ respectively denote the number of drones experiencing LOS and NLOS propagation (i.e., $\mathsf{K} = \mathsf{K}_{\mathrm{LOS}} + \mathsf{K}_{\mathrm{NLOS}}$) at time $t = 0$. Based on the stationary property of PPP, and Slivnyak's theorem [7], the analysis is carried for a typical user o assumed to be located at the origin.

The drones, equipped with M transmit antennas, are considered to be serving a user per resource block. Let $\boldsymbol{h}_b^T \sim \mathcal{CN}(\mathbf{0}, \mathbf{I}_{\mathsf{M}})$ denote the small-scale fading channel between the typical user o and its serving drone b. Here, we utilize the Nakagami-m distribution, which can capture a wide range of small-scale fading conditions through tuning of the parameter m. Considering the drones apply conjugate-beamforming (CB), the intended (from drone b) small-scale fading channel power gains under LOS and NLOS propagation are distributed as $g_b \sim \mathcal{G}\left(m\mathsf{M}, \frac{1}{m}\right)$ and $g_b \sim \mathcal{G}\left(\mathsf{M}, \frac{1}{m}\right)$, respectively. The interfering (from drone i) small-scale fading channel power gains under LOS and NLOS propagation are distributed as $g_i \sim \mathcal{G}\left(m, \frac{1}{m}\right)$ and $g_i \sim \mathcal{G}\left(1, 1\right)$, respectively [8].

The path-loss function is defined as

$$L(r) = \max\left(\beta_0, \beta_1 r^{-\alpha}\right) \tag{1}$$

where r is the distance, β_0 is the minimum coupling loss, β_1 is a constant parameter of the path-loss function, and α is the path-loss exponent. All channels undergo free-space path-loss, i.e., $\alpha = 2$. Moreover, $\beta_1 = \frac{1}{\epsilon}\left(\frac{c}{4\pi f_c}\right)^2$, where $c = 3 \times 10^8$ m/s is the light speed, and f_c is the carrier frequency. For LOS ($L_{\mathrm{LOS}}(r)$) and NLOS ($L_{\mathrm{NLOS}}(r)$) links, $\epsilon = \epsilon_{\mathrm{LOS}} = 1$ dB and $\epsilon = \epsilon_{\mathrm{NLOS}} = 20$ dB, respectively [9].

We utilize the following function for the probability of LOS propagation

$$\Pr\left[\mathrm{LOS}, r_{(.)} = r\right] = \begin{cases} 1 & \text{if } r \in [0, \mathsf{D}) \\ 0 & \text{if } r \in [\mathsf{D}, \mathsf{R}] \end{cases} \tag{2}$$

where D denotes the critical distance. Under this model, given the drones follow from a homogeneous PPP Φ, the average number of LOS and NLOS drones in a 3D ball of radius R are respectively $\bar{\mathsf{K}}_{\mathrm{LOS}} = \frac{4}{3}\pi\mathsf{D}^3\lambda$ and $\bar{\mathsf{K}}_{\mathrm{NLOS}} = \frac{4}{3}\pi\left(\mathsf{R}^3 - \mathsf{D}^3\right)\lambda$.

We consider the cellular association strategy in which the typical user o connects to the drone b which provides the greatest received SINR.

Lemma 1. Considering there are $\mathsf{K}_{\mathrm{LOS}}$ LOS drones (i.e., $\mathsf{K}_{\mathrm{LOS}} > 0$) uniformly-deployed on the finite 3D ball of radius D, the CDF and PDF of the distance between the typical user o and its serving LOS drone, $r_{b,\mathrm{LOS}}$, are respectively given by

$$\mathcal{F}_{r_{b,\mathrm{LOS}}}(r) = 1 - \left(1 - \frac{r^3}{\mathsf{D}^3}\right)^{\mathsf{K}_{\mathrm{LOS}}}, \quad 0 \le r \le \mathsf{D} \tag{3}$$

$$\mathcal{P}_{r_{b,\mathrm{LOS}}}(r) = \frac{3r^2\mathsf{K}_{\mathrm{LOS}}}{\mathsf{D}^3}\left(1 - \frac{r^3}{\mathsf{D}^3}\right)^{\mathsf{K}_{\mathrm{LOS}}-1}, \quad 0 \le r \le \mathsf{D}. \tag{4}$$

Proof: The result follows from [10, Theorem 2.1] with $n = 1$, $d = 3$.

Lemma 2. Considering there are $\mathsf{K}_{\mathrm{NLOS}}$ NLOS drones (i.e., $\mathsf{K}_{\mathrm{LOS}} = 0$) uniformly-deployed on the finite 3D ball double-bounded by radii D ($< \mathsf{R}$) and R, the CDF and PDF of the distance between the typical user o and its serving NLOS drone, $r_{b,\mathrm{NLOS}}$, are respectively given by

$$\mathcal{F}_{r_{b,\mathrm{NLOS}}}(r) = 1 - \left(1 - \frac{r^3 - \mathsf{D}^3}{\mathsf{R}^3 - \mathsf{D}^3}\right)^{\mathsf{K}_{\mathrm{NLOS}}}, \quad \mathsf{D} \le r \le \mathsf{R} \tag{5}$$

$$\mathcal{P}_{r_b,\text{NLOS}}(r) = \frac{3r^2 \mathsf{K}_{\text{NLOS}}}{\mathsf{R}^3 - \mathsf{D}^3} \left(1 - \frac{r^3 - \mathsf{D}^3}{\mathsf{R}^3 - \mathsf{D}^3}\right)^{\mathsf{K}_{\text{NLOS}}-1}, \quad \mathsf{D} \leq r \leq \mathsf{R}. \tag{6}$$

Proof: The proof, omitted due to space limitations, follows from the probability distribution of a double-bounded random process given in [11, Eqn. (3)].

3 Unified Framework

In this section, we present a stochastic geometry-based model for performance analysis of drone mobile networks. By introducing Slivnyak's theorem, the analysis is carried for a typical user o assumed to be located at the origin.

3.1 3D Brownian Motion

The drones are considered to be mobile according to a 3D Brownian motion (BM). At time $t > 0$, the movement of an arbitrary drone can be captured through the following stochastic differential equation (SDE) [12].

$$\mathrm{d}l(t) = \sigma \, \mathrm{d}b(t) \tag{7}$$

where $l(t) = \{l_x(t), l_y(t), l_z(t)\}$ is a vector for the Cartesian coordinates at time t, $b(t) = \{b_x(t), b_y(t), b_z(t)\}$ represents the standard BM (i.e., Wiener process) vector at time t, and σ is a positive constant (e.g., representing average velocity). Here, we consider $b_x(t), b_y(t), b_z(t) \sim \mathcal{N}(0, t)$. The corresponding Euclidean distance with respect to the origin at time t can be formulated as $\hat{r}(t) = \sqrt{l_x^2(t) + l_y^2(t) + l_z^2(t)}$. Here, the mobility model should account for the finite volume of the 3D ball as well as the LOS/NLOS propagation conditions. Hence, we consider the case where (i) the mobile LOS drone cannot be at a distance larger than D with respect to the origin at any given time, i.e., $r_{\text{LOS}}(t) = \max(0, \min(\mathsf{D}, \hat{r}(t)))$, and (ii) the mobile NLOS drone cannot be at a distance smaller than D and larger than R with respect to the origin at any given time, i.e., $r_{\text{NLOS}}(t) = \max(\mathsf{D}, \min(\mathsf{R}, \hat{r}(t)))$.

Lemma 3. With the 3D BM mobility model under consideration, the CDF and PDF of the mobile LOS drone distance with respect to the origin at time t (i.e., $r_{\text{LOS}}(t) = \max(0, \min(\mathsf{D}, \hat{r}(t)))$) are respectively given by

$$\mathcal{F}_{r_{\text{LOS}}(t)}(w) = \left(1 + \frac{2}{\sqrt{\pi}} \Gamma\left(\frac{3}{2}, \frac{w^2}{4\sigma t}\right)(\mathcal{H}(w - \mathsf{D}) - 1)\right)\mathcal{H}(w) \tag{8}$$

and

$$\mathcal{P}_{r_{\text{LOS}}(t)}(w) = \left(1 + \frac{2}{\sqrt{\pi}} \Gamma\left(\frac{3}{2}, \frac{w^2}{4\sigma t}\right)(\mathcal{H}(w - \mathsf{D}) - 1)\right)\delta(w) + \\ \left(\frac{2}{\sqrt{\pi}} \Gamma\left(\frac{3}{2}, \frac{w^2}{4\sigma t}\right)\delta(w - \mathsf{D}) - \frac{w^2}{2\sqrt{\pi}\,(\sigma t)^{\frac{3}{2}}} \exp\left(-\frac{w^2}{4\sigma t}\right)(\mathcal{H}(w - \mathsf{D}) - 1)\right)\mathcal{H}(w). \tag{9}$$

Lemma 4. With the 3D BM mobility model under consideration, the CDF and PDF of the mobile NLOS drone distance with respect to the origin at time t (i.e., $r_{\mathrm{NLOS}}(t) = \max\left(\mathsf{D}, \min\left(\mathsf{R}, \hat{r}(t)\right)\right)$) are respectively given by

$$\mathcal{F}_{r_{\mathrm{NLOS}}(t)}(w) = \left(1 + \frac{2}{\sqrt{\pi}}\Gamma\left(\frac{3}{2}, \frac{w^2}{4\sigma t}\right)\left(\mathcal{H}(w-\mathsf{R}) - 1\right)\right)\mathcal{H}(w-\mathsf{D}) \tag{10}$$

and

$$\begin{aligned}\mathcal{P}_{r_{\mathrm{NLOS}}(t)}(w) = &\left(1 + \frac{2}{\sqrt{\pi}}\Gamma\left(\frac{3}{2}, \frac{w^2}{4\sigma t}\right)\left(\mathcal{H}(w-\mathsf{R}) - 1\right)\right)\delta(w-\mathsf{D}) \\ &+ \left(\frac{2}{\sqrt{\pi}}\Gamma\left(\frac{3}{2}, \frac{w^2}{4\sigma t}\right)\delta(w-\mathsf{R})\right)\mathcal{H}(w-\mathsf{D}) \\ &- \left(\frac{1}{2\sqrt{\pi}w}\left(\frac{w^2}{\sigma t}\right)^{\frac{3}{2}}\exp\left(-\frac{w^2}{4\sigma t}\right)\left(\mathcal{H}(w-\mathsf{R}) - 1\right)\right)\mathcal{H}(w-\mathsf{D}).\end{aligned} \tag{11}$$

Next, we aim to characterize the distribution of the reference transmitter-receiver distance based on the 3D BM mobility and LOS/NLOS propagation models under consideration.

Corollary 1. The closest transmitter-receiver distance for the serving LOS drone becomes equivalent to the 3D BM mobility model with the following drones spatial density [13]

$$\lambda = \frac{3\log\left(1 - \frac{\mathcal{H}(w)}{\sqrt{\pi}}\left(2\mathcal{H}(w-D)\Gamma\left(\frac{3}{2}, \frac{w^2}{4t\sigma}\right) - 2\Gamma\left(\frac{3}{2}, \frac{w^2}{4t\sigma}\right) + \sqrt{\pi}\right)\right)}{4\pi D^3\log\left(1 - \frac{w^3}{D^3}\right)} \tag{12}$$

Corollary 2. The closest transmitter-receiver distance for the serving NLOS drone becomes equivalent to the 3D BM mobility model with the following drones spatial density

$$\lambda = \frac{3\log\left(1 - \frac{\mathcal{H}(w-\mathsf{D})}{\sqrt{\pi}}\left(2\mathcal{H}(w-\mathsf{R})\Gamma\left(\frac{3}{2}, \frac{w^2}{4t\sigma}\right) - 2\Gamma\left(\frac{3}{2}, \frac{w^2}{4\sigma t}\right) + \sqrt{\pi}\right)\right)}{4\pi\left(\mathsf{R}^3 - \mathsf{D}^3\right)\log\left(1 - \frac{w^3 - \mathsf{D}^3}{R^3 - \mathsf{D}^3}\right)}. \tag{13}$$

3.2 Deterministic Motion

Next, we consider the case where the movement of the drones is deterministic such that they move at a constant speed towards a target. At time $t > 0$, the corresponding Euclidean distance with respect to the origin is given by

$$\hat{r}(t) = r_0 - vt \tag{14}$$

where r_0 is the distance at time $t = 0$ and v is the constant speed, respectively. Note that a negative value for v indicates movement in the opposite direction and vice versa. Here, we need to account for the finite volume of the 3D ball as well as the LOS/NLOS propagation conditions. Hence, we consider the case where (i) the mobile LOS drone cannot be at a distance larger than R with respect to the origin at any given time, i.e., $r_{\mathrm{LOS}}(t) = \max(0, \min(\mathsf{D}, \hat{r}(t)))$, and (ii) the mobile NLOS drone cannot be at a distance smaller than D and larger than R with respect to the origin at any given time, i.e., $r_{\mathrm{NLOS}}(t) = \max(\mathsf{D}, \min(\mathsf{R}, \hat{r}(t)))$.

Lemma 5. With the deterministic mobility model under consideration, the CDF and PDF of the mobile LOS drone distance with respect to the origin at time t (i.e., $r_{\mathrm{LOS}}(t) = \max(0, \min(\mathsf{D}, \hat{r}(t)))$) are respectively given by

$$\mathcal{F}_{r_{\mathrm{LOS}}(t)}(w) = \mathcal{H}(w)\left(1 + (\mathcal{H}(w - \mathsf{D}) - 1)\left(1 - \frac{(vt+w)^3}{\mathsf{D}^3}\right)^{\mathsf{K}_{\mathrm{LOS}}}\right) \tag{15}$$

and

$$\mathcal{P}_{r_{\mathrm{LOS}}(t)}(w) = \delta(w) + (\mathcal{H}(w)\delta(w-\mathsf{D}) + (\mathcal{H}(w-\mathsf{D}) - 1)\delta(w))\left(1 - \frac{(vt+w)^3}{\mathsf{D}^3}\right)^{\mathsf{K}_{\mathrm{LOS}}}$$
$$- \mathcal{H}(w)(\mathcal{H}(w-\mathsf{D}) - 1)\frac{3\mathsf{K}_{\mathrm{LOS}}(vt+w)^2}{\mathsf{D}^3}\left(1 - \frac{(vt+w)^3}{\mathsf{D}^3}\right)^{\mathsf{K}_{\mathrm{LOS}}-1}. \tag{16}$$

Lemma 6. With the deterministic mobility model under consideration, the CDF and PDF of the mobile NLOS drone distance with respect to the origin at time t (i.e., $r_{\mathrm{NLOS}}(t) = \max(\mathsf{D}, \min(\mathsf{R}, \hat{r}(t)))$) are respectively given by

$$\mathcal{F}_{r_{\mathrm{NLOS}}(t)}(w) = \mathcal{H}(w-\mathsf{D})\left(1 + (\mathcal{H}(w-\mathsf{R}) - 1)\left(\frac{\mathsf{R}^3 - (vt+w)^3}{\mathsf{R}^3 - \mathsf{D}^3}\right)^{\mathsf{K}_{\mathrm{NLOS}}}\right) \tag{17}$$

and

$$\mathcal{P}_{r_{\mathrm{NLOS}}(t)}(w) = \delta(w-\mathsf{D})\left(1 + (\mathcal{H}(w-\mathsf{R}) - 1)\left(\frac{\mathsf{R}^3 - (vt+w)^3}{\mathsf{R}^3 - \mathsf{D}^3}\right)^{\mathsf{K}_{\mathrm{NLOS}}}\right)$$
$$+ \mathcal{H}(w-\mathsf{D})\left(\delta(w-\mathsf{D})\left(\frac{\mathsf{R}^3 - (vt+w)^3}{\mathsf{R}^3 - (\mathcal{H}(w-\mathsf{R}) - 1)\mathsf{D}^3}\right)^{\mathsf{K}_{\mathrm{NLOS}}}\right)$$
$$- \mathcal{H}(w-\mathsf{D})\left(\frac{3\mathsf{K}_{\mathrm{NLOS}}(vt+w)^2}{\mathsf{R}^3 - \mathsf{D}^3}\left(\frac{\mathsf{R}^3 - (tv+w)^3}{\mathsf{R}^3 - \mathsf{D}^3}\right)^{\mathsf{K}_{\mathrm{NLOS}}-1}\right) \tag{18}$$

3.3 SINR Formulation

The received SINR at the reference user o is given by

$$\mathrm{SINR} = \frac{X}{I + \sigma^2} \tag{19}$$

where

$$X = pg_b L(r_b) \tag{20}$$

and

$$I = \sum_{i \in \Phi \setminus \{b\}} pg_i L(r_i) \tag{21}$$

with p and σ^2 respectively used to denote the transmit power and noise variance.

3.4 SE Formulation

Theorem 1. The average rate (in nat/s/Hz) of the typical user is given by

$$\mathbb{E}\left[\log\left(1 + \mathrm{SINR}\right)\right] = \int_0^{+\infty} \int_0^{+\infty} \frac{1 - \mathcal{F}_{\mathrm{SINR}|r_b=r}[\gamma]}{1+\gamma} \, \mathrm{d}\gamma \, \mathcal{P}_{r_b}(r) \, \mathrm{d}r \tag{22}$$

where $\mathcal{F}_{\mathrm{SINR}|r_b=r}[\gamma]$ and $\mathcal{P}_{r_b}(r)$ denote the CDF of the SINR conditioned on $r_b = r$ and the PDF of the transmitter-receiver distance (given in *Lemma 1*), respectively.

4 Numerical and Simulation Results

In this section, we evaluate the performance of 3D deterministic random (Brownian) motion model. To confirm our framework, we use MC methods to obtain simulation result in different scenarios.

4.1 Impact of Number of Antennas

To gain insight into the effect of different number of antennas, we provide results using deterministic model, stationary model and 3D Brownian model via MC simulation in Fig. 1. A key point to note is that due to the deterministic model is the most ideal movement model and the 3D Brownian model is the most worst movement model, so in Fig. 1 the deterministic model curves are the upper bound of reality and the 3D Brownian model curves are the lower bound of reality. Also, SE increases as the number of antennas increases and we can obtain the better performance of drone mobile networks by adjusting the number of antennas.

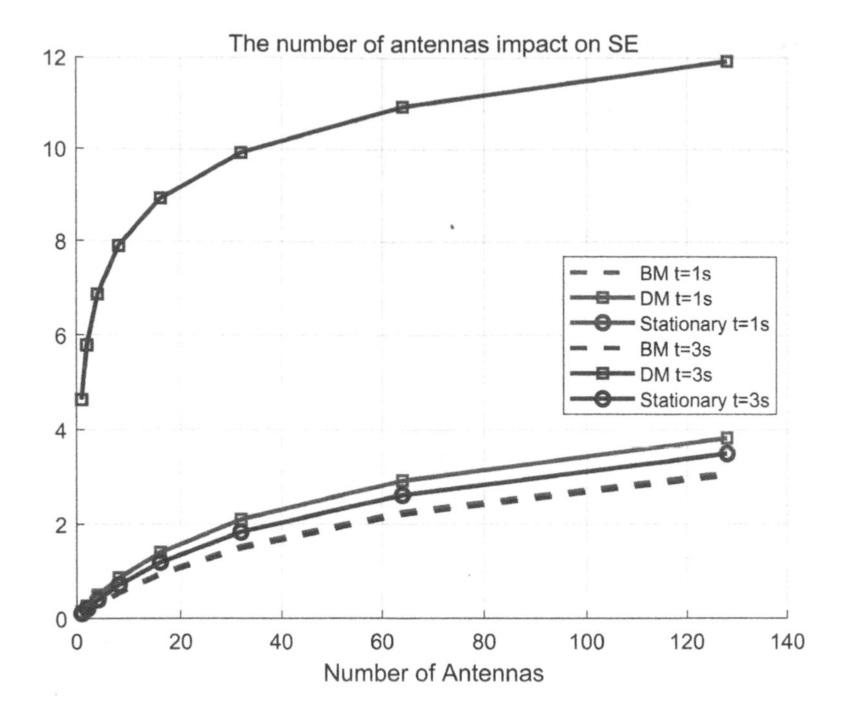

Fig. 1. Spectral efficiency against different number of antennas, m = 2, R = 100 m, D = 18 m, $\lambda = 10^{-5}/m^3$, v = 3 m/s, p = 20 W.

4.2 Impact of Nakagami-M Fading Parameter

We evaluate the Nakagami-m fading parameter's impact on spectral efficiency. Figure 2 shows the drone mobile networks performance with different Nakagami-m fading parameter. We observe that when the Nakagami-m fading parameter increase s, some curve's SE decrease and the other curve's SE increase. The reason is that when the serving drone move three second based on deterministic movement model, the channel of serving drone b becomes LOS, but other curve's channels are still NLOS (i.e., the channel function is different), and we can find that the performance of SE at time t = 3 is better than time t = 0. We can derive the SE of drone mobile networks in different channel cases.

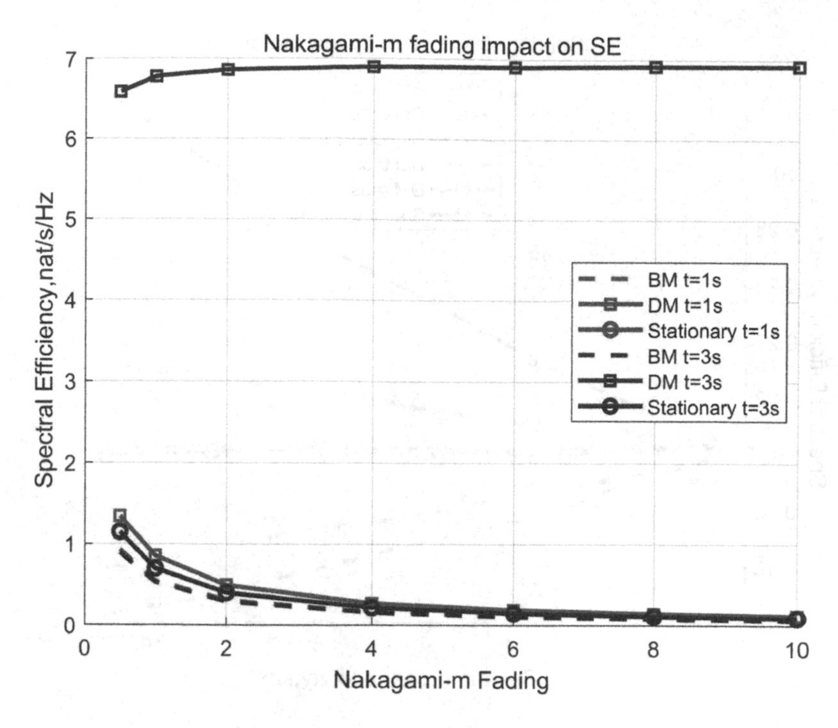

Fig. 2. The Nakagami-m fading impact on Spectral efficiency, M = 4, R = 100 m, D = 18 m, $\lambda = 10^{-5}/m^3$, v = 3 m/s, p = 20 W.

4.3 Impact of Different Drone Velocity

We explore the influence of different drone velocity under different movement strategies at time $t = 0$ and time $t = 3$ in Fig. 3. A key point we can find that the drone velocity has a huge impact on SE, especially on deterministic movement model and 3D Brownian movement model. So we can obtain the performance boundaries when we change the velocity of drone, but also need to consider the security issues at high velocity.

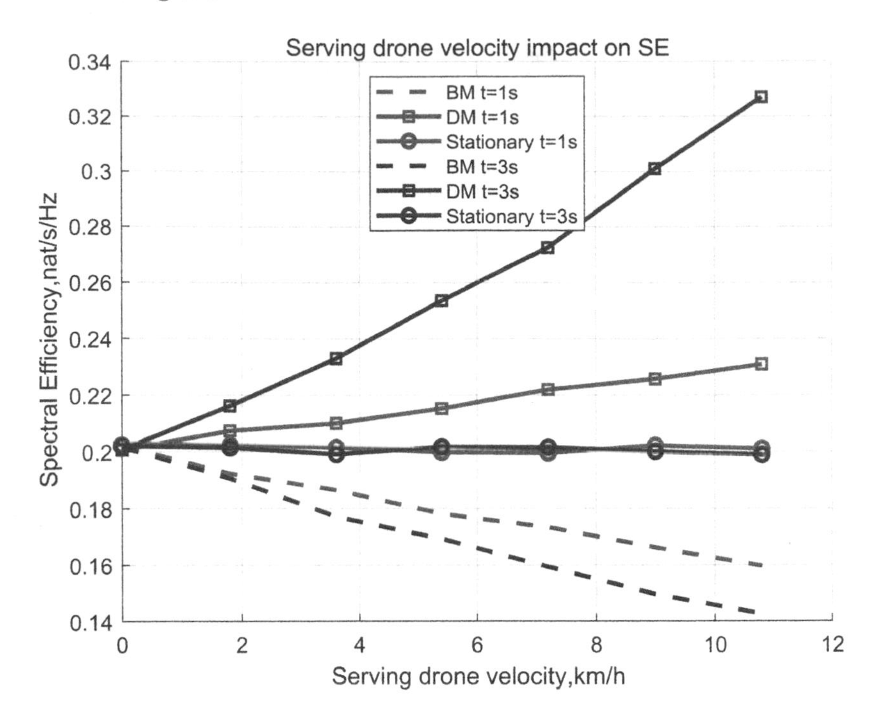

Fig. 3. The drone velocity impact on Spectral efficiency, M = 4, m = 2, R = 100 m, D = 18 m, $\lambda = 10^{-5}/m^3$, p = 20 W.

4.4 Impact of Different Deployment Density

We examine the influence of drone deployment density in improving drone mobile networks under various movement model and different time. Since the deployment density of drones largely determines the economic cost and network performance of drone mobile network, it is necessary to study the impact of deployment density on system performance. Figure 4 represents the performance of SE in different drone deployment density and we can find that there is a best deployment density around $\lambda = 3 \times 10^{-5}/\text{m}^3$. Therefore, in practical applications, we can optimize the system performance and cost by adjusting the deployment density.

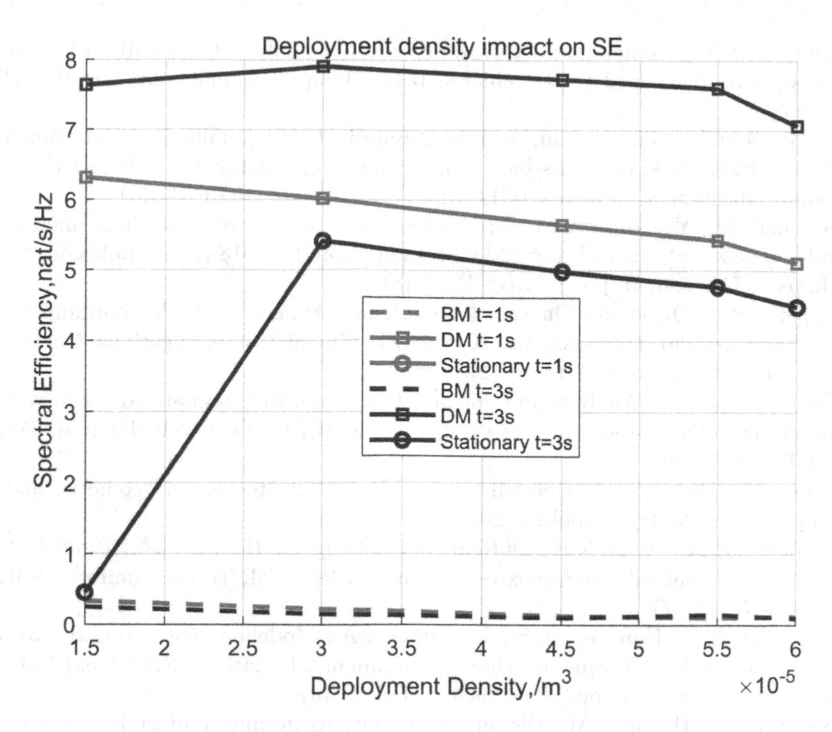

Fig. 4. The serving drone deployment density impact on Spectral efficiency, M = 4, m = 2, R = 100 m, D = 18 m, v = 3 m/s, p = 20 W.

5 Conclusion

This paper provides a system-level analysis of the drone mobile networks. In order to make the drone mobile networks model closer to the actual application scenario, this paper explores the feasibility and performance of the deterministic random (Brownian) motion model, and also explicitly account for some parameters of the drone networks, such as the impact of drone mobility based on deterministic random (Brownian) motion. Then we derive the expressions of SINR, average rate through the mathematical tools provided by stochastic geometry theory. We also validated the theoretical derivation by Monte Carlo simulation. The simulation results confirm the feasibility of the deterministic random (Brownian) motion model. In the follow-up study, we will use this as a basis to analyze the performance of drone mobile networks after the introduction of MIMO.

References

1. Turgut, E., Gursoy, M.C.: Downlink analysis in unmanned aerial vehicle (UAV) assisted cellular networks with clustered users. IEEE Access **6**, 36313–36324 (2018)

2. Chetlur, V.V., Dhillon, H.S.: Downlink coverage analysis for a finite 3-D wireless network of unmanned aerial vehicles. IEEE Trans. Commun. **65**(10), 4543–4558 (2017)
3. Yi, W., Liu, Y., Nallanathan, A., Karagiannidis, G.K.: A unified spatial framework for clustered UAV networks based on stochastic geometry. In: 2018 IEEE Global Communications Conference (GLOBECOM), pp. 1–6. IEEE (2018)
4. Alzenad, M., Yanikomeroglu, H.: Coverage and rate analysis for unmanned aerial vehicle base stations with LoS/NLoS propagation. In: 2018 IEEE Globecom Workshops (GC Wkshps), pp. 1–7. IEEE (2018)
5. López-Pérez, D., et al.: On the downlink performance of UAV communications in dense cellular networks. In: 2018 IEEE Global Communications Conference (GLOBECOM), pp. 1–7. IEEE (2018)
6. Zhang, S., Liu, J.: Analysis and optimization of multiple unmanned aerial vehicle-assisted communications in post-disaster areas. IEEE Trans. Veh. Technol. **67**(12), 12049–12060 (2018)
7. Chiu, S.N., Stoyan, D., Kendall, W.S., Mecke, J.: Stochastic Geometry and Its Applications. Wiley, Hoboken (2013)
8. Shojaeifard, A., Wong, K.K., Di Renzo, M., Zheng, G., Hamdi, K.A., Tang, J.: Massive MIMO-enabled full-duplex cellular networks. IEEE Trans. Commun. **65**(11), 4734–4750 (2017)
9. Al-Hourani, A., Kandeepan, S., Jamalipour, A.: Modeling air-to-ground path loss for low altitude platforms in urban environments. In: 2014 IEEE Global Communications Conference, pp. 2898–2904. IEEE (2014)
10. Srinivasa, S., Haenggi, M.: Distance distributions in finite uniformly random networks: theory and applications. IEEE Trans. Veh. Technol. **59**(2), 940–949 (2010)
11. Kumaraswamy, P.: A generalized probability density function for double-bounded random processes. J. Hydrol. **46**(1–2), 79–88 (1980)
12. Park, J., Jung, S.Y., Kim, S.L., Bennis, M., Debbah, M.: User-centric mobility management in ultra-dense cellular networks under spatio-temporal dynamics. In: 2016 IEEE Global Communications Conference (GLOBECOM), pp. 1–6. IEEE (2016)
13. Athanasios. Papoulis. Probability, random variables, and stochastic processes (1985)

Modulation Recognition with Alpha-Stable Noise Over Fading Channels

Lingfei Zhang[1,2], Mingqian Liu[3(✉)], Jun Ma[2], and Chengqiao Liu[2]

[1] College of Physics and Electronic Information Engineering, Qinghai Nationalities University, Xining 810007, China
[2] School of Computer, Qinghai Normal University, Xining 810003, China
[3] State Key Laboratory of Integrated Service Networks, Xidian University, Xi'an 710071, China
mqliu@mail.xidian.edu.cn

Abstract. This paper proposes a method based on kernel density estimation (KDE) and expectation condition maximization (ECM) to realize digital modulation recognition over fading channels with non-Gaussian noise in the cognitive radio networks. A compound hypothesis test model is adopt here. The KDE method is used to estimate the probability density function of non-Gaussian noise, and the improved ECM algorithm is used to estimate the fading channel parameters. Numerical results show that the proposed method is robust to the noise type over fading channels. Moreover, when the GSNR is 10 dB, the correct recognition rate for the digital modulation recognition under non-Gaussian noise is more than 90%. Gaussian noise, and the improved ECM algorithm is used to estimate the fading channel parameters. Numerical results show that the proposed method is robust to the noise type over fading channels. Moreover, when the GSNR is 10 dB, the correct recognition rate for the digital modulation recognition under non-Gaussian noise is more than 90%.

Keywords: Cognitive radio · Modulation recognition · Fading channel · Non-Gaussian noise · Alpha stable distribution

1 Introduction

With the rapid development of communication technology and the increasing tension in spectrum resources, cognitive radio technology has become one of the key technologies to solve these problem [1]. In the cognitive radio network (CRN), spectrum sensing technology and spectrum access technology are particularly important. Among them, spectrum sensing technology not only needs to accurately detect the occurrence of authorized user signals, but also needs to identify the modulation type of authorized user signals [2]. Then we can determine the authorized user information, such as the type of service, the strength of the service, and so on. CRN technology is easily affected by fading channels, inter-user interference and electromagnetic pulse noise. This interference and noise often highlights that the probability density function is a thick tail of non-Gaussian distribution [3]. Therefore, it has practical engineering significance to

X. Wang et al. (Eds.): 6GN 2020, LNICST 337, pp. 323–332, 2020.
https://doi.org/10.1007/978-3-030-63941-9_24

study the digital modulation signal identification method under non-Gaussian fading channel in CRN.

Several algorithms have been reported to solve modulation recognition based on non-Gaussian noise. Some scholars have studied the digital modulation recognition under the alpha stable distributed noise model. Some have studied the modulation recognition under the mixed Gaussian noise distribution. In [4], a novel modulation classification method was proposed by using cyclic correntropy spectrum (CCES) and deep residual neural network (ResNet). CCES is introduced to effectively suppress non-Gaussian noise through the designated Gaussian kernel. In [5], T. Dutta et al. proposed a cyclostationary (CS) property based on FB classification technique under non-Gaussian impulsive noise condition. CS features perform well at low signal to noise ratio (SNR). But the performances of the classifiers degrade due to the presence of the impulse noise. In [6], Hu Y H et al. used fractional low-order wavelet packet decomposition and neural network recognition method, but this method has poor recognition performance under low GSNR. In [7], under the selective channel, the Gaussian mixture model is used to model the noise, and the identification problem of the amplitude phase modulation signal is studied. In the actual communication process, the signal is affected by the fading of channel besides a lot of non-Gaussian noise. In [8], D. E. Kebiche et al. investigated the performance of the Rao-test based detector for wideband spectrum sensing under non-Gaussian noise in a multi-carrier transmission framework. It showed that the Rao-test based detector combined with the universal filtered multicarrier (UFMC) outperforms the traditional OFDM based system in a realistic non-Gaussian noise environment. In [9], a signal identification method based on normalized fractional low-order cumulants for alpha-stable distributed noise fading channel was proposed. When the GSNR is low, the classification and recognition performance is poor. In [10], S. Hu, Y. Pei et al. proposed a low-complexity blind data-driven modulation classifier which operates robustly over Rayleigh fading channels under uncertain noise condition modeled using a mixture of three types of noise, namely, white Gaussian noise, white non-Gaussian noise and correlated non-Gaussian noise. The performance of proposed classifier approaches that of maximum likelihood classifiers with perfect channel knowledge.

In view of the above problems, this paper proposes a new method of modulated signal recognition under the alpha stable distributed noise fading channel. Firstly, the kernel density estimation algorithm is used to estimate the probability density function of the alpha stable distribution noise. The improved Expectation Conditional Maximization algorithm is then used to estimate the fading channel parameters. Finally, the modulation type of the signal is identified according to the composite hypothesis test model. The simulation results show that when the characteristic index of alpha stable distribution noise is 1.5 and the GSNR is 10 dB, the recognition rate of the signal in the fading channel is more than 90%. It can be seen that this method is effective and feasible.

2 System Model

Let B be the number of classes of all possible modulation types in the set of signals to be identified. The transmitted signal of class b can be expressed as:

$$s_b(t) = \sum_{m=-\infty}^{\infty} s_{mb} g(t - nT) \tag{1}$$

where T is the symbol period of the transmitted signal, $g(t)$ is the shaping filter function, and $s_{mb} \in s_b$ is the signal value of a certain modulation method. In CRN, the model of the received signal $r(t)$ can be described as:

$$r(t) = \sum_{l=0}^{L-1} h_l s_b(t - \tau_l) + \omega(t) \tag{2}$$

where L is the number of paths in the fading channel, τ_l is the time delay of each path, h_l is the attenuation range of each path, and $\omega(t)$ is the alpha stable distribution noise. The received signal is sampled at the receiving end by the sampling period T_s We assume that the time delay τ_l generated by each path is an integer times the sampling period and that the number of sampling points of the time delay after sampling is still represented by τ_l, the sampled received signal can be expressed as:

$$r[k] = \sum_{l=0}^{L} h_l s_b[kT_s - \tau_l] + \omega[k], k = 1, 2, \cdots, k \tag{3}$$

where K is the signal length, $\{h_l\}_{l=0}^{L-1}$ and $\{\tau_l\}_{l=0}^{L-1}$ are unknown channel parameter. Since the alpha stable distribution does not have a uniform closed PDF, $\omega(t)$ is usually described by a feature function [11], which is expressed as:

$$\phi(\theta) = \exp\{j\mu\theta - \gamma|\theta|^{\alpha}[1 + j\beta \text{sgn}(\theta)\omega(\theta, \alpha)]\} \tag{4}$$

where $\text{sgn}(\theta)$ is a symbolic function,

$$\text{sgn}(\theta) = \begin{cases} 1, & \theta > 0 \\ 0, & \theta = 0 \\ -1, & \theta < 0 \end{cases}, \quad \omega(\theta, \alpha) = \begin{cases} \tan(\alpha\pi/2), & \alpha \neq 1 \\ \frac{2}{\pi}\log|t|, & \alpha = 1 \end{cases}$$

$\gamma \geq 0$ is the dispersion coefficient, also known as the scale coefficient, α is called the characteristic index and its value range is $0 < \alpha \leq 2$. When $\alpha = 2$, the alpha stable distribution noise is converted into Gaussian noise. The parameter $\beta(-1 \leq \beta \leq 1)$ is a symmetrical parameter that determines the symmetry of the distribution. When $\beta = 0$, the alpha stable distribution is called the $s\alpha s$ distribution. The parameter mu is called the positional parameter. For the $s\alpha s$ distribution, if $0 < \alpha < 1$, μ is the median. If $1 < \alpha \leq 2$, μ is the mean. If $\mu = 0$ and $\gamma = 1$ are satisfied, the alpha stable distribution is called the standard alpha stable distribution.

3 Modulation Identification Under Alpha Stable Noise and Fading Channel

3.1 Probability Density Function Estimation of Alpha Stable Distributed Noise

We assume that each sample in the sample set $X = \{X_1, X_2, \cdots, X_N\}$ is independent and obeys the distribution of PDF as $p(X)$. The PDF is estimated using the Kernel Density Estimation (KDE) [12].

First we estimate the probability density in a small area. If a small area R is in the space where the sample is located, the probability that a random variable X is in the area R is

$$P_R = \int_R P(X)dX \tag{5}$$

known by the definition of binomial distribution, the probability that k of the N independent and identically distributed samples of the sample set are in the area R is:

$$P_k = C_N^k P_R^k (1 - P_R)^{N-k} \tag{6}$$

where C_N^k represents the number of combinations of k values randomly taken from N values. Then we get the expectation of $k : E[k] = k = NP_R$.Therefore, the estimate of P_R can be expressed as:

$$P_R = \frac{k}{N} \tag{7}$$

When $p(x)$ is continuous and the volume of R is sufficiently small, it can be considered that $p(x)$ in R is a constant, then Eq. (5) can be expressed as:

$$P_R = \int_R p(x)dx = p(x)V \tag{8}$$

where V is the volume of R. Substituting (7) into (8), we have:

$$\hat{p}(x) = \frac{k}{NV} \tag{9}$$

So we have completed the probability density estimation in a small area. But the probability density function estimated in this way is not continuous. Therefore, this paper uses the KDE method. When the small volume is fixed, the small volume of sliding is used to estimate the probability density of each point.

We assume that X is a d-dimensional vector, each small volume is a hypercube, each dimension has an edge length of h' and each small volume has a volume of $V = h'^d$ The d-dimensional unit window function is used to calculate the number of samples falling into each small volume:

$$\phi(u_1, u_2, \cdots, u_d)\begin{cases} 1, & |u_i| \leq \frac{1}{2}, i = 1, 2, \cdots, d \\ 0, & \text{other} \end{cases} \tag{10}$$

At this time, $\phi\left(\frac{x-x_i}{h'}\right)$ can be used to determine whether the sample X_i is in a cube whose center is X and whose edge length is h'. Let the number of observation points be N, then the samples falling in the above cube can be expressed as follows:

$$k_N = \sum_{i=1}^{N} \phi\left(\frac{x - x_i}{h'}\right) \tag{11}$$

Substituting Eq. (11) into Eq. (9), a PDF estimate at point x is:

$$\hat{P}(x) = \frac{1}{N}\sum_{i=1}^{N} \frac{1}{V}\phi\left(\frac{x - x_i}{h'}\right) \tag{12}$$

From a nuclear perspective, we can define the kernel function as:

$$K(x, x_i) = \frac{1}{V}\phi\left(\frac{x - x_i}{h}\right) \tag{13}$$

The estimate of the PDF can be regarded as using the kernel function to perform interpolation operation on the sample within the range of values. Its expression is:

$$K(x) = \frac{1}{2\sqrt{\pi}} \exp\left(-\frac{1}{2}x^2\right) \tag{14}$$

In short, a kernel density estimation process needs to move the kernel function to the position of each observation. And then select the global bandwidth to control the smoothness of the probability density function and the expansion of the kernel function. The revised evaluation formula is:

$$\hat{f}(x) = \frac{1}{N}\sum_{i=1}^{N} \frac{1}{h}K\left(\frac{x - x_i}{h}\right) \tag{15}$$

3.2 Estimation of Fading Channel Parameters

In this paper, based on the principle of ECM algorithm [13], combined with the model of modulation recognition under non-Gaussian fading channel, the parameter estimation steps of the proposed fading channel are as follows:

1) E-step: Under the assumption $H_b(b = 1, 2, \cdots, B)$ corresponding to each modulated signal, we calculate the conditional expectation of the logarithm likelihood function of the complete data. We have:

$$\begin{aligned} Q\left(\theta, \theta^P | H_b|\right) &= E\left[\log(p(c|\theta, H_b)) | \mathrm{o}, \theta^P, H_b\right] \\ &= \sum_{h} \log(p(c|\theta, H_b)) P\left(h|\mathrm{o}, \theta^P, H_b\right) \end{aligned} \tag{16}$$

where B is the number of modulation modes in the alternative set. Considering that the distribution of $\{r[k]_{k=1}^{K}\}$ and $\{s_b[k]\}_{k=1}^{K}$ is different, $\log(p(c|\theta, H_b))$ can be written as:

$$\begin{aligned}
&\log(p(c|\theta, H_b)) \\
&= \log\left(\prod_{k=1}^{K} p(r[k], s_b[k], H_b)\right) \\
&= \sum_{k=1}^{K} \log\left(\frac{1}{N_b} p(r[k]|s_b[k], \theta)\right)
\end{aligned} \tag{17}$$

where:

$$\begin{aligned}
&p(r[k]|s_b[k], \theta) \\
&= \frac{1}{N}\sum_{n=1}^{N} \frac{1}{h\lambda_n} K\left(\frac{r[k] - \sum_{l=0}^{L-1} h_l s_b[kT_s - \tau_l] - \omega_n}{h\lambda_n}\right)
\end{aligned} \tag{18}$$

By Bayesian theory:

$$\begin{aligned}
&p(h|o, \theta p, H_b) \\
&= \prod_{i=1}^{K} P\left(s_b[k]|r[k], \theta^P, H_b\right) \\
&= \prod_{i=1}^{K} \frac{1}{N_b} \frac{p(r[k]|s_b[k], \theta^p)}{P(r[k]|\theta^p, H_b)}
\end{aligned} \tag{19}$$

where:

$$\begin{aligned}
&p\left(r[k]|s_b[k], \theta^p\right) \\
&= \frac{1}{N}\sum_{n=1}^{N} \frac{1}{h\lambda_n} K\left(\frac{r[k] - \sum_{l=0}^{L-1} h_l^p s_b\left[kT_s - \tau_l^p\right] - \omega_n}{h\lambda_n}\right) \\
&p\left(r[k]|\theta^p, H_b\right) \\
&= \frac{1}{N_b}\sum_{s_b[k]}\left[\frac{1}{N}\sum_{n=1}^{N} \frac{1}{h\lambda_n} K\left(\frac{r[k] - \sum_{l=0}^{L-1} h_l^p s_b\left[kT_s - \tau_l^p\right] - \omega_n}{h\lambda_n}\right)\right]
\end{aligned} \tag{20}$$

Finally, $Q\left(\theta, \theta^P\right)$ can be obtained:

$$Q\left(\theta, \theta^p|H_b\right)$$

$$= \sum_{k=1}^{K} \sum_{s_b[k]} \left[\log\left(\frac{1}{N_b} p(r[k]|s_b[k], \theta) \right) \frac{1}{N_b} \frac{p(r[k]|s_b[k], \theta^p)}{p(r[k]|\theta^p, H_b)} \right] \tag{21}$$

2) M-step: This process maximizes the $Q(\theta, \theta^P)$ in the E-step to find a new estimate of the unknown parameter and then substitutes it as the known quantity into the next iteration.

$$\hat{\theta} = \arg\max_{\theta} Q(\theta, \theta^p | H_b) \tag{22}$$

The E-step of the ECM algorithm is the same as the EM algorithm, and each M-step is replaced with several simpler conditions to maximize the CM-step. For a certain hypothesis, the unknown parameter vector $\theta = \{h_1, \cdots h_L, \tau_1, \cdots, \tau_L\}$ is divided into $\theta_1 = \{h_1, \cdots, h_L\}$ and $\theta_2 = \{\tau_1, \cdots, \tau_L\}$. That is, the unknown channel parameter vector is $\theta = \{\theta_1, \theta_2\}$, $s = 2$ Each unknown parameters is given an initial value before the iterative process begins. In the iteration, keeping θ_2^p of the current iteration unchanged, we derivate $Q(\theta, \ \theta^p | H_b)$ to θ_1 and let its derivative be 0. We get the estimation θ_1^{p+1} of θ_1 in the next iteration. Then keeping the value of θ_1^{p+1} unchanged, we derivate $Q(\theta, \theta^P | H_b)$ to θ_2 and let its derivative be 0. We get the estimate θ_2^{p+1} of θ_2 in the next iteration. When the set convergence condition is satisfied, the iteration stops. The obtained parameter estimation value is the final unknown parameter estimation value of the iteration under the p + 1th assumption.

3.3 Modulation Recognition Based on Compound Hypothesis Test

The recognition method based on the likelihood function describes the recognition process as a compound hypothesis test process. The modulation mode corresponding to the maximum hypothesis in the (logarithmic) likelihood function of the received signal is determined as the modulation mode of the signal [14].

We use the KDE method to estimate the PDF of alpha stable noise distribution. Combing

$$p(r[1], \cdots, r[k] | H_b) = \prod_{k=1}^{K} \sum_{i=1}^{N_b} p(r[k] | s_{bi}[k]) P(s_{bi}[k] | H_b) \tag{23}$$

$$P(s_{bi}[k] | H_b) = 1 / N_b \tag{24}$$

we obtain:

$$p(r[1], \cdots, r[K] | H_b) = \prod_{k=1}^{K} \frac{1}{N_b} \sum_{i=1}^{N_b} p(r[k] | s_{bi}[k]) \tag{25}$$

where N_b represents the possible number of values of the amplitude of the bth modulation signal. It can be seen that the probability density distribution of the received signal is equal to the probability density distribution of the noise when the transmitted signal is known. This knowledge can be obtained by the estimation method of the fading channel parameters. Therefore, the process of modulation recognition under non-Gaussian fading channels is expressed as:

$$\hat{H} = \underset{H_b}{\arg\max} \log(r[1], \cdots, r[K] | H_b) \tag{26}$$

where $H_b(b = 1, 2, \cdots, B)$ indicates the modulation type of the received signal S_b.

4 Simulation Results and Analysis

In order to verify the effectiveness of the method, simulation experiments are carried out by MATLAB simulation software. The simulation parameters are set as follows: the signal set to be identified is BPSK, QPSK, 16QAM and 64QAM, which are four commonly used digital modulated signals. The noise is additive standard τ_l distributed noise. The carrier frequency of the modulated signal is 10 kHz. The code element rate is 1200 baud. The roll-off factor of the shaping filter is 0.35. The sampling frequency is 40 kHz. The number of signal sampling points is 3000 and the multipath channel is ITU_V_B channel of Rec.ITU-RM.225. The number of Monte Carlo simulations is 2000.

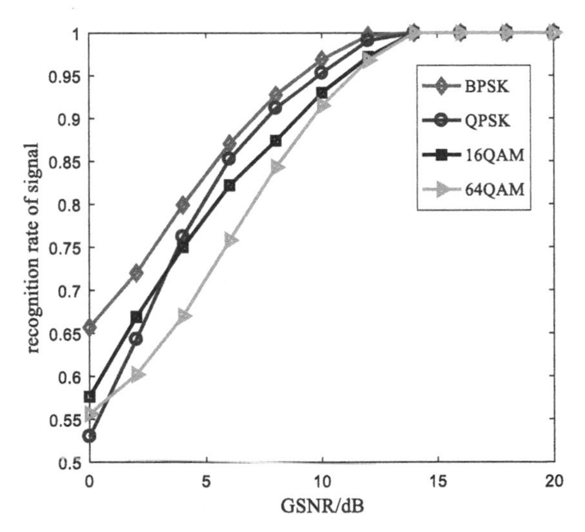

Fig. 1. Identification performance of signals under different GSNR

When $\alpha = 1.5, h_{opt} = 0.7$, the recognition performance of the signal at different GSNR is shown in Fig. 1. It can be seen from Fig. 1 that the recognition performance of different signals under the fading channel increases as the GSNR increases. When the GSNR is 10 dB, the recognition accuracy of different signals is more than 90%. It can be seen that the proposed identification method has good recognition performance under the fading channel. It is effective and feasible.

When $h_{opt} = 0.7$, $GSNR = 10\,\text{dB}$, the recognition performance of the signals under different characteristic indices is shown in Fig. 2. It can be seen from Fig. 2 that the recognition performance of different signals under the fading channel increases as the characteristic indices increases. When α is greater than 1, the correct recognition rate of different signal recognition is above 85%. When $\alpha = 2$, the correct recognition rate of different signals under the fading channel is 100%. It is shown that the proposed method is not only suitable for non-Gaussian noise fading channels, but also has good recognition performance under the environment of Gaussian noise fading channels. Therefore, the proposed method is robust to different noise types.

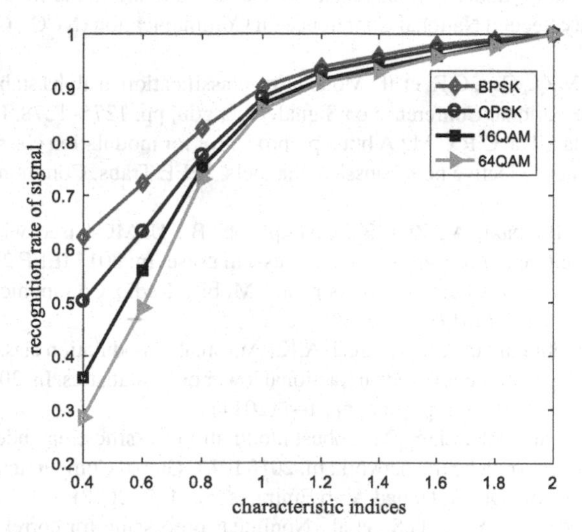

Fig. 2. Identification performance of signals under different characteristic indices

5 Conclusion

Aiming at the problem of modulation recognition in non-Gaussian fading channel in cognitive networks, this paper proposes a modulation recognition model based on compound hypothesis test suitable for this scenario. Under this model, the probability density function of non-Gaussian noise is estimated by the kernel density estimation method. The parameters of the fading channel are estimated using an improved expectation condition maximization algorithm. The simulation results show that the proposed identification method is effective and feasible under non-Gaussian noise fading channels and robust to different noise types.

Acknowledgments. The authors acknowledge the financial support of the Key Projects of R&D and Achievement Transformation in Qinghai Province (Grant: 2018-NN-151), the National Natural Science Foundation of China (Grant: 62071364 and 61761040).

References

1. Wang, D.Y., Zhang, N., Li, Z., et al.: Leveraging high order cumulants for spectrum sensing and power recognition in cognitive radio networks. IEEE Trans. Wireless Commun. **17**(2), 1298–1310 (2018)
2. Zhang, M., Diao, M., Guo, L.M.: Convolutional neural networks for automatic cognitive radio waveform recognition. IEEE Access **5**, 11074–11082 (2017)
3. Yang, G.S., Wang, J., Zhang, G.Y., et al.: Joint estimation of timing and carrier phase offsets for MSK signals in alpha-stable noise. IEEE Commun. Lett. **22**(1), 89–92 (2018)
4. Ma, J., Lin, S., Gao, H., Qiu, T.: Automatic modulation classification under non-gaussian noise: a deep residual learning approach. In: ICC2019 - 2019 IEEE International Conference on Communications (ICC), Shanghai, China, pp. 1-6 (2019)
5. Dutta, T., Satija, U., Ramkumar, B., Manikandan, M.S.: A novel method for automatic modulation classification under non-Gaussian noise based on variational mode decomposition. In: 2016 Twenty Second National Conference on Communication (NCC), Guwahati, pp. 1–6 (2016)
6. Hu, Y.H., Liu, M.Q., Cao, C.F., et al.: Modulation classification in alpha stable noise. In: 2016 IEEE 13th International Conference on Signal, Chengdu, pp. 1275–1278. IEEE (2016)
7. Amuru, S.D., da Silva, C.R.C.M.: A blind preprocessor for modulation classification applications in frequency-selective non-Gaussian channels. IEEE Trans. Commun. **63**(1), 156–169 (2015)
8. Kebiche, D.E., Baghaki, A., Zhu, X., Champagne, B.: UFMC-based wideband spectrum sensing for cognitive radio systems in non-Gaussian noise. In: 2017 IEEE 28th Annual International Symposium on Personal, Indoor, and Mobile Radio Communications (PIMRC), Montreal, QC, pp. 1–7 (2017)
9. Vinod, A.P., Madhukumar, A.S., Krishna, A.K.: Automatic modulation classification for cognitive radios using cumulants based on fractional lower order statistics. In: 2011 URSI General Assembly and Scientific Symposium, pp. 1–4 (2011)
10. Hu, S., Pei, Y., Liang, P.P., Liang, Y.: Robust modulation classification under uncertain noise condition using recurrent neural network. In: 2018 IEEE Global Communications Conference (GLOBECOM), Abu Dhabi, United Arab Emirates, pp. 1–7 (2018)
11. Zhang, G.Y., Wang, J., Yang, G.S., et al.: Nonlinear processing for correlation detection in symmetric alpha-stable noise. IEEE Signal Process. Lett. **25**(1), 120–124 (2018)
12. Pardo, A., Real, E., Krishnaswamy, V., et al.: Directional kernel density estimation for classification of breast tissue spectra. IEEE Trans. Med. Imaging **36**(1), 64–73 (2017)
13. Meng, X.L., Rubin, D.B.: Maximum likelihood estimation via the ECM algorithm: a general framework. Biometrika **80**(2), 267–278 (1993)
14. Carvajal, R., Rivas, K., Agüero, J.C.: On maximum likelihood estimation of channel impulse response and carrier frequency offset in OFDM systems. In: 2017 IEEE 9th Latin-American Conference on Communications (LATINCOM), Guatemala City, pp. 1–4 (2017)

An Improved Linear Threshold Model

Xiaohong Zhang[1], Nanqun He[1], Kai Qian[1], Wanquan Yang[2], and Jianji Ren[1](✉)

[1] College of Computer Science and Technology, Henan Polytechnic University, Jiaozuo 454000, China
{xh.zhang,renjianji}@hpu.edu.cn, 211809010012@home.hpu.cn, qkai237@126.com

[2] Henan College of Survey and Mapping, Zhengzhou 451464, China
ywanquan@126.com

Abstract. Linear threshold model is one of the widely used diffusion models in influence maximization problem. It simulates influence spread by activating nodes for an iterative way. However, the way it makes activation decisions limits the convergence speed of itself. In this paper, an improvement is proposed to speed up the convergence of Linear threshold model. The improvement makes activation decisions with one step ahead considering the nodes which will be activated soon. To assist in activation decision making, the improvement introduces a new state and updates state transition rules. The experiment results verify the performance and efficiency of the improvement.

Keywords: Linear threshold model · Performance · Influence maximization · Convergence speed

1 Introduction

Social network platforms such as Facebook, WeChat, Twitter, etc., have become widely used mediums that people communicate with each other. As one of the most important problems in social network analysis, the problem of influence maximization has attracted tremendous attentions [15,20,26]. It aims to find a small set of influential nodes so that the influence of those nodes can be spread most widely. Although the problem is first raised in the field of marketing, it exists in many other fields such as political movements [16], rumor controlling [29,31], and so on.

Diffusion models are critical for solving the influence maximization problem since they can simulate the spread of influence, Linear threshold model (LT model) [17] is one of the most widely used diffusion models. It simulates influence spread by activating nodes in an iterative way. In each iteration, a activation decisions are made only according to the nodes activated in previous iterations. The nodes activated in current iteration are not counted for any activation decision made in the same iteration, which limits the convergence speed

X. Wang et al. (Eds.): 6GN 2020, LNICST 337, pp. 333–344, 2020.
https://doi.org/10.1007/978-3-030-63941-9_25

of LT model. Although many efforts have been made on the improvement of LT model [2,7,23,27], how to improve the convergence speed of LT model is still an open issue.

In this work, we focus on the problem of the convergence speed of LT model. To deal with the problem, an improvement is proposed to activate nodes with one step ahead considering the nodes which will be activated. Concretely, the improvement makes activation decisions according to not only active nodes but also the inactive nodes to be activated soon. To identify those inactive nodes, the improvement introduces a new state for them, that is, ready-active state, and new state transition rules.

The rest of this paper is organized as follows: related work is reviewed in Sect. 2 followed by motivation in Sect. 3. The proposed approach is elaborated in Sect. 4 and evaluated in Sect. 5. Finally, the paper is concluded in Sect. 6.

2 Related Work

Influence maximization is one of the most important problems in the areas of social network analysis, and has attracted much attentions in recent years. Domingos and Richardson [8,24] first study the influence maximization problem in probabilistic environments. Kempe et al. [17] prove that the influence maximization problem is an NP-hard problem, and propose LT model to simulate the process of information diffusion. In LT model, each node of a social network is only in one of the following two states: inactive state and active state. LT model starts with the assumption that all the nodes are in the inactive state except the set of seed nodes which are initialized in the active state.

Many researches have been carried out to select the set of seed nodes. Kempe et al. [17] propose a simple greedy algorithm approximating the optimum with a factor of $(1-1/e)$. Various variants of the greedy algorithm, e.g. CELF [18], CELF++ [12], constrained greedy algorithm [33], etc., have been proposed to improve the efficiency of the simple greedy algorithm. In addition to greedy algorithms, many different algorithms, e.g., centrality based algorithms [9,10,25], community based algorithms [3,19,28,32], influence estimation based algorithms [21,22], etc., are also exploited to identify the set of seed nodes. Besides the work on the selection of seed nodes, many efforts are put on the improvement of LT model.

Various researches concern on competitive environments in which more than one player competes with each other to influence the most nodes. He et al. [13] proposed a competitive LT model to block the influence of competing products. Bozorgi et al. [3] extend LT model to give each node a decision-making ability about incoming influence spread. Galhotra et al. [11], Zhang et al. [30] and Yang et al. [29] concern on the improvement of LT model to allow a node to be either positive or negative in the speed process of information or influence. Zhang et al. [14,34] and Caliò et al. [5] revise LT model to spread influence according to trust/distrust relationships. Chan et al. [6] propose the non-progressive LT model to deal with the case in which active nodes may become inactive. Although much efforts have been made on the improvement of LT model, the convergence problem of the model is still an open problem.

For the convenience of presentation, all the notations in this work are described in Table 1.

Table 1. Notations

Notations	Description
$\boldsymbol{G}$	A directed graph
$\boldsymbol{E}$	The edge set of $\boldsymbol{G}$
$\boldsymbol{V}$	The vertex set of $\boldsymbol{G}$
v	A vertex, e.g., v_i represents the i_{th} vertex of $\boldsymbol{V}$
θ_a	Activation threshold of a node
$\boldsymbol{N}_i^a$	The active neighbor set of v_i
$\boldsymbol{N}_i^r$	The ready-active neighbor set of v_i
$State(v_i)$	The state of node v_i
Inf^{To}	The influence to a node
Inf_p^{To}	The potential influence to v_i
$Inf(v_i, v_j)$	The influence of v_i to v_j

3 Motivation

LT model is proposed to simulate the spread of influence by Kempe et al. [17]. In LT model, a social network is denoted as a directed graph $G = (V, E)$, where V represents the set of nodes, and E expresses the set of directed edges. Each node can only be in one of the following two states: the active state and the inactive state. $((\forall v_i) v_i \in V)$, v_i can transmit from the inactive state to the active state if function (1) is satisfied. Otherwise, it stays in the inactive state.

$$\Sigma_{v_j \in N_{v_i}^a} Inf(v_j, v_i) \geq \theta_a \tag{1}$$

LT model starts with a small set of seeds. All the nodes regarded as seeds are initialized in the active state while all the other nodes in inactive state. LT model updates the states of nodes in an iterative way and converges if no inactive nodes are activated any more. It decides whether an inactive node can be converted to the active state according to the influence from its neighbors which are activated in previous iterations. The influence from the nodes being activated in the current iteration is not counted, which limits the convergence speed of LT model.

In this work, we aim to make an improvement on LT model to accelerate its convergence. The improvement makes activation decisions according to the influence from not only active neighbors but also the inactive neighbors which will change into the active state.

4 LT Model with One Step Forward Looking

In this section, we elaborate the improvement of LT model. The improvement aims to speed up the convergence speed of LT model with one step ahead considering the nodes which will be activated. Concretely, in each iteration, the improvement calculates the influence to an inactive node according to its active neighbors and inactive neighbors which will be activated soon. If the calculated influence overpasses θ_a, the inactive node changes into the active state.

4.1 State Transition

To activate a node, the improvement calculates the influence from its active neighbors, and some of its inactive neighbors. Here, the inactive neighbors indicate the inactive nodes which will change from the inactive state to the active state in the next iteration. To distinguish those inactive neighbors form other inactive neighbors, we introduce the ready-active state.

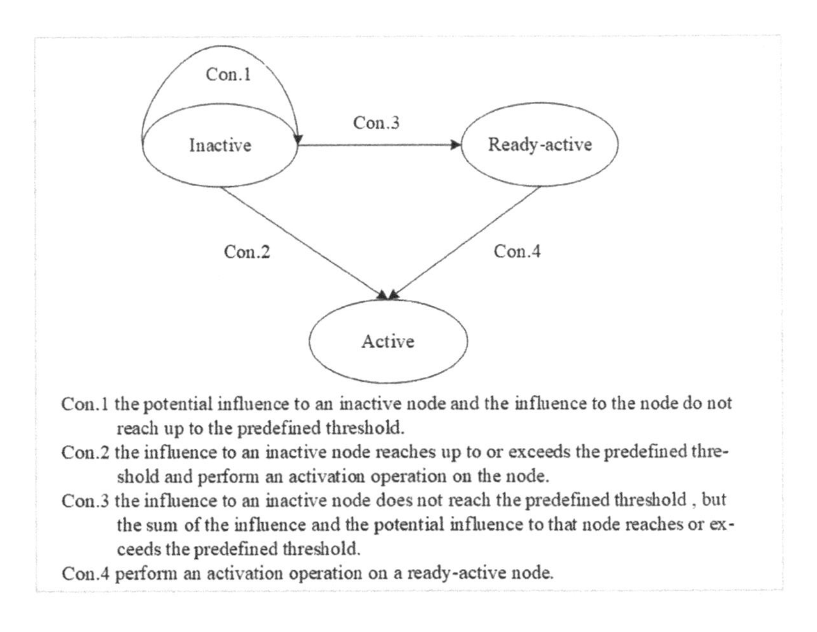

Con.1
Inactive
Con.3
Ready-active
Con.2
Con.4
Active
Con.1 the potential influence to an inactive node and the influence to the node do not reach up to the predefined threshold.
Con.2 the influence to an inactive node reaches up to or exceeds the predefined threshold and perform an activation operation on the node.
Con.3 the influence to an inactive node does not reach the predefined threshold , but the sum of the influence and the potential influence to that node reaches or exceeds the predefined threshold.
Con.4 perform an activation operation on a ready-active node.

Fig. 1. Node state transition

Definition 1. ***Ready-active state.*** $(\forall v_i)((v_i \in V)$ *and* $State(v_i) = inactive)$, v_i *is in the ready-active state if* v_i *satisfies the following conditions: (1) the influence from all the active neighbors of* v_i *reaches up to or overpasses the predefined threshold, and (2) no activation operation has been performed on* v_i.

Definition 2. *Potential influence.* *$(\forall v_i)((v_i \in V)$ and $State(v_i) = inactive)$, the potential influence of v_i indicates the influence from its ready-active neighbors. It is calculated by function (2).*

$$Inf_p^{To} = \Sigma_{v_j \in N_i^r} Inf(v_j, v_i) \tag{2}$$

In the function, $inf(v_j, v_i)$ represents the influence from v_j to v_i, N_i^r represents the ready-active neighbor set of v_i. It can be calculated according to the existing influence models. After adopting the ready-active state, state transition is performed according to the following rules:

Rule 1. Perform an activation operation on an inactive node if the influence from the active neighbors of that node reaches or overpasses an predefined threshold, and change that node into the active state.

Rule 2. Change an inactive node into the ready-active state if the influence from its active neighbors does not reach the predefined threshold, but the influence from its active neighbors and the potential influence from its ready-active neighbors reach or exceed the predefined threshold.

Rule 3. Change the state of a ready-active node into the active state after performing an activation operation on that node.

Figure 1 shows the state transition in the improvement. According to the figure, an inactive node can be transited into the ready-active state or active state. It can also keep being in the inactive state. If Con.1 is satisfied, it keeps being in the inactive state. If Con.2 is satisfied, it changes into the active state. If Con.3 is satisfied, it changes into the ready-active state. A ready-active state can only change into the active state only if Con.4 is satisfied.

4.2 Influence Propagation

The improved model propagates influence in an iterative way as LT model does. However, it is different from LT model by considering ahead the inactive node to be activiated soon, that is, the actve-ready nodes. It first initializes seed nodes in the active state and all other nodes in the inactive state and starts iterations. In each iteration, it calculates the influence and the potential influence to each inactive node, and performs state transition according to Rules 1 to 3. The iteration converges if the difference in the number of the nodes activated during two adjacent iterations is less than a predefined threshold.

Algorithm 1 shows the influence propagation in the improved model. We take Fig. 2 as an example to illustrate the influence spread with the improved model. In the initial state, v_2 and v_5 are regarded as seed nodes, the states of the two nodes are active, and the remaining nodes are inactive.

In the first iteration, v_2 has an influence on v_1 exceeds θ_a and v_1 transitions to an active state. Similarly, v_4 also becomes active. The sum of influence of v_2 on v_3 and the potential influence of v_4 on v_3 exceeds θ_a, and v_3 is changed into

Algorithm 1. Influence propagation in the improved model

Input:: seeds, G(V,E)
Output: total number of the nodes activated

```
1: actives ← seeds
2: do
3:     num=actives.size();
4:     inactives ← obtain the inactive neighbors of the nodes in seeds
5:     for each node in inactive do
6:         calculate the influence to that node, that is Inf^{To};
7:         if Inf^{To} ≥ θ_a then
8:             activate the node and updata the state of that node to be active;
9:             insert that node to actives
10:        else
11:            calculate the potentional influence to that node, that is Inf_p^{To}
12:            if Inf^{To} + Inf_p^{To} ≥ θ_a then
13:                activate v_i
14:                insert v_i to actives
15:            end if
16:        end if
17:    end for
18: while (actives.size()–num ≥ ξ)
19: return num
```

the ready-active state, the activation operation is performed on v_3, and v_3 is changed into the active state. The remaining nodes keep being inactive.

In the second iteration, the influence to v_6 comes from v_3 overpasses θ_a, and v_6 is changed into the active state. The sum of the influence of node v_5 on v_7 and the potential influence of v_6 on v_7 exceeds θ_a, so v_7 changes into the ready-active state, the activation operation is performed on v_7, and v_7 is changed into an active state. The remaining nodes keep being inactive.

In the third iteration, v_7 has an influence on v_8 that exceeds θ_a, so v_8 changes into the active state. Similarly, v_9 is also activated. The other node states remain unchanged.

In the fourth iteration, no nodes can be activated, so the iteration converges.

5 Experiment and Evaluation

In this section, we elaborate the abundant experiments conducted to evaluate the improved model by comparing with LT model, and discuss the experiment results.

5.1 Data Sets and Environment

We use six-real-world data sets in our experiments. All these data sets are collected from Stanford Large Network Dataset Collection [1]. All these data sets are in different scales and expressed in directed graphs. The details of those data sets are described below.

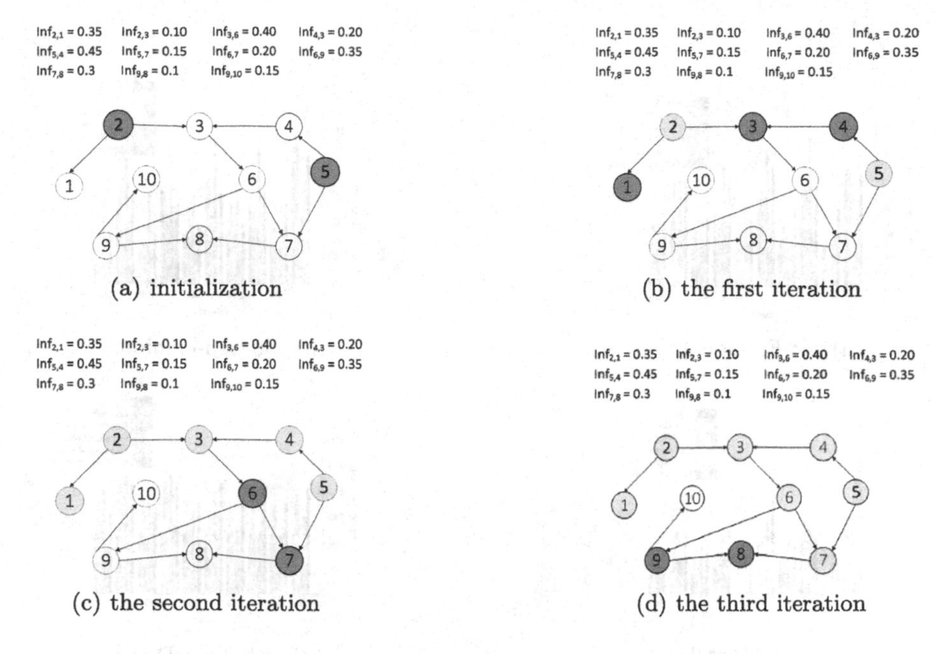

(a) initialization (b) the first iteration (c) the second iteration (d) the third iteration

Fig. 2. Information dissemination based on the improved model

- soc-Epinions1 data set: This data is collected according to the trust relationship interaction on the website. It includes 75879 nodes and 508837 edges. Each vertex describes a reviewer and each edge denotes the trust relationship between two reviewers.
- soc-sign-epinions data set: This data set is extracted from Epinions.com. It contains about 131828 nodes and 841372 edges. Each vertex represents a user, and edge describes one user described by one vertex trusting the other user described by the other vertex.
- email-EuAll data set: This network is generated according to the email data from a large European research institution. It includes 265214 nodes and 420045 edges. Each node corresponds to an email address. Each edge describes that at least one email is sent from one node to the other node.
- web-NotreDame data set: This data set is collected from the web site of the University of Notre Dame. It includes 325729 nodes and 1497134 edges. Nodes represent pages from the web site and edges represent hyperlinks between those pages.
- wiki-Talk data set: This data set is collected from Wikipedia which is a free encyclopedia written collaboratively by volunteers around the world. It includes 2394385 nodes and 5021410 edges. A node represents a Wikipedia user. An edge represents one user at least edits a talk page of the other node.
- soc-LiveJournal1 data set: This data set is collected from a free online community with almost 10 million members. It includes 4847571 nodes and 6899373 edges. Nodes represent members and edges represent friendship of members.

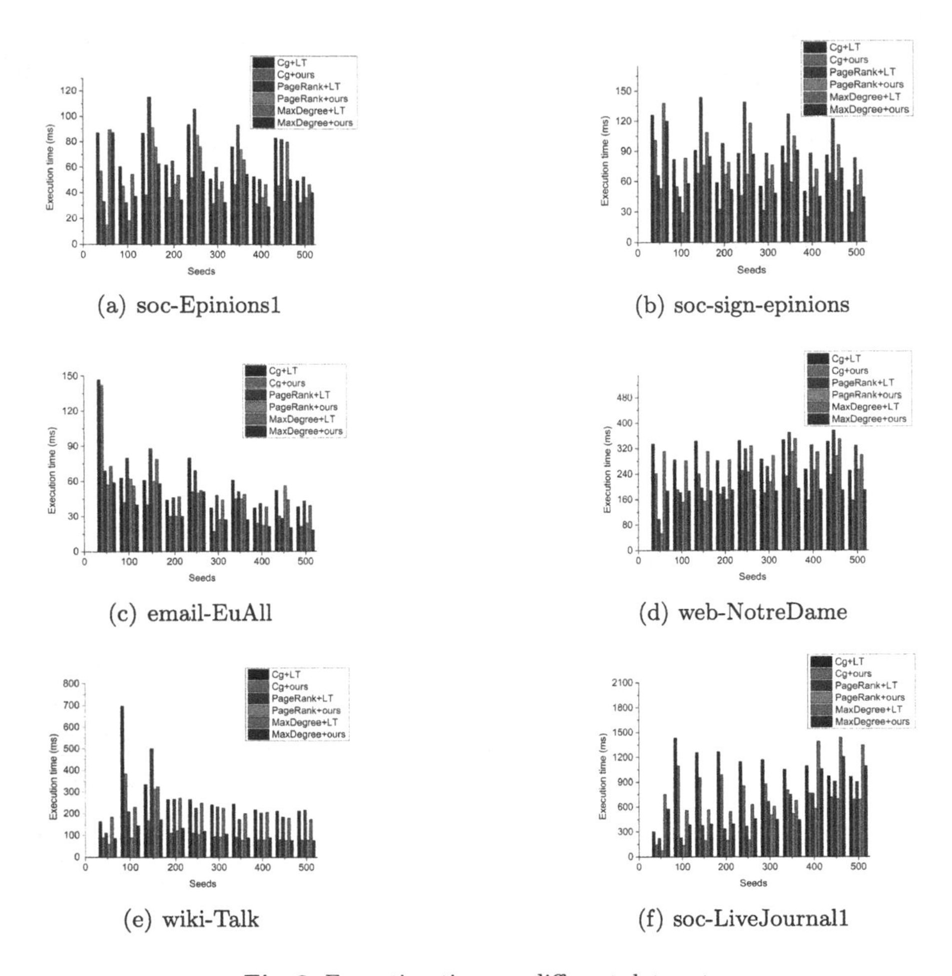

(a) soc-Epinions1

(b) soc-sign-epinions

(c) email-EuAll

(d) web-NotreDame

(e) wiki-Talk

(f) soc-LiveJournal1

Fig. 3. Execution time on different data sets

Table 2. Configuration of the big data platform

Hardware layer		HDFS		Spark	
Audit/Node	24	File block	256	Parallel tasks/actuators	7
Memory/Node	32G	Number of copies	3	Working node A	19
Hard disk/Node	3TB	Data node	19	Actuator number/work node	1
Tocal nodes	20	Nodes	19	Worker	1

All the experiments are conducted on the big data platform which is deployed on the cluster consisting of 20 severs. The details of the platform are described in Table 2. We choose execution time and the total number of activated nodes as the metrics to evaluate our work.

5.2 Result Discussion

The evaluation of our work is performed by comparing with LT model when different seed selection algorithms are adopted. Cg denotes the seed selection algorithm based on influence spread in a constrained greedy way algorithms [33]. MaxDegree and PageRank describe the algorithm selecting seeds according to Max degree policy [17] and PageRank algorithm [4], respectively.

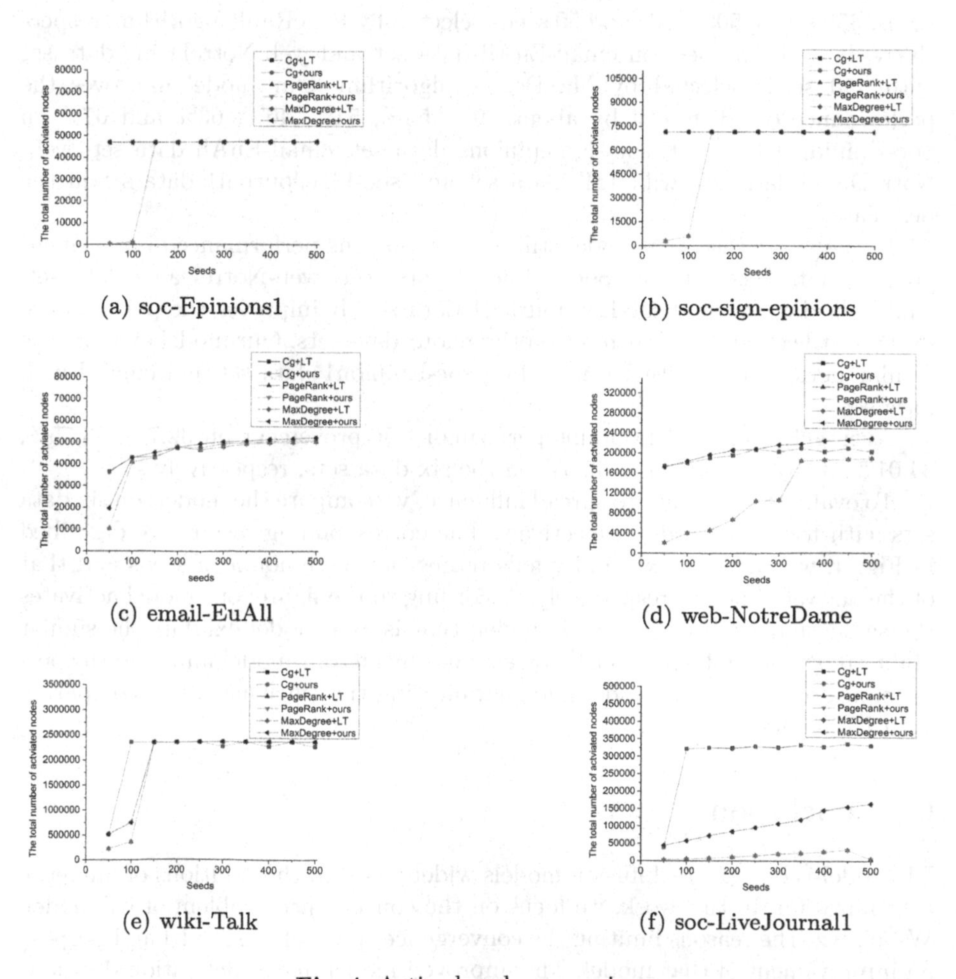

(a) soc-Epinions1

(b) soc-sign-epinions

(c) email-EuAll

(d) web-NotreDame

(e) wiki-Talk

(f) soc-LiveJournal1

Fig. 4. Actived node comparison

Cg+ours, MaxDegree+ours and PageRank+ours denote the results of the corresponding seed selection algorithms combined with our model, respectively. Cg+LT, MaxDegree+LT and PageRank+LT describe the results of those algorithms with LT model, respectively. All the results shown in this subsection are the average results of 100 independent runnings.

Figure 3 shows the execution time of our model and LT model on different data sets. In the figure, the x axis represents the total number of seeds selected by different algorithms. The y axis expresses the execution time of our model and LT model spreading influence with the selected seeds. According to the figure, our model converges faster than LT model on all data sets no matter which algorithm is used to select seeds.

More details, our model performs best on Soc-Epinions1 data set, soc-sign-epinions data set, wiki-Talk data set and soc-LiveJournal1 data set with 450 seeds, 350 seeds, 500 seeds and 50 seeds selected by PageRank algorithm, respectively. It performs best on email-EuAll data set and web-NotreDame data set with 450 seeds selected by MaxDegree algorithm. Our model improves the performance of LT model by about 59%, 53%, 54%, 46%, 63% and 67% on Soc-Epinions1 data set, soc-sign-epinions data set, email-EuAll data set, web-NotreDame data set, wiki-Talk data set and soc-LiveJournal1 data set in the best case.

In the worst case, our model still obtains obvious performance improvement on four data sets, that is, Soc-Epinions1 data set, web-NotreDame data set, wiki-Talk data set and soc-LiveJournal1 data set. It improves the performance by 13% at least and 37% at most on these four data sets. Our model exhibits the similar performance with LT model on soc-Epinion1 data set and email-EuAll data set.

Averagely, our model obtains performance improvement on 32.7%, 33.75%, 34.01%, 31.6%, 54.38% and 32.3% on the six data sets, respectively.

To evaluate the ability to spread influence, we compare the models on six data sets with the same seeds, respectively. The corresponding results are described in Fig. 4, where the x axis and y axis represents the amount of seeds and that of the activated nodes, respectively. According to the figure, our model activates the same number of nodes as LT model, that is, our model exhibits the similar ability to spread influence to LT model. That means our model improves the performance of influence spread while guaranteeing the efficiency when comparing with LT model.

6 Conclusion

LT model is one of the diffusion models widely used in the solutions of influence maximization. In this work, we focus on the convergence problem of LT model. We analyze the reasons limiting the convergence speed of the model and propose an improvement of that model. The improved model makes activation decisions according to not only active nodes but also the inactive nodes to be activated soon. The experiment results on abundant datasets verify the performance and efficiency of the improved model. In the future, we will research the application of Linear threshold model in competitive environments.

References

1. http://snap.stanford.edu/data/ (2020)
2. Borodin, A., Filmus, Y., Oren, J.: Threshold models for competitive influence in social networks, pp. 539–550 (2010)
3. Bozorgi, A., Samet, S., Kwisthout, J., Wareham, T.: Community-based influence maximization in social networks under a competitive linear threshold model. Knowl. Based Syst. **134**, 149–158 (2017)
4. Brin, S., Page, L.: The anatomy of a large-scale hypertextual web search engine. Comput. Netw. ISDN Syst. **30**, 107–117 (1998)
5. Caliò, A., Tagarelli, A.: Complex influence propagation based on trust-aware dynamic linear threshold models. Appl. Netw. Sci. **4**(1), 1–41 (2019). https://doi.org/10.1007/s41109-019-0124-5
6. Chan, T.H.H., Ning, L.: Influence maximization under the non-progressive linear threshold model. arXiv Social and Information Networks (2015)
7. Chen, W., et al.: Influence maximization in social networks when negative opinions may emerge and propagate, pp. 379–390 (2011)
8. Domingos, P., Richardson, M.: Mining the network value of customers. In: Proceedings of the Seventh ACM SIGKDD International Conference on Knowledge Discovery and Data Mining, pp. 57–66. ACM (2001)
9. Freeman, L.C.: A set of measures of centrality based on betweenness. Sociometry **40**(1), 35–41 (1977)
10. Freeman, L.C.: Centrality in social networks conceptual clarification. Social Netw. **1**(3), 215–239 (1978)
11. Galhotra, S., Arora, A., Roy, S.: Holistic influence maximization: combining scalability and efficiency with opinion-aware models, pp. 743–758 (2016)
12. Goyal, A., Lu, W., Lakshmanan, L.V.S.: CELF++: optimizing the greedy algorithm for influence maximization in social networks, pp. 47–48 (2011)
13. He, X., Song, G., Chen, W., Jiang, Q.: Influence blocking maximization in social networks under the competitive linear threshold model, pp. 463–474 (2012)
14. Hosseinipozveh, M., Zamanifar, K., Naghshnilchi, A.R.: Assessing information diffusion models for influence maximization in signed social networks. Expert Syst. Appl. **119**, 476–490 (2019)
15. Huang, H., Shen, H., Meng, Z., Chang, H., He, H.: Community-based influence maximization for viral marketing. Appl. Intell. **49**(6), 2137–2150 (2019). https://doi.org/10.1007/s10489-018-1387-8
16. Katz, E., Lazarsfeld, P.F., Roper, E.: Personal influence : the part played by people in the flow of mass communications. Am. Sociol. Rev. **17**(4), 357 (1956)
17. Kempe, D., Kleinberg, J., Tardos, E.: Maximizing the spread of influence in a social network. In: Proceeding of the Ninth ACM SIGKDD International Conference on Knowledge Discovery and Data Mining (2003)
18. Leskovec, J., Krause, A., Guestrin, C., Faloutsos, C., Vanbriesen, J.M., Glance, N.: Cost-effective outbreak detection in networks, pp. 420–429 (2007)
19. Li, X., Cheng, X., Su, S., Sun, C.: Community-based seeds selection algorithm for location aware influence maximization. Neurocomputing **275**, 1601–1613 (2018)

20. Liu, W., Chen, X., Jeon, B., Chen, L., Chen, B.: Influence maximization on signed networks under independent cascade model. Appl. Intell. **49**(3), 912–928 (2018). https://doi.org/10.1007/s10489-018-1303-2
21. Lu, W., Zhou, C., Wu, J.: Big social network influence maximization via recursively estimating influence spread. Knowl. Based Syst. **113**, 143–154 (2016)
22. Lu, Z., Fan, L., Wu, W., Thuraisingham, B., Yang, K.: Efficient influence spread estimation for influence maximization under the linear threshold model. Comput. Soc. Netw. **1**(1), 1–19 (2014). https://doi.org/10.1186/s40649-014-0002-3
23. Pathak, N., Banerjee, A., Srivastava, J.: A generalized linear threshold model for multiple cascades, pp. 965–970 (2010)
24. Richardson, M., Domingos, P.: Mining knowledge-sharing sites for viral marketing. In: Proceedings of the Eighth ACM SIGKDD International Conference on Knowledge Discovery and Data Mining, pp. 61–70. ACM (2002)
25. Sabidussi, G.: The centrality index of a graph. Psychometrika **31**(4), 581–603 (1966)
26. Saxena, B., Kumar, P.: A node activity and connectivity-based model for influence maximization in social networks. Soc. Netw. Anal. Min. **9**(1), 1–16 (2019). https://doi.org/10.1007/s13278-019-0586-6
27. Trpevski, D., Tang, W.K., Kocarev, L.: Model for rumor spreading over networks. Phys. Rev. E **81**(5), 056102 (2010)
28. Wang, Y., Cong, G., Song, G., Xie, K.: Community-based greedy algorithm for mining top-k influential nodes in mobile social networks, pp. 1039–1048 (2010)
29. Yang, L., Li, Z., Giua, A.: Containment of rumor spread in complex social networks. Inf. Sci. **506**, 113–130 (2020)
30. Zhang, H., Dinh, T.N., Thai, M.T.: Maximizing the spread of positive influence in online social networks, pp. 317–326 (2013)
31. Zhang, R., Li, D.: Identifying influential rumor spreader in social network. Discrete Dyn. Nat. Soc. **2019**, 1–10 (2019)
32. Zhang, X., Zhu, J., Wang, Q., Zhao, H.: Identifying influential nodes in complex networks with community structure. Knowl. Based Syst. **42**, 74–84 (2013)
33. Zhang, X., Li, Z., Qian, K., Ren, J., Luo, J.: Influential node identification in a constrained greedy way. Physica A **557**, 124887 (2020). https://doi.org/10.1016/j.physa.2020.124887
34. Zhang, Y., Wang, Z., Xia, C.: Identifying key users for targeted marketing by mining online social network, pp. 644–649 (2010)

Performance Analysis for Caching in Multi-tier IoT Networks with Joint Transmission

Tianming Feng[1], Shuo Shi[1(✉)], Xuemai Gu[1], and Zhenyu Xu[2]

[1] Harbin Institute of Technology, Harbin 150001, China
{fengtianming,crcss,guxuemai}@hit.edu.cn
[2] Huizhou Engineering Vocational College, Huizhou 516023, China
hitusa@126.com

Abstract. The rapid growth of the number of IoT devices in the network has brought huge traffic pressure to the network. Caching at the edge has been regarded as a promising technique to solve this problem. However, how to further improve the successful transmission probability (STP) in cache-enabled multi-tier IoT networks (CMINs) is still an open issue. To this end, this paper proposes a base station (BS) joint transmission scheme in CMIN where the nearest BS that stores the requested files in each tier is selected to cooperatively serve the typical UE. Based on the proposed scheme, we derive an integral expression for the STP, and optimize the content caching strategy for a two-tier network case. The gradient projection method is used to solve the optimization problem, and a locally optimal caching strategy (LCS) is obtained. Numerical simulations show that the LCS achieves a significant gain in STP over three comparative baseline strategies.

Keywords: IoT networks · Cache-enabled networks · Joint transmission

1 Introduction

The rapidly growing number of mobile IoT devices on the network has led to explosive growth in the demand for mobile data traffic. Deploying multi-tier edge base station (BS) to form a heterogeneous network (HetNet) has become an effective solution to meet these demands [1]. However, densely deployed BSs will inevitably cause high inter-cell interference. As one of the downlink coordinated multipoint transmission (CoMP) transmission techniques, joint transmission (JT) [2], where a user is simultaneously served by several BSs, can effectively mitigate the interference. Based on the recent observation that a large portion of the traffic is caused by repeatedly downloading a few popular contents, caching the popular contents at the edge of the IoT network, such as edge BSs, IoT devices, can greatly reduce the traffic burden of IoT network.

X. Wang et al. (Eds.): 6GN 2020, LNICST 337, pp. 345–356, 2020.
https://doi.org/10.1007/978-3-030-63941-9_26

The concept of caching technique has been extensively studied in the recent years. The authors of [3] design an optimal tier-level caching strategy for HetNets. This work is extended to a K-tier multi-antenna multi-user HetNets scenario by [4]. In [5], the authors design an optimal caching strategy in heterogeneous IoT networks to improve the offloading rate for backhaul links. [6] jointly optimizes content placement and activation densities of BSs of different tiers to reduce the energy consumption for heterogeneous industrial IoT networks. [7] investigates the influence of JT on caching strategy and obtains a locally optimal solution in the general case and a globally optimal solution in some special cases. Based on this work, the authors of [8] study the advantage of introducing local channel state information into JT and design an algorithm for locally optimal caching strategy in IoT networks. [9] studies the tradeoff between the content diversity gain and the cooperative gain and proposes an optimal caching strategy to balance the tradeoff.

However, all the aforementioned works that jointly consider JT and caching technique in multi-tier networks only limit the content caching strategy design to the single-tier BSs. That is, only one tier of BSs in the network have cache capacity, the BSs of other tiers do not have the ability to cache. Moreover, the BS cooperation is just limited to the BSs from the same tier. In this paper, the content caching strategy with BS JT is studied in a cache-enabled multi-tier IoT network (CMIN). Different from the existing literature, every tier of BSs in our model has cache capacity to store a limited number of contents. JT scheme is adopted in this work, under which the user is jointly served by the nearest BSs that cache the requested content from each tier. So that the BS cooperation is "vertical" (i.e., cross-tier) rather than "horizontal" (i.e., in-tier) considered in the existing literature. Based on the JT and caching model, integral expressions for the successful transmission probability (STP) are derived for an M-tier IoT network case and a special case of two-tier network. Then, a locally optimal caching strategy (LCS) for the two-tier IoT network is obtained by maximizing the STP of the network using the gradient projection method (GPM). Finally, simulation results demonstrate the LCS achieves significant gains in STP over several comparative baselines with BS JT.

2 System Model

2.1 Network and Caching Model

We consider a downlink large-scale CMIN consisting of M tiers of BSs, where the set of BS tiers is denoted as $\mathcal{M} = \{1, 2, \cdots, M\}$. The BSs in the m-th tier are spatially distributed as an independent homogeneous Poisson point process (PPP) Φ_m, with density λ_m and transmission power P_m, $m \in \mathcal{M}$. The locations of the BSs of all tiers are denoted by Φ, i.e., $\Phi = \bigcup_{m \in \mathcal{M}} \Phi_m$. We focus on the analysis of a typical user equipment (UE) u_0, which is assumed to locate at the origin without loss of generality. When the typical UE is served by a BS located at $x_m \in \mathbb{R}^2$ from the m-th tier, the signal power it receives can be expressed as $P_m \|x_m\|^{-\alpha_m} |h_{x_m}|^2$, where $\|x_m\|^{-\alpha_m}$ and $|h_{x_m}|^2$ correspond to large-scale fading and small-scale fading, respectively; $\alpha_m > 2$ denotes the path-loss exponent; h_{x_m} models the Rayleigh fading between the typical UE and the BS, i.e., $h_{x_m} \overset{d}{\sim} \mathcal{CN}(0,1)$ or channel power $|h_{x_m}|^2 \overset{d}{\sim} \text{Exp}(1)$ [1].

The CMIN considered in this paper has a content database denoted by $\mathcal{N} = \{1, 2, \cdots, N\}$. There are $N \geq 1$ files in the database, whose sizes are assumed to be equal and normalized to one for ease of analysis. The popularity of file n is denoted as $a_n \in [0, 1]$. Here, we assume the popularity distribution follows a Zipf distribution [3], i.e.,

$$a_n = \frac{n^{-\gamma}}{\sum_{n \in \mathcal{N}} n^{-\gamma}}, \quad \text{for } \forall n \in \mathcal{N}, \tag{1}$$

where $\gamma \geq 0$ is the Zipf exponent. Each UE randomly requests one file according to the file popularity distribution in one time slot.

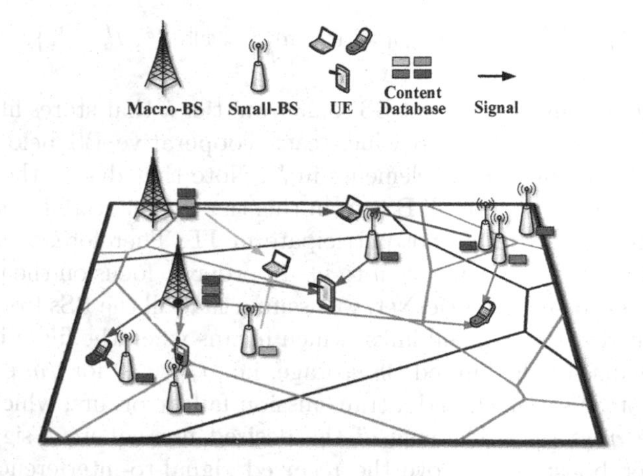

Fig. 1. Illustration of a two-tier cache-enabled IoT network with joint transmission scheme. There are four different files in the database, which are indicated by four different colors. The colors of the UEs represent the files they request. In this scenario, $N = 4$, $C_1 = 3$, $C_2 = 1$.

In tier m, each BS has a limited cache size C_m which can store at most C_m different files out of N. In this paper, we consider a *random caching* scheme [3,10], in which the file n is stored in a BS from tier m randomly with probability $t_{mn} \in [0, 1]$, which is called the *caching probability*. Denote by $\mathbf{T}_m = [t_{m1}, \cdots, t_{mN}]$ the caching probability vector of all N files for tier m, which is identical for all BSs in tier m. Given $\mathbf{T}_m$, one BS from tier m can randomly choose C_m files to store, using the probabilistic content caching policy proposed in [10]. Let $\mathbf{t}_n = [t_{1n}, \cdots, t_{Mn}]^T$ be the caching distribution of file n, and $\mathbf{T} = [\mathbf{t}_1, \cdots, \mathbf{t}_N]$ be the caching probability matrix for all files and all BS tiers. The rows and columns of $\mathbf{T}$ correspond to $\mathbf{T}_m$ and $\mathbf{t}_n$ and further correspond to different tiers and different files, respectively. Due to the limited cache size, we have [3,10]:

$$0 \leq t_{mn} \leq 1, \quad \forall m \in \mathcal{M}, n \in \mathcal{N} \tag{2}$$

$$\sum_{n \in \mathcal{N}} t_{mn} \leq C_m, \forall m \in \mathcal{M} \tag{3}$$

Note that the BSs that cache file n from the m-th tier can also be denoted as a homogeneous PPP $\Phi_{m,n}$, with density $\lambda_{m,n} = t_{mn}\lambda_m$, and the remaining BSs that do not cache file n from tier m are denoted as $\Phi_{m,-n}$, with density $\lambda_{m,-n} = (1 - t_{mn})\lambda_m$, according to the thinning theorem for PPP.

2.2 Joint Transmission

Assume that the typical UE u_0 requests file n in current time slot. We adopt a joint transmission scheme in which the typical UE u_0 is served jointly by the *nearest* BSs that cache the requested file from each tier, as shown in Fig. 1. Therefore, the cooperative BS set denoted by $\mathcal{C}_n$ can be defined as

$$\mathcal{C}_n \triangleq \{\{x_{k,n,0}\} | \, x_{k,n,0} = \arg \max_{x \in \Phi_{k,n}} \|x\|^{-\alpha_k}, \forall k \in \mathcal{K}\}, \tag{4}$$

where $x_{k,n,0}$ represents the nearest BS to u_0 from tier k that stores file n; $\mathcal{K} \subseteq \mathcal{M}$ is the set of indexes of tiers to which each cooperative BS belongs, and let $K \triangleq |\mathcal{K}|$ denote the number of elements in $\mathcal{K}$. Note that due to the effect of $\mathbf{T}$, there may be a tier m so that all BSs from this tier do not store file n if $t_{mn} = 0$, i.e., the BSs from this tier do not participate in JT. Therefore, there are up to $K \leq M$ BSs to jointly transmit file n to u_0. In order to focus on the performance analysis of the cache-enabled HetNet, we assume that all the BSs from every tiers are not equipped with backhaul links, which means when the file n is not stored in any BS because of the limited BS storage, i.e., $t_{mn} = 0$ for $\forall m \in \mathcal{M}$, the file request can not be satisfied, and a transmission failure occurs, which is referred to as a *cache miss case*. The sum of the desired non-coherent signal yields a received power boost to improve the received signal-to-interference-plus-noise ratio (SINR). Based on the JT scheme described above, the received signal at u_0 when requesting file n can be written as

$$y_n = \underbrace{\sum_{x \in \mathcal{C}_n} P_{\nu(x)}^{1/2} \|x\|^{-\alpha_{\nu(x)}/2} h_x X_n}_{\text{desired signal}} + \underbrace{\sum_{x \in \Phi \backslash \mathcal{C}_n} P_{\nu(x)}^{1/2} \|x\|^{-\alpha_{\nu(x)}/2} h_x X_x}_{\text{interference}} + Z, \tag{5}$$

where $\nu(x)$ returns the index of tier to which a BS located at x belongs, i.e., $\nu(x) = m$ iff $x \in \Phi_m$; X denotes the symbol jointly sent by the cooperative BSs, which is the desired symbol of u_0; X_x denotes the symbol sent by the BSs located outside $\mathcal{C}_n$, which is regarded as interference symbol to u_0; $Z \overset{d}{\sim} \mathcal{CN}(0, N_0)$ models the background thermal noise.

2.3 Performance Metric

Since the strength of interference in HetNets is usually much stronger than that of the thermal noise, it is reasonable to neglect the impact of the thermal noise and just consider the interference-limited network, i.e., $N_0 = 0$. The SIR of the typical UE u_0 requesting file n is given by

$$\mathrm{SIR}_n = \frac{\left| \sum\limits_{x \in \mathcal{C}_n} P_{\nu(x)}^{1/2} \|x\|^{-\alpha_{\nu(x)}/2} h_x \right|^2}{\sum\limits_{x \in \Phi \backslash \mathcal{C}_n} P_{\nu(x)} \|x\|^{-\alpha_{\nu(x)}} |h_x|^2} = \frac{\left| \sum_{k \in \mathcal{K}} P_k^{1/2} \|x_{k,n,0}\|^{-\alpha_k/2} h_{k,n,0} \right|^2}{\sum\limits_{m=1}^{M} P_m I_m}, \tag{6}$$

where $I_m = \sum_{x \in \Phi_m \backslash \{x_{m,n,0}\}} \|x\|^{-\alpha_m} |h_x|^2$ is the interference caused by all BSs from tier m normalized by the transmission power P_m. In this paper, we employ the STP as the system performance metric. When u_0 requests file n, the transmission will succeed if the received data rate at the UE exceeds a given threshold r [bps/Hz], i.e., $\log_2(1 + \mathrm{SIR}_n) \geq r$. Furthermore, the STP is defined as

$$q(\mathbf{T}) \triangleq \Pr[\mathrm{SIR} \geq \tau] = \sum_{n \in \mathcal{N}} a_n q_n(\mathbf{t}_n), \tag{7}$$

$q_n(\mathbf{t}_n) \triangleq \Pr[\mathrm{SIR}_n \geq \tau]$ denotes the STP when u_0 requests file n, and the second equality holds due to the total probability theorem. For notational simplicity, we define the ratios of transmission power and BS density as $P_{ij} \triangleq P_i/P_j$ and $\hat{\lambda}_{ij} \triangleq \lambda_i/\lambda_j$, respectively.

3 Analysis of Performance Metric

In this section, we first derive the expression of STP for a given caching probability matrix $\mathbf{T}$. Then verify the obtained expression using Monte Carlo simulation.

For ease of notation, we first have the following definitions:

$$F(\alpha, x) = {}_2F_1\left(-\frac{2}{\alpha}, 1; 1 - \frac{2}{\alpha}; -x\right) - 1, \tag{8}$$

where ${}_2F_1(a, b; c; d)$ denotes the Gauss hypergeometric function.

Theorem 1. *The STP for the joint transmission scheme considered in this work is given by*

$$q(\mathbf{T}) = \sum_{n \in \mathcal{N}} a_n q_n(\mathbf{t}_n), \tag{9}$$

where $q_n(\mathbf{t}_n)$ *is expressed as*

$$q_n(\mathbf{t}_n) = \int_0^\infty \int_0^\infty \cdots \int_0^\infty q_{n,\mathbf{R}_0}(\mathbf{t}_n, \mathbf{r}) \prod_{k \in \mathcal{K}} f_{R_{k,0}}(r_k) \,\mathrm{d}\mathbf{r}. \tag{10}$$

$\mathbf{R}_0 = [R_{1,0}, \cdots, R_{k,0}, \cdots], k \in \mathcal{K}$ *is the distance vector between the typical UE and its serving BSs, i.e.,* $R_{k,0} = \|x_{k,n,0}\|$, *and* $\mathbf{r}$ *is a realization of* $\mathbf{R}_0$; $f_{R_{k,0}}(r_k)$ *is the probability density function (PDF) of* $R_{k,0}$, *which can be given by*

$$f_{R_{k,0}}(r_k) = 2\pi\lambda_k t_{kn} r_k e^{-\pi\lambda_k t_{kn} r_k^2}. \tag{11}$$

$q_{n,\mathbf{R}_0}(\mathbf{t}_n, \mathbf{r})$ *is the STP conditioned on* $\mathbf{R}_0 = \mathbf{r}$ *and is given by*

$$q_{n,\mathrm{R}_0}(\mathbf{t}_n, \mathbf{r}) = \prod_{m\in\mathcal{K}} \exp\left(-F\left(\alpha_m, \frac{\tau}{\sum_{k\in\mathcal{K}} P_{km}\frac{r_m^{\alpha_m}}{r_k^{\alpha_k}}}\right)\pi\lambda_m t_{mn} r_m^2\right) \times \prod_{m=1}^{M} \exp\left(-\frac{\pi\lambda_m(1-t_{mn})}{\mathrm{sinc}(2/\alpha_m)}\left(\frac{\tau}{\sum_{k\in\mathcal{K}} P_{km} r_k^{-\alpha_k}}\right)^{\frac{2}{\alpha_m}}\right), \tag{12}$$

where, $\mathrm{sinc}(x) = \frac{\sin(\pi x)}{\pi x}$.

Proof. According to the total probability theorem, we have (9). Next, we calculate $q_n(\mathbf{t}_n)$. As illustrated before, the interference can be categorized into two types: 1) the interference caused by the BSs in $\Phi_{m,n}$ which are farther away from u_0 than the serving BS from tier m; 2) the interference caused by the BSs in $\Phi_{m,-n}$ which may be closer to u_0 than the serving BS from tier m. For the second case, if there are no BSs that store file n, i.e., $t_{mn} = 0$, all the BSs in the tier m are interfering BSs. Therefore, the interference from tier m can be rewritten as $I_m = \mathbb{1}(t_{mn} > 0)I_{m,n} + I_{m,-n}$, where $\mathbb{1}(\bullet)$ is the indicator function; $I_{m,n} \triangleq \sum_{x\in\Phi_{m,n}\setminus\{x_{m,n,0}\}} \|x\|^{-\alpha_m} |h_x|^2$ and $I_{m,-n} \triangleq \sum_{x\in\Phi_{m,-n}} \|x\|^{-\alpha_m} |h_x|^2$. The received signal power of u_0 is $S = \left|\sum_{k\in\mathcal{K}} P_k^{1/2} \|x_{k,n,0}\|^{-\alpha_k/2} h_{k,n,0}\right|^2$. Conditioning on $\mathbf{R}_0 = \mathbf{r}$, we can obtain the conditional STP as

$$\begin{aligned} q_{n,\mathbf{R}_0}(\mathbf{t}_n, \mathbf{r}) &= \Pr\left[\mathrm{SIR}_n \geq \tau \middle| \mathbf{R}_0 = \mathbf{r}\right] \\ &= \mathbb{E}_{I_m}\left[\Pr\left[S \geq \tau \sum_{m=1}^{M} P_m I_m \middle| \mathbf{R}_0 = \mathbf{r}\right]\right] \\ &\overset{(a)}{=} \mathbb{E}_{I_{m,n}, I_{m,-n}}\left[e^{-\frac{\tau \sum_{m=1}^{M} P_m(\mathbb{1}(t_{mn}>0)I_{m,n} + I_{m,-n})}{\sum_{k\in\mathcal{K}} P_k r_k^{-\alpha_k}}}\right] \\ &\overset{(b)}{=} \prod_{m\in\mathcal{K}} \mathbb{E}_{I_{m,n}}\left[e^{-\frac{\tau P_m I_{m,n}}{\sum_{k\in\mathcal{K}} P_k r_k^{-\alpha_k}}}\right] \prod_{m=1}^{M} \mathbb{E}_{I_{m,-n}}\left[e^{-\frac{\tau P_m I_{m,-n}}{\sum_{k\in\mathcal{K}} P_k r_k^{-\alpha_k}}}\right] \\ &\triangleq \prod_{m\in\mathcal{K}} \mathcal{L}_{I_{m,n}}(sP_m, \mathbf{r}) \prod_{m=1}^{M} \mathcal{L}_{I_{m,-n}}(sP_m, \mathbf{r})\Bigg|_{s=\frac{\tau}{\sum_{k\in\mathcal{K}} P_k r_k^{-\alpha_k}}}, \end{aligned} \tag{13}$$

where (a) follows from that $S \overset{d}{\sim} \mathrm{Exp}\left(\frac{1}{\sum_{k\in\mathcal{K}} P_k r_k^{-\alpha_k}}\right)$; (b) is due to the independence of the homogeneous PPPs; $\mathcal{L}_{I_{m,n}}$ and $\mathcal{L}_{I_{m,-n}}$ represent the Laplace transforms of the interference $I_{m,n}$ and $I_{m,-n}$, respectively, which can be derived as follows

$$\begin{aligned}\mathcal{L}_{I_{m,n}}(sP_m,\mathbf{r}) &= \mathbb{E}_{\Phi_{m,n},|h_x|}\left[e^{-sP_m\sum_{x\in\Phi_{m,n}\setminus\{x_{m,n,0}\}}\|x\|^{-\alpha_m}|h_x|^2}\right]\\ &\overset{(a)}{=} \mathbb{E}_{\Phi_{m,n}}\left[\prod_{x\in\Phi_{m,n}\setminus\{x_{m,n,0}\}}\mathbb{E}_{|h_x|}\left[e^{-sP_m\|x\|^{-\alpha_m}|h_x|^2}\right]\right]\\ &\overset{(b)}{=} \mathbb{E}_{\Phi_{m,n}}\left[\prod_{x\in\Phi_{m,n}\setminus\{x_{m,n,0}\}}\frac{1}{1+sP_m\|x\|^{-\alpha_m}}\right]\\ &\overset{(c)}{=} e^{-2\pi\lambda_{m,n}\int_{r_m}^{\infty}\left(1-\frac{1}{1+sP_m r^{-\alpha_m}}\right)r\mathrm{d}r}\\ &= e^{-F\left(\alpha_m,\frac{sP_m}{r_m^{\alpha_m}}\right)\pi\lambda_m t_{mn} r_m^2},\end{aligned} \tag{14}$$

where (a) is due to the independence of the channels; (b) follows from $|h_x|^2 \overset{d}{\sim}$ Exp(1); (c) is from the probability generating functional for a PPP and converting from Cartesian to polar coordinates. Similarly, we have

$$\begin{aligned}\mathcal{L}_{I_{m,-n}}(sP_m,\mathbf{r}) &= \mathbb{E}_{\Phi_{m,-n},|h_x|}\left[e^{-sP_m\sum_{x\in\Phi_{m,-n}}\|x\|^{-\alpha_m}|h_x|^2}\right]\\ &= e^{-2\pi\lambda_{m,-n}\int_0^{\infty}\left(1-\frac{1}{1+sP_m r^{-\alpha_m}}\right)r\mathrm{d}r}\\ &= e^{-\frac{\pi\lambda_m(1-t_{mn})}{\mathrm{sinc}(2/\alpha_m)}(sP_m)^{(2/\alpha_m)}}.\end{aligned} \tag{15}$$

Then, remove the condition $\mathbf{R}_0 = \mathbf{r}$, $q_n(\mathbf{t}_n)$ can be obtained as

$$q_n(\mathbf{t}_n) = \int_0^{\infty}\int_0^{\infty}\cdots\int_0^{\infty} q_{n,\mathbf{R}_0}(\mathbf{t}_n,\mathbf{r})\prod_{k\in\mathcal{K}} f_{R_{k,0}}(r_k)\,\mathrm{d}\mathbf{r}. \tag{16}$$

The proof of Theorem 1 is completed.

From Theorem 1, we can know that even though the expression of the STP is in integral form, the Gaussian hypergeometric function can be effectively calculated with a numerical method when the number of tiers of the network M is small. Figure 2 plots the STP $q(\mathbf{T})$ versus τ with and without JT, respectively. From Fig. 2, we can see that the derived analytical expression of STP with JT matches the corresponding Monte Carlo results perfectly. Besides, the STP of the system with BS JT has been greatly improved compared to the non-cooperation scheme, which confirms the effectiveness of the BS JT.

Next, a special case is considered, where the tier number $M = 2$.

Corollary 1 (STP for the Two-Tier HetNets). *For the two-tier network where $M = 2$, let $\alpha_1 = \alpha_2 = \alpha$, the STP is given by*

$$q^t(\mathbf{T}) = \sum_{n\in\mathcal{N}} a_n \tilde{q}_n(\mathbf{t}_n), \tag{17}$$

where

$$\tilde{q}_n(\mathbf{t}_n) = \int_{\mathbb{R}^{2+}} Q_{1,2}(u_1,u_2,t_{1n})Q_{2,1}(u_2,u_1,t_{2n})\mathrm{d}\mathbf{u}, \tag{18}$$

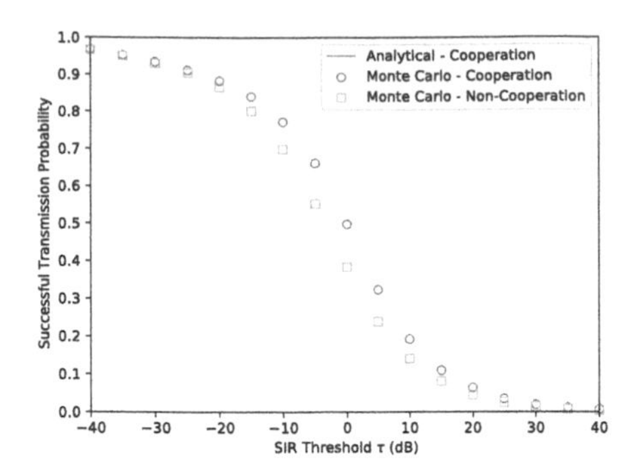

Fig. 2. STP $q(\mathbf{T})$ versus τ. $M = 3$, $N = 9$, $C_1 = 7$, $C_2 = 5$, $C_3 = 3$, $\gamma = 1$, $\alpha_1 = 4.3$, $\alpha_2 = 4$, $\alpha_3 = 3.8$, $\lambda_1 = 1/(200^2\pi)\,\mathrm{m}^{-2}$, $\lambda_2 = 5/(100^2\pi)\,\mathrm{m}^{-2}$, $\lambda_3 = 1/(50^2\pi)\,\mathrm{m}^{-2}$, $P_1 = 46\,\mathrm{dBm}$, $P_2 = 32\,\mathrm{dBm}$, $P_3 = 23\,\mathrm{dBm}$, $\mathbf{T}_1 = [1,1,1,1,0.8,0.7,0.6,0.5,0.4]$, $\mathbf{T}_2 = [1,1,0.8,0.7,0.6,0.5,0.4,0,0]$, $\mathbf{T}_3 = [0.8,0.7,0.6,0.5,0.4,0,0,0,0]$. In the Monte Carlo simulations, the BSs are deployed in a square area of $1,000 \times 1,000\,\mathrm{m}^2$, and the results are obtained by averaging over 10^5 independent realizations.

$$Q_{i,j}(x,y,z) = z\exp\left(-xz - zA_{i,j}(x,y) - (1-z)B_{i,j}(x,y)\right), \tag{19}$$

$$A_{i,j}(x,y) = xF\left(\alpha, \frac{\tau}{1 + P_{ji}\left(\hat{\lambda}_{ji}\frac{x}{y}\right)^{\frac{\alpha}{2}}}\right), \tag{20}$$

$$B_{i,j}(x,y) = \frac{x}{\mathrm{sinc}(2/\alpha)}\left(\frac{\tau}{1 + P_{ji}\left(\hat{\lambda}_{ji}\frac{x}{y}\right)^{\frac{\alpha}{2}}}\right)^{\frac{2}{\alpha}}. \tag{21}$$

From Corollary 1, we can observe that the STP is not only influenced by the caching probability t_n, the path-loss exponent α, and the SIR threshold τ but also affected by the ratios of transmission power P_{ij} and BS density $\hat{\lambda}_{ij}$, respectively, when $M = 2$.

4 STP Maximization

In this section, we maximize the STP by optimizing the placement probability matrix $\mathbf{T}$ for the two-tier IoT network. The optimization problem for this case can be formulated as

Problem 1 (Maximization of STP for Two-Tier IoT network).

$$\max_{\mathbf{T}} q^t(\mathbf{T})$$
$$\text{s.t. } (2), (3),$$

where the objective function is given by Corollary 1. Let $\mathbf{T}^{\star} = [\mathbf{T}_1^{\star T}, \mathbf{T}_2^{\star T}]^T$ be the optimal solution of this problem. From Problem 1, we can observe that the constraint set is convex, whereas the convexity of the objection function is hard to determine because of its complicated expression. Therefore, a locally optimal solution can be obtained using the gradient projection method (GPM) [11], the procedure of which is summarized in Algorithm 1. In Step 3 of Algorithm 1, the stepsize $s(k)$ satisfies

Algorithm 1. Locally Optimal Solution to Problem 1

1: Initialize $\epsilon = 10^{-6}$, $k = 0$, $k_{\max} = 10^6$, and $t_{mn}(0) = \frac{C_m}{N}$, for $\forall n \in \mathcal{N}, m \in \{1, 2\}$.
2: **repeat**
3: For $\forall n \in \mathcal{N}, m \in \{1,2\}$, compute $\bar{t}_{mn}(k+1) = t_{mn}(k) + s(k)\frac{\partial q^t(\mathbf{T}(k))}{\partial t_{mn}(k)}$, where $s(k)$ satisfies (22).
4: For $\forall n \in \mathcal{N}, m \in \{1,2\}$, compute the projection $t_{mn}(k+1) = [\bar{t}_{mn}(k+1) - u_m^{\star}]_0^1$, where the scalar $u_m^{\star}$ satisfies $\sum_{n \in \mathcal{N}} [\bar{t}_{mn}(k+1) - u_m^{\star}]_0^1 = C_m$, $[x]_0^1 = \max\{\min\{1, x\}, 0\}$.
5: $k \leftarrow k + 1$.
6: **until** $|t_{mn}(k+1) - t_{mn}(k)| < \epsilon$, for $\forall n \in \mathcal{N}, m \in \{1,2\}$ or $k > k_{\max}$.

$$\lim_{k \to \infty} s(k) = 0, \quad \sum_{k=0}^{\infty} s(k) = \infty, \tag{22}$$

and the partial derivative $\frac{\partial q^t(\mathbf{T}(k))}{\partial t_{mn}(k)} = a_n \frac{\partial \tilde{q}(\mathbf{t}_n(k))}{\partial t_{mn}(k)}$ is given by

$$\begin{aligned} \frac{\partial \tilde{q}_n(\mathbf{t}_n)}{\partial t_{in}} =& \frac{\tilde{q}_n^0(t_{1n}, t_{2n})}{t_{in}} + \int_{\mathbb{R}^{2+}} \left(-A_{i,j}(u_i, u_j) + B_{i,j}(u_i, u_j) - u_i\right) \\ & \times Q_{i,j}(u_i, u_j, t_{in}) Q_{j,i}(u_j, u_i, t_{jn}) \, \mathrm{d}\mathbf{u}, \qquad t_{1n} > 0, t_{2n} > 0, \end{aligned} \tag{23}$$

where $i, j \in \{1, 2\}$, $i \neq j$. By using Algorithm 1, a locally optimal caching strategy (LCS) can be obtained.

5 Numerical Results

In this section, some simulations are conducted to compare the LCS obtained by Algorithm 1 with three baseline strategies, i.e., MPC (most popular caching) [12], IIDC (i.i.d. caching) [13] and UDC (uniform distribution caching) [14]. Note that the three baselines also adopt a joint transmission scheme. We focus on a two-tier IoT network with randomly deployed macro BSs and small BSs. Unless otherwise noted, we set $M = 2$, $N = 100$, $C_1 = 30$, $C_2 = 20$, $\gamma = 1$, $\alpha_1 = \alpha_2 = 4$, $\lambda_1 = 5/(200^2\pi)\,\mathrm{m}^{-2}$, $\lambda_2 = 1/(50^2\pi)\,\mathrm{m}^{-2}$, $P_1 = 43\,\mathrm{dBm}$, $P_2 = 23\,\mathrm{dBm}$, $\tau = 0\,\mathrm{dB}$.

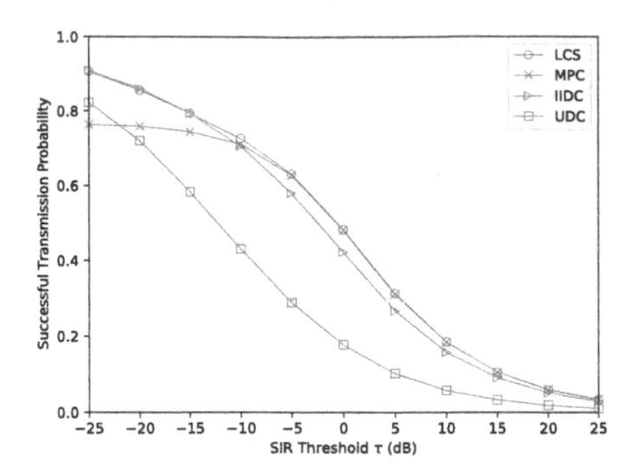

Fig. 3. Comparison of the LCS with three baseline strategies.

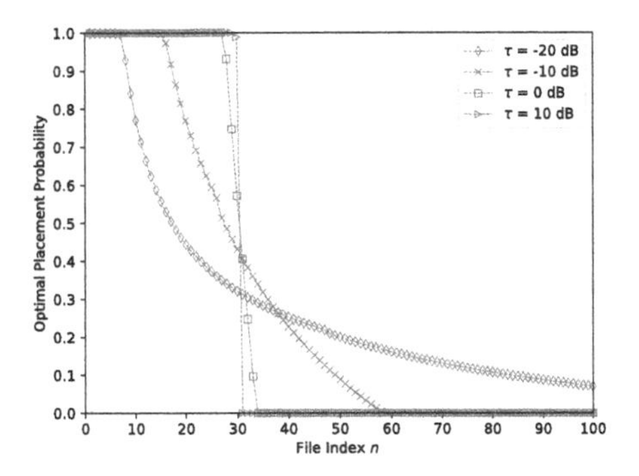

Fig. 4. Locally optimal caching probability vector $\mathbf{T}_1^{\star}$ versus file index n.

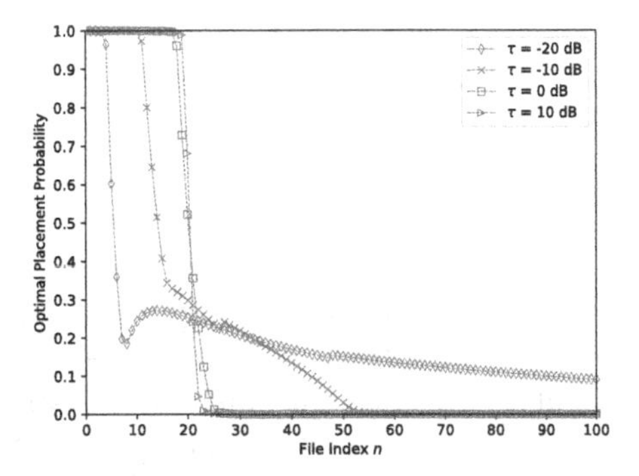

Fig. 5. Locally optimal caching probability vector $\mathbf{T}_2^{\star}$ versus file index n.

Figure 3 shows the comparison between the LCS and the three baseline strategies. From Fig. 3, we can observe that the LCS outperforms all the three baselines. In particular, when τ is small, the LCS is consistent with the IIDC; however, when τ is large, the LCS is consistent with the MPC. This is because that when τ is small, caching more different files can provide higher file diversity; and when τ is large, caching the most popular files can guarantee the most frequent file requests, which is more meaningful for improving the STP.

Figure 4 and Fig. 5 depict the locally optimal caching vector $\mathbf{T}_1^\star$ and $\mathbf{T}_2^\star$, respectively. From the figures, one can see that files of higher popularity have higher probability to be cached at BSs. Furthermore, there exist a situation where we have $1 \leq N_1 < N_2 \leq N$ that satisfies $t_1^\star = t_2^\star = \cdots = t_{N_1}^\star = 1$, $t_i^\star \in (0, 1)$ for all $i \in (N_1, N_2)$ and $t_{N_2}^\star = t_{N_2+1}^\star = \cdots = t_N^\star = 0$ for each tier, as illustrated in [8, Remark 2].

6 Conclusion

In this paper we studied the caching probability optimization in cache-enabled multi-tier IoT networks with base station joint transmission (JT). We developed a JT scheme and derived the STP expression for the K-tier cache-enabled IoT networks when adopting random caching scheme. Then the locally optimal caching strategy (LCS) was obtained by using the gradient projection method. Simulations were conducted to verify the established STP model and compare the LCS with three baseline strategies, i.e., MDP, IIDC and UDC. The results showed that LCS outperforms all the three baselines in terms of STP.

References

1. Nigam, G., Minero, P., Haenggi, M.: Coordinated multipoint joint transmission in heterogeneous networks. IEEE Trans. Commun. (2014). https://doi.org/10.1109/TCOMM.2014.2363660
2. Nigam, G., Minero, P., Haenggi, M.: Spatiotemporal cooperation in heterogeneous cellular networks. IEEE J. Sel. Areas Commun. **33**(6), 1253–1265 (2015). https://doi.org/10.1109/JSAC.2015.2417017
3. Wen, J., Huang, K., Yang, S., Li, V.O.: Cache-enabled heterogeneous cellular networks: optimal tier-level content placement. IEEE Trans. Wireless Commun. (2017). https://doi.org/10.1109/TWC.2017.2717819
4. Kuang, S., Liu, X., Liu, N.: Analysis and optimization of random caching in K -Tier Multi-antenna multi-user HetNets. IEEE Trans. Commun. **67**(8), 5721–5735 (2019). https://doi.org/10.1109/TCOMM.2019.2913378
5. Zhang, S., Liu, J.: Optimal probabilistic caching in heterogeneous IoT networks. IEEE Internet Things J. **7**(4), 3404–3414 (2020). https://doi.org/10.1109/jiot.2020.2969466
6. Yang, J., Ma, C., Jiang, B., Ding, G., Zheng, G., Wang, H.: Joint optimization in cached-enabled heterogeneous network for efficient industrial IoT. IEEE J. Sel. Areas Commun. **8716**(c), 1 (2020). https://doi.org/10.1109/jsac.2020.2980907

7. Wen, W., Cui, Y., Zheng, F.C., Jin, S., Jiang, Y.: random caching based cooperative transmission in heterogeneous wireless networks. IEEE Trans. Commun. **66**(7), 2809–2825 (2018). https://doi.org/10.1109/TCOMM.2018.2808188
8. Feng, T., Shi, S., Gu, S., Xiang, W., Gu, X.: Optimal content placement for cache-enabled IoT networks with local channel state information based joint transmission. IET Commun. (2020, to be published). https://doi.org/10.1049/iet-com.2020.0167
9. Chae, S.H., Quek, T.Q., Choi, W.: Content placement for wireless cooperative caching helpers: a tradeoff between cooperative gain and content diversity gain. IEEE Trans. Wireless Commun. **16**(10), 6795–6807 (2017). https://doi.org/10.1109/TWC.2017.2731760
10. Blaszczyszyn, B., Giovanidis, A.: Optimal geographic caching in cellular networks. In: IEEE International Conference on Communications (2015). https://doi.org/10.1109/ICC.2015.7248843
11. Bertsekas, D.P.: Nonlinear Programming. Athena Scientific, 2nd edn. (1999)
12. Baştuğ, E., Bennis, M., Kountouris, M., Debbah, M.: Cache-enabled small cell networks: modeling and tradeoffs. EURASIP J. Wireless Commun. Netw. **2015**(1), 41–47 (2015)
13. Bharath, B.N., Nagananda, K.G., Poor, H.V.: A learning-based approach to caching in heterogenous small cell networks. IEEE Trans. Commun. **64**(4), 1674–1686 (2016). https://doi.org/10.1109/TCOMM.2016.2536728
14. Tamoor-ul-Hassan, S., Bennis, M., Nardelli, P.H.J., Latva-Aho, M.: Modeling and analysis of content caching in wireless small cell networks. In: Proceedings of IEEE ISWCS, Bussels, Belgium, pp. 765–769, August 2015. https://doi.org/10.1109/ISWCS.2015.7454454

Fast Estimation for the Number of Clusters

Xiaohong Zhang, Zhenzhen He, Zongpu Jia, and Jianji Ren(✉)

College of Computer Science and Technology, Henan Polytechnic University, Jiaozuo 45400, Henan, China
{xh.zhang,jiazp,renjianji}@hpu.edu.cn, hezzedu@163.com

Abstract. Clustering analysis has been widely used in many areas. In many cases, the number of clusters is required to been assigned artificially, while inappropriate assignments affect analysis negatively. Many solutions have been proposed to estimate the optimal number of clusters. However, the accuracy of those solutions drop severely on overlapping data sets. To handle the accuracy problem, we propose a fast estimation solution based on the cluster centers selected in a static way. In the solution, each data point is assigned with one score calculated according to a density-distance model. The score of each data point does not change any more once it is generated. The solution takes the top k data points with the highest scores as the centers of k clusters. It utilizes the significant change of the minimal distance between cluster centers to identify the optimal number of the clusters in overlapping data sets. The experiment results verify the usefulness and effectiveness of our solution.

Keywords: Clustering · The number of clusters · Density · Distance

1 Introduction

Clustering analysis is one of the ways to perform unsupervised analysis [1]. It is dedicated to dividing data into clusters with the goal of the similarity between data within the cluster and minimizing the similarity between data between clusters [2]. It has been widely used in many areas such as image processing, bioinformatics, in-depth learning, pattern recognition and so on. Clustering analysis can be classified into partition-based clustering, density-based clustering [3], grid-based clustering, hierarchical clustering [4] and so on [5,6]. However, in many cases, the number of clusters must be assigned artificially, while inappropriate assignments affect analysis results negatively. If the number of clusters is much larger than the actual number of clusters, the resulting clustering results will be very complicated and the characteristics of the data cannot be analyzed. If the number of clusters is much smaller than the actual number of clusters, some valuable information will be lost in the clustering results. The loss of this information leads to the inability to obtain valuable information in later data mining.

X. Wang et al. (Eds.): 6GN 2020, LNICST 337, pp. 357–370, 2020.
https://doi.org/10.1007/978-3-030-63941-9_27

Therefore, many solutions have been proposed to determine the optimal number of clusters. Some solutions utilize cluster validity Indexes, e.g., DB Index [7], I Index [8] and Xie-Beni Index [9], to determine the optimal number. Some solutions exploit heuristics to deduce the number. For example, the solution of Laio et al. [10] is one of those solutions. It estimates the optimal number according to density and distance (the density of data and the distance between). However, the heuristic still needs to input the number of clusters artificially, and cannot fully cluster automatically. Recently, Gupta et al. [11] propose a solution to identify the optimal number according to the last leap and the last major leap of the minimal distances between cluster centers. However, when running on overlapping data sets, the accuracy of most of those solutions drops severely.

In order to solve the problem of poor estimation of cluster numbers on overlapping data sets, we propose an algorithm that focuses on cluster number estimation on overlapping data sets. And this method has a very fast speed. The solution selects cluster centers in a static way. It generates a score for each data point according to a density-distance model. The score of each data point does not change any more once it is generated. The solution takes the top k data points with the highest scores as the centers of k clusters. It utilizes the significant change of the minimal distance between cluster centers to identify the optimal number of the clusters in overlapping data sets.

The rest of this paper is organized as follows: we review the relevant published work in Sect. 2. After analyzing the estimation problem of the number of the clusters in overlapping data sets in Sect. 3, we elaborate our solution in Sect. 4. The experimental results are discussed in Sect. 5. Finally, the paper concludes in Sect. 6.

2 Related Work

It is very important to determine the number of clusters which data points are grouped into, especially for partition-based clustering solutions. However, it is not easy to estimate the optimal value of the number. Fortunately, cluster validity Indexes [12–16] provide a useful tool for the estimation. Davies and Bouldin [7] proposed DBI index based on inter-cluster similarities to obtain the best cluster number. Xie and Beni [9] put forward Xie-Beni Index based on intra cluster compactness and inter cluster separation, and utilize the index to determine the optimal number. Bensaid [17] and Ren et al. [18] improve the solution of Xie and Beni to enhance the reliability and robustness of that solution, respectively. Some other validity Indexes [19], e.g., Bayesian Information Criterion (BIC) [20], diversity [21], intra-cluster coefficient and inter-cluster coefficient [22], are also exploited to estimate the number of clusters. To obtain better estimation, some solutions even apply multiple indexes in the estimation [5].

In addition to cluster validity indexes, some other factors are also utilized for the estimation. The solutions proposed by Wang et al. [23] and Laio et al. [10] perform the estimation according to the factors related to density. Concretely, the solution of Laio et al. calculates a produce of density and distance for each data point and estimations the number of clusters based on those products. Because the user is required to input the number of clusters according to the visualization. Therefore, it is impossible to process data in batches. Recently, Gupta et al. [11] observed that the last significant drop of the distance between cluster centers indicated the natural number of clusters. Based on the observation, they proposed the Last Leap solution (LL) and the Last Major Leap solution (LML) to estimate the natural number of clusters.

Many algorithms already have a very high accuracy for determining the number of clusters in some simple data sets. However, when running on the data sets with overlapping clusters, the accuracy of those solutions drops severely. Here, the overlapping data set refers to a data set with no obvious boundary between clusters. For example, Fig. 1 shows a non-overlapping two-dimensional data set, and Fig. 2 shows an overlapping two-dimensional data set. At the same time, a detailed explanation of the notions mentioned below is shown in Table 1.

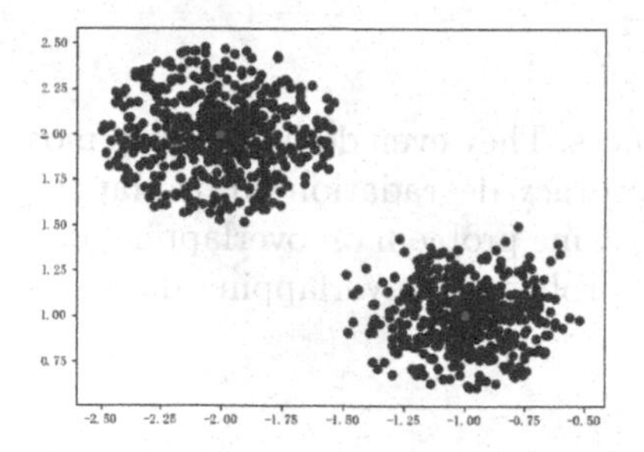

Fig. 1. Non-overlapping

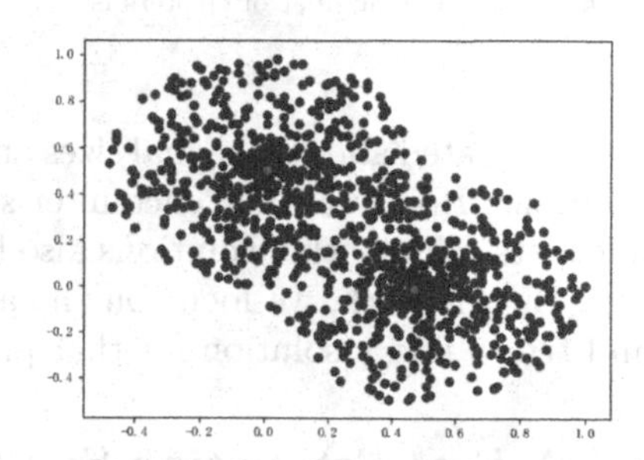

Fig. 2. Overlapping

3 Motivation

Given a data set-P, $P = \{p_1, p_2, \cdots, p_m\}$, where $(\forall p_i) p_i \in \mathbb{R}^d$, partition-based solutions try to divide P into k subsets noted as C_1, C_2, ..., C_k. Each of those subsets is known as a cluster and identified by a cluster center. Partition-based solutions require users to offer the number of clusters, i.e., the number of k, which indicates that users are involved in the procedure of clustering somehow. To make sure that clustering is truly unsupervised, clustering solutions should be equipped with the ability of estimating the optimal number of clusters. The objective of this work is to search for the optimal number from a set-$K = \{k_i | k_i \in N^* and\ k_i < k_{max}\ and\ k_{max} = \sqrt{m}\}$.

Gupta et al. observed that the last significant drop of the minimal distances between cluster centers indicates the optimal number of clusters. Based on the observation, they proposed the Last Leap solution (LL) and the Last Major Leap solution (LML). LL and LML work well on the data sets in which the clusters are

Table 1. Notions.

Notions	Description
P	Data set
C	Center set
M	Data set size
K	Number of clusters
k_{max}	Maximum number of clusters
Weight	Density weight
$P_{p_i}^h$	Relative high-density point set
$dist$	Density-bound minimum distance
$score_{d \cdot d}$	Density-distance score
cls	Cluster center closeness
k_{break}	The break value describes the minimum value of k which satisfies that $cls(k) > 1$
f	Calculate the change of minimum distance between center points
$\overline{\rho}$	Density mean of the dataset
k_i	The number of clusters is i

well-separated and have equal sizes and variances. They even do better than most solutions. However, they encounter severe accuracy degradation on overlapping data sets. Many other solutions also have the same problem on overlapping data sets. In this work, we focus on the accuracy problem on overlapping data sets, and try to find a solution for that problem.

4 A Fast Estimation Solution

In this section, we elaborate a fast estimation solution for the number of clusters in overlapping data sets. The minimum distance between the center points is used to measure the change in the degree of separation between clusters. The solution exploits a density-distance model to select cluster centers in a static way. In order to realize the automatic determination of the number of clusters, it is necessary to use the formula to determine the degree of change in distance. However, when the value of k is greater than the optimal value, the following situations are likely to occur. On the whole, the distance between the center points has not changed much at this time. But from a local perspective, it is a big change. This leads to misjudgment. Therefore, we use the tightness of the center point to narrow the K value range to avoid the distance between the center points being too small. Finally, we take the number satisfying the constraint of the minimum distances as the optimal number of clusters.

4.1 Selecting Cluster Centers

Density-based clustering solutions have the ability of selecting global optimal points as cluster centers without iterations. In order to the performance of clustering, Laio et al. proposed a fast clustering algorithm based on the cluster centers selected based on the products of the density and distance of each data point.

Here, the distance of a data point represents the minimum distance between that point and any other point which has higher density than that point. The algorithm has the ability which can select proper cluster centers in non-sphere and strongly overlapping data sets.

In order to avoid the influence of outliers on judging the change in minimum distance between the center points, we introduce the density-distance model designed based on density weight. Density weight is defined to measure the importance of the density of a data point.

Definition 1. ***Density weight.*** *$(\forall p_i)p_i \in P$, the density weight of p_i describe the importance of $\rho(p_i)$ in deciding whether to accept p_i as a cluster center. It is calculated by Function 1.*

$$weight(p_i) = \frac{\rho(p_i)}{\overline{\rho}} \tag{1}$$

Definition 2. ***Relative high-density point set.*** *$(\forall p_i)p_i \in P$, the relative high-density point set of p_i consists of the points which have a higher density than p_i. It is noted as $P^h_{p_i}$, and described as*
$P^h_{p_i} = \{p_j | \exists (p_j \in P \text{ and } \rho(p_j) > \rho(p_i))\}$.

Definition 3. ***Density-bound minimum distance.*** *$(\forall p_i)p_i \in P$, the density-bound minimum distance represents the minimum distance from p_i to any point with higher density. It is denoted as $dist_{min \cdot \rho}(p_i)$, and calculated by Function 2, where $dist(p_i, p_k) = (p_i - p_k)^2$.*

$$dist_{min \cdot \rho}(p_i) = min_{(p_k \in P^h_{p_i})} dist(p_i, p_k) \tag{2}$$

Definition 4. ***Density-distance score.*** *$(\forall p_i)p_i \in P$, the density-distance score of p_i is defined to measure whether p_i is suitable for a cluster center. It is denoted as $score_{d \cdot d}(p_i)$.*

The density-distance score is calculated by a density-distance model.

The score can be calculated according to Function 3. Considering Function 1, Function 3 can be transformed into function 4.

$$score_{d \cdot d}(p_i) = \rho(p_i) \cdot weight(p_i) \cdot dist_{(min \cdot \rho)}(p_i) \tag{3}$$

$$score_{d \cdot d}(p_i) = \frac{\rho(p_i)^2}{\overline{\rho}} \cdot dist_{min \cdot \rho}(p_i) \tag{4}$$

The Points with high density-distance scores are more suitable for being cluster centers than the points with low scores. Therefore, the interference of outliers is reduced by assigning lower scores.

To select cluster centers quickly, our approach calculates a density-distance score for each data point, and sorts all these data points in the descending orders of density-distance scores. $(\forall k_i)k_i \in K$, it takes the top k_i data points as the centers of k_i clusters.

4.2 Adjusting the Search Space of the Optimal Number

The aim of our solution is to find the optimal number of the clusters in an overlapping data set from a search space described by $S = \{1, 2, \ldots, k_{max}\}$. The size of the search space is decided by k_{max}. If k_{max} is much larger than the optimal number of clusters, some relatively close data points are probably selected as cluster centers. Those data points lead to the misjudgement on the significant changes of the minimum distances between cluster centers. This results in the erroneous estimation of the optimal number of clusters. Furthermore, the too large value of k_{max} increases additional search cost.

To avoid the erroneous estimation and the additional search cost, we introduce the definition of cluster center closeness. Cluster center closeness is introduced to assist the proper assignment of k_{max}. It describes the tightness among cluster centers. The lower value of the closeness indicates the sparser distribution of the cluster centers, and vice versa.

Definition 5. ***Cluster center closeness.*** *The cluster center closeness of k clusters describes the adjacency of those centers. It is noted as* $cls(k)$ *and calculated by Function 5.*

$$cls(k) = \frac{dist_{m \cdot f \cdot c}(\overline{dist}, count_{\rho < \overline{\rho}})}{\min\limits_{c_i \in C_k, c_j \in C_k \, and \, i \neq j} dist(c_i, c_j)} \tag{5}$$

In Function 5, $dist_{m \cdot f \cdot c}(\overline{dist}, count_{\rho < \overline{\rho}})$ describes the minimum distance of the centers. It is calculated by Function 6, where $count_{\rho < \overline{\rho}}$ denotes the total number of points of which the densities are smaller than the average density.

$$dist_{m \cdot f \cdot c}(\overline{dist}, count_{\rho < \overline{\rho}}) = \frac{\overline{dist} \cdot count_{\rho < \overline{\rho}}}{m} \tag{6}$$

The case that $cls(k_i)$ exceeds 1 indicates that some centers of the k_i clusters from the same cluster. This means that k_i overpasses the optimal number of clusters. In this situation, it is unnecessary to search the optimal number from k_i to k_{max}. Here, we define break value to describe k_i in this situation.

Definition 6. ***Break value.*** *The break value describes the minimum value of k which satisfies that* $cls(k) > 1$. *It is denoted as* k_{break} *and calculated by Function 7*

$$k_{break} = \min_{k_i \in K and \; cls(k_i) > 1} k_i \tag{7}$$

To improve performance and avoid misjudgement, it is necessary to exclude the values from k_{break} to k_{max} from the search space for the optimal number of clusters. The new search space is described as $S' = \{k_i \mid k_i \in N^* and \; k_i < k_{break}\}$. Consequently, the cluster center set is also adjusted to be consistent with the change of the search space. Actually, it is narrowed to only include $(k_{break} - 1)$ points with the highest density-distance scores.

4.3 Identifying the Optimal Number of Clusters

Cluster centers are selected only according to density-distance scores. Concretely, The top k data points with the highest scores are taken as the centers of k clusters, while the top $k+1$ data points with the highest scores are taken as the centers of $k+1$ clusters. Therefore, cluster centers can be selected ahead. Correspondingly, the minimum distance between cluster centers is fixed for a given number of clusters.

Based on the observation of Gupta et al., the significant change of the minimum distance between cluster centers is exploited to identify the optimal number of clusters. Furthermore, the optimal number is identified according to the last significant change. Here, we utilize Function 8 to discover any significant change.

$$f(C_k, C_{k+1}) = \frac{\min\limits_{c_i \in C_k, c_j \in C_k \text{ and } i \neq j} dist(c_i, c_j)}{\min\limits_{c_i \in C_{k+1}, c_j \in C_{k+1} \text{ and } i \neq j} dist(c_i, c_j)} \tag{8}$$

If $f(C_k, C_{k+1})$ reaches up to or overpass a predefined threshold, the significant change of the minimum distances occurs when k increases to $(k+1)$. All the significant changes of the minimum distances can be discovered by calculating the values of $f(C_k, C_{k+1})$ when the number of cluster increases from 1 to k_{break}. The number related to the last significant change of the minimum distances is taken as the optimal number of clusters.

The framework to estimate the optimal number of clusters is described as following:

Step 1. $k_{max} \leftarrow \sqrt{m}$;

Step 2. *Calculate a density-distance score for each data point according to Function 4;*

Step 3. *Select the top K_{max} points with the highest density-distance scores as centers of K_{max} clusters;*

Step 4. *Calculate a $cls(k)$ for each k from 1 to k_{max} according to Function 5;*

Step 5. *Calculate k_{break}, and reconstruct the search space of the optimal number as S';*

Step 6. *Calculate all the significant change of the minimum distances between cluster centers according to S' and Function 8;*

Step 7. *Take the element related to the last significant change of the minimum distances in S' as the optimal number.*

5 Experiments and Discussion

In this section, we present extensive experiments on artificial data sets and real data sets to evaluate our approach. In the experiments, we compare our approach with ten different solutions. Before going into details, we introduce the data sets and performance metrics used in the experiments.

5.1 Data Sets and Metrics

To evaluate our approach, we carried out extensive experiments on thirteen data sets. This includes nine overlapping data sets and four non-overlapping data sets. The overlapping data sets are SM2, Flame, SM5, SM6, SM8, Wine, Seeds, Iris and Ionoshpere. Five of those data sets are artificial data sets and the rest of the data sets are real data sets. There are also four non-overlapping data sets, namely A-N-O-1, A-N-O-2, A-N-O-3 and A-N-O-4. The details of these data sets are described in Table 2. Nine artificial data sets are shown in Fig. 3.

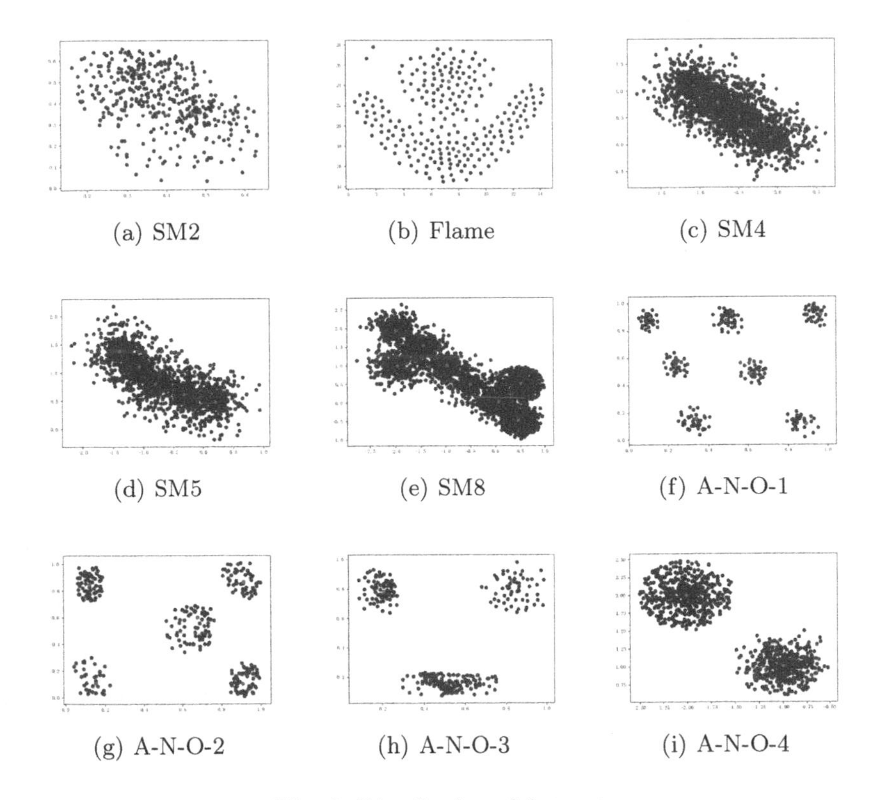

Fig. 3. Distribution of data sets

Accuracy and execution time are adopted as two metrics used to evaluate our approach. Accuracy is exploited to measure the effectiveness of our approach while execution time is utilized to measure the performance.

Table 2. Details of dataset.

Number	Dataset	Features	Clusters	Instances
1	SM2	2	2	403
2	Flame	2	2	240
3	SM4	2	4	3200
4	SM5	2	5	2600
5	SM8	2	8	4400
6	Iris	4	3	150
7	Seeds	7	3	210
8	Wine	13	3	178
9	Ionoshpere	34	2	351
10	A-N-O-1	2	7	264
11	A-N-O-2	2	5	306
12	A-N-O-3	2	3	280
13	A-N-O-4	2	2	1100

In evaluation, our approach is compared with ten different approaches showed in Table 3. All those approaches are executed 20 times on all of the data sets and all the results discussed in the rest of this section are the average results of the 20 executions.

Table 3. Approaches to be compared with ours

Approaches	Selection criteria for k	Min no. of clusters
PC [28]	max	2
ZXF [30]	Knee	2
LL	max & 1	1
LML	max & 1	1
I Index [8]	max	2
BIC	max	2
CH Index [24]	max	2
CE [25]	min	2
FHV [26]	min	1
Jump [27]	max	2
Our approach	max & 1	1

5.2 Experiment Results and Analysis

We track the 20 executions of each approach on the overlapping data sets and record the estimated number of clusters in Table 4. If an approach gets two different numbers of clusters in a data set during 20 executions, the results are record as the two number separated by a slash. If an approach obtains multiple numbers of clusters, the results are described as two numbers connected by a dash. One of the two number is the obtained minimal number, and the other is the obtained maximal number.

According to the Table 4, LML estimates the number of clusters correctly only on the Iris data set. BIC does better than LML. It obtains the correct estimation on SM2 and SM8 data sets. Jump and I also exhibit relatively high accu-

Table 4. Estimation of the optimal number of clusters in overlapping data sets

Approaches	Iris	Seeds	Wine	Ionoshpere	SM2	Flame	SM4	SM5	SM8
PC	2	2	2	2	2	2	2	2	2
ZXF	6	6	6	4–7	6	4	10/11	11–14	9/11
LL	2	2	2	1–9	1	4	1/2	2	8
LML	3	2	2	1–9	1	4	3/55	2	8–12
I	3	3	7	2	2	4	3	3	2
BIC	8–12	14	13	15–18	2	4	3	3/15	8
CH	3	3	13	2	2	8	2	2	8
CE	2	2	2	2	2	2	2	2	2
FHV	2	1	1	1	1	1	1	1	2
Jump	3	3	10–13	2	12–18	4	2–56	2–50	8
Our approach	3	3	3	2	2	2	4	5	8

Table 5. Estimate of the optimal number of clusters in non-overlapping data sets

Approaches	A-N-O-1	A-N-O-2	A-N-O-3	A-N-O-4
PC	7	5	3	2
ZXF	7	5	3	10/11
LL	7	5	3	2
LML	7	5	3	2–32
I	7	5	3	2
BIC	7	11	8	2
CH	7	11/12	4	2
CE	7	5	3	2
FHV	7	5	3	2
Jump	7	5	3	2
Our approach	7	5	3	2

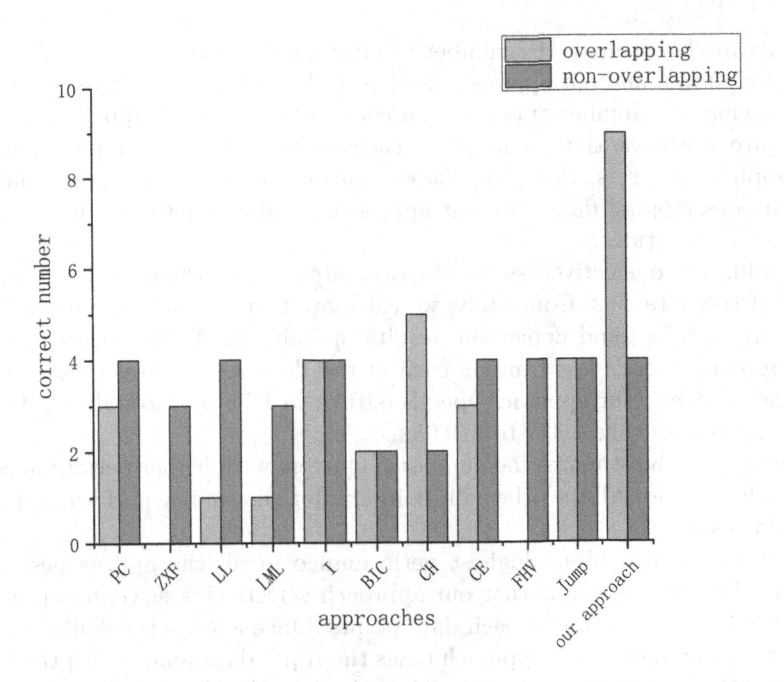

Fig. 4. The correct number of each method on the overlapping data set and the non-overlapping data set.

Table 6. Execution time of approaches

Approaches	A-N-O-1	A-N-O-2	A-N-O-3	A-N-O-4	Iris	Seed	Wine	Ionoshpere	SM2	Flame	SM4	SM5	SM8
PC	2.23	2.40	2.47	18.16	1.37	2.05	1.58	4.17	13.49	2.26	145.44	91.73	279.05
ZXF	2.22	2.30	2.50	18.21	1.37	2.03	1.57	4.13	13.63	2.24	148.9	92.75	259.20
LL	2.25	2.29	2.44	18.11	1.36	2.04	1.61	3.98	13.79	2.26	151.75	92.85	260.87
LML	2.22	2.29	2.44	18.14	1.40	2.04	1.58	3.98	13.67	2.34	152.10	93.35	260.65
I	2.23	2.30	2.45	18.14	1.37	2.13	1.58	4.06	13.74	2.25	153.12	94.22	261.63
BIC	2.23	2.28	2.42	18.17	1.35	2.02	1.57	3.96	13.79	2.41	143.81	92.52	261.83
CH	2.21	2.28	2.44	18.15	1.35	2.04	1.56	3.94	13.44	2.41	143.91	93.85	263.74
CE	2.18	2.23	2.39	18.21	1.32	1.99	1.53	3.99	13.72	2.39	144.75	93.40	262.61
FHV	2.23	2.29	2.44	18.28	1.40	2.09	1.62	4.06	13.76	2.29	143.83	93.84	272.38
Jump	2.28	2.34	2.48	18.27	1.39	2.08	1.60	4.05	13.58	2.26	143.19	92.83	269.28
Our approach	0.027	0.033	0.028	0.20	0.015	0.033	0.032	0.066	0.049	0.034	1.03	0.724	1.70

*Each execution time is measured in second.

racy. They estimate the number of clusters correctly on 4 overlapping datasets. They are followed by CE, and PC which estimate the number of clusters correctly on 3 data sets. CH can correctly estimate the number of clusters on the five data sets. Among those solutions, our approach does best. It estimates the number of clusters correctly on all the overlapping data sets.

We also evaluate the accuracy of our approach on non-overlapping data sets, and describe the results in Table 5. According to the table, BIC and CH estimate the number of clusters correctly on two data sets, while LML and ZXF do better.

They correctly estimate the number of clusters on three data sets. PC, LL, I, FHV, CE, Jump and our approach does best. They obtain the results consistent with the natural number of clusters on each of the non-overlapping data sets.

Figure 4 shows the accuracy of each method on overlapping and non-overlapping data sets. Based on the estimation on both overlapping data sets and non-overlapping data sets, our approach exhibit highest accuracy than all of the other solutions.

In addition to effectiveness, we also evaluate the performance of our approach on all of the data sets. Concretely, we calculated the average execution time of all the approaches and depict the results in Table 6. According to the Table, our approach spends less time on each of the data sets than each of the other approaches does. Our approach spends 0.015 s to 1.7 s on those data sets, while other approaches take 1.32 s to 279.05 s.

Our approach estimates the number of clusters with higher performance than other solutions do. All the other solutions exhibit the similar performance on the same data set.

Our approach exhibits highest performance in all the approaches for two reasons. The first reason is that our approach selects cluster centers in a static way. It calculates a score for each data points. Once a score is calculated, it will not change any more. Our approach takes the top k data points with the highest scores as the centers for k clusters. The other way to choose a cluster center is through iteration. When certain conditions are met, the iteration will stop and the center point will be obtained. The second reason is that our approach narrow the search space of the optimal number of cluster, and hence degrading search cost.

6 Conclusion

In this work, we focused on the problem to estimate the optimal number of clusters in overlapping data sets. To deal with the problem, we proposed a fast estimation approach. The approach selects cluster centers in a static way according to density and distance. It utilizes the significant change of the minimal distance between cluster centers to identify the optimal number of clusters. The experimental result demonstrated the usefulness and effectiveness of our approach. In the future, we will conduct research on estimating the optimal number of the clusters in more complex data sets.

Acknowledgement. This work is supported by the National Natural Science Foundation of China (61602156 and 61433012), the project of the Scientific and Technological in Henan province (172102310677), the project of the Basic and Frontier Technology in Henan province (142300 410147) and the PhD foundation of Henan Polytechnic university (B2012-099).

References

1. Anil, K.: Data clustering: 50 years beyond K-means. Pattern Recogn. Lett. **31**(8), 651–666 (2010)

2. He, Z., Jia, Z., Zhang, X.: A fast method for estimating the number of clusters based on score and the minimum distance of the center point. Information **11**, 16 (2020)
3. Chen, Z.W., Chang, D.X.: Automatic clustering algorithm base on density difference. J. Softw. **29**(4), 935–944 (2018)
4. Jia, R.Y., Li, Z.: The level of K-means clustering algorithm base on minimum spanning tree. Microelectron. Comput. **33**(3), 86–93 (2016)
5. Ünlü, R., Xanthopoulos, P.: Estimating the number of clusters in a dataset via consensus clustering. Expert Syst. Appl. **125**, 33–39 (2019)
6. Bai, L., Cheng, X., Liang, J., Shen, H., Guo, Y.: Fast density clustering strategies based on the k-means algorithm. Pattern Recogn. **71**, 375–386 (2017)
7. Davies, D.L., Bouldin, D.W.: A cluster separation measure. IEEE Trans. Pattern Anal. Mach. Intell. **1**(2), 224–227 (1979)
8. Maulik, U., Bandyopadhyay, S.: Performance evaluation of some clustering algorithms and validity indices. IEEE Trans. Pattern Anal. Mach. Intell. **24**(12), 1650–1654 (2002)
9. Beni, G., Xie, X.: A validity measure for fuzzy clustering. IEEE Trans. Pattern Anal. Mach. Intell. **13**(8), 841–847 (1991)
10. Rodriguez, A., Laio, A.: Machine learning clustering by fast search and find of density peaks. Science **344**(619), 1492 (2014)
11. Gupta, A., Datta, S., Das, S.: Fast automatic estimation of the number of clusters from the minimum inter-center distance for k-means clustering. Pattern Recogn. Lett. **116**, 72–79 (2018)
12. He, L., Wu, L.D., Cai, Y.C.: Survey of clustering algorithms in data mining. Appl. Res. Comput. **71**, 375–386 (2017)
13. Macqueen, J.: Some methods for classification and analysis of multivariate observations. In: Fifth Berkeley Symposium on Mathematical Statistics and Probability, vol. 1, no. 14, pp. 281–297. California Press, Berkely (1967)
14. Zhai, D.H., Yu, J., Gao, F.: K-means text clustering algorithm based on initial cluster centers selection according to maximum distance. Appl. Res. Comput. **31**(3), 713–719 (2014)
15. de Amorim, R.C., Hennig, C.: Recovering the number of clusters in data sets with noise features using feature rescaling factors. Inf. Sci. **324**, 126–145 (2015)
16. Teklehaymanot, F.K., Muma, M., Zoubir, A.M.: A novel Bayesian cluster enumeration criterion for unsupervised learning. IEEE Trans. Signal Process **66**(20), 5392–5406 (2018)
17. Bensaid, A.M., Hall, L.O., Bezdek, J.C.: Validity-guided (re)clustering with applications to image segmentation. IEEE Trans. Fuzzy Syst. **4**, 112–123 (1996)
18. Ren, M., Liu, P., Wang, Z., Yi, J.: A self-adaptive fuzzy c-means algorithm for determining the optimal number of clusters. Comput. Intell. Neurosci. 3–15 (2016)
19. Sweeney, T.E., Chen, A.C., Gevaert, O.: Combined mapping of multiple clustering algorithms (COMMUNAL): a robust method for selection of cluster number, K. Sci. Rep. **5**, 16971 (2015)
20. Wang, M., Abrams, Z.B., Kornblau, S.M.: Thresher: determining the number of clusters while removing outliers. BMC Bioinformatics **19**(1), 9 (2018)
21. Kingrani, S.K., Levene, M., Zhang, D.: Estimating the number of clusters using diversity. Artif. Intell. Res. **7**(1), 15 (2018)
22. Doan, H., Nguyen, D.: A method for finding the appropriate number of clusters. Int. Arab J. Inf. Technol. **15**(4), 675–682 (2018)
23. Wang, Y., Shi, Z., Guo, X., Liu, X., Zhu, E., Yin, J.: Deep embedding for determining the number of clusters. In: AAAI (2018)

24. Caliński, T., Harabasz, J.: A dendrite method for cluster analysis. Commun. Stat. **3**(1), 1–27 (1974)
25. Bezdek, J.C.: Mathematical models for systematics and taxonomy. In: Eighth International Conference on Numerical Taxonomy, vol. 3, pp. 143–166 (1975)
26. Dave, R.N.: Validating fuzzy partitions obtained through c-shells clustering. Pattern Recognit. Lett. **17**(6), 613–623 (1996)
27. Sugar, C.A., James, G.M.: Finding the number of clusters in a dataset: an information-theoretic approach. J. Am. Stat. Assoc. **98**(463), 750–763 (2003)
28. Bezdek, J.C.: Cluster validity with fuzzy sets. J. Cybernet. **3**(3), 58–73 (1973)
29. Pakhira, M.K., Bandyopadhyay, S., Maulik, U.: Validity index for crisp and fuzzy clusters. Pattern Recognit. **37**(3), 487–501 (2004)
30. Zhao, Q., Xu, M., Fränti, P.: Sum-of-squares based cluster validity index and significance analysis. In: Kolehmainen, M., Toivanen, P., Beliczynski, B. (eds.) ICANNGA 2009. LNCS, vol. 5495, pp. 313–322. Springer, Heidelberg (2009). https://doi.org/10.1007/978-3-642-04921-7_32

Cyber Security and Privacy

Research on Detection Method of Malicious Node Based on Flood Attack in VANET

Yizhen Xie[1], Yuan Li[2](✉), and Yongjian Wang[3]

[1] Beijing University of Posts and Telecommunications, Beijing 100876, China
[2] Southwest University of Science and Technology, Mianyang 621010, Sichuan, China
Liyuan_3033@163.com
[3] National Internet Emergency Center, Beijing 100029, China

Abstract. For the reason of the variability and mobility of VANET's topology, it is easy to be attacked by attackers. This paper proposes two detection methods for TCP Synchronize Sequence Numbers (SYN) flood attacks and UDP traffic flood attacks in combination with the requirements of the rapid and real-time detection for malicious nodes in the security of the 5G VANET, and to enhance the ability to identify and detect malicious nodes. For the TCP SYN flood attack, the number of access requests with the same node ID is limited, and the number of semi-connections in Road Side Unit (RSU) is adjusted. For the UDP traffic flood attack, the RSU is used to monitor and analyze the data traffic of each node in VANET. Through the feasibility analysis, the above two detection methods can effectively detect flood attacks in VANET, thus assisting the network defense mechanism to ensure network security. Thereby providing guarantee for the security of VANET in the 5G communication environment.

Keywords: VANET security · 5G · Flood attack · TCP SYN flood · UDP traffic flood

1 Introduction

As an intelligent system for the application of the Internet of Things in transportation systems, VANET is an integrated application and extension of related technologies such as computers, Internet, mobile communication networks, and Internet of Things [1]. VANET plays an important role in easing urban traffic pressure, reducing traffic accident rates, improving road utilization, and unmanned driving. 5G VANET uses 5G network-based C-V2X technology, which mainly includes four major communication scenarios: vehicle-Cloud communication: vehicle and VANET service platform interaction information through 5G network; vehicle-vehicle communication: vehicle to vehicle via LTE/5G-V2X technology interaction information; vehicle-road communication: vehicle and road infrastructure facilities use LTE/5G-V2X technology to interaction information; vehicle-people communication: vehicle and user intelligent terminals interaction information through 5G networks. The emergence of 5G communication technology has helped the development of VANET. The characteristics of low latency

X. Wang et al. (Eds.): 6GN 2020, LNICST 337, pp. 373–383, 2020.
https://doi.org/10.1007/978-3-030-63941-9_28

and high speed will improve the performance and reliability of VANET. However, due to the open network environment of the VANET, the mobility of nodes, and the dynamic structure of the topology, it is vulnerable to be attacked by attackers and causes serious security problems. Therefore, VANET security is an important part of the research field of VANET.

The security threats of wireless communication in the 5G VANET scenario can be divided into three categories: traffic attacks cause network overload, identity forgery causes information leakage or misjudgment, and malicious interference causes system errors. Flood attacks are the main aspect in traffic attacks. Through the research on the flood attack problem faced by VANET, this paper mainly detects the two attack modes of TCP SYN flood attack and UDP traffic flood attack. The main research contents are as follows:

For the architecture of VANET adopting the TCP protocol, when the vehicle node initiates a connection request to the RSU and initializes, the RSU uses the SYN header to perform a "Triple handshake" connection with the vehicle nodes, The attacker can quickly initiate a large number of connections in a short time. The requests cause the RSU to fail to respond, thus consuming a large amount of RSU computation processing resources until the service is denied. For this TCP SYN flood attack, the node's ID restriction method is used for detection and defense. When the node sends a access request, the RSU records the node's ID. When the same ID sends the access request multiple times in a short time, the node's requests should be restricted. For the node's ID that has already existed in the network, its request packet will no longer be accepted. At the same time, the RSU shortens the TCP handshake half-connection suspension timeout period and closes the request connection initiated by the duplicate ID.

For the architecture of VANET using the UDP protocol, when a normal node in the network is hacked into a malicious node and initiates a traffic attack, the RSU will be targeted. A malicious node will frequently broadcast data packets to other nodes. The destination address of the data packet is the RSU, and a large number of data packets occupy network bandwidth resources and RSU's computing processing capability in a short time. For such flood attacks, RSU is used to monitor the data traffic of nodes in the network in real time. RSU counts the data traffic of each node, calculates the traffic average value according to the time period, and records the source address of the received data packet. If the node data traffic average value is higher than the normal value and the data packet source address frequency is too high in a certain period of time, the node may be determined to initiate a flood attack and isolate it. In addition, UDP packet verification is added to analyze the availability of the data packet and improve the detection accuracy.

5G technology protects the safety of connected cars. With its powerful mobile bandwidth, 5G communication technology can reach a peak rate of 20 Gbit/s, support lower latency ($\leq$10 ms), higher reliability (>99.99%), and more terminal connections (1 million terminals can be connected per square kilometer) [2], which can meet the safety communication needs of VANET In the detection of SYN TCP flood attacks, the high reliability of 5G can improve the detection performance; in the detection of UDP flood attacks, 5G can guarantee the real-time nature of traffic detection with low latency characteristics.

The second section of this paper introduces the current research status of VANET security and flooding attack detection. The third section introduces TCP SYN flood and UDP traffic flood attack methods and corresponding detection methods; the feasibility analysis is carried out in the fourth section; finally, summarize the full text.

2 Related Work

VANET is defined by the China IoT School-Enterprise Alliance. VANET is a network that integrates information such as vehicle location, vehicle speed, and vehicle travel path. It can collect the environment and status information of the vehicles and their surroundings through wireless devices such as GPS, radio frequency, sensors and cameras [3]. Then the related information is transmitted through the Internet technology, and finally the relevant analysis technology is used to process the information, thereby more effectively realizing the traffic of the vehicles and improving the overall traffic efficiency of the city. Since the rise of VANET in recent years, relevant security defense measures have not been improved in time. At present, there are still many security problems in VANET, which have occurred from malicious attacks by attackers, with serious consequences.

In 2010, researchers at Rutgers University in Southern Carolina demonstrated how to crack the car's internal network and counterfeit the tire pressure sensors' information of some car brands. Even destroyed the TPMS system over 40 meters away through wirelessly interfering. At the Las Vegas hacking conference in 2013, two hackers demonstrated how to attack the Toyota Prius and Ford Mavericks control systems to achieve a series of operations such as sudden braking, including high-speed driving, brake failure, and steering of the steering wheel [4]. In 2015, two security researchers demonstrated a vulnerability in the car's "Uconnect" feature of Chrysler's Jeep Cherokee, remotely invading and controlling the target vehicle, and freely controlling the entertainment system, wipers, steering wheel, engine, etc. At the beginning of 2017, the safety weather van of VANET had been urgently transferred to the data security and privacy of customers. In June, a database of dealers in the United States was attacked, involving sales data leaks from more than 10 million vehicles. In December, Nissan Motor officially announced that its financial company database data information was stolen by hackers, customers' personal information and loan information were all stolen [5]. MegamosCrypto protection systems from Audi, Porsche, Bentley and other Volkswagen brands were also breached.

At present, many scholars have conducted research on the detection of malicious nodes in VANET. How to meet the requirements of low-latency, high reliability, high speed, large capacity, and high security in high-density vehicle scenarios is the challenge faced by 5G VANET and the focus of related research. However, most of the existing literature research on flood attack detection is based on traditional wired networks and wireless sensor networks, and few researchers have studied the flood attacks in VANET. The current research status of flood attack detection is as follows:

In [6], for SYN Flood attack on the target computer, the detection function is set on the target computer. If it finds that receives SYN packets for the local machine without responding, it is considered to be subjected to a SYN flood attack and refuses to connect with it. However, this method is not suitable for the scenario where vehicle nodes in VANET frequently access the network and is likely to cause misjudgment.

The most obvious feature of UDP traffic flood attack is that the traffic is greatly increased. It is also a common method based on traffic changing detection. The real-time defense mechanism (DDM) of DHCP flooding attack is proposed in [7]. The dynamic peak estimation model is established by using two key parameters of real-time DHCP traffic average speed and IP pool margin to evaluate whether the port is attacked. If it is attacked, the mitigation model starts defense. In the mitigation model, the IP pool cleaning is performed by using the response feature of the Address Resolution Protocol (ARP). The intra-period interception mechanism is designed to intercept the attack source, which can reduce the blocking and minimize the interception to the user. The impact of normal use. The shortcoming of this method is that the computational processing capability of the node and the RSU is relatively high, otherwise the real-time performance of the defense cannot be met.

Literature [8] proposed a method for constructing a new SYN-agent that uses the TCP header reservation flag to inform the server of a complete the triple TCP handshake. If it is a SYN-attack, there should be no further ACK after this. After a short period of time, the half-open TCP connection is removed from the proxy. Therefore, the TCP SYN flood attack is avoided to make a large number of TCP semi-connections occupy resources. This method is applicable to a scenario where the nodes are relatively fixed. In VANET, the vehicle enters and exits the network at any time, and the detection is difficult.

Literature [9] proposed a SYN Flood attack mitigation method based on supervised learning classification method, which identifies and blocks SYN Floods before they reach their targets, thus preventing resource consumption and performance loss. This method selected a classifier and adjusts parameters according to the policy and change characteristics of the SYN Flood attack, but the learning attack strategy occupies a large amount of computing resources of the RSU.

Literature [10] proposed an effective method to detect and defend against UDP flood attacks under IP spoofing types. This method utilizes a Bloom filter-based storage efficient data structure and an IP Wedge Reference Detection Method. It achieves higher detection rate while defending against UDP flood attacks with IP spoofing, and has lower storage and computational costs. However, this method does not consider the availability of attack packets.

The detection method for traditional network flood attacks is not applicable to VANET. Considering the actual network environment of VANET, combined with the low-latency and high-reliability characteristics of 5G communication, this paper proposes a detection method for TCP SYN and UDP traffic flood attacks in VANET. 5G communication technology can meet the needs of single-vehicle uplink and downlink data rates greater than 10 Mbit/s, 50 Mbit/s in some scenarios, latency of 3-50 ms, and reliability greater than 99.99%. At the same time, it can meet the real-time interactive data of vehicles, roads and pedestrians, and the high data transmission demand of up to TB-level per day [11]. The application of 5G communication technology in the low-latency and high-mobility VANET scenario solves many problems and challenges faced by the current VANET, and enables the VANET to obtain better performance under high-speed movement [12].

3 Principles and Methods of Flood Attacks

3.1 TCP SYN Flood

In the process of establishing a connection with TCP, a "Triple handshake" is required. As shown in the figure, first handshake: When the TCP network establishes a connection, the client sends a SYN packet to the server and enters the SYN_SENT state, waiting for the server to confirm. The second handshake: After receiving the SYN packet, the server confirms the SYN of the client and sends a SYN+ACK packet to the client. The server enters the SYN_RECV state, which is the half-connected state [13]. The third handshake: After receiving the SYN+ACK packet of the server, the client feeds back the ACK response packet to the server. After receiving the response packet, the server completes the successful TCP connection, as shown in Fig. 1.

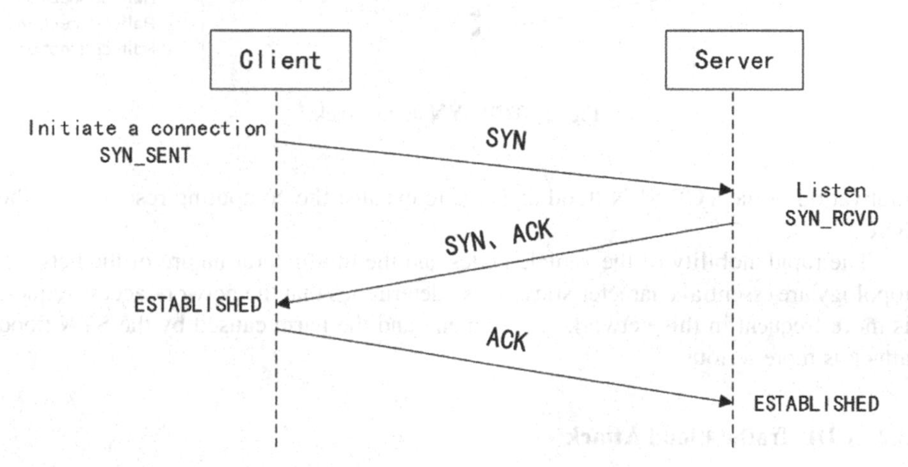

Fig. 1. Triple handshake.

The attacker uses the defect of the TCP "Triple handshake" connection mechanism to initiate a SYN attack. The attacker controls malicious node to initiate a connection and attempts to access VANET. The malicious node masquerades as a normal node to initiate a connection request to the RSU as a client. After receiving the SYN+ACK packet of the server in the second handshake phase, the malicious node does not continue to complete the third handshake, and does not return the ACK response packet to the server, so that the server remains half-connected. After the attacker initiates multiple connection requests, the server will suspend the corresponding number of half-connected states. As a VANET relay server, the RSU will gradually be exhausted by a large number of useless half-connected state processing resources, rejecting other connection services, resulting in normal legal vehicle nodes unable to access VANET, as shown in Fig. 2. The attack

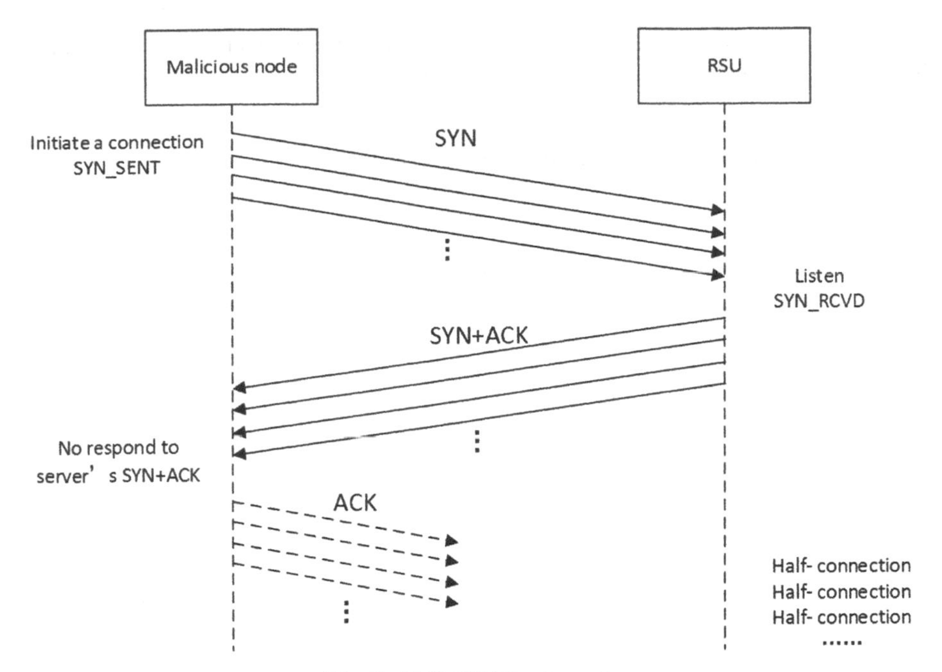

Fig. 2. TCP SYN flood attack.

motivation of the TCP SYN flood attack is to exhaust the computing resources of the RSU.

The rapid mobility of the vehicle nodes and the multilateral nature of the network topology are essential characteristics. It also determines that the network access request is more frequent in this network environment, and the harm caused by the SYN flood attack is more serious.

3.2 UDP Traffic Flood Attack

UDP is a connectionless protocol and does not require any connection to be made to transmit data. The attacker sends a large number of UDP packets to the RSU [14]. The RSU is busy processing the UDP packets and cannot process normal packet requests or responses. A large number of useless UDP packets carry a large amount of network traffic quickly occupying network bandwidth resources, causing network congestion to reject other normal services, as shown in Fig. 3.

In addition, in the UDP flood attack, the attacker sends a large number of UDP packets or malformed UDP packets with fake IP addresses. These fake IP addresses do not exist in current VANET, and the destination IP address of the packets will never be available. So these packets will always be forwarded within the network, resulting in continuous attack traffic, and the target node does not get back information to cause system resources to run out or even crash. The malformed UDP packet will not be parsed after the node receives it. Therefore, after the node receives the packet, it detects whether it is normal. The incomplete abnormal packet will be discarded, but the detection and

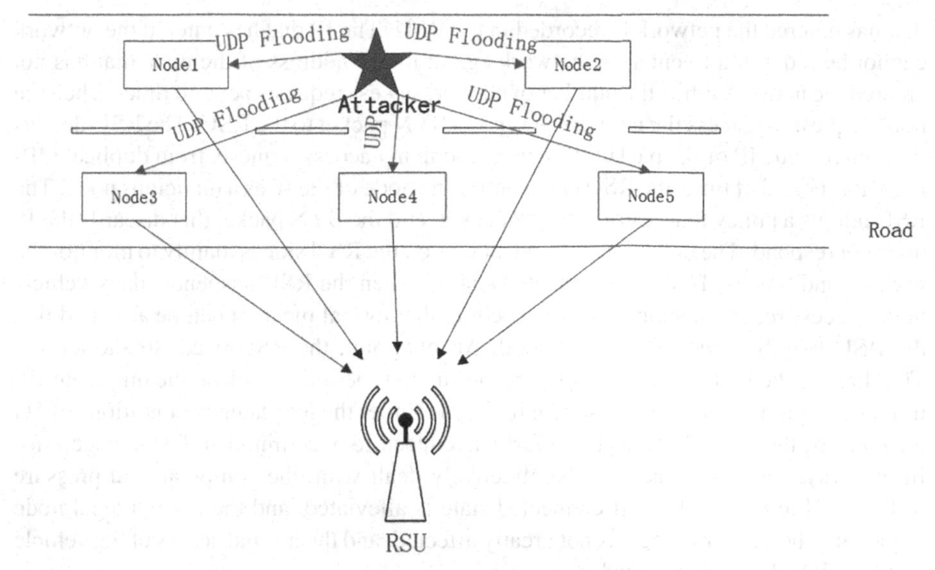

Fig. 3. UDP traffic flood attack.

discarding of the packet requires the node to calculate the processing resource. Many malformed data packets will occupy large detection computing resources of the node until the computing power of the node is exhausted, causing the target node to denial of service.

The attacking mode is that the attacker places the malicious node in VANET in advance, or attacks normal nodes in the network and capture control to become malicious nodes. Since UDP is based on a connectionless protocol, the node and the RSU do not continuously maintain a connection, and the member nodes of VANET change constantly. So it is difficult to detect the malicious node causing the UDP flood attack in VANET.

4 Detection Method

4.1 Detection for TCP SYN Flood Attack

Because the TCP SYN flood attack utilizes the TCP "Triple handshake" process, the server will suspend the half-connection and wait for the client to acknowledge the ACK packet. Therefore, one of the keys to detecting the SYN flood attack is to detect the authenticity and availability of the client's SYN packet. If a client continuously sends a SYN packet request to the network, but does not respond to the ACK packet to the server, causing the server to suspend a large number of half-connected states, it can detect that the client initiates a TCP SYN flood attack. Secondly, limiting the number of server-side TCP half-connections and the timeout period can alleviate the SYN flood attacks sent by attackers.

In the practical application environment of VANET based on 5G technology, the IP of the vehicle node requesting to access the network is detected, and the IP of the node that has entered the network is recorded in the RSU. The IP that has entered the network cannot be requested to enter the network again; the IP address of the node that has not entered the network limits the number of network access requests per unit time. When the node requests to access the network, it sends a SYN packet to the RSU. The RSU detects the source node IP of the packet. If there is too many access requests from duplicate IPs in a short period of time, the RSU can identify the node of the IP as a malicious node. The RSU adopts a policy that restricts the IP packet, and the SYN packet that discards the IP does not respond. The detection of flood attacks on the RSU side is mainly to monitor the number and time of TCP half-connected states. When the RSU suspends many vehicle nodes access requests, significantly exceeding the normal mean, it can be assumed that the RSU may be under TCP SYN flood. At this point, the RSU needs to shorten the TCP handshake half-connection suspension timeout period and close the duplicate IP-initiated request connection, as shown in Fig. 4. Under the low-latency conditions of 5G technology, the timeout needs to be further reduced to accommodate fast connections. In this way, the flood attack can be effectively dealt with, the computational pressure of the RSU to handle the half-connected state is alleviated, and the normal legal node requesting the network access is not greatly affected, and the normal access of the vehicle nodes in VANET is guaranteed.

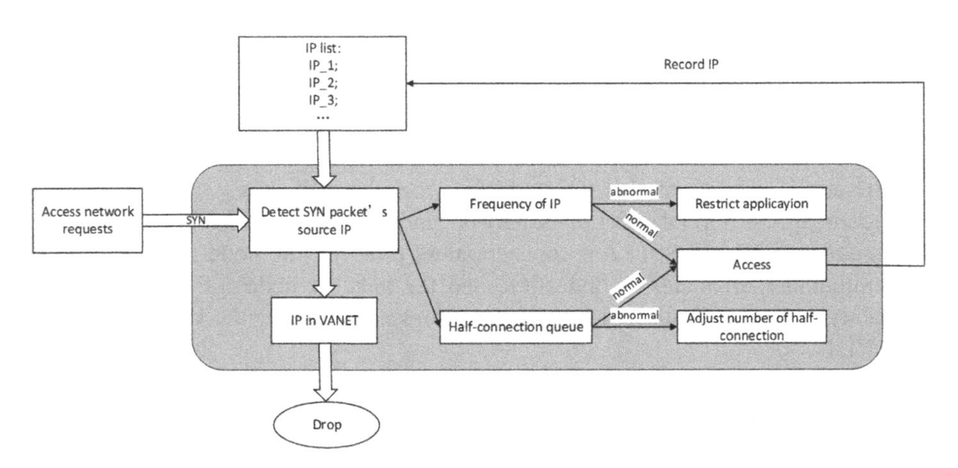

Fig. 4. TCP SYN flood detection method.

4.2 Detection for UDP Traffic Flood Attack

Due to the variability of the composition of the vehicle network node, the identity of the node is difficult to authenticate. Before the vehicle node sends the UDP data packet, the legality of the node cannot be known. Therefore, in the case of UDP traffic flood attacks, it is better to monitor node's traffic. The RSU monitors the data traffic of each node in the network to analyze whether the traffic is abnormal. There are also two ways to monitor traffic: First, the RSU directly monitors the data traffic sent and received by each node in the network, and performs traffic change statistics. However, in this way, each time the node sends and receives data, the RSU must monitor and sample it, which will occupy a large amount of computing resources of the RSU and occupy much network bandwidth. Second, the vehicle node counts the source IP address of the data packet when receiving the data packet, and generates a traffic statistics data packet, where the packet includes the data source IP address, the destination IP address, and the data packet size. Each node performs traffic statistics packets and uploads them to the RSU. The RSU collates and counts the traffic statistics of each node. Because the UDP packet of the node is sent to multiple nodes, the traffic condition of the source node is counted by multiple nodes, thus avoiding the contingency difference of the individual nodes and improving the credibility. After the RSU counts the traffic changes of each node according to the time interval, it can be compared with the normal ones. The difference is due to the increase in the amount of data reported by the vehicle or the surge in data traffic caused by the flooding attack. Due to the high-bandwidth and low-latency characteristics of 5G, the average value of traffic needs to be adjusted in real time according to the different states of ordinary VANET communication and 5G communication to adapt to different network conditions. Therefore, the RSU can detect a malicious node that initiates a UDP flood attack inside VANET. The detection method is shown in Fig. 5.

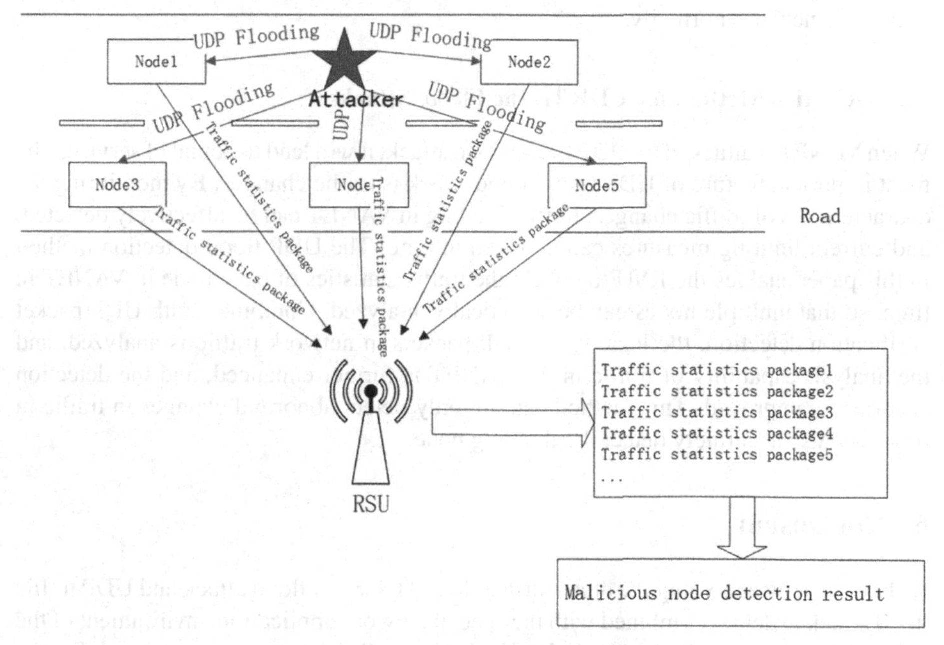

Fig. 5. UDP traffic attack detection mode.

In addition, an attacker may send many UDP packets with a fake destination IP address. At this condition, the destination IP address connectivity of the UDP packet in the network is detected and analyzed. If the destination IP does not exist in the RSU routing list or is not connectable, the forwarding of the packet is stopped. It is also necessary to detect the integrity and availability of UDP packets to avoid the large amount of malformed data packets occupying the computing power of the vehicle nodes or RSUs.

5 Experimental Analysis

5.1 Detection Method for TCP SYN Flood Attack

In the VANET application environment, once the RSU is attacked by the TCP SYN flood, other legitimate vehicles cannot enter the network normally. The road conditions and vehicle emergency information on this road cannot be accepted in time. The RSU analyzes all the captured SYN packets, and obtains the statistics of the SYN packets sent by each node. The RSU detects the suspended half-connection status in real time, and detecting the number of these TCP half-connections can effectively determine whether it has been attacked by SYN flood. RSU reduces the half-connection waiting time can alleviate the calculation and buffering pressure, and effectively prevent the same vehicle ID from launching a network access request multiple times to maliciously attack the RSU to occupy network access request resources. While detecting and mitigating TCP SYN flood attacks, this method can also ensure that other legitimate vehicle nodes can access the network normally.

5.2 Detection Method for UDP Traffic Flood Attacks

When VANET is attacked by UDP traffic flood attack, it will lead to denial of service. The most important feature of UDP traffic flood attack is traffic changes. By monitoring the characteristics of traffic changes, traffic flooding in VANET can be effectively detected, and current limiting measures can be taken in time. The UDP flood detection method in this paper enables the RSU to collect the traffic statistics of each node in VANET in time, so that multiple nodes can be statistically analyzed. Combined with UDP packet verification detection, the legality of UDP packets in network traffic is analyzed, and the analysis capability of traffic in the VANET is further enhanced, and the detection accuracy is improved. This method can not only detect abnormal changes in traffic in time, but also accurately detect the flooding node.

6 Conclusion

In this paper, by analyzing the attack principle of TCP SYN flood attack and UDP traffic flood attack in detail, combined with the special network application environment of the 5G VANET, the detection methods for the above two flood attacks are proposed. For the TCP SYN flood attack, the method of restricting the repeated network access request and adjusting the number and time of TCP half-connections is adopted, which effectively alleviates the flooding of the SYN packet and ensures that the legal vehicle node normally enters the network; for the UDP traffic flood attack, the whole network node is adopted. The method of traffic monitoring and UDP packet check validity effectively mitigates UDP traffic flood and improves detection accuracy.

The two types of flood attacks in this article also have limitations. The detection method for the TCP SYN flood attack will be affected by the network changes. The average traffic in the detection method for UDP flood attacks will be affected by 5G network conditions, and the detection performance of both will be affected by mobility of the vehicles. When multiple nodes attack at the same time, the detection effect may not be ideal.

References

1. Zhou, H.: Research on traffic information collection and processing methods in the Internet of Vehicles. Jilin University (2013)
2. Xu, C.: Impact of 5G connected vehicles on the development of autonomous driving technology. Inf. Commun. 46–47 (2018)
3. Lin, T.: Research on multi-path TCP transmission and control in vehicle networking. Dalian University of Technology (2017)
4. Nan, Y., Rongbao, K.: Analysis and protection of vehicle network security threats. Commun. Technol. **48**(12), 1421–1426 (2015)
5. Chen, L.: Security threats and research status of Internet of Vehicles. http://wemedia.ifeng.com/73941858/wemedia.shtml. Accessed 16 Aug 2018
6. Huang, Y., Wan, L., Li, X.: SYN flooding attack based on IP spoofing. Comput. Technol. Dev. **18**(12), 159–161,165 (2008)
7. Zou, C., Liu, P., Tang, X.: Real-time defense of DHCP flood attack in SDN network. Comput. Appl. http://knsCnki.net/kcms/detail/51.1307.TP.20181203.1628.008.html. Accessed 16 Aug 2018
8. Liu, P., Sheng, Z.: Defending against TCP SYN flooding with a new kind of SYN-agent. In: 2008 International Conference on Machine Learning and Cybernetics, Kunming, pp. 1218–1221 (2008)
9. Degirmencioglu, A., Erdogan, H.T., Mizani, M.A., Yılmaz, O.: A classification approach for adaptive mitigation of SYN flood attacks: preventing performance loss due to SYN flood attacks. In: NOMS 2016–2016 IEEE/IFIP Network Operations and Management Symposium, Istanbul, pp. 1109-1112. IEEE (2016)
10. Verma, K., Hasbullah, H., Kumar, A.: An efficient defense method against UDP spoofed flooding traffic of denial of service (DoS) attacks in VANET. In: 2013 3rd IEEE International Advance Computing Conference (IACC), Ghaziabad, pp. 550–555. IEEE (2013)
11. Chen, W., Li, Y., Liu, W.: Analysis of the progress and key technologies of the connected car industry ZTE. http://kns.cnki.net/kcms/detail/34.1228.TN.20200217.1748.004.html. Accessed 21 Feb 2020
12. Liangmin, W., Xiaolong, L., Chunxiao, L., Jing, Y., Weidong, Y.: Prospects of 5G telematics. J. Netw. Inf. Secur. **2**(06), 1–12 (2016)
13. Tang, H., Zeng, Y.: Detection of SYN flood attack based on semi-join list. Comput. Eng. **37**(19), 135–137,144 (2011)
14. Houmer, M., Hasnaoui, M.L., Elfergougui, A.: Security analysis of vehicular ad-hoc networks based on attack tree. In: 2018 International Conference on Selected Topics in Mobile and Wireless Networking (MoWNeT), Tangier, pp. 21–26. IEEE (2018)

Two Attacking Strategies of Coordinated Cyber-Physical Attacks for Cascading Failure Analysis in Smart Grid

WenJie Kang[1,2,3,4](✉)

[1] Information Technology (Internet Supervision) Department, Hunan Police Academy, Changsha, China
kangwenjiebishen@126.com
[2] School of Electronic Information and Electrical Engineering, Changsha University, Changsha, China
[3] College of Systems Engineering, National University of Defense Technology, Changsha, China
[4] Key Laboratory of Hunan Province for New Retail Virtual Reality Technology, Hunan University of Technology and Business, Changsha, China

Abstract. As a classic Cyber-Physical System (CPS), smart grids often suffer from various types of attacks, one of which the most threatening attacks is Coordinated Cyber-Physical Attack (CCPA). In order to improve the robustness of the smart grid under CCPA, two attack strategies are proposed to analyze the cascading failure of smart grids. Firstly, we define attack goals (AGs) function to identify important cyber and physical nodes as possible targets. Secondly, based on these targets, the algorithm of optimal attack and saturation attack strategy is designed and applied to CCPA for analyzing the effect of those attack strategies on smart grids. Finally, node loss is used as an evaluation index to compare the attack effect of CCPA, Cyber Attack (CA) and Physical Attack (PA). The experimental results show that when the same proportion of nodes are removed, the CCPA has more node losses than the CA and PA, regardless of based on the optimal attack strategy or the saturation attack strategy.

Keywords: Smart grids · Attack goals (AGs) · Coordinated Cyber-Physical Attack · Attacking strategy

Supported by National Natural Science Foundation of China under Grant Nos. 61501482, 61572514, 61903049 and 61702539, Scientific Research Fund of Hunan Provincial Education Department (19C0160 and 20B057), Open Fund of Key Laboratory of Hunan Province (2017TP1026), Hunan Key Laboratory Open Research Fund Project 2017TP1026, Hunan Provincial Natural Science Foundation of China under Grant No. 2018JJ3611, Changsha Science and Technology Program (Grant K1705007) and NUDT Research Project under Grant No. ZK-18-03-47.

X. Wang et al. (Eds.): 6GN 2020, LNICST 337, pp. 384–396, 2020.
https://doi.org/10.1007/978-3-030-63941-9_29

1 Introduction

The fifth-generation (5G) contributions to meet the requirements of data transmission and real-time state perception of industrial control network. Moreover, the next-generation (B5G/6G) will address the major opportunities and face new challenges, not only to create more new technologies, protocols and applications, but also to bring more security risks, such as coordinated cyber-physical attacks, cross-layer attacks etc. On the one hand, attackers can use acquired knowledge of target network to carry out cross-layer attacks on the Cyber-Physical Systems from the information domain to the physical domain; On the other hand, attackers can cooperate with each other and use their own resources to launch multi-directional and multi-target cyber-physical attacks; The study of cross-layer attack has completed in [1]. This paper will analyze the impact of CCPA combined with different attack strategies on the cascading failures of smart grid.

An incident in 2015 involving Ukrainian power systems was seen as a coordinated cyber-attack (CCA), where attackers injected malicious commands into the cyber domain, causing a breakdown in the physical domain for several hours [2]. Continuous attacks on Venezuelan power grid on March 7 and 8, 2019 resulted in power supply interruptions in 18 states across the country, which has once again caught the attention of scientists. As a major threat to the ICT infrastructure of power systems [3], CCAs are described as an organized cyber disruption, in which the attackers may have a well-organized plan to launch multiple cyber-attacks intended to compromise the same target [2].

With the increasing prevalence of terrorism and sabotage activities, the power grid is becoming more vulnerable to various kinds of cyber and physical attacks [4]. In the future, the attackers may launch cyber and physical attacks at the same time and collaborate to finish the task by sharing the same (multiple) targets and attacking them simultaneously. Coordinated cyber-physical attacks (CCPAs) on smart grids could lead to undetectable line outages, leading to a need for topology preservation and load-redistribution attacks that trigger cascading failures [5]. An exploration of potential attack goals (AGs) will contribute to clarifying the target, rather than blindly anticipating attack strategies. A coordinated cyber-physical attack following these AGs will maximize the destruction.

Liu et al. [6] proposed a framework that models a class of cyber-physical switching attack in smart grid systems to demonstrate how attack construction on a linearized version of the system still executes on nonlinear and realistic models of the system. Deng et al. [7] proposed CCPAs in smart grid by utilizing cyber attacks to mask physical attacks which can lead to power outages and potentially cause cascading failures. CCPAs used a false data injection attack vector based on phasor measurement unit (PMU) to avoid physical attacks being detected. The mathematical model of locally coordinated cyber-physical attacks is proposed to use incomplete network information in order to cause undetectable transmission line outages [8]. Liu et al. [9] developed coordinated cyber-physical attack based on variable structure systems theory to enable large-scale power system disturbances. Since CCPAs on its critical infrastructure can cause disastrous human and economic losses, a stochastic game-theoretic approach is proposed to generate the optimal strategies that defenders can adopt to protect the smart grid against CCPAs [10]. Tian et al. [11] investigated Multilevel Programming-Based CCPAs and the countermeasure with one leader and multiple followers in smart grid. Lakshminarayana et al. [12] proposed a moving target defense (MTD) strategy to detect coordinated cyber-physical attacks (CCPAs) that consists of a physical attack and followed by a coordinated cyber attack.

The rest of the paper is organized as follows. Section 1 introduces the model of coordinated cyber-physical attacks that contains identification evaluation of cyber and physical attack goal, CCPA based on optimal attack strategy and CCPA based on saturation attack strategy. Section 2 shows the experimental results and analyzes the reasons for the results. Finally, Sect. 3 draws relevant conclusions and presents future work.

1.1 Identification of Cyber and Physical Attack Goals (AGs) [13]

In order to generate an attack sequence of CCPA, the first thing to do is to identify the cyber AGs and physical AGs. As a characteristic of the power grid, power flow can cause the redistribution of voltage and frequency following the breakdown or failure of a substation. Therefore, physical AGs not only rely on the power flow, but also depends on the characteristic of network structure. Due to the coupling relationship, cyber AGs are mainly related to its coupled physical AGs, degree and dependency.

When analyzing the power-flow process, we ignore the internal complex changes in the power system and directly analyze the results by using active power P and reactive power Q as the load of the substations. When the load $L_i = \sqrt{P_i^2 + Q_i^2}$ of a substation i is over a certain threshold range $[(1-\alpha)*L_i, (1+\alpha)*L_i]$, it will fail due to overload. However, the threshold value relies on the capacity of the network and reflects the robustness of the network itself. This means that an overloaded or underloaded substation will malfunction, triggering the load redistribution again. By simulating attacker behavior, it is possible to reveal which substations are likely to cause more substation failure; in this way, these substations can easily be highlighted as AGs. The malfunction condition of a substation occurs when the substation's load exceeds the network's capacity.

The impact of nodes is used to describe the importance of each substation. A larger impact represents a failed node can cause more node failures. Therefore, Failure Node Set (FNS) is defined as a collection of failed nodes caused by the failure of a substation k, which is used to evaluate the impact of the failed substation k. Differences in parameter α may result in a different FNS. We use Formula 1 to assess the impact of the substations and adopt the average of IM under all tolerance parameters in order to evaluate physical AGs.

$$IM_i^{\alpha_j} = \begin{cases} n(FNS_i^{\alpha_j}), FNS_i^{\alpha_j} = FNS_{Max}^{\alpha_j} \\ n(FNS_i^{\alpha_j}) - n(FNS_{Max}^{\alpha_j} \bigcap FNS_i^{\alpha_j}), otherwise \end{cases} \tag{1}$$

where $n(FNS_i^{\alpha_j})$ represents the size of FNS of substation i under tolerance parameter α_j. $FNS_{Max}^{\alpha_j} \bigcap FNS_i^{\alpha_j}$ is the intersection of $FNS_{Max}^{\alpha_j}$ and $FNS_i^{\alpha_j}$. If FNS of substation i is contained by a maximum FNS, it is insignificant and will not be used as an attack goal. It means that a substation with a larger $n(FNS_i)$ and a smaller $n(FNS_{Max} \bigcap FNS_i)$ has a bigger probability of node i being attacked. Hence, the probability of nodes being physical AGs is described as:

$$Prob_i^P = \mu * \frac{ID_i^+}{ID_i^-} * \frac{1}{M} \sum_{j=1}^{M} Im_i^{\alpha_j} \tag{2}$$

where $Prob_i^P$ denotes the probability that substation i will be selected as an AG for the power grid, and ID_i^+ and ID_i^- represent dependence out-degree and dependence in-degree of node i, respectively. α_j denotes a tolerance parameter j that reflects the capacity of the network to deal with overload. M denotes the number of tolerance parameters.

The ultimate goal of the attackers is to destroy physical devices, and they may select cyber AGs in order to control the failure of these physical AGs. Because the coupling relationship is the same, we will use Formula 1 as the function of AGs in the communication network.

Due to the coupling relationship, cyber nodes that control key substations become more important. The large-degree nodes are usually transfer stations for information collection and data transmission via the communication network and are of great significance to network security. As such, these factors should be taken into account when calculating the probability of nodes being cyber AGs in a communication network:

$$Prob_i^C = \mu * \frac{ID_i^+}{ID_i^-} * \frac{D_i}{D_{Max}} * \sum_{CR_{ij}=1} Prob_j^P \tag{3}$$

where $Prob_i^C$ represents the probability that cyber node i will be an AG in a communication network. D_i denotes the degree of node i, while D_{Max} is the maximum value of the degree of all nodes. $CR_{ij} = 1$ represents a coupling link from cyber node i to physical node j.

Based on the probability of nodes being AGs, the attacker may choose the cyber and physical nodes with higher probability as the target of CCPA. By simulating the CCPA scenarios in real situations, we design two attack strategies: optimization attack strategy (OAS) and saturation attack strategy (SAS). The OAS is to select the least AGs to maximize the attack effect when attacking the same number of AGs. The SAS is to cover every cyber and physical AGs without redundant attack, and select the least number of AGs to make the attack achieve the effect of attacking all AGs. Cyber attack (CA) refers to the invasion, attack and destruction of important nodes in information system by means of information technology. However, Physical attack (PA) refers to the use of violent means, special tools or weapons to destroy important power stations in the power network one by one.

1.2 CCPA Based on Optimal Attack Strategy

The optimal attack strategy for coordinated cyber physical attacks is to find the optimal attack sequence and achieve the effect of exceeding cyber attacks or physical attacks. The CCPA based on optimal attack strategy algorithm is designed to generate attack sequences by searching for cyber or physical AGs. From the perspective of the attack effect, the number of nodes in attack sequence is less than that of cyber attack or physical attack, and its attack effect is better than that of cyber attack and physical attacks.

We design the objective function $Max\{*\}$ of the OAS to satisfy the conditions of formulas (5)–(9). The purpose of formula (4) is to find the attack sequence with the best attack effect when attacking the same number of AGs. Formula (5) represents the same number of cyber and physical nodes being attacked, which is a constant.

$$Max(OA(t^P + t^C)) = Max(f^P + f^C) \tag{4}$$

s.t.

$$num(t^P + t^C) - m \tag{5}$$

$$0 < num(t^P) \leq N \tag{6}$$

$$0 < num(t^C) \leq N \tag{7}$$

$$OA(t^P + t^C) \geq PA(t_1^P), num(t_1^P) = m \tag{8}$$

$$OA(t^P + t^C) \geq CA(t_1^C), num(t_1^C) = m \tag{9}$$

Where f^P and f^C represent the number of failed physical nodes and failed cyber nodes, respectively. t^P and t^C denote the set of the physical and cyber AGs, respectively. m is the number of cyber and physical AGs. $num(*)$ denotes the number of $*$. $OA(*)$, $CA(*)$ and $PA(*)$ represent the attack effect of CCPA Based on optimal attack strategy, cyber attack and physical attack of $*$, respectively. t_1^* represents a set of $*$ different from t^*.

The main steps of Algorithm 1 are as follows:

Step 1: Initialization. The first N cyber AGs are taken as the cyber candidate sequence CyberAGs in order of degree. The first L physical AGs are taken as the physical candidate sequence PhysicalAGs, and the coupling relationship matrix CR_{ij} is obtained.
Step 2: Traverse all nodes of the physical candidate sequence. If there are cyber nodes coupled with it that belong to the information candidate sequence CyberAGs, then remove these nodes from the CyberAGs, named as delete-CyberAGs( ).
Step 3: Traversing the cyber candidate sequence CyberAGs, if the CyberAGs contains nodes AG_i^P and meets the condition of $CR_{AG_i^C AG_j^P} = 1$, and then removing them from the physical candidate sequence PhysicalAGs.
Step 4: Cyber physical attack sequence: attacksequence is equal to Cyber-AGs+PhysicalAG

Algorithm 1: The CCPA based on optimal attack strategy is used to identify the attack sequence of cyber and physical AGs.

Input: $AG^P = (p_1, p_2, ..., p_m)$,$D^C$,$CR_{ij}$,$L$
Output: attacksequence

```
deletecyberAGs ← null
deletephysicalAGs ← null
cyberAGs ← null
physicalAGs ← null
attacksequence ← null
for i = 1;i < N;i++ do
    cyberAGs.add(D_i^C)
    physicalAGs.add(AG_i^P)
end
for i = 1;i < N;i++ do
    for i = 1;i < L;i++ do
        if CR_ij == 1 then
            deletecyberAGs.add(AG_j^C)
        end
    end
end
cyberAGs.removeAll(deletecyberAGs)
for i = 0;i < N;i++ do
    for j = 0;j < L;j++ do
        if cyberAGs.contain(AG_j^C) && CR_ji == 1 then
            deletephysicalAGs.add(AG_i^P)
        end
    end
end
physicalAGs.removeAll(deletephysicalAGs)
attacksequence ← cyberAGs + physicalAGs
```

1.3 CCPA Based on Saturation Attack Strategy

A saturation attack strategy of CCPA involves an attack sequence of cyber and physical AGs. Saturation attacks strategy cover the whole range of AGs, excluding repeated attacks. Such an attack sequence does not contain redundant attacks where the same AG is repeatedly attacked or where two AGs are attacked to produce the same attack effect.

We design the objective function $Min\{*\}$ of the SAS to satisfy the conditions of formulas (11)–(15). The purpose of formula (10) is to find the attack sequence of the least number of AGs to make the attack achieve the effect of attacking all AGs without redundant attack.

$$Min(num(t^P + t^C)) \tag{10}$$

s.t.

$$SA(t^P + t^C) = PA(t_2^P) + CA(t_2^C) \tag{11}$$

$$0 < t^P \leq N \tag{12}$$

$$0 < t^C \leq N \tag{13}$$

$$num(t_2^P) = m \tag{14}$$

$$num(t_2^C) = m \tag{15}$$

Where f^P and f^C represent the number of failed physical nodes and failed cyber nodes, respectively. t^P and t^C denote the set of the physical and cyber AGs, respectively. m is the number of cyber and physical AGs. $num(*)$ denotes the number of $*$. $SA(*)$, $CA(*)$ and $PA(*)$ represent the attack effect of CCPA Based on saturation attack strategy, cyber attack and physical attack of $*$, respectively. t_2^* represents a set of $*$ different from t^*.

CCPA based on saturation attack strategy is an attack mode that covers all attack goals, and it eliminates redundant attacks and repeated attacks. A redundant attack means that two different attack targets produce the same attack effect, and a repeated attack means that one target is attacked multiple times. Algorithm 2 describes the attack sequence generation process for a saturated attack. The steps are as follows:

Step 1. Initialization n cyber AGs, cyber candidate sequence CyberAGs=null, the first n physical AGs, physical candidate sequence PhysicalAGs=null, the coupling relationship matrix CR_{ij} is obtained.
Step 2. Traversing all the nodes of the cyber candidate sequence CyberAGs, traversing all the nodes of the physical candidate sequence PhysicalAGs, if the condition meets $CR_{AG_i^C AG_j^P} = 1$, removing the physical node j from the deleteAGs and removing failure node set getFNS(AG_j^P) that contains physical node j from the node set deleteAGs.
Step 3. Traverse all the remaining nodes of the physical candidate sequence PhysicalAGs. If the deleteAGs does not contain nodes AG_j^C, the physical candidate sequence PhysicalAGs will be added.
Step 4. Cyber physical attack sequence: attacksequence is equal to $CyberAGs + PhysicalAGs$.

2 Experiments and Analysis

In order to verify the effectiveness of CCPA based on different attack strategies, we used a fraction of the practical smart grid as experimental data. The smart grid consists of power grid and communication network, in which cyber nodes are coupled with physical nodes by two-way coupling link with one-to-one corresponding. Figure 1(a) shows that the power grids is composed of 154 substations and 192 transmission lines. Here, square nodes represent generators and circular nodes represent substations. The communication network is constructed by 154 cyber nodes and 153 communication lines in Fig. 1(b), in which the control centers are represented by square nodes and monitoring/controlling nodes are represented by circular nodes.

Algorithm 2: The CCPA based on saturation attack strategy is used to identify the attack sequence of cyber and physical AGs.

```
Input: AG^P = (p_1, p_2, ..., p_m), AG^C = (c_1, c_2, ..., c_n),
CR_ij
Output: attacksequence
deletephysicalAGs ← null
cyberAGs ← null
physicalAGs ← null
attacksequence ← null
for i = 1;i < AG^C.length;i ++ do
    cyberAGs.add(D_i^C)
    for j = 0;j < AG^P.length;j ++ do
        if CR_ij == 1 then
            deletephysicalAGs.add(AG_j^P)
            deletephysicalAGs.addAll(getFNS(AG_j^P))
        end
    end
end
for j = 0;j < AG^P.length;j ++ do
    if !deletephysicalAGs.contain(AG_j^P) then
        physicalAGs.add(AG_j^P)
    end
end
attacksequence ← cyberAGs + physicalAGs
```

No matter what attack strategy or method is adopted, the first thing an attacker has to do is to identify the attack goals. In order to achieve the desired attack effect, the appropriate attack goals should be chosen based on different attack strategies. According to the formula (2), we can get the probability of substations being physical AGs in Fig. 2(a). The greater the probability of the node, the easier it is to be selected as the target of attack. Similarly, the probability of cyber nodes being AGs can be computed by the formula (3) in Fig. 2(b). In the case of limited attack resources, an attacker may select a certain proportion of nodes as attack goals according to his own situation.

The optimal attack strategy is to find an optimal attack sequence when the same proportion of nodes are attacked, so that the attack effect of CCPA is better than other attacks under the same conditions. Figure 3(a) shows that the network layers and distribution of physical and cyber AGs. Labels "C" and "P" represent cyber and physical AGs, respectively. The x-axis represents the distribution of cyber or physical AGs between nodes 1 and 154. We assume that 10% of nodes are selected as targets in the case of limited resources, so 15 cyber AGs and 15 physical AGs can be found by Algorithm 1. The attack sequence consists of cyber AGs and Physical AGs. When the same proportion of nodes are attacked, the CCPA based on OAS has a significantly better attack effect than PA and CA, as shown in Fig. 3(b).

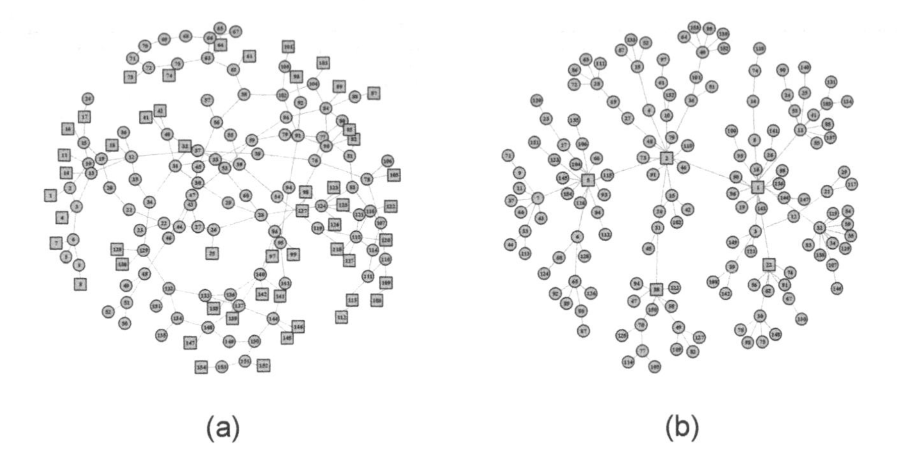

Fig. 1. The network structure diagram of smart grid. (a) Power grid. (b) Communication network.

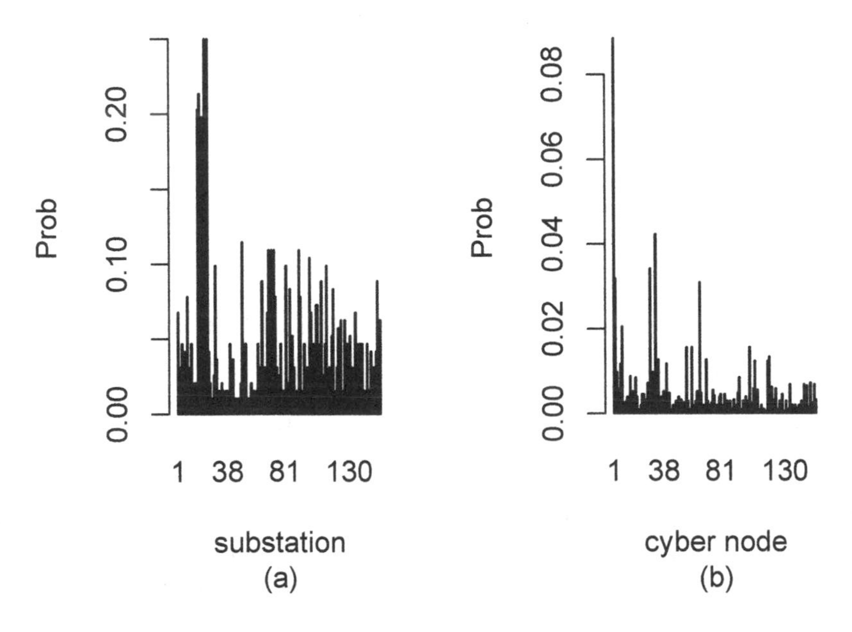

Fig. 2. (a) The probability of substations being physical AGs. (b) The probability of nodes being cyber AGs.

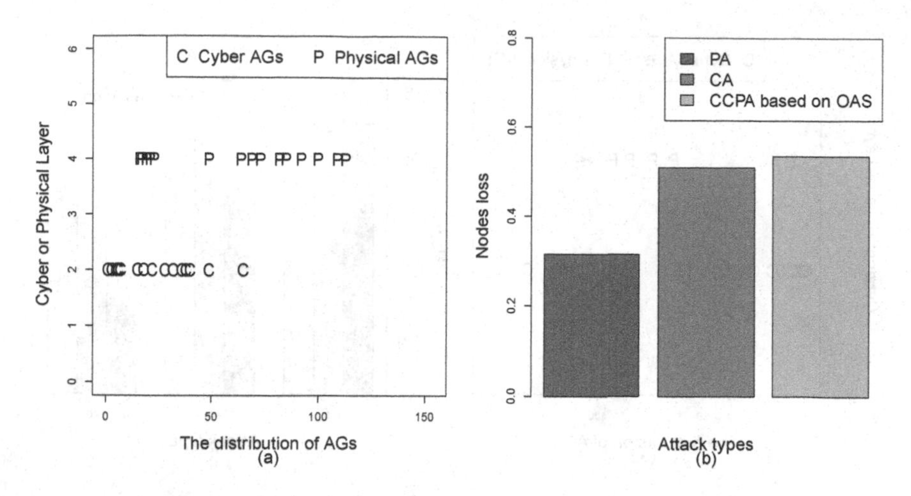

Fig. 3. The CCPA based optimal attack strategy. (a) The network layers and distribution of AGs. (b) Comparison of attack effects of cyber-attack (CA), physical attack (PA) and CCPA based on OAS.

The saturation attack strategy is to cover each node as much as possible without considering resource consumption or having sufficient resources, while avoiding redundant attacks. According to Algorithm 2, 30 cyber AGs and 9 physical AGs can be identified in Fig. 4(a), which has the same attack effect with 30 cyber AGs and 30 physical AGs. The tolerance α has a greater impact on the robustness of smart grid, so we calculate node loss of CCPA based on SAS to compare with PA and CA. It is clear that CCPA based on SAS has better attack effect than CA and PA regardless of $\alpha = 0.3$,$\alpha = 0.4$ and $\alpha = 0.5$ in Fig. 4(b).

In order to better show the node loss under different attack types, we show the curve of the cascading failure process by attacking the nodes one by one in Fig. 5. The black curve, red curve and blue curve represent the process of node loss under CCPA based on SAS, CA and PA, respectively. It is easy to see from Fig. 5(a), (b) and (c) that the ranking of node loss is CCPA based on SAS > CA > PA regardless of $\alpha = 0.3$, $\alpha = 0.4$ and $\alpha = 0.5$. This means that CCPA may become an attack mode that attackers are willing to choose, because it is difficult to defend and has higher attack effects.

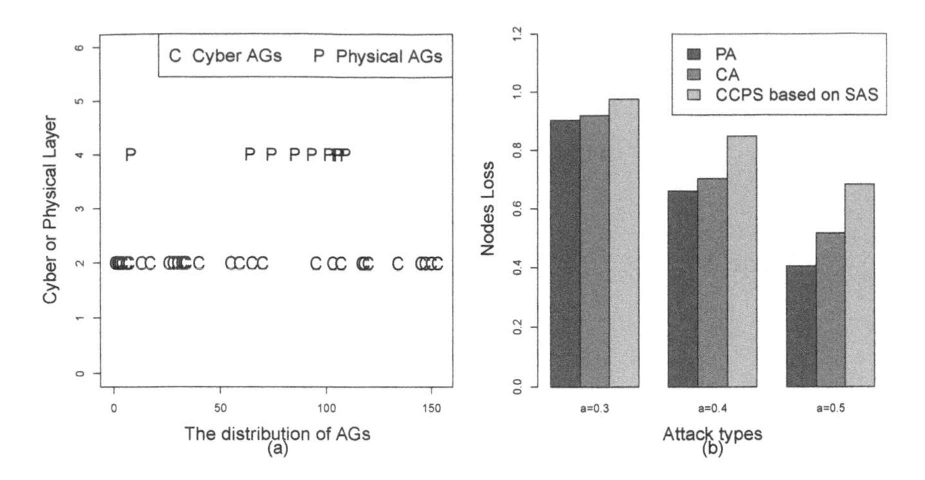

Fig. 4. The CCPA based saturation attack strategy (a) The network layers and distribution of AGs. (b) Comparison of attack effects of cyber-attack (CA), physical attack (PA) and CCPA based on SAS under different tolerances α.

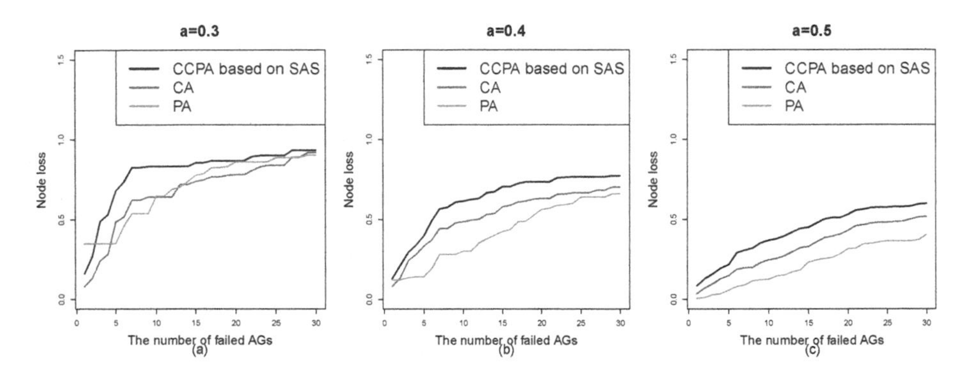

Fig. 5. The comparison of node loss of CA, PA and CCPA based on SAS. (a) $\alpha = 0.3$. (b) $\alpha = 0.4$. (c) $\alpha = 0.5$

3 Conclusion

In this paper, the optimal attack strategy and saturation attack strategy are proposed and applied to CCPA. Through experiments, we can draw the following conclusions: 1)When the same number of nodes are attacked, a set of attack sequence can always be identified by OAS to achieve a higher attack effect; 2) In order to achieve the same attack effect, the saturated attack strategy can find the attack sequence of fewer nodes regardless of $\alpha = 0.3$, $\alpha = 0.4$ and $\alpha = 0.5$; 3) whether based on OAS or SAS, the CCPA has a better attack effect than CA and PA.

In fact, Coordinated Cyber-Physical Attacks not only include the coordination of different attack strategies and means, but also the coordination of the multi-directional and multi-targets. This require us to find an effective integrated defense mechanism or strategy to deal with CCPAs. At the same time, we also found an unconventional phenomenon from the experimental results, that is, different attack sequences composed of the same cyber and physical nodes have different attack effects on the cascading failure of smart grid.

In the future, there are a few points worthy of our in-depth study as follows:1) Research on analyzing the effect of different attack sequences on cyber-physical systems; 2) Research and identification of multiple types of attack strategies; 3) Research on cooperative defense model against CCPA.

References

1. Kang, W.J., Zhu, P.D., Hu, G., Hang, Z., Liu, X.: Cross-layer attack path exploration for smart grid based on knowledge of target network. In: Liu, W., Giunchiglia, F., Yang, B. (eds.) KSEM 2018. LNCS (LNAI), vol. 11061, pp. 433–441. Springer, Cham (2018). https://doi.org/10.1007/978-3-319-99365-2_38
2. Smith, R.: Assault on California power station raises alarm on potential for terrorism. Wall Street J. http://online.wsj.com/news/articles/SB10001424052702304851104579359141941621778
3. Yang, Q., Yang, J., Yu, W., An, D., Zhang, N., Zhao, W.: On false data-injection attacks against power system state estimation: modeling and countermeasures. IEEE Trans. Parallel Distrib. Syst. **25**(3), 717–729 (2014)
4. Xiang, Y., Wang, L., Yu, D., Liu, N.: Coordinated attacks against power grids: load redistribution attack coordinating with generator and line attacks. In: IEEE Power Energy Society General Meeting, pp. 1–5 (2015)
5. Li, Z., Shahidehpour, M., Alabdulwahab, A., Abusorrah, A.: Bilevel model for analyzing coordinated cyber-physical attacks on power systems. IEEE Trans. Smart Grid **7**(5), 2260–2272 (2016)
6. Liu, S., Mashayekh, S., Kundur, D.: A framework for modeling cyber-physical switching attacks in smart grid. IEEE Trans. Emerg. Top. Comput. **1**(2), 273–285 (2013)
7. Deng, R., Zhuang, P., Liang, H.: CCPA: coordinated cyber-physical attacks and countermeasures in smart grid. IEEE Trans. Smart Grid **8**(5), 2420–2430 (2017)
8. Li, Z., Shahidehpour, M., Abdulwhab, A., et al.: Analyzing locally coordinated cyber-physical attacks for undetectable line outages. IEEE Trans. Smart Grid **9**(1), 35–47 (2017)
9. Liu, S., Feng, X., Kundur, D., et al.: Switched system models for coordinated cyber-physical attack construction and simulation. In: 2011 IEEE First International Workshop on Smart Grid Modeling and Simulation (SGMS), pp. 49-54. IEEE (2011)
10. Wei, L., Sarwat, A., Saad, W., et al.: Stochastic games for power grid protection against coordinated cyber-physical attacks. IEEE Trans. Smart Grid **9**(99), 684–694 (2016)
11. Tian, M., Cui, M., Dong, Z., et al.: Multilevel programming-based coordinated cyber physical attacks and countermeasures in smart grid. IEEE Access **7**, 9836–9847 (2019)

12. Lakshminarayana, S., Belmega, E.V., Poor, H.V.: Moving-target defense for detecting coordinated cyber-physical attacks in power grids. IEEE Access **7**, 9836–9847 (2019)
13. Kang, W., Zhu, P., Liu, X.: Integrated defense mechanism based on attack goals against three attack strategies in smart grid. In: IEEE INFOCOM 2020 - IEEE Conference on Computer Communications Workshops (INFOCOM WKSHPS), Toronto, ON, Canada, pp. 1027–1032 (2020)

Private Cloud in 6G Networks: A Study from the Total Cost of Ownership Perspective

Yuanfang Chi[1], Wei Dai[2], Yuan Fan[1], Jun Ruan[1], Kai Hwang[2,3], and Wei Cai[2,3](✉)

[1] Alibaba Group, Beijing, China
{cyf170506,fanxi.fy,jun.ruanj}@alibaba-inc.com
[2] The Chinese University of Hong Kong, Shenzhen, Shenzhen, China
[3] Shenzhen Institute of Artificial Intelligence and Robotics for Society, Shenzhen, China
weidai@link.cuhk.edu.cn, {hwangkai,caiwei}@cuhk.edu.cn

Abstract. Security and privacy concerns are increasingly important when massive data processing and transferring becomes a reality in the era of the Sixth Generation (6G) Networks. Under this circumstance, it becomes a trend that the enterprises tend to host their data and services on private clouds dedicated to their own use, rather than the public cloud services. However, in contrary to the well-investigated total cost of ownership (TCO) for public clouds, the analytic research on the cost of purchase and operation for private clouds is still a blank. In this work, we first review the state-of-the-art TCO literature to summarize the models, tools, and cost optimization techniques for public clouds. Based on our survey, we envision the TCO modeling and optimization for private clouds by comparing the differences of features between public and private clouds.

Keywords: Cloud computing · Total cost of ownership · Case study

1 Introduction

It took ten years for the network infrastructure to evolve from 4th generation (4G) to 5th generation (5G). According to Moore's Law, the expected waiting time for the 6th generation (6G) network will be even shorter. According to predictions, the bandwidth will be increased to terabytes, with which the users are able to exchange a large volume of data in a short time. Under this circumstance, security and privacy are increasingly important, since the risks of data leakage and the potential damages to service hijack are dramatically grown at the same time.

This work was supported by Project 61902333 supported by National Natural Science Foundation of China, by the Shenzhen Institute of Artificial Intelligence and Robotics for Society (AIRS).

X. Wang et al. (Eds.): 6GN 2020, LNICST 337, pp. 397–413, 2020.
https://doi.org/10.1007/978-3-030-63941-9_30

Meanwhile, cloud computing has been widely adopted by startups and well-established enterprises, thanks to its prominent elastic and on-demand features. The flexibility of on-demand usage reduces development, service deployment, and maintenance costs. Huge data computation, backup, and recovery tasks have become easier with the cloud. The cloud computing market is expanding. Many large, medium, and small enterprises will choose to invest in cloud services [24] to enjoy the benefits of cloud computing, but risks often accompany the investment. For a company that provides cloud services or uses cloud services, how to stabilize development and revenue is also a key success factor. On the one hand, operating an enterprise with cloud-based IT infrastructure can improve the stability of the enterprise and bring intuitive benefits. A number of articles in the financial and business domain have reported the cloud computing business models, such as the VE model [46] and Cloud Business Model Framework (CBMF) [69]. More discussions and studies about business models can be found at [5,7,13,16,38,40,48]. On the other hand, analyzing Return On Investment (ROI) can also increase the value of investment. ROI analysis and modeling methods in cloud computing have been discussed in [4,37,64]. Cost analysis and optimization modeling can effectively help decision-making and increase profit under the premise of guaranteeing service quality [39].

Nevertheless, the privacy and security problems of cloud services are never fully solved. The tenants have no choice but surrender their data and programs to the public cloud providers, who may not be trustworthy if the data is extremely sensitive. To address this concern, cloud providers propose private cloud service, which is typically deployed inside the organization and is typically behind the firewall [74]. Private cloud enhances the enterprise customers' confidence to embrace the cloud since it is an isolated solution that can prevent data from being transmitted to the public network. It maximizes the customers' control over their data, provides improved security and service quality [52]. In addition, the local network within a private cloud also reduces network latency and improves transmission quality. The isolation, low latency, and security of private cloud make many applications possible to deploy on the cloud, such as the Industrial Internet of Things (IIoT) [9], Federal Learning [29], and Cluster Learning [35].

From the users' perspective, the cost efficiency is always a key metric when considering cloudization. On the other hand, cloud providers also need to evaluate their profits in the cloud business. Hence, the total cost of ownership (TCO), a standard approach in analyzing the cost of purchase and operation of an asset, is a critical topic for both parties. By definition, the TCO in cloud computing covers the capital and operational costs of building, using, and maintaining a cloud data center. It involves all direct and indirect costs, such as the cost of cloud service providers purchasing, deploying, operating, and maintaining assets, and the cost of cloud users renting cloud resources. Due to the importance of TCO, the analysis and optimization methods for public cloud and hybrid cloud have been well investigated. However, few work has been devoted to the cost analysis associated with private clouds. In this work, we survey the literature on public cloud TCO analysis methods and the optimization solutions that minimize the TCO of a cloud service. We compare and contrast the similarities and differences between public and private clouds from a TCO perspective and provide an in-depth discussion on what else can be further optimized to accomplish a lower TCO in a private cloud.

The remainder of the paper is organized as follows. We present an overall review of existing works on TCO in Sect. 2. Afterwards, we compare the public and private cloud from TCO optimization perspective in Sect. 3. Section 4 concludes the work and envision the TCO research for private clouds.

2 Literature Review

TCO was first being studied in the business domain by Ellram since 1993 [15]. She claimed that TCO modeling is constructive for business parchment decision making, ongoing supplier management, and understanding of indirect costs. However, she also pointed out that the quantification and measurement of TCO is complexity [14]. In 1999, Milligan suggested that accurate total cost measurement is elusive due to the lack of analysis methods [43]. Thus, one of the leading research directions in TCO is to find an accurate modeling and measurement method. Later, Degraeve and Roodhooft illustrated the application of TCO in the supply chain domain. They split purchasing activities into three levels corresponding to three different expenses. Then, they built a mathematical model to optimally select suppliers such that the total cost of ownership is minimized [72]. As the IT industry rising in 2004, few scholars began to study TCO in the information technology domain. They have worked on topics such as revenue analysis in IT by taking advantage of TCO [4], how to utilize TCO to decide whether to adopt open-source software in IT companies [31,45], and optimization models to decrease the cost of IT infrastructure [1]. After cloud computing entered the market, some researchers analyzed the business value of cloud computing and mentioned that cloud providers should pay attention to the cost [27], while few studies took the modeling of TCO in cloud computing in to consideration. Although Patel and Shah presented a cost model to calculate the cost of building a data center, they did not take indirect costs such as maintenance cost, operation cost, and labor cost into consideration [50]. TCO in cloud computing was finally studied by Li et al. in 2009. A cloud cost amortization model is built to make it possible to calculate the direct and indirect cost of cloud computing infrastructures. Their model can be divided into eight parts: server cost, software cost, network cost, support and maintenance cost, power cost, cooling cost, facilities cost, and real-estate cost. They also implemented an interactive tool for cloud providers to calculate cloud TCO [33]. Scholars begin to pay much more attention to models and tools to analyze TCO in cloud computing and the optimization to reduce TCO.

In order to give the readers better comprehension on the landscape of the cloud computing TCO research, we classified the surveyed papers into groups as shown in Fig. 1. The selected papers are mainly divided into two groups: models and tools for cloud TCO, and cost optimization methods. For the papers related to models and tools, we will further divide into two subclasses from the perspective of cloud computing providers and cloud computing users respectively. For the papers related to cost optimization methods, we will further classify them into three topics: task scheduling, resource scheduling, and heterogeneous resource optimization. More details on the classification system are in the following.

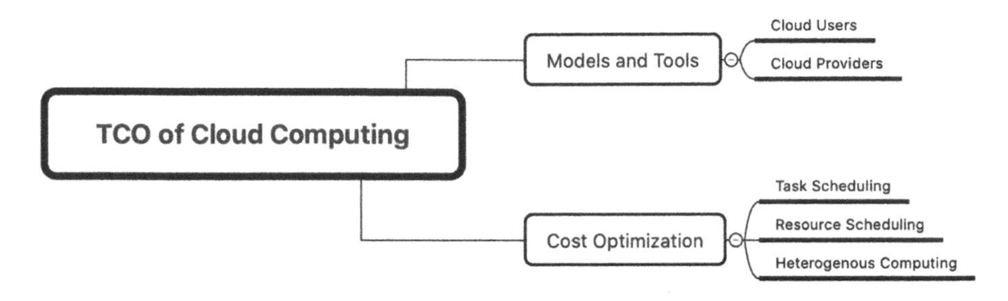

Fig. 1. Our proposed classification system of cloud computing TCO papers.

- Models and Tools for Cloud Computing TCO (Sect. 2.1): We consider papers that provide modeling methods or tools to measure or calculate TCO for cloud computing providers or users. These studies can be further categorized into two classes based on their viewpoints: models and tools for cloud providers and models and tools for cloud users.
- Cost optimization (Sect. 2.2): We consider papers that optimize cloud computing TCO by optimizing one aspect of resources in cloud computing. Each work focus on different optimization directions. Such studies can be further categorized into three groups: task scheduling, resource scheduling, and heterogeneous resource optimization.

2.1 Models and Tools

In cloud computing, a model and tool that measures the total cost of ownership can help analyze investment risk, manage cloud resources, and improve the effectiveness of decision making. The modeling and tools related papers are roughly categorized into two classes: (i) Modeling and tools for Cloud Providers and (ii) Modeling and tools for Cloud Users. They are surveyed in the following.

Models and Tools for Cloud Providers. The cost has been known as one of the most important factors for cloud providers. Moreover, there is significant capital consumption, such as cloud data center construction, service software development, maintenance. So, cost evaluation will be helpful for revenue estimation, investment risk analysis, and decision making. From the perspective of cloud providers, the research of cloud TCO modeling and tools can be further classified into two subclasses, (i) TCO breakdown and (ii) energy cost models and tools.

TCO Break Down. The primary purpose of cloud total cost of ownership breakdown research is to provide cloud providers with a measurement of cost. It focuses on the modeling approaches of server cost, software cost, network cost, support and maintenance cost, power cost, cooling cost, facilities cost, and real-estate cost.

In 2009, Li et al., as the pioneers, first solved the problem that the lack of a method to measure TCO. They proposed a cloud cost amortization model to calculate the total cost, and further use the model to find VM utilization cost given the number of running VMs [33]. Later, researchers studied cloud TCO breakdown under different business models. In 2015, Filiopoulou et al. applied the system of system (SoS) method to reorganize each part of the total cost. They regraded each type of cost as a subsystem of cloud computing and illustrated a cost modeling framework for each subsystem [17]. In 2016, Simonet et al. demonstrated a TCO breakdown modeling method for distributed cloud computing (DCC). Besides, they categorized DCC actors into five levels and presented related cost models [56]. It provides a detailed reference for cloud providers who intend to invest distributed could computing.

Through the above studies, researchers showed an overview of cloud TCO breakdown. However, they either overlooked some detailed cost measurements for some terms in the formulation or overlooked the cost calculation under some specific business scenarios. In real business, the cost can be significantly altered by different business scenarios and other dynamic factors such as application migration, timely changed resource utilization rate, and satisfactory network dependability and availability. In the rest of the subsection, we will review researches that focus on cost modeling for one aspect of TCO breakdown.

In 2011, Mencer et al. demonstrated a cost model for software development in cloud computing. They believed that the computational efficiency and development cost of the software would be affected by different abstraction level programming languages. By analyzing software development cost and TCO, they concluded that TCO is not necessarily minimized by minimizing programming effort because programming efforts might be related to conditions of servers [41]. In 2013, Sun and Li illustrated a labor cost model to calculate labor efforts in a service migration scenario. They quantified the skill level of employees and built a probabilistic model to estimate the number of persons needed per day to finish migration [61]. Omana et al. proposed an analyzing method for cloud providers to determine whether to replace aged assets. By measuring the cost of power, cooling, physical space, asset attachments, and IT support, they utilized statistical methods to examine the relationship between asset cost and resource capacities, such as the number of CPU cores and memory. They finally mined some useful results to help decision making and device management [24].

Thanakornworakij et al. did a cloud TCO research on the premise of ensuring system availability and satisfying service-level agreement (SLA), and quality of service (QoS). They modeled the cost of servers, network, software, power, cooling, facilities, maintenance, and availability. Then, they obtained the relationship between the number of devices needed and revenue under 99% system availability, which is beneficial for cloud providers to define the right size of the data center [62]. Sousa's research team presented a modeling strategy under a similar scenario. They introduced Mean Time to Failure (MTTF) and Mean Time to Repair (MTTR) to build a detailed maintenance cost model. They further derived a Stochastic Model Generator for Cloud Infrastructure Planning (SMG4CIP) which concerned both dependability and cost requirements [58,59].

Furthermore, some studies presented modeling methods for dynamic factors in cloud TCO such as resource utilization cost. Although [33] tried to use the ratio of running VMs to model utilization cost at a given time point, they can only get a rough result because the cost of running VMs depends on other frequently changed variables like CPU rate. As a follow-up study of [33], in 2013, Vrček and Brumec pointed out that many cloud TCO related studies ignored some hidden variables to build cost models, such as CPU utilization rate, data transmission rate, system load. They utilized CPU rate per hour to build a cost model and analyzed how the CPU rate affects the total cost in cloud computing [66]. Their study offered some vital insight into cloud computing TCO. Since time dimension can be added into the cost model, cloud providers can generate flexible strategy by analyzing per hour cost. Molka and Byrne had a similar idea that an accurate prediction model of resource utilization rate can derive an accurate utilization cost calculation. They demonstrated a prediction model using an online nonlinear autoregressive method and then formulated a model to calculate the real-time cost using CPU utilization data [44]. If there are some more accurate prediction model that can be applied and plugged in [44], cloud providers can obtain more accurate information to make decisions. Some research about resource prediction can be found in [54]. In 2017, Singh et al. also took account of the resource utilization rate to calculate utilization cost. They captured the effect of the relinquishment of cost and revenue [57].

Energy Cost Modeling and Tools. With the development of cloud computing and its convenience, the energy and electricity power consumed by cloud data center per year increase dramatically [51], which leads to a higher expense to operate a data center. An investigation indicated that energy cost account for 42% of all expenditure per month in data center until 2011 [22]. Thus, energy cost, as part of the total cost of ownership, gradually becomes a focus of cloud providers' concern. We survey some representative energy cost modeling methods and tools as follows.

In 2010, Yu and Bhatti thought [33] ignored approaches to gain specific values in the energy cost formulas. They pointed out there is a lack of an information model that collects energy usage and resource usage data for devices for cloud providers to manage assets. They proposed a Scalable Energy Monitor (SEM) architecture to measure energy consumption and further to calculate energy cost [71]. Later in 2012, Uchechukwu et al. studied a detailed energy cost modeling method. They divided energy into static and non-static parts where energy consumed by storage, computation, and communication is modeled. By analyzing cost using their measurement, they found it is possible to save energy costs [65]. In the same year, Chen et al. had a similar idea. They built a model with a static part and a dynamic part, and further developed an analytic tool to measure and summarize energy consumed by different tasks [8]. Recently, some researchers focus on establishing more advanced models. In 2018, Jawad et al. illustrated a smart Power Management Model (PMM) to schedule power by adapting a Nonlinear Autoregressive Network with Exogenous Inputs and Neural Network based forecasting algorithm (NARX-NN) [26].

Models and Tools for Cloud Users. In general, cloud users are small companies or startups who rent cloud services from cloud providers to operate their businesses. There are three types of service in cloud computing: Infrastructure as a Service (IaaS) that provides computation power and data storage, Platform as a Service (PaaS) that provides developers platforms and Software as a Service (SaaS) that allows users access software service by light-weight clients [40]. Besides, there are different cloud providers in the market with different pricing models [55], such as Amazon EC2, Microsoft Azure, Google Compute Platform. It is not trivial for cloud users to choose the optimal leasing solution from the intricate market. Several representative papers are selected to illustrate methods and tools to analyze TCO from the perspective of cloud users.

In 2009, Kondo et al. first found monthly cost can be decreased drastically by deploying tasks on cloud servers instead of on Volunteer Computing (VC) platform [28]. Later in 2011, Han introduced the concept of cloud TCO from the perspective of cloud users. He presented a detailed analysis of TCO by comparing cloud service with local storage and servers [23]. Inspired by [23,28], in 2012, Martens et al. first contributed the TCO measurement for cloud users. They summarized cost variables to be considered by cloud users when renting cloud services, and proposed cost models for IaaS, PaaS and SaaS [39]. In 2013, Martens and his team continued their research on TCO models. They demonstrated a deployment planning strategy for cloud users and presented more detailed cost models for IaaS, PaaS and SaaS [67].

Additionally, some researchers focused on developing tools and methods that compare different cloud providers and estimate resources acquired by users' services or applications to further provide an optimal leasing plan. Liew and Su proposed a tool called CouldGuide which can predict the cloud computing resources required, and helps users select the most suitable leasing scheme from multiple cloud service products with different pricing modes according to user policy, such as maximizing performance or minimizing cost [34]. Aniceto et al. gave a more detailed deployment strategy for small IT companies as cloud users. They abstracted the resource into the number of instances, and built a statistical model using historical data to predict the number of instances demanded at the moment. The authors further categorized cloud products into reserved instances and on-demand instances according to the pricing model. They finally reduced 32% cost compared with only adapting on-demand deployment by formulating a mixed deployment strategy [47]. In 2018, Ghule and Gopal further proposed a comparison framework for cloud users to analyze the difference between IaaS cloud providers by defining suggestive comparison parameters, such as reliability, performance, serviceability [18].

2.2 Cost Optimization

In above sections, we reviewed modeling methods and tools to measure TCO. It may be not enough for providers or users to make an optimal decision. In order to sustain a thriving business in enterprise, it is an effective way to maximize profit by optimizing and reducing the total cost of ownership. In cloud computing, many papers studied cost optimization from different aspect. We survey and summarize several representative papers from the aspect of task scheduling, resource scheduling and heterogeneous computing.

Task Scheduling. As cloud providers, all users' operation requests will be executed in provider back end. Effectively scheduling tasks requested by users can improve service quality, reduce system latency, and reduce costs. We select several typical studies. They adopted different optimization methods in different application scenarios and finally achieved certain results.

Pandey et al. optimized the general task scheduling problem in cloud computing, and solved the Task-Resource Scheduling Problem by using the Particle Swarm Optimization (PSO)-based Heuristic optimization method. They minimized the cost of task execution to reduce the operating costs of the TCO. The proposed PSO method can achieve lower cost than BRS (best resource selection) algorithm, and the PSO can converge faster than the GA (Genetic Algorithm) [49]. Zhang et al. further considered the scheduling strategy when tasks can be executed parallelly. They proposed the DeCloud architecture which is responsible for scheduling users' data requests from a centralized data storage center. They presented Generic Searching Algorithm (GSA) and Heuris Searching Algorithm (HSA) to schedule parallel and non-parallel tasks respectively. Finally, their method is better than greedy search and random search [73].

Resource Scheduling. As a typical resource scheduling problem in cloud computing, elastic computing has always been a concern of cloud providers. Over-provisioning will lead to lower resource utilization, thus increasing cloud computing operating costs. Under-provisioning will result in unmet user requirements and reduction of service quality. We survey two cost-related resource scheduling studies as below.

Each research team has different focuses and directions on this issue. Wu et al. demonstrated an optimization model to schedule resource in SaaS. They minimized the cost by minimizing resources allocated to VMs while meeting SLA and QoS requirements. Their proposed algorithm can reduce the cost by 50% compared to the base algorithm, ProfminVio [70]. Mao et al. viewed VM (Virtual Machine) as the unit of recourse, and considered several factors including the type of VM, the startup latency of the VM, and the deadline of the task to construct an auto scaling schedule strategy. They performed different types of tasks by scheduling different kinds of VMs, and finally got a lower cost compared with using fixed VM type under the condition of satisfactory performance and deadline [36].

Heterogeneous Computing. Companies may encounter business situations that require a mix of local computing resources and exogenous public cloud computing resources, which is common in hybrid clouds. Some studies have found that it is possible to optimize the cost of reducing the amount of investment required by the enterprise by optimally allocating the workload or task quota for local and heterogeneous resources.

In 2010, Trummer et al. utilized the COP (Constraint optimization problem) method to minimize the rental cost of external cloud services such that the cost of the enterprise is minimized [63]. However, their research did not analyze the cost of local resources. Thus, it cannot give a global optimal cost solution. Bittencourt and Madeira proposed HCOC (Hybrid Cloud Optimized Cost) scheduling algorithm. It enables computing tasks in a hybrid cloud to be dynamically assigned to local resources (private clouds) or external sources (public clouds). Their method can effectively improve the efficiency of task completion, and reduce the cost compared to greedy schedule algorithm [3]. In 2015, Laatikainen et al. focused on the optimization of the storage costs of hybrid clouds. They first formulated the measurement of hybrid cloud storage costs and then analyzed the impact of refining the reassessment interval on the cost savings attainable by using hybrid cloud storage. Finally they obtained that shortening reassessment interval and the acquisition of public cloud storage capacity allows the volume variability to be reduced, yielding a reduction of the overall costs [30].

3 Overall Comparison Between Public and Private Cloud

Many cloud service providers start to deliver the entire data center as a whole cloud computing solution to customers, such as Alibaba Cloud Apsara Stack [21], Dell EMC [25], HPE Helion Open-Stack [11], Microsoft Azure Stack [42], which deploys public cloud software in a smaller private cloud, providing customers with customized, secure, low-latency private cloud services.

In the reviewed works above, most of them are related to cost modeling and optimization methods in public or hybrid clouds while private cloud is seldomly mentioned. In this chapter, we hope that by comparing and contrasting the public and private cloud environment, readers can draw a better understanding of the differences and similarities between public and private cloud, thus better understand the total cost of owning a private cloud and possible ways that can be attempted to optimize the cost. Table 1 gives a brief comparison of characteristic of public and private cloud.

As shown in the table above, both public and private clouds can provide IaaS, PaaS, and SaaS services, but they have significant differences in the other

Table 1. Comparison of characteristics between public cloud and private cloud.

Characteristics	Private cloud	Public cloud
Service Type	IaaS, PaaS, SaaS	IaaS, PaaS, SaaS
Service Users	Large Organizations	General Public or Small Startups
Device Ownership	Users	Cloud Service Providers
Deployment Location	User's Data Center	Cloud Provider's Data Center
Quality of Service	Stable	Unstable
Data Security	High	Low
Use Cost	High	Low
Scalability	Scale-out Only	Elastic Computation
Maintenance	Technician/Service Provider	Public Cloud Service
Procurement	High Customization	Low Customization

nine areas. The users of private cloud services are large organizations, communities and enterprises, such as government departments while public cloud orients to general public users, small companies and startups. In general, private cloud users need to pay more in the initial stage because they need to purchase a complete cloud computing infrastructure, while public cloud users only pay for the rental service in on demand price. For private cloud users, high costs usually are accompanied by stable cloud service quality and high data security. Since the private cloud is deployed in the user's data center, the user can have an isolated network built within the domain of an intranet to stably run the service. Meanwhile, data security is guaranteed because users can manage data autonomously. In contrast, since the public cloud is deployed in the data center of the service provider, the public cloud user needs to access the cloud service through the public network (Internet), and the quality of service often depends on the network quality between the user side and the public cloud data center. Since public cloud users don't know where their data is physically stored, there is a hidden risk of data leakage. In terms of procurement, since private cloud users may need to carry some personalized services, their infrastructure will also need to be customized. Servers in the private cloud can be highly customized where users can select the model specification according to the business requirement, while the public cloud server has a low degree of customization. Generally public providers purchase servers suitable for the multi-tenant technology.

3.1 The Difference from Cloud Provider's Perspective

In this section, we will analyze the similarities and differences between public cloud providers and private cloud providers in analyzing TCO from two aspects.

Procurement, Deployment and Maintenance. When creating a data center strategy, it is hard for public cloud providers to predict the exact user demands. So they need to adopt high-performance computing servers, implement scalable storage and networks to meet the increasing demand for compute-intensive, storage-intensive, or IO bandwidth-intensive workloads. Private clouds, on the other hand, can get their customer requirements directly from their customers so the demands are more predictable than public clouds. However, to better meet each customer's business objects, each private cloud data center has to be customized in a certain level. This customization leads to huge personnel costs for private cloud providers. Furthermore, public cloud is deployed in carefully chosen sites that can be managed and continuously improved by providers, whereas the private cloud is deployed on customers' sites. In order to maintain the quality of service (QoS) and meet the Service Level Agreement (SLA), both public and private cloud service providers need to provide sufficient support on hardware maintenance, disaster recovery or the ability of service scale-in/scale-out, etc. On-site supports are offered by private cloud providers, which also increases the cost of owning a cloud data center for cloud provider's aspect. Therefore, a more flexible planning tools and automated deployment and maintenance tools are crucial for private cloud providers to lower their TCO, compared to public cloud providers.

Resource Scheduling. Public cloud service providers have realized resource pooling and on-demand cloud instances by leveraging virtualization, multi-tenant and elastic computing technology. By optimizing the resource scheduling algorithms, resource utilization is evidently increased. Equipment idle time and equipment operating costs can be reduced. Virtualization technology is also the fundamental in private clouds. Recently, many companies use light-weighted virtual machine, dockers, as the container to carry their applications [2]. Private cloud providers still need to develop resource scheduling strategies to optimally allocate resources to each docker carrying user business applications. The effectiveness of the resource scheduling strategy becomes even more crucial for private cloud since any waste of resource will reflect directly on the customer bill.

3.2 The Similarities and Differences from Cloud User's Perspective

As summarized in Table 1, there is a significant difference between public cloud users and private cloud users in nature. Generally, public cloud users are general public, small companies, and startups, and they usually cannot build a data center of its own due to the lack of funds. The public cloud's pay-as-you-go mode can meet the user's demand for resource on-demand, so that such users can quickly deploy their own business. In contrast, most of private cloud users are governments, large and medium-sized enterprises and communities. They have higher requirements for network service quality and data security, and initially have sufficient funds to purchase data centers. Compared with public cloud users, the cost spent by private cloud users in a short period of time would be much greater than that of public cloud users by nature. According to some studies, in the case of Amazon, if users need to operate their business for more than 18 months, the invested cost of public cloud users may be equal to or more than that of private cloud users [10].

Use and Maintenance. Both public cloud users and private cloud users need to develop their own commercial applications on the top of infrastructure using IaaS. The cost of application development depends on the technology stack and technology architecture used by the users. Some PaaS cloud services enable users to quickly complete application development or allows users to adapt the outsourcing API interface to complete product release. Thus, the development costs can be further reduced. For example, Google App Engine [20] can help users quickly establish a web application development; Cloud Healthcare API [19] can provide an intelligent healthcare interface that enables users to complete the development related to smart healthcare. On the SaaS side, public cloud users can use the software on the cloud to complete business tasks directly, such as Dropbox [12] to complete the task of cloud storage. PaaS and SaaS private cloud users may need to work harder if they choose to develop their own PaaS and SaaS applications for security reasons.

From the perspective of operation and maintenance, public cloud users need to have certain operation and maintenance knowledge of servers and networks. For private cloud users, there are usually professionals from private cloud providers side to sustain the basic operation and maintenance, which greatly reduces the cost of learning and training professionals.

Security and Network Reliability. Security and confidentiality are long-term issues in the public cloud [60]. Since public cloud users' data is stored in remote data centers, users may have concerns about confidentiality of data. If the data center crashes, users will also face data loss issues. Although public cloud multi-tenant technology increases the utilization of hardware resources through virtualization technology, multi-tenant environments may lead to potential data leakage caused by side-channel attacks [53]. The public cloud's latency may also affect the user's profit, if the user's business is latency sensitive [68]. Compared to the public cloud, private cloud users can manage and back up data autonomously because the infrastructure of the private cloud is completely owned by users. Since the network is relatively independent and the network traffic does not go through the public network, the delay will be much lower. Therefore, the risk of private cloud users in terms of security, network reliability and latency will be much lower than that of public cloud users.

Cloud Evaluation. For public cloud users and private cloud users, due to different charging standards of different providers, the choice of cloud service providers directly affects the cost of their investment. For public cloud users, there are some tools to help them choose the service provider and rental solution that best suits their requirements, such as [18,32,34]. However, there are no tools to provide users with a comparison of private cloud providers. This is not a trivial problem because private cloud users need to compare private cloud service offerings in terms of investment costs, business value benefits, security and stability, and IT cost savings. On the other hand, cloud users need to estimate the amount of computing and storage resources consumed by their own business before adopting cloud computing solutions. Public cloud and private cloud users can control their costs by evaluating and predicting the resource requirements. Rodrigo N. Calheiros et al. proposed EMUSIM that can help users evaluate and predict the resources that will be consumed by their own business after migrate to cloud [6].

4 Conclusion

Nowadays, cloud computing is widely adopted for personal or organization use. How to help users and providers better understand their costs of owning a cloud service, so they can further optimize their usage or ways to build the service to lower the overall TCO has become the next question. Many have provided breakdowns of TCO calculation for public cloud, but few has discussed how TCO is different for the private cloud. In this work, by discussing the similarities and differences between public and private clouds in terms of server procurement and deployment, software, operations and maintenance and resource allocation algorithms, we hope to shade a light on the unique points that people can look at for TCO optimization of private clouds.

References

1. Ardagna, D., Francalanci, C., Trubian, M.: A cost-oriented approach for infrastructural design. In: SAC (2004)
2. Bhimani, J., Yang, Z., Leeser, M., Mi, N.: Accelerating big data applications using lightweight virtualization framework on enterprise cloud, 09 2017. https://doi.org/10.1109/HPEC.2017.8091086
3. Bittencourt, L.F., Madeira, E.R.M.: HCOC: a cost optimization algorithm for workflow scheduling in hybrid clouds. J. Internet Serv. Appl. **2**, 207–227 (2011)
4. Brocke, J.V., Lindner, M.: Service portfolio measurement - a framework for evaluating the financial consequences of out-tasking decisions, pp. 203–211, 01 2004. https://doi.org/10.1145/1035167.1035197
5. Buyya, R., Yeo, C.S., Venugopal, S., Broberg, J., Brandic, I.: Cloud computing and emerging it platforms: vision, hype, and reality for delivering computing as the 5th utility. Future Gener. Comput. Syst. **25**, 599–616 (2009)
6. Calheiros, R., Netto, M., De Rose, C., Buyya, R.: EMUSIM: an integrated emulation and simulation environment for modeling, evaluation, and validation of performance of cloud computing applications. Softw. Pract. Exp. **43**, 595–612 (2013). https://doi.org/10.1002/spe.2124
7. Chang, V.I., Wills, G.B., Roure, D.D.: A review of cloud business models and sustainability. In: 2010 IEEE 3rd International Conference on Cloud Computing, pp. 43–50 (2010)
8. Chen, F., Schneider, J., Yang, Y., Grundy, J., He, Q.: An energy consumption model and analysis tool for cloud computing environments. In: 2012 First International Workshop on Green and Sustainable Software (GREENS), pp. 45–50, June 2012. https://doi.org/10.1109/GREENS.2012.6224255
9. Choo, K.R., Gritzalis, S., Park, J.H.: Cryptographic solutions for industrial Internet-of-Things: research challenges and opportunities. IEEE Trans. Industr. Inf. **14**(8), 3567–3569 (2018). https://doi.org/10.1109/TII.2018.2841049

10. Dantas, J., Matos, R.R.M., Araujo, J., Maciel, P.R.M.: Eucalyptus-based private clouds: availability modeling and comparison to the cost of a public cloud. Computing **97**, 1121–1140 (2015)
11. Development, H.P.E.: HPE Helion OpenStack (2019). https://www.hpe.com/us/en/product-catalog/detail/pip.hpe-helion-openstack-cloud-software.1010838414.html. Accessed 15 Oct 2019
12. Dropbox: Dropbox (2019). https://www.dropbox.com/. Accessed 15 Oct 2019
13. Centre for Economics and Business Research Ltd.: The cloud dividend: part one. The economic benefits of cloud computing to business and the wider EMEA economy-France, Germany, Italy, Spain and the UK. Business & Information Systems Engineering (2010)
14. Ellram, L.M.: Total cost of ownership: a key concept in strategic cost management decisions (1993)
15. Ellram, L.M.: Total cost of ownership: elements and implementation (1993)
16. Etro, F.: The economic consequences of the diffusion of cloud computing (2010)
17. Filiopoulou, E., Mitropoulou, P., Tsadimas, A., Michalakelis, C., Nikolaidou, M., Anagnostopoulos, D.: Integrating cost analysis in the cloud: a SoS approach. 2015 11th International Conference on Innovations in Information Technology (IIT), pp. 278–283 (2015)
18. Ghule, D., Gopal, A.: Comparison parameters and evaluation technique to help selection of right IaaS cloud. In: 2018 5th IEEE Uttar Pradesh Section International Conference on Electrical, Electronics and Computer Engineering (UPCON), pp. 1–6, November 2018. https://doi.org/10.1109/UPCON.2018.8597059
19. Google: Cloud Healthcare API (2019). https://cloud.google.com/healthcare/. Accessed 15 Oct 2019
20. Google: Google App Engine (2019). https://cloud.google.com/appengine/. Accessed 15 Oct 2019
21. Group, A.: Apsara Stack (2009). https://www.alibabacloud.com/product/apsara-stack. Accessed 15 Oct 2019
22. Hamilton, J.: Cooperative expendable micro-slice servers (CEMS): low cost, low power servers for internet-scale services, 01 2009
23. Han, Y.: Cloud computing: case studies and total cost of ownership (2011)
24. Iglesias, J.O., Perry, P., Stokes, N., Thorburn, J., Murphy, L.: A cost-capacity analysis for assessing the efficiency of heterogeneous computing assets in an enterprise cloud. In: IEEE/ACM 6th International Conference on Utility and Cloud Computing, UCC 2013, Dresden, Germany, 9–12 December 2013, pp. 107–114. IEEE Computer Society (2013). https://doi.org/10.1109/UCC.2013.32
25. Inc., D.: Dell EMC (2019). https://www.dellemc.com/en-us/solutions/cloud/vmware-cloud-on-dellemc.htm#scroll=off. Accessed 15 Oct 2019
26. Jawad, M., et al.: A robust optimization technique for energy cost minimization of cloud data centers. IEEE Trans. Cloud Comput. 1 (2018). https://doi.org/10.1109/TCC.2018.2879948
27. Klems, M., Nimis, J., Tai, S.: Do clouds compute? a framework for estimating the value of cloud computing. In: WEB (2008)
28. Kondo, D., Javadi, B., Malecot, P., Cappello, F., Anderson, D.P.: Cost-benefit analysis of cloud computing versus desktop grids. In: 2009 IEEE International Symposium on Parallel Distributed Processing, pp. 1–12, May 2009. https://doi.org/10.1109/IPDPS.2009.5160911
29. Konecný, J., McMahan, H.B., Yu, F.X., Richtárik, P., Suresh, A.T., Bacon, D.: Federated learning: strategies for improving communication efficiency. CoRR abs/1610.05492 (2016). http://arxiv.org/abs/1610.05492

30. Laatikainen, G., Tyrväinen, P.: Cost efficiency of hybrid cloud storage cost efficiency of hybrid cloud storage: shortening acquisition cycle to mitigate volume variation, 12 2015
31. Larsen, M.H., Holck, J., Pedersen, M.K.: The challenges of open source software in it adoption: enterprise architecture versus total cost of ownership (2004)
32. Li, A., Yang, X., Kandula, S., Zhang, M.: CloudCmp: comparing public cloud providers. In: Internet Measurement Conference (2010)
33. Li, X., Li, Y., Liu, T., Qiu, J., Wang, F.: The method and tool of cost analysis for cloud computing. In: 2009 IEEE International Conference on Cloud Computing, pp. 93–100, September 2009. https://doi.org/10.1109/CLOUD.2009.84
34. Liew, S., Su, Y.Y.: CloudGuide: helping users estimate cloud deployment cost and performance for legacy web applications, pp. 90–98, 12 2012. https://doi.org/10.1109/CloudCom.2012.6427577
35. Liu, Q., Cheng, L., Ozcelebi, T., Murphy, J., Lukkien, J.: Deep reinforcement learning for IoT network dynamic clustering in edge computing. In: 2019 19th IEEE/ACM International Symposium on Cluster, Cloud and Grid Computing (CCGRID), pp. 600–603, May 2019. https://doi.org/10.1109/CCGRID.2019.00077
36. Mao, M., Li, J., Humphrey, M.: Cloud auto-scaling with deadline and budget constraints. In: 2010 11th IEEE/ACM International Conference on Grid Computing, pp. 41–48, October 2010. https://doi.org/10.1109/GRID.2010.5697966
37. Skilton, M., Director, C., Cloud Business Artifacts Project, et al.: Building return on investment from cloud computing. White Paper (2009)
38. Marston, S., Li, Z., Bandyopadhyay, S., Zhang, J., Ghalsasi, A.: Cloudcomputing - the business perspective. Decis. Support Syst. **51**(1), 176–189 (2011). https://doi.org/10.1016/j.dss.2010.12.006
39. Martens, B., Walterbusch, M., Teuteberg, F.: Costing of cloud computing services: a total cost of ownership approach. In: 2012 45th Hawaii International Conference on System Sciences, pp. 1563–1572 (2012)
40. Mell, P., Grance, T.: The NIST definition of cloud computing. Natl. Inst. Stand. Technol. **53**(6), 50 (2009)
41. Mencer, O., Vynckier, E., Spooner, J., Girdlestone, S., Charlesworth, O.: Finding the right level of abstraction for minimizing operational expenditure. In: WHPCF@SC (2011)
42. Microsoft: Azure Stack (2009). https://docs.microsoft.com/en-us/azure-stack/operator/azure-stack-overview?view=azs-1908. Accessed 15 Oct 2019
43. Milligan, B.: Tracking total cost of ownership proves elusive. Purchasing **127**, 22–23 (1999)
44. Molka, K., Byrne, J.: Towards predictive cost models for cloud ecosystems: poster paper. In: IEEE 7th International Conference on Research Challenges in Information Science (RCIS), pp. 1–2, May 2013. https://doi.org/10.1109/RCIS.2013.6577736
45. Moyle, K.: Total cost of ownership and open source software (2004)
46. Mvelase, P., Sibiya, G., Dlodlo, N., Oladosu, J., Adigun, M.: A comparative analysis of pricing models for enterprise cloud platforms. In: 2013 Africon, pp. 1–7, September 2013. https://doi.org/10.1109/AFRCON.2013.6757870
47. Orbegozo, I.S.A., Moreno-Vozmediano, R., Montero, R.S., Llorente, I.M.: Cloud capacity reservation for optimal service deployment. In: CLOUD 2011 (2011)
48. Pal, R., Hui, P.: Economic models for cloud service markets: pricing and capacity planning. Theoret. Comput. Sci. **496**, 113–124 (2013)

49. Pandey, S., Wu, L., Guru, S.M., Buyya, R.: A particle swarm optimization-based heuristic for scheduling workflow applications in cloud computing environments. In: 2010 24th IEEE International Conference on Advanced Information Networking and Applications, pp. 400–407, April 2010. https://doi.org/10.1109/AINA.2010.31
50. Patel, C.D., Shah, A.: Cost model for planning, development and operation of a data center (2005)
51. BONE Project: WP 21 tropical project green optical networks: report on year 1 and unpdate plan for activities. No. FP7-ICT-2007- 1216863 BONE project (2009)
52. Qing, L., Boyu, Z., Jinhua, W., Qinqian, L.: Research on key technology of network security situation awareness of private cloud in enterprises. In: 2018 IEEE 3rd International Conference on Cloud Computing and Big Data Analysis (ICCCBDA), pp. 462–466, April 2018. https://doi.org/10.1109/ICCCBDA.2018.8386560
53. Ren, K., Wang, C., Wang, Q.: Security challenges for the public cloud. IEEE Internet Comput. **16**(1), 69–73 (2012). https://doi.org/10.1109/MIC.2012.14
54. da Rosa Righi, R., et al.: A survey on global management view: toward combining system monitoring, resource management, and load prediction. J. Grid Comput. **17**(3), 473–502 (2019). https://doi.org/10.1007/s10723-018-09471-x
55. Sharma, U., Shenoy, P., Sahu, S., Shaikh, A.: A cost-aware elasticity provisioning system for the cloud. In: 2011 31st International Conference on Distributed Computing Systems, pp. 559–570, June 2011. https://doi.org/10.1109/ICDCS.2011.59
56. Simonet, A., Lebre, A., Orgerie, A.: Deploying distributed cloud infrastructures: who and at what cost? In: 2016 IEEE International Conference on Cloud Engineering Workshop (IC2EW), pp. 178–183, April 2016. https://doi.org/10.1109/IC2EW.2016.48
57. Singh, S., Aazam, M., St-Hilaire, M.: RACE: relinquishment-aware cloud economics model. In: 2017 24th International Conference on Telecommunications (ICT), pp. 1–6, May 2017. https://doi.org/10.1109/ICT.2017.7998279
58. Sousa, E., Lins, F., Tavares, E., Cunha, P., Maciel, P.: A modeling approach for cloud infrastructure planning considering dependability and cost requirements. IEEE Trans. Syst. Man Cybern. Syst. **45**(4), 549–558 (2015). https://doi.org/10.1109/TSMC.2014.2358642
59. Sousa, E., Maciel, P., Medeiros, L., Lins, F., Tavares, E., Medeiros, E.: Stochastic model generation for cloud infrastructure planning. In: 2013 IEEE International Conference on Systems, Man, and Cybernetics, pp. 4098–4103, October 2013. https://doi.org/10.1109/SMC.2013.699
60. Subashini, S., Kavitha, V.: A survey on security issues in service delivery models of cloud computing. J. Netw. Comput. Appl. **34**, 1–11 (2011)
61. Sun, K., Li, Y.: Effort estimation in cloud migration process. In: 2013 IEEE Seventh International Symposium on Service-Oriented System Engineering, pp. 84–91, March 2013. https://doi.org/10.1109/SOSE.2013.29
62. Thanakornworakij, T., Nassar, R., Leangsuksun, C., Paun, M.: An economic model for maximizing profit of a cloud service provider. In: 2012 Seventh International Conference on Availability, Reliability and Security, pp. 274–279 (2012)
63. Trummer, I., Leymann, F., Mietzner, R., Binder, W.: Cost-optimal outsourcing of applications into the clouds. In: 2010 IEEE Second International Conference on Cloud Computing Technology and Science, pp. 135–142, November 2010. https://doi.org/10.1109/CloudCom.2010.64
64. Tsalis, N., Theoharidou, M., Gritzalis, D.: Return on security investment for cloud platforms. In: 2013 IEEE 5th International Conference on Cloud Computing Technology and Science, vol. 2, pp. 132–137 (2013)

65. Uchechukwu, A., Li, K., Shen, Y.: Improving cloud computing energy efficiency. In: 2012 IEEE Asia Pacific Cloud Computing Congress (APCloudCC), pp. 53–58 (2012)
66. Vrek, N., Brumec, S.: Role of utilization rate on cloud computing cost effectiveness analysis. In: International Conference on Information Society (i-Society 2013), pp. 177–181, June 2013
67. Walterbusch, M., Martens, B., Teuteberg, F.: Evaluating cloud computing services from a total cost of ownership perspective. Manage. Res. Rev. **36**, 613–638 (2013)
68. Wan, Z.: Cloud computing infrastructure for latency sensitive applications. In: 2010 IEEE 12th International Conference on Communication Technology, pp. 1399–1402, November 2010. https://doi.org/10.1109/ICCT.2010.5689022
69. Weinhardt, C., et al.: Cloud computing - a classification, business models, and research directions. Bus. Inf. Syst. Eng. **1**, 391–399 (2009)
70. Wu, L., Garg, S.K., Buyya, R.: SLA-based resource allocation for software as a service provider (SaaS) in cloud computing environments. In: 2011 11th IEEE/ACM International Symposium on Cluster, Cloud and Grid Computing, pp. 195–204, May 2011. https://doi.org/10.1109/CCGrid.2011.51
71. Yu, Y., Bhatti, S.N.: Energy measurement for the cloud. In: International Symposium on Parallel and Distributed Processing with Applications, pp. 619–624 (2010)
72. Z Degraeve, F.R.: Improving the efficiency of the purchasing process using total cost of ownership information: the case of heating electrodes at Cockerill Sambre SA (1999)
73. Zhang, P., Han, Y., Zhao, Z., Wang, G.: Cost optimization of cloud-based data integration system. In: 2012 Ninth Web Information Systems and Applications Conference, pp. 183–188, November 2012. https://doi.org/10.1109/WISA.2012.13
74. Zheng, L., Hu, Y., Yang, C.: Design and research on private cloud computing architecture to support smart grid. In: 2011 Third International Conference on Intelligent Human-Machine Systems and Cybernetics, vol. 2, pp. 159–161, August 2011. https://doi.org/10.1109/IHMSC.2011.109

Research on Risk Transmission Process and Immune Strategy of Mine Electric Power Information Network

Caoyuan Ma[1(✉)], Qi Chen[1], Wei Chen[2], Long Yan[1], and Xianqi Huang[1]

[1] China University of Mining and Technology, Beijing, China
{Mcaoyuan,ts18130025a31,ts18130210p31,
ts18060170p3me2}@cumt.edu.cn
[2] Jiangsu Normal University, Xuzhou, China
chenwei@jsnu.edu.cn

Abstract. The power information network is becoming more and more important in the safe and efficient production operation of the mine power system. Meanwhile, the power information network may be subject to security risks, such as malicious virus attacks, which poses challenges to mine safety production. Based on the complex network theory, this paper proposes a complex network model of the power information network. Aiming at the possible attack risk of the power information network, the SIR epidemic model is used to analyze and research on the evolution process of the power information network risk. On this basis, two immunization strategies are proposed to suppress the continuous propagation of power information network security risks. The immunization process of the power information network is simulated to verify the significance of the immunization strategy in the process of power information network security risk transmission.

Keywords: Mine power information network · Complex network · Risk propagation · Infectious disease model · Immune strategy

1 Introduction

As a special information communication network of the power system, the power information network takes responsibility to the power system production and management, as well as plays an important role in the safe and stable operation of the power system [1]. With the development of smart grid, the application of power information network is more and more widely used in mine power production operation. The power information network system for smart mines is a comprehensive information platform, which is proposed to meet the demand of data exchange and informatization. The construction and implementation of the system is able to solve the problem of data sharing and integration between the application systems of the mine power grid and at the same time

Electronic supplementary material The online version of this chapter (https://doi.org/10.1007/978-3-030-63941-9_31) contains supplementary material, which is available to authorized users.

X. Wang et al. (Eds.): 6GN 2020, LNICST 337, pp. 414–423, 2020.
https://doi.org/10.1007/978-3-030-63941-9_31

provide global graphics, global data permissions and data exchange services. Therefore, the information islands problem is solved for many application systems within the mine power grid. Various data resources are able to be interconnected among application systems promoting the informatization of mine power grids.

However, in the process of production operation, the safety margin of the power information network is also continuously reduced, and the power information network has potential hidden dangers. If effective prevention and control strategies are not adopted during the risk contagion, it is easy to cause the local failure at the beginning. Even it develops into an avalanche-like cascading failure, which causes the network to collapse and seriously affects the safe production operation of the power system. This poses a challenge to the construction of a safe and reliable power information network. Therefore, issues of security risks of the power information network should be paid attention to.

From some early analysis of power information network security risk propagation process, the power information network is a typical complex network. The analysis method of complex network added to the analysis process has the better revelation of the overall dynamic propagation behavior process of the information network system. In 1999, Barabási and Albert constructed a complex network model of scale-free networks [2], whose scale-free feature widely existing in various real networks is a typical feature of complex networks. Similarly, the research [3] shows that the power dispatch communication data network is also a scale-free network. Based on the above theory, literature [4] studied the influence of power information network on power system network. As well as the cascading failure propagation process of power information network is analyzed based on the complex network model of scale-free network. The above analysis of the communication of security risks in the information and communication network does not take into account the source and form of the risk nor the propagation law of the risk. Literature [5] did research on the hidden dangers of the power information network originated from malicious attacks, such as hackers, computer viruses, Trojan horses, etc. Therefore, the security protection against these malicious attacks is needed. Literatures [6] studied the spreading of computer viruses in communication networks. Authors found that the spreading of computer viruses and biological viruses are similar, proposed a computer virus SIR infectious disease model. However, the two literatures above analyze the communication network without considering combination of the complex network theory. The topology of the communication network itself is also not taken into account in the process of virus propagation. In addition, the electric power information network has a complex network structure without scale characteristics. The intrusion and spreading of viruses in the network are also random and accidental. These characteristics have an impact on the risk propagation process. However, research of these aspects is not considered in the above literatures. The index of the network reliability, vulnerability and other indicators are the evaluation method of network system security [7]. For the power information network, in order to effectively resist various malicious attacks and ensure the safe and stable operation of the power system, it is necessary to screen out the relatively vulnerable position of the power information network. However, these problems are not quite well presented only depending on the reliability analysis of the network. Therefore, it is necessary to take corresponding measures to conduct vulnerability analysis on the power information network. Literature [8] pointed out that node importance measurement is of great significance for studying the vulnerability of complex networks. Literature [9] and literature [8] proposed corresponding evaluation methods of node importance of complex networks, which provided new ideas for vulnerability analysis and further immune optimization of complex networks.

This paper analyzes the topology of this particular complex network of power information network. As well as combined with the propagation characteristics of risk, the authors study the SIR model of power information network risk propagation and the optimized immune strategy of power information network. In the end, the simulation is constructed and analyzed.

2 The Foundation of Complex Network Model of Electric Power Information Network

The power information network can be regarded as a complex network which is composed of nodes and lines. A complex network model is established by abstractly simplifying an actual power information network. Therefore, it is easy to analyze the topology structure and risk propagation process of the power information network. At present, complex networks are used to evaluate several basic statistical attributes of their characteristics, including degree, clustering coefficient and shortest path.

(1) The degree of the node. In the network, the degree refers to the number of nodes which are directly connected to the node i. The value of degree is represented by ki. The average degree is the average value of all nodes in the complex network, the value of <k>, shown as,

$$\langle k \rangle = \frac{1}{N} \sum\nolimits_{i=1}^{N} k_i \tag{1}$$

(2) Degrees distribution. The degree distribution refers to the number of nodes with a degree value of k accounts for the proportion of the total number of nodes in the entire complex network, defined as Pi(k).
(3) The shortest path length. There are usually multiple paths between any two nodes i, j in a complex network. The path with the fewest number of connected edges is defined as the shortest path length, which is denoted by dij. The average value of all the shortest path lengths is the average shortest path length L of this complex network, shown as,

$$L = \frac{1}{N(N-1)} \sum\nolimits_{ij}^{n} d_{ij} \tag{2}$$

(4) Clustering coefficient C. The clustering coefficient C of complex networks is an important parameter evaluating the aggregation degree of nodes in complex networks. The size of the clustering coefficient indicates the degree of small grouping within the network. The aggregation coefficient of the entire network is:

$$C = \frac{1}{N}\sum\nolimits_{i=1}^{N} C_i \tag{3}$$

3 Research on the Spatiotemporal Evolution of Security Risks in Electric Power Information

After the nodes of the power information network are randomly invaded by malicious viruses, the risk propagation process of the entire network is usually propagated from a single attacked failure node to its neighbors. Therefore, the neighboring nodes may also have failure. The risk further spreads to its neighboring nodes and gradually spreads to more nodes. If effective preventive measures are not adopted in time, this spreading trend is possible to be very serious. As a result, most nodes have failure due to malicious attacks in the end. Even the entire power information network is systemic collapsed. Since the power network depends on the control of the power information network, there may be a large-scale power outage in the end.

After being maliciously attacked, the nodes of the power information network can return to normal by manually repairing. This scenario is similar to the SIR model in infectious disease theory. Therefore, the SIR model can be used for analogy and fitting when establishing the risk propagation model of the power information network. The SIR model can be used to describe the transmission process, that is, the infected person has immunity after returning to health. Authors study the power information network with a total of N nodes. During the propagation process, N keeps constant. After a node is attacked, it will not be delayed for a long time regardless of a failure or continuing spreading to other nodes, which is similar to the incubation period of infectious diseases. In this case, the nodes of the power information network can be divided into three categories: S-type nodes, I-type nodes and R-type nodes. S-type nodes, which have not been infected, represent susceptible nodes in the power information network. Type I nodes represent nodes that have been infected in the power information network. The risk can continue to spread to other type S nodes from type I nodes. R-type nodes indicate repaired nodes. These nodes have immunity and will not continue to be infected by I-type nodes for a certain period of time. At time t in the propagation process, S(t) is defined as the proportion of S-type nodes to the total number of nodes N. Similarly I(t) and R(t) have the same definition. β represents the probability of infected S-type nodes. γ represents the probability that the type I node returns to normal. The differential equation is:

$$\begin{cases} \frac{dS(t)}{dt} = -\beta S(t)I(t) \\ \frac{dI(t)}{dt} = \beta S(t)I(t) - \gamma R(t) \\ \frac{dR(t)}{dt} = \gamma I(t) \end{cases} \tag{4}$$

The spatio-temporal evolution of risk transmission is analyzed in the complex network of electric power information network by using the SIR infectious disease model. This is certain to obtain the whole process of the network from infection to gradual cure. Figure 1 shows the change process with the time of the proportion of three nodes in the network.

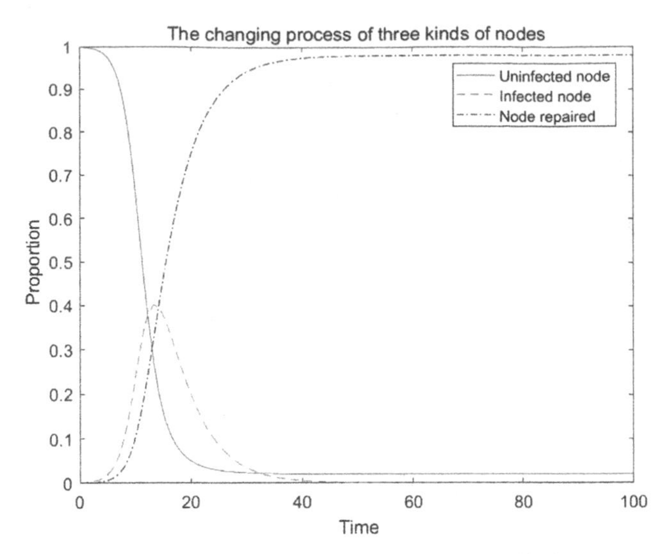

Fig. 1. Three kinds of nodes change curve with time

4 Research on Optimization of Immune Strategy for Power Information Network

For the spread of infectious diseases, the cost-effectiveness of pre-vaccination is much higher than that of post-treatment. There is a similar principle for the risk propagation of the power information network. That is to say, the nodes are selected for pre-immune strengthening, which can greatly reduce the repair cost after the failure of the power information network. Literature [10] mentioned this pre-immunization strategy. If it is an infectious disease of people in the society, the process of vaccination will be affected by the individual's subjective willingness. What should be noticed is that individuals may not be vaccinated in time, even people refuse to vaccinate. So it is not convenient to apply simply the infectious disease immunity model. However, there is no such a complex problem in the power information network. Each node does not have the same subjective willingness as the general population. This ideal objective scenario is convenient for us to apply infectious disease immunity strategies to the power information network.

With the above theoretical conditions, the specific infectious disease immunization strategies are considered. At present, there are three effective immunization strategies: random immunization, acquaintance immunization and targeted immunization.

Random immunization randomly selects some nodes from the network nodes for immunization. There is no additional condition for this kind of immunization. First, acquaintance immunization randomly selects a proportion of nodes from a complex network with a total number of N nodes, and then neighbor nodes are randomly selected from each selecting node to be immunized. Targeted immunization is specific to specific complex network structures. Some nodes play an important role in the transmission process of infectious diseases. If nodes are infected, the intensity of the infectious disease will eventually be stronger. While eventually the spread intensity will be relatively weaker if they are not infected. This scenario inspires us to identify these important nodes according to the relevant indicators of the nodes. This is the idea of targeted immunity.

Literature [11] pointed out that, in contrast, targeted immunity is a good idea provided that we have mastered the index information of each node of the network. In this paper, the information acquired of the relevant power information network are qualified to use the targeted immune method. Literature [10] presented that these important nodes can be better identified by selecting the node degree as an index. Literature [8] further introduces the topological coincidence degree of neighbor nodes, which are integrated with the node degree to better identify these important nodes.

This paper comprehensively considers the node importance evaluation algorithm in terms of the node degree and the neighboring node's topological coincidence degree. The specific algorithm is as follows:

It is generally acknowledged that the larger the node degree, the more important the node is in the network [12]. However, the importance of a node in a complex network not only depends on the degree of the node, but also depends on the degree of dependence of the neighbor node on the node. What is called neighbor node refers to the low-order neighbor node within two hops. If there is no other connection between the two nodes b and c which are connected to the node a, the information can only be transmitted through the node a. On the condition that the node a fails, the information cannot be transmitted. If there is a common neighbor node d between b and c with the exception of the node a, the central position of the node a weakens, and the robustness of the system increases.

Through the above discussion, the similarity of the node domain can be defined. The higher the similarity of the node domain, the lower the dependence of the entire complex network on the node is. This means that the importance of the node is relatively low. The similarity is defined as sim(b, c). If there is no connection between nodes b and c, it is the equivalent of the result of the first formula. If there is a connection, then it is equal to the second result, which is the value 1. The formula is as follows:

$$sim(b,c) = \begin{cases} \frac{|n(b) \cap n(c)|}{|n(b) \cup n(c)|} \\ 1 \end{cases} \tag{5}$$

A node importance evaluation index LLS(i) based on domain similarity is proposed by combining the degree of the node. The formula is as follows:

$$LLS(i) = \sum\nolimits_{b,c \in n(i)} (1 - sim(b,c)) \tag{6}$$

n(i) represents the neighbor node of the node i. The LLS index comprehensively considers the similarity between the degree of the node and the neighbor node. The larger the LLS value, the more important the node is.

The above are the two power information network immunization strategies. The algorithm flow chart of the two strategies is as follows (Fig. 2):

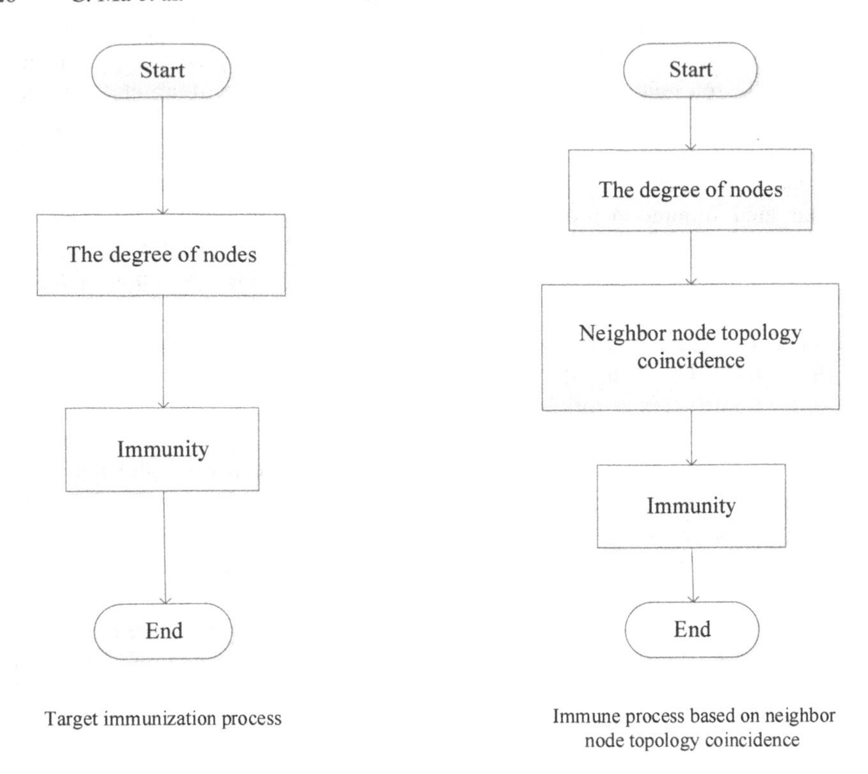

Fig. 2. Comparison of two immunization strategies

5 Analysis of Examples of Electric Power Information Network

Based on the above model and immune strategy, 54-node power information network [13] and 26-node power information network [1] were used for simulation verification separately.

First, the first algorithm of the target immune strategy is applied to the power information network. Degrees are regarded as indicators, the nodes are arranged in order of degree. As well as the degree of each node is obtained. So that the first five nodes from largest to smallest are selected to be immunized (Figs. 3 and 4).

Then the second algorithm is applied to the 54-node power information network. That is to say, the second algorithm is a node importance evaluation algorithm that comprehensively considers the degree of overlap between the node degree and the neighbor node topology. The LLS(i) value of each node is calculated. And the largest five Nodes are to be selected to be immunized.

The non-immunized model and the model after these two immunizations are compared and simulated. The comparison results are shown in the figure (Figs. 5 and 6):

In the figure, the number of initial infected nodes will gradually increase. With optimization of the two immunization strategies, the number of infected nodes is at a lower level and no longer growing. Therefore, the infection process is controlled. The

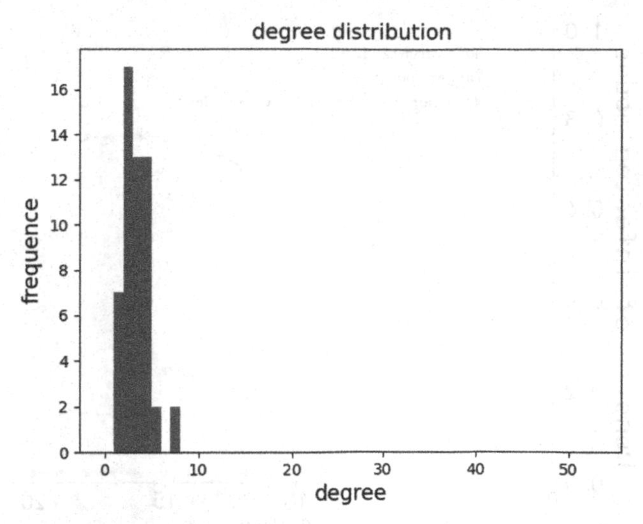

Fig. 3. 54 node degree distribution diagram

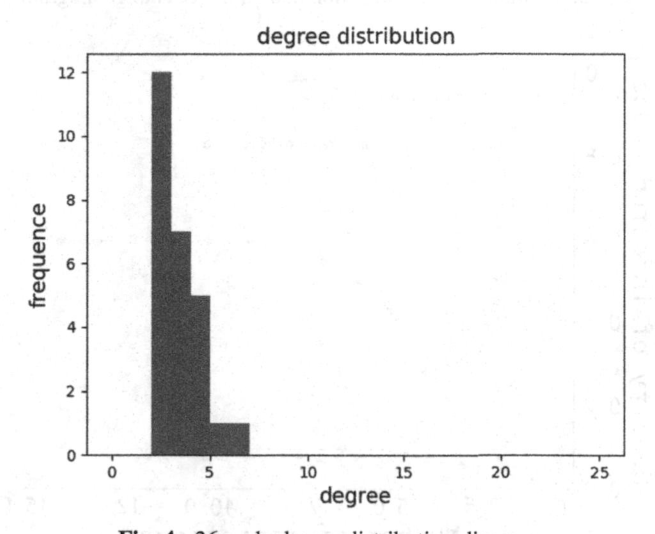

Fig. 4. 26 node degree distribution diagram

total number of infected nodes is effectively reduced by both strategies. An immune strategy that comprehensively considers the degree of node and topological overlapping degree of the neighbor node is better than the target immune strategy, which validates the previous theoretical ideas.

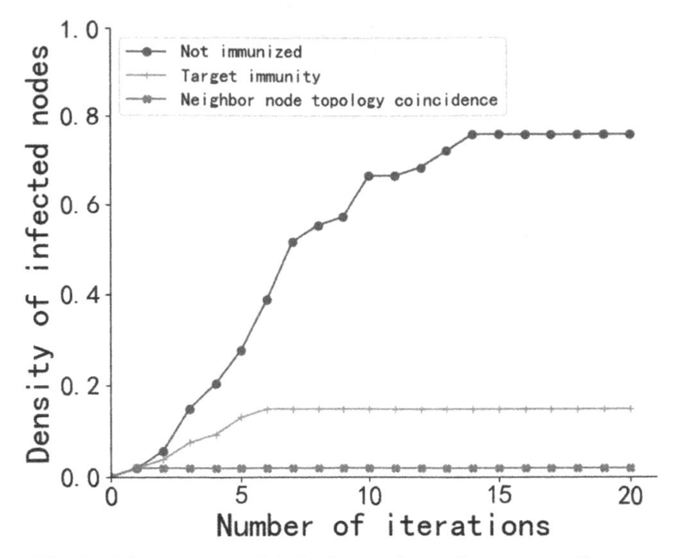

Fig. 5. 54 node network infection node number change diagram

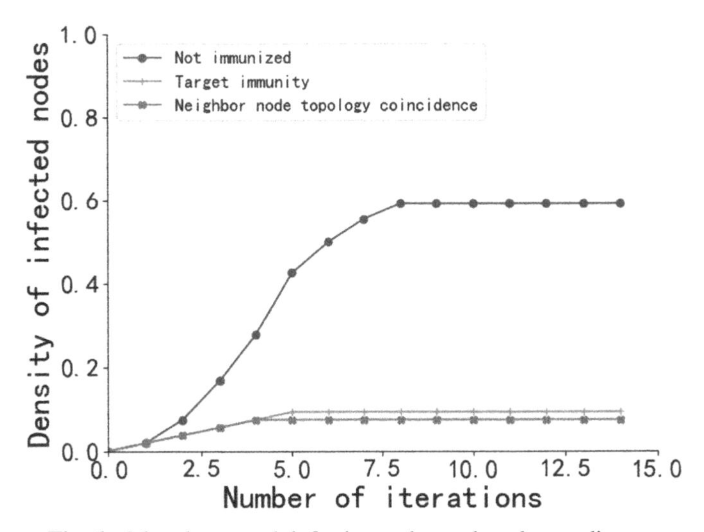

Fig. 6. 26 node network infection node number change diagram

6 Conclusion

Based on the complex network theory, the SIR epidemic model is adopted to analyze and study the evolution process of power information network risk in this paper. The target immune strategy and the immune strategy are used, which comprehensively considers the degree of overlap between the node degree and the neighbor node topology. The power information network immune process is simulated to verify the effectiveness of these two immune strategies in the power information network security risk propagation process. When the most important nodes are identified more accurately for immune protection, the scale of infection can be further reduced.

References

1. Sun, J., Cui, L.: Business-based vulnerability analysis and evaluation method for power communication network. Power Syst. Prot. Control **45**(24) (2016)
2. Albert, B.: Emergence of scaling in random networks. Science **286**(5439) (1999)
3. Hu, J., Li, Z., Duan, X.: Structural feature analysis of the electric power dispatching data network. Proc. CSEE **29**(4), 53–59 (2009)
4. Cao, Y., Zhang, Y.: Analysis of cascading failures under the influence of power system and communication network interaction. Electr. Power Autom. Equip. **33**(1) (2013)
5. Xie, F.: Design and research of safety protection of electric power monitoring system in new energy power plant. Appl. Energy Technol. **40**(3) (2018)
6. Guo, X.: Spatial propagation mechanism of a kind of SIR computer virus model. J. Anhui Norm. Univ. **40**(3) (2017)
7. Yu, Q.: Research on Risk Spreading Behavior of Complex Networks Based on Risk Spreading Paths and Nodes. Lanzhou University (2017)
8. Ruan, Y.: Evaluation algorithm of node importance of complex network based on domain similarity. Acta Phys. Sin **66**(3) (2017)
9. Xia, L.: Research on Information Propagation Dynamic Modeling and Immune Strategy on Complex Network. Nanjing University of Posts and Telecommunications (2017)
10. Jiang, W.: Overview of precaution and recovery strategies for cascading failures in multilayer networks. Acta Phys. Sin **69**(8) (2020)
11. Liu, X.: Research on Infectious Disease Transmission and Immunization Strategy on Complex Network. Lanzhou University (2015)
12. Liu, W.: Evolution of self-organized critical state of power grid based on weighted network topological entropy. Proc. CSEE **35**(22) (2015)
13. Zhang, K.: Analysis of Influence of Communication Network Based on Dependent Network Theory on Power Network Robustness. Southwest Jiaotong University (2014)

Workshop on Intelligent Computing for Future Wireless Network and Its Applications (ICFWNIA)

Design and Implementation a Smart Pillbox

Chin-Chia Hsu[1], Tzer-Long Chen[2], I-Fang Chang[2], Zhen-Yu Wu[3], and Chia-Hui Liu[4](✉)

[1] Department of Leisure Sports and Health Management, St. John's University, New Taipei City, Taiwan
[2] Department of Finance, Providence University, Taichung, Taiwan
[3] Department of Information Management, National Penghu University of Science and Technology, Penghu, Taiwan
[4] Department of Applied Mathematics, Chinese Culture University, Taipei, Taiwan
ljh34@ulive.pccu.edu.tw

Abstract. Nowadays, Taiwan has been moving closer to aging society. An increasing number of elderly needs related medical equipment to assist their lives, such as medicine boxes and walking aids. Many aging people suffer from chronic diseases and they take pills and nutritional products. However, the elderly sometimes forgets to take medicines, such as hypertension medicines and other medicines since their memory degradation. If older person forgets to take pills, it may cause consequence diseases such as stroke. In addition, it is important for patients with chronic diseases to follow doctor's orders. However, there are too many elderly people with "medication adherence", this research uses a smart phone with a smart pill box to supervise and remind the elderly to take medicine. Considering that most of the users are elderly people who are not familiar with the operation process of the mobile APP in the smart phone, a novel real-time transmission of electronic drug orders was developed in the system. It supports a simple and useful user interface for elderly. In the proposed system, the doctor completed the diagnosis and sent the electronic drug list from Near-field communication (NFC) to the APP of the elderly through the doctor's mobile phone APP at first. Then, the information such as setting the medication time and the number of days to return to the doctor are set. When it is time to take the medicine for elderly, the mobile phone APP starts the alarm to remind the user to take the medicine. In the proposed smart pillbox, we use the Arduino UNO development board to design, control, and add a time RTC clock module on the board to control the time. When it is time to take medicine, the servo motor will open/close the pillbox's medicine port.

Keywords: Internet of Things · Smart pillbox · Technology assistant tool · Near-field communication (NFC)

1 Introduction

With the advent of an aging society, many problems have been entrained. According to the research and analysis of the prevalence of multiple chronic patients, the prevalence

X. Wang et al. (Eds.): 6GN 2020, LNICST 337, pp. 427–432, 2020.
https://doi.org/10.1007/978-3-030-63941-9_32

of chronic conditions among the elderly (the population over 65 years old) is 73.39%, and the prevalence of multiple chronic conditions 62.63%. In other words, the average elderly person suffers from one or more chronic diseases. Therefore, to compliance with medical order is particularly important for patients with chronic diseases. This article mainly develops a multi-functional smart pills box based on a smart phone. Since Taiwan has become an aging society, many elders need to take chronic disease drugs. Due to the different time and number of drugs, the elders may be confusing or forgetting to take medicine. In the proposed smart pill box system, it is mainly used to remind the elderly to take medicine and record the time of medication.

The main functions include: 1. To remind the elderly to use medicine. 2. To send the doctor's electronic medicine list to the smart mobile device APP for the elderly. 3. To supervise the medication for the elderly. In this paper, designing the alarm function of the mobile APP reminds the elderly to take medicine on time since they sometimes forget to take the medicine. In addition, the elderly people are not suitable for operating mobile device applications. This paper uses NFC technology to transfer the medicine information to the elderly mobile devices. The elderly people do not need to perform related APP operations. It reminds the elderly people to take medicine. The smart pillbox will automatically open the medicine intake port when the medicine is taken. When the medicine intake port is not closed, it will automatically record that the elderly person does not take medicine. The APP device proposed in this article will automatically calculate the time to reclaim the medicine, reminding the elderly to remember to take the medicine. This article considers the power consumption of the smart pillbox. The relevant calculations are mainly based on mobile devices to avoid that the pillbox is disable since the power problems.

We use the Arduino UNO development board as the control of the smart pillbox and network transmission communication. In the proposed smart pillbox, we use the micro switch and the servo control to control it to open, and use Bluetooth technology in the communication part [10] and NFC [9] for transmission, using Bluetooth and NFC technology for communication transmission can reduce the power consumption of the transmission. In addition, this paper also designs a mobile device APP to facilitate the connection with the smart pillbox and design the APP interface for doctors. It is used to facilitate the transmission of electronic drug orders with the elderly. The system architecture proposed in this paper has been implemented.

2 Related Works

In [1], remote medical treatment is mainly carried out through the concept of the Internet of Things. It is assumed that each medicine has an RFID tag. The elder wears a wearable device to detect physiological information. The doctor can review the health of the elders through the physiological information. Elders can sense what kind of medicine they take through wearable devices. This plan mainly uses the concept of Internet of Things to capture the physiological information of elders.

In the literature [2, 3, 5], the main function of the smart pill box is to remind people to take the medication. The elderly will be reminded to take the medicine since they do not take the medication. However, the elderly's physiological information is not recorded and analysis, so that the doctor cannot understand the adaptation of the elderly to medication.

In [4], physiological information is mainly retrieved through a wearable device. If the physiological information is collected and then analyzed effectively, it will be great help to the use of chronic diseases or acute drugs. It also uses physiological information analysis to understand the elders 'physical condition after taking medication.

In [6], the main purpose is to detect the aching state of the body through the wearable device. The long-term detection of physiological information through the wearable device can effectively make the doctor easier to grasp the disease.

In literature [7] mainly provides the concept of intelligent medical care. Patients with wearable devices can monitor physiological information for a long time, which allows doctors to diagnose the condition more accurately. Doctors can also conduct remote medical consultations to reduce medical costs. Because wearable devices belong to Low power consumption devices can increase the life span of power.

In the literature [8], it mainly evaluates the needs of the elderly for smart kits. In the research results, it is found that the elderly has a positive preference for technological assistance, but the ability to use technology products is low. When using a pill box, it is necessary to consider the interface operated by the elderly.

3 Proposed Methods

3.1 The System Architecture

There are three functions in the proposal system:

(1) Medication reminder: After the seniors are treated, the doctor will set the medication time through the mobile device APP and set the consultation time. When the medication time expires, the smart pillbox will automatically open and the mobile device will remind the medication to be used. If it has not been turned off and the time for taking the medicine is exceeded, the alarm signal will be sent to the smartphone, and the smartphone will record that the pill has not been taken.
(2) Medication time setting: There will be a doctor's order when the elderly goes to the doctor. It can be setting on the smart pillbox APP. Consider that most elderly people are not familiar with the operation process of APP, the real-time transmission of electronic medicine slips are designed in the proposed system. Therefore, the elderly just goes home and put the medicine into the medicine box according to the daily points without setting and operating the APP, as shown in Fig. 1.
(3) Supervise medication status: This article will record the medication status of the elderly and will remind the elderly to re-take the medicine.

3.2 The Development for Doctors

In this article, the function of the doctor-side APP is mainly to record patient data, transmit the time of taking medicine and the number of days of taking medicine. In the proposed smart pillbox system, it uses the NFC to transmit the data and the database to

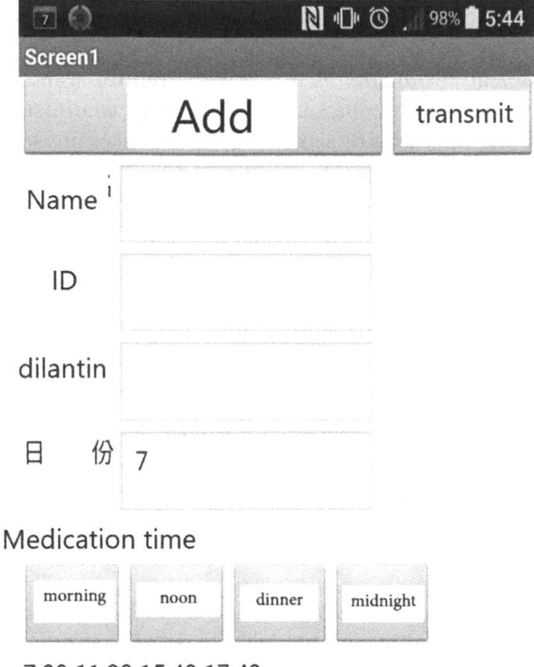

Fig. 1. Doctor's order

record the retrieved data. It will display the personal information section of the elderly and display the name of the medicine used by the elderly and the number of days the medicine is used. In addition, the time for the elderly to use the medicine can also be set. The doctor-side APP will use NFC to send relevant information to the database of the elderly app and delete the information of the last medication to facilitate the use of database space. In addition, the doctor-side APP will also set the reminder time of the elderly app. This function is to reduce the incompatibility of the elderly in operating IT.

3.3 The Mobile Application Development for Elderly

The designed APP includes the following functions: It displays the current time and sets alarm to remind the elders to take medication. It reminds elders before they should go to the doctor. It records the situation of the pillbox using Bluetooth and the time to take the medication by NFC. In addition, they can also close the pill box through the APP and set the time of the last medication. In the proposed system, most settings have been completed by the doctor's APP through NFC. Therefore, the APP interface for the elders do not require additional settings, reducing the complexity of the elderly operating the APP.

4 The Implementation Result

This article implements the smart pillbox and the APP program for doctors and seniors. App inventor 2 is used in the APP development software, and the Arduino development is used in the hardware part. Table 1 shows the hardware and software devices developed in this system. Figure 1 and Fig. 2 are the experimental result.

Table 1. The component of software and hardware.

Hardware
1. Arduino UNO R3 2. Servo Motor 3. RTC clock module 4. Bluetooth module Component: 1. Resistance 2. Microswitch
Software
1. Development of the Arduino with C 2. App Inventor

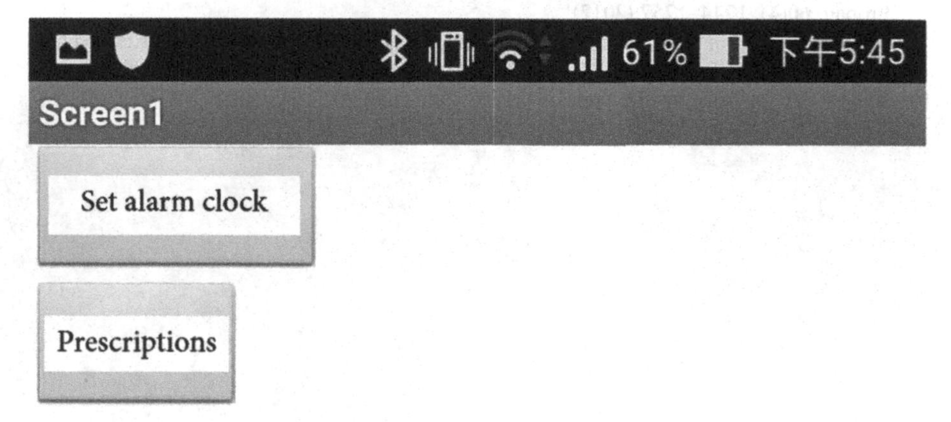

Fig. 2. Implementation result

5 Conclusion

With the convenience of a smart phone, the proposed smart pillbox system can easily set the time for taking medication. When the time is up, the smart phone will automatically remind, and the pillbox can only be opened. The proposed smart pillbox can improve the problem that the senior patients often forget the plight of taking medicine and taking the wrong medicine. The system can be further improved by increasing sensor modules to retrieve the physiological status of the elderly. In addition, elderly people's families can also know about the elderly's physical condition and medication status through remote programs. We will also further consider power management and operability to make the system more complete in the future.

References

1. Yang, G., et al.: A health-IoT platform based on the integration of intelligent packaging, unobtrusive bio-sensor, and intelligent medicine box. IEEE Trans. Indust. Inform. **10**(4) (2014)
2. HERO Smart Pillbox. https://read01.com/Rm7JjO.html
3. i-mami Smart Pillbox. http://www.i-mami.com.tw/web/static/med_advantage#column
4. Andreu-Perez, J., Leff, D.R., Ip, H.M.D., Yang, G.-Z.: From wearable sensors to smart implants—toward pervasive and personalized healthcare. IEEE Trans. Biomed. Eng. **62**(12) (2015)
5. MivaTek Smart Pillbox. https://www.mivatek.com.tw/%E5%95%86%E5%93%81/%E6%99%BA%E6%85%A7%E8%97%A5%E7%9B%92%E5%A5%97%E7%B5%84/
6. Bonato, P.: Wearable sensors and systems. IEEE Eng. Med. Biol. Mag. **29**(3), 25–36 (2010)
7. Islam, S.K., Fathy, A., Wang, Y., Kuhn, M., Mahfouz, M.: Hassle-free vitals. IEEE Microwave Mag. **15**(7), 15–33 (2014)
8. Smith, S.L., Archer, J.W., Timms, G.P., Smart, K.W., Barker, S.J., Hay, S.G., Granet, C.: A millimeter-wave antenna amplitude and phase measurement system. IEEE Trans. Antennas Propag. **60**(4), 1744–1757 (2012)
9. Fischer, J.: NFC in cell phones:the new paradigm for an interactive world. IEEE Commun. Mag. **47**(6), 22–28 (2009)
10. Bluetooth Core Specification Version 4.1, 03 December 2013, Bluetooth SIG, Inc.

Multi-function Electric Scooter Assistant System for Seniors

Tzer-Long Chen[1], Chien-Yun Chang[2], Zhen-Yu Wu[3], and Chia-Hui Liu[4](✉)

[1] Department of Finance, Providence University, Taichung, Taiwan
[2] Department of Fashion Business and Merchandising, Ling Tung University, Taichung, Taiwan
[3] Department of Information Management, National Penghu University of Science and Technology, Penghu, Taiwan
[4] Department of Applied Mathematics, Chinese Culture University, Taipei, Taiwan
ljh34@ulive.pccu.edu.tw

Abstract. Many countries are moving towards an aging society. Many elderly people live alone, and they need to take themselves. The purpose of this article is to increase the safety when elderly people ride on a scooter alone. When an accident occurs, their family can clearly know the location of the accident and can record the entire accident process. This paper uses the IFTTT (If This Then That) system. When the accident occurs, IFTTT is used to send the location to the family LINE. The family can know the location of the elderly through remotely monitor the real-time behavior of the monitored person. Therefore, they can immediately understand the situation when the accident occurs. In addition, they can also know whether the elderly is riding the scooter through the pressure sensor. It can transmit GPS location immediately and start the camera to record.

Keywords: IoT · Aging society · Elderly

1 Introduction

With the rapid growth of the elderly population, today's long-term care needs, family care responsibilities are becoming increasingly heavy. In order to build a long-term care system that meets the needs of the elderly and people with physical and mental disabilities. The current changes in the world's population structure along with the advent of an aging society have brought many problems. Many elderly people use wheelchair while go out alone since the disease and degradation. Accidents are prone to happen for elderly's inconvenient activities and their relatives unable to know that. It easily causes regrets. Many elder people use a scooter as their mobility equipment since they are inconvenient to walk. However, the scooter does not have any warning devices or driving recorders. More sensors and instant messaging are required. In this paper, we use the IFTTT system (If This Then That). When an accident occurs, IFTTT is used to send the location to the relatives' LINE application. The relatives can know the location of the elderly through the line, and remotely monitor the real-time behavior of the monitored person when the accident occurs. They can immediately realize the situation.

X. Wang et al. (Eds.): 6GN 2020, LNICST 337, pp. 433–438, 2020.
https://doi.org/10.1007/978-3-030-63941-9_33

The also know whether the elderly are riding a scooter by using the pressure sensor in the proposed system. The proposed system can effectively determine whether the scooter has an accident and send the GPS position and start the camera to record immediately.

2 Related Works

Through the new concept of smart health care, this article discusses how to combine smart health care with mobile health. The literature [1] designed mobile medical system to monitor the patient's medication, and whether to use the medicine according to the doctor's order to avoid the patient's wrong medication. In the literature [2] is design a smart city to arrange many sensor devices in the city. The main goal is to make the city more intelligent, such as: smart grid, smart vehicles, etc., to effectively use public resources. In the literature [3–5], function of the smart pill box is to remind the medication if there is no medication, the elderly will be reminded to take the medicine, but the elderly's physiological information and physiological information analysis are not recorded, so that the doctor cannot understand the elderly adaptability of medication. In the literature [6], remote medical treatment carried out through the concept of the Internet of Things. It is assumed that each medicine has an RFID tag, and the elder is wearing a wearable device to detect physiological information. The elders can sense which medicine to take through the wearable device. This project uses the concept of Internet of Things to capture the elders' physiological information. This article proposes that the multifunctional elderly scooter assist system uses the Internet of Things as the basis for various accident sensing and send a message to notify the location of the car accident in real time. When the accident occurs, the camera will be activated to shoot. Whether the rider is riding on a scooter, this article carries out practical development and actual testing proves that the method of this article is feasible.

3 The Proposed Scheme

3.1 Riding Situation Determination

The proposed system's flow chart shows in the Fig. 1. Detecting whether the elderly is riding on the electric scooter is the first process in the proposed system. Next, it retrieves the data of the pressure sensor around the vehicle. If the data indicate that is under pressure and the elderly is also sitting on the scooter. It means that the accident impact is slight. If the elderly is not on the scooter and the pressure the value of the sensor is large, indicating that the impact force is large.

3.2 Activate the Monitor Device

In order to be able to monitor the impact of the scooter in real time, the system immediately activates the relevant photographic equipment after the sensor detects it as shown in Fig. 2. At the same time, the system will also perform GPS position detection when the system performs photo recording and GPS position detection.

The relevant information will be sent to the Line of friends and relatives. When the elderly is in good condition, he can press a button to send a message by himself to inform his relatives and friends.

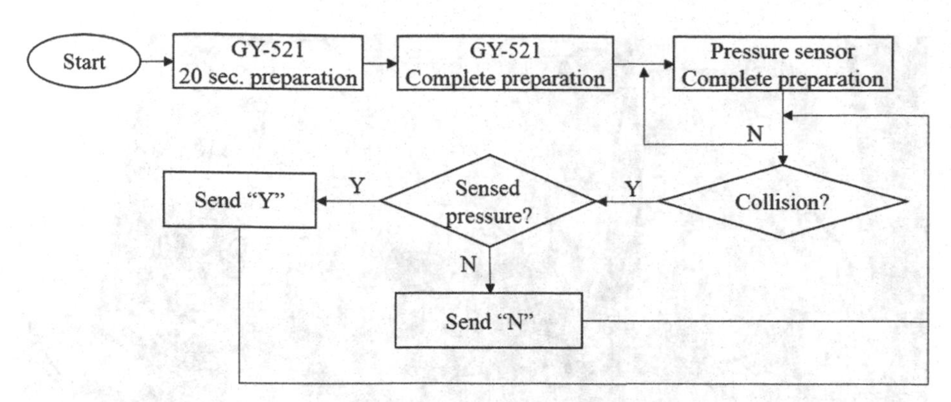

Fig. 1. Flowchart of the system

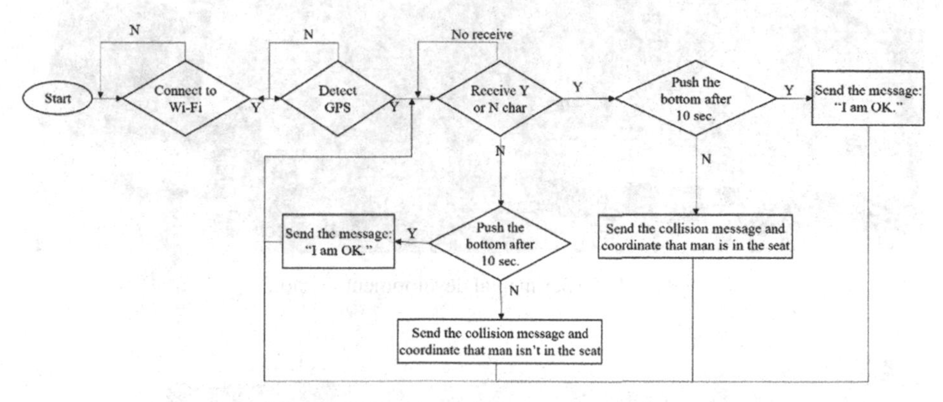

Fig. 2. Flow chart of start monitor device

4 Experiment Result

The experimental results of this paper are shown in Figs. 3, 4 and 5. Figure 3 is the actual experimental development version implemented in this paper. The development version is equipped with sensors and Wi-Fi communication. Figure 4 is the real-time image after collision and Fig. 5 is the instant communication transmission. The current position of elderly people can be known from the experimental results. The method proposed in this paper is feasible and can also be practically applied to scooters.

Fig. 3. Experimental development version

Fig. 4. Real-time image

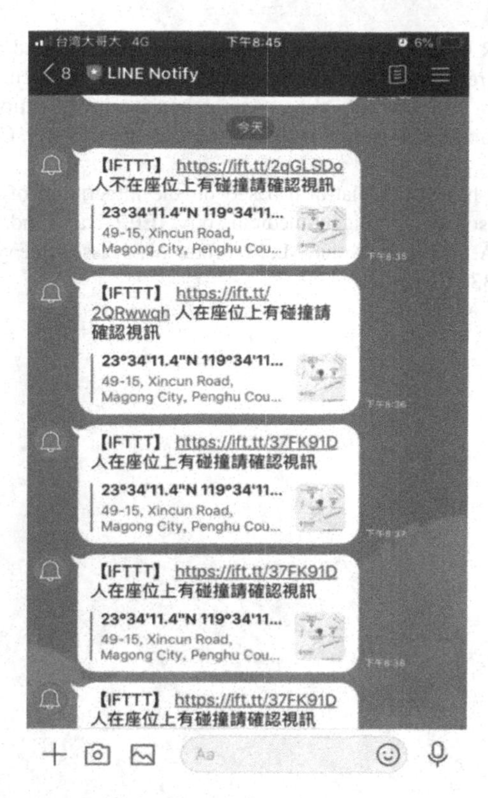

Fig. 5. Instant messaging

5 Conclusion

This system can be located at the location of the accident. Due to the relationship between technology and capability, currently only the collision location can be sent during a collision, and the image processing ability is poor, so the image will be delayed; In the future, we hope that the map can be opened when the electric vehicle moves. It shows the location of the monitored person and the route they walked, so that the monitor can be seen briefly. The proposed system in this paper can be applied to the elderly scooter. It can effectively propose driving safety, and the experimental results can know that the method proposed in this article is feasible.

References

1. Krishna, A.M., Tyagi, A.K.: Intrusion detection in intelligent transportation system and its applications using blockchain technology. In: 2020 International Conference on Emerging Trends in Information Technology and Engineering (2020)
2. Rwibasira, M., Suchithra, R.: A survey paper on consensus algorithm of mobile-healthcare in blockchain network. In: 2020 International Conference on Emerging Trends in Information Technology and Engineering (2020)

3. Marangappanavar, R.K., Kiran, M.: Inter-planetary file system enabled blockchain solution for securing healthcare records. In: 2020 Third ISEA Conference on Security and Privacy (2020)
4. Sahu, M.L., Atulkar, M., Ahirwal, M.K.: Comprehensive investigation on IoT based smart healthcare system. In: 2020 First International Conference on Power, Control and Computing Technologies (2020)
5. Yang, G., et al.: A health-IoT platform based on the integration of intelligent packaging, unobtrusive bio-sensor, and intelligent medicine box. IEEE Trans. Ind. Inf. **10**(4), 1 (2014)
6. Islam, S.K., Fathy, A., Wang, Y., Kuhn, M., Mahfouz, M.: Hassle-free vitals. IEEE Microwave Mag. **15**(7), S25–S33 (2014)

On the Security Policy and Privacy Protection in Electronic Health Information System

Hsuan-Yu Chen[1], Yao-Min Huang[2], Zhen-Yu Wu[3](✉), and Yin-Tzu Huang[4]

[1] Department of Radiology, Tri-Service General Hospital, National Defense Medical Center, Taipei, Taiwan

[2] Department of Management Sciences, National Chiao Tung University, Hsinchu, Taiwan

[3] Department of Information Management, National Penghu University of Science and Technology, Penghu, Taiwan

zywu@gms.npu.edu.tw

[4] Department of Electrical Engineering, National Taiwan University, Taipei, Taiwan

Abstract. The integration of information technology and medical techniques could benefit expanding the time and location for medical services as well as provide better medical quality that it has become a trend in medical domain. For this reason, this study proposes to develop an integrated healthcare information system. Furthermore, based on user privacy and system information security, the key in a smart national health insurance card or a patient's fingerprint together with a doctor's group signature and public key are utilized for retrieving electronic medical records from the system so as to protect the confidential information in the system and guarantee patients' privacy. This system covers other relevant functions of medical record mobility, data link, information security protection, and drug conflict avoidance. In the integrated healthcare information system, a patient's privacy could be protected through hiding. The retrieval location of medical records is always the same that doctors from different hospitals could access to the medical records through authorization. When dealing with an urgent case, the real-time medical record retrieval could effectively enhance the recovery rate. The information in the system procedure broadly covers diagnoses of patients, insurance claims, and drug collection in order to prevent doctors from prescribing wrong medicine, avoid the troublesome of insurance claim application, and reduce patients' problem about collecting receipts.

Keywords: Healthcare information system · Group signature · Electronic medical records · Privacy · Protection

1 Introduction

For several years of network development, the application has become complicated and the relevant technology is getting diversified. Digitalization in daily life has become a major trend in modern technology development. Common e-commerce, e-medical treatment, e-banking, g-government, and online community application are the applications of network.

X. Wang et al. (Eds.): 6GN 2020, LNICST 337, pp. 439–444, 2020.
https://doi.org/10.1007/978-3-030-63941-9_34

Relative to medical application, the meaning of information technology lies in expanding the clinical facilitating function of medical information. The technology covers health insurance IC card device, digital certificate and signature, electronic medical records, and electronic prescriptions [1, 5]. The introduction of information technology to medical systems presents obvious performance on medical management. However, the development and maintenance costs for medical information systems is large that it is simply tried in large medical centers, e.g. National Taiwan University Hospital, Vetcrans General Hospital, and Chang Guang Medical Foundation, in the beginning of the development. Besides, it is merely one-way information feed-in, rather than interactive information feedback, such as assisting medical personnel in teaching, inquiry, aid, and alert with built-in professional medical knowledge. After the practice of national health insurance in 1995, novel and mature system development technology has been emerging to cope with frequently changing health insurance payment reporting and accelerate the application of medical information to small and medium medical institutions. With the application of new technology, the development of programs becomes faster, more flexible, and more easily maintained. After the provision with direct operation and application for doctors, medical information systems are developed the comprehensive function to avoid the delivery of manual document, acquire real-time revision of users' feed-in information, maintain the timeliness of information in database, and get rid of the time difference in information update in traditional batch processing. It therefore could reduce and even avoid possible human errors of nurses, pharmacists, and technicians. An information system is also an inevitable infrastructure in medical management for the improvement of medical quality.

Research on the data structure of health insurance IC cards and digital certificates, electronic medical records, and the data format of electronic prescriptions is mature. However, the integration of cross-medical institution electronic medical record format, cross-department secure patient record information exchange agreement [2], telemedicine, and caregiver authentication mechanisms still require improvement. These are research on cross-medical institution electronic medical record format exchange agreement and medical information system transfer format.

The reinforcement and integration of medical information systems could assist in the promotion of medical quality and efficiency, where the integration of function and technology is the key to implement medical informatization.

First, reinforce the function of health insurance IC cards and the compatibility with integrated medical information systems. The generally used health insurance IC cards are wafer cards with small memory capacity, bad computing function, and not being able to support digital signature or encryption/decryption requirements in actual applications that it is not convenient for the verification and acquisition of electronic medical records, prescriptions, and examination forms. The selection of the saving medium of a health insurance IC card and the wafer material, e.g. ROM, RAM, or EPROM, should be carefully considered and matched the system according to the characteristics. Furthermore, the built-in data could be graded according to the importance of data or segmented the necessity according to emergency use. The reading and security mechanism of built-in data should be well planned and designed in advance.

Second, establish electronic prescription system with complete functions and the integration with medical information systems. As the rapid development of electronic medical records, prescriptions are inclining to electronic [3]. In terms of current medical systems and health insurance systems, the assistance of digital signature allows pharmacies and patients verifying the correctness, integrity, and non-repudiation of electronic prescriptions; meanwhile, pharmacies could complete the health insurance payment reporting through online verification mechanisms. It could enhance the integrity of medical information networks and is worth of development in the future.

Third, protect patients' and doctors' privacy. Confidentiality in medical practice is the basic element to establish good doctor-patient relationship. Patients have the right to request the confidentiality of personal medical information and doctors have the obligation to respect patients' medical privacy. In the essence of law, privacy is a limited right which is passively restricted to balancing the conflict among public health benefits, third party benefits, and personal privacy benefits [4]. Moreover, under current medical system with referral, consultation, and health insurance IC cards, medical division and medical teams are the trend as well as the major challenge to maintain patients' privacy. In this case, authorized access of patients' medical records could implement the protection of both doctors' and patients' privacy.

An integrated medical information system should be a secure, convenient, and complete system with following characteristics.

1.1 Portability of Recent Medical Records

A patient's recent medical records are logged in the health insurance IC card. With authentication, a doctor could read the patients' medical information in other medical institutions and rapidly access the patient's medical history to promptly make correct diagnoses. It would enhance doctors' diagnosis correctness and efficiency.

1.2 Medicine Collection Function

The function stresses on the convenience for patients collecting medicine. An agent's identity and authentication information are integrated with the medical information system and registered in the patient's health insurance IC card. When the patient is not able or busy to collect medicine by himself/herself, especially the one with physical & mental disabilities or with difficulty in moving, could entrust a legal agent, through authority mechanism, to collect medicine.

1.3 Linkability and Privacy Protection

In the digital medical system, patients' and doctors' medical privacy is extremely emphasized. In the beginning of the system development, the following points should be drawn.

Anonymity: Patients and doctors use pseudonyms in an integrated medical information system.

Patient identity linkability: Merely the health insurance reporting unit could access a patient's real identity. Pharmacies' linkability to patient identity is simply the holder of electronic prescriptions, but not the real identity.

Doctor identity linkability: Merely specific just medical associations could grasp a doctor's real identity. The linkability of Bureau of National Health Insurance to doctor identity is simply the prescription writer, but not the real identity.

Non-linkability of doctor identity: Pharmacies could not link a doctor's identity from electronic prescriptions.

1.4 Avoiding Patients' Repeated Medicine Collection

After the comprehensively electronic health science and technology, the access to electronic prescriptions might be easier than to written prescriptions. To avoid a patient repeatedly collecting medicine in different pharmacies with the copy of electronic prescriptions to result in medical resource waste, a healthy integrated medical information system should present the function to prevent such similar behaviors.

2 Research Method

The proposed Integrated Medical Information System (IMIS), including the architecture, physical mechanism, and execution process, are introduced in this section. IMIS integrates e-patient records and e-prescriptions to simplify the originally complicated and time-consuming medical process, e.g. diagnosis, inspection, medicine collection, emergency, and insurance payment. It even applies cryptography, e.g. encryption/decryption and digital signatures, to protect patients' and doctors' privacy, and prevent illegal benefit acquisition. Besides, when there are medical malpractice claims, the just third party in the system would inspect the message and signature to calm the dispute.

The architecture and operation process of IMIS are shown in Fig. 1. The entire mechanism could be divided into registration phase, diagnosis phase, collecting medicine phase, and subvention phase. The physical mechanism contains insurers, Bureau of National Health Insurance, pharmacies, patients, doctors, and agents. Bureau of National Health Insurance is responsible for national medical businesses, covering from doctors to patients' medicine collection. A patient is first offered a national health insurance IC card used for medical institutions; doctors and pharmacies are provided consultation and medicine authentication, including collecting, storing, and updating patients' electronic medical records and prescriptions; and, an agent is verified the qualification to collect medicine. What is more, it is also responsible for subsidizing medical expenses and partial or full diagnosis and treatment expenses of patients, issuing group signature and certificate for doctors diagnosing and treating patients, as well as assisting in possible medical malpractice claims among pharmacies, insurer, doctors, and patients. A pharmacy is responsible for verifying prescriptions and the correctness of signature in prescriptions as well as provides medicine for patients or agents. Patients, doctors, and agents play primary roles in IMIS.

Based on protecting patients' and doctors' privacy, a patient, at registration phase, should apply for an anonymous national health insurance IC card from Bureau of

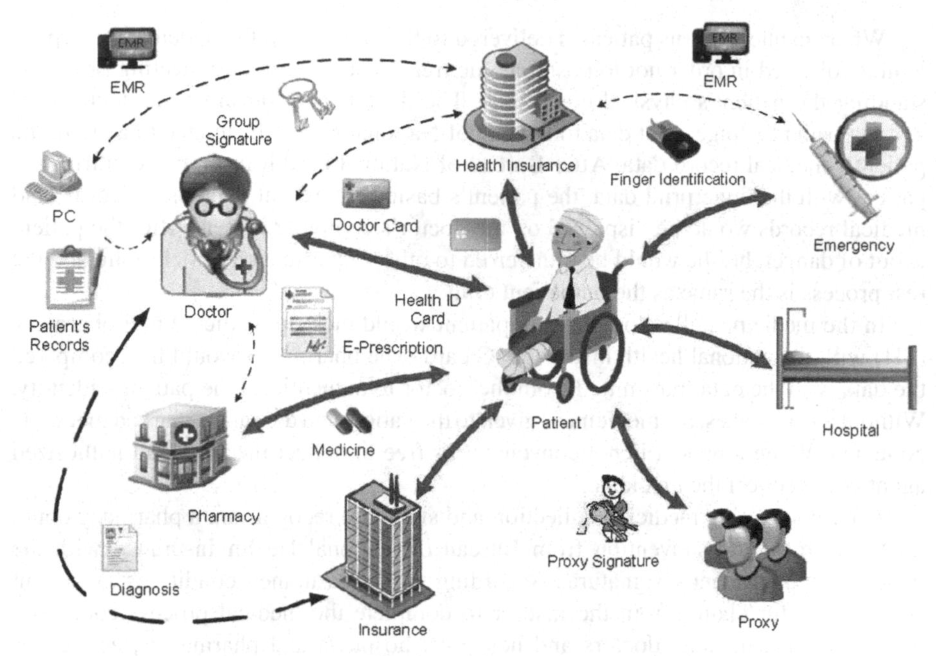

Fig. 1. IMIS process

National Health Insurance (NHI), while a doctor needs to apply for personal group signature secret key and doctor card. The characteristic of group signature could satisfy the requirement for anonymity. Besides, there is the function in IMIS of an agent collecting medicine for patients; the proxy signature secret key would be saved in the card.

At diagnosis phase, when a patient sees a doctor with the national health insurance IC card, the doctor would first insert the doctor card and the national health insurance IC card to the machine and transmit the patient data to Bureau of National Health Insurance for authenticating the patient's signature. When the patient is confirmed as the card holder, the patient's basic data, medical records, and recent diagnosis records would be displayed on the doctor's computer screen to help the doctor understand the patient's physical conditions and outpatient situations.

After the outpatient, the doctor would update the patient's medical record data and fill in prescriptions for the patient collecting medicine in a pharmacy. Such motions require the doctor using the personal group secret key for the signature to be responsible for the diagnosis. Moreover, the doctor would write the diagnosis records and prescription data in the patient's national health insurance IC card.

When a patient needs to apply for insurance, he/she could simply asks the doctor transferring the diagnosis content into a diagnosis certificate with the doctor's group secret key signature. The certificate is then transmitted to the insurer to complete the integration of insurance and medical treatment.

When an unconscious patient is delivered to the emergency, the patient's fingerprint is first collected in order not to postpone the treatment because of collecting or understanding the patient's physical conditions. The doctor then transmits the doctor card and the patient's fingerprint data to Bureau of National Health Insurance to acquire the patient's medical record data. After Bureau of National Health Insurance confirms the patient with the fingerprint data, the patient's basic data, recent diagnosis records, and medical records would be displayed on the doctor's computer screen. After the patient is out of danger, he/she would be transferred to other department and sickroom, and the rest process is the same as the outpatient clinic.

In the medicine collection phase, a patient would pick up medicine in a pharmacy (PH) with the national health insurance IC card. The pharmacist would first compared the data with the data transmitted from the doctor to authenticate the patient's identity. Without any mistakes, the medicine is given to the patient and the card is marked medicine collected. When a patient is not convenient or free to collect medicine, an authorized agent could collect the medicine.

After completing medicine collection and signature recognition, a pharmacy could apply for medicine subvention from Bureau of National Health Insurance with the patient's or the agent's signature. According to the insurance conditions, a patient could apply for claims from the insurer to complete the medical process composed of patients and doctors, doctors and hospitals, hospitals and pharmacies, as well as insurance companies and patients.

3 Conclusion

A integrated medical information system with security policy and privacy protection, which combines e-patient records, e-prescriptions, modified smart cards, and fingerprint identification systems and applies proxy signature and group signature, is proposed in this study.

References

1. Dai, C.-y.: Secure Access Control of Personal Health Records in Cloud Computing Using Attribute-Based Encryption. Department of Information Science and Engineering, National Chiao Tung University, Hsinchu City (2013)
2. Huang, K.-H., Hsieh, S.-H., Chang, Y.-J., Lai, F., Hsieh, S.-L., Lee, H.-H.: Application of portable CDA for secure clinical-document exchange. J. Med. Syst. **34**(4), 531–539 (2010)
3. Yang, Y., Han, X., Bao, F., Deng, R.H.: A Smart-card-enabled privacy preserving E-prescription system. IEEE Trans. Inf. Technol. Biomed. **8**(1), 47–58 (2004)
4. Stallings, W.: Cryptography and Network Security: Principal and Practices, 7th edn. Prentice Hall, Boston (2016)
5. Takeda, H., Matsumura, Y., Kuwata, S.: Architecture for networked electronic medical record systems. Int. J. Med. Inform. **60**(2), 161–167 (2000)

The Development of Smart Home Disaster Prevention System

Zhan-Ping Su[1], Chin-Chia Hsu[2], Zhen-Yu Wu[3], and Chia-Hui Liu[4](✉)

[1] Department of Business Management, Chungyu University of Film and Arts, New Taipei City, Taiwan
[2] Department of Leisure Sports and Health Management, St. John's University, New Taipei City, Taiwan
[3] Department of Information Management, National Penghu University of Science and Technology, Penghu, Taiwan
[4] Department of Applied Mathematics, Chinese Culture University, Taipei, Taiwan
ljh34@ulive.pccu.edu.tw

Abstract. Building smart home within Internet of Things (IoT) technology can bring residents living convenience. A smart home infrastructure needs to be built with the Internet of things and wireless network. In this article, using the raspberry Pi to build a smart home disaster prevention system. The system functions proposed in this article are: 1. Real-time monitoring of home environment and safety issues, 2. Indoor temperature and humidity sensing, 3. Pot temperature and humidity sensing, 4. Disaster detection, 5. Fish tank water level detection. The proposed method helps to provide a comfortable home environment. It also reduces the damage caused by home disasters. Using the concept of home automation, we propose the smart home disaster prevention system which can reduce the pressure on residents to maintain the home environment.

Keywords: IoT · Smart home · Disaster prevention system

1 Introduction

To improve the living quality of life, we can build smart home within Internet of Things (IoT) technology. It can bring residents living convenience. It monitors remotely any data or home appliances in the home by anytime, anywhere. A smart home infrastructure needs to be built with the Internet of things and wireless network. In this article, using the raspberry Pi to build a smart home disaster prevention system. This article uses various sensors such as flame sensors to monitor various dangerous areas in the home to avoid fires. The system functions proposed in this article are: 1. Real-time monitoring of home environment and safety issues, 2. Indoor temperature and humidity sensing, 3. Pot temperature and humidity sensing, 4. Disaster detection, 5. Fish tank water level detection. The proposed method helps to provide a comfortable home environment. It also reduces the damage caused by home disasters. Users can check the changes of the

X. Wang et al. (Eds.): 6GN 2020, LNICST 337, pp. 445–451, 2020.
https://doi.org/10.1007/978-3-030-63941-9_35

home environment through the proposed platform, which helps to adjust the energy-saving appliances in the home. Smart homes are based on the technology of the IoT, wireless networks and Internet. It can control lights, windows, temperature and humidity, audio-visual equipment and home appliances. It can further monitor environmental factors in the home. These factors are carried out through relevant algorithms. According to the analysis, the home environment can be controlled at the optimal temperature and humidity through home appliance control. n recent years, the government has raised the green energy requirements. Smart home system can monitor the use of home appliances in the home for power control. As aging society comes, smart homes also need to be included in disaster prevention monitoring, such as: Kitchen smoke, temperature, etc., to ensure that elderly people forget to turn off the fire source to cause a fire when cooking. Many families now have fish farming or potted plants, and smart homes also need to monitor fish tanks and potted plants. Smart homes mainly focus on improving the quality of living. Using the concept of home automation, we propose the smart home disaster prevention system which can reduce the pressure on residents to maintain the home environment.

2 Related Works

In [1], it mainly proposed that smart electric lamps need to be judged together with the external environment and sunlight in order to reduce the waste of electricity, and the smart power management system needs to be common to different household appliances. In [2], it mainly proposes a network security mechanism to ensure smart appliances, it was attacked by hackers, which led to accidents. In [2], it mainly proposed identity verification and private encryption mechanism of messages, so that legal users can access and control home appliances. In [3], algorithms are mainly used to judge households. The temperature fluctuation is reasonable, and the fire detection is further conducted through the algorithm. In [4], mainly judge the indoor and outdoor ventilation status, and further start the relevant home appliances, so that the temperature of the home can be reduced to the most suitable state. In [5], mainly design the distribution of sensors and servers, and need to consider the use of electricity and the network to reduce the waste of electricity in smart home appliances. In [6], the security of smart home networks is mainly proposed. Since smart homes mainly use Wi-Fi networks, Wi-Fi networks will have security problems. Therefore, the literature [6] mainly uses 4G mobile communication for network connection. The literature [7] mainly proposes energy theft detection. The literature [7] mainly uses algorithms to estimate the household electricity consumption compared with the electricity consumption of electricity meters. To determine whether power has been stolen.

3 The Proposed Scheme

This paper proposed a smart home disaster prevention system based on the IoT and wireless networks. In addition to improving the quality of family life, it also performs disaster detection functions. The functions of the smart home disaster prevention system are as follows: 1. Real-time monitoring of the home environment And safety issues, 2.

indoor temperature and humidity sensing, 3. pot temperature and humidity sensing, 4. disaster detection, 5. fish tank water level detection. From the experimental results, the proposed system is implemented. It can also improve the quality of life.

3.1 System Model

The function diagram of the system is shown in Fig. 1. The proposed system controls the equipment through Bluetooth and relays and construct an IoT environment within wireless networks. The main functions of this article are indoor temperature and humidity detection, watering prompt, abnormal gas concentration sensing and fish tank reminders for adding water. The system proposed in this article is based on normal sensors. Therefore, the price is reasonable and easy to deploy the system. In addition, this system can detect whether there is a disaster through abnormal gas concentration sensing and flame sensor to ensure the safety and life of the family quality.

3.2 The Proposed Scheme

This article mainly uses Bluetooth and relays to control home appliances. The functions are: 1. Real-time monitoring of home environment and safety issues, 2. Indoor temperature and humidity sensing, 3. Pot temperature and humidity sensing, 4. Disaster detection, 5. Fish tank water level detection. The temperature and humidity sensors are installed in the system, if the temperature and humidity are higher than the threshold, household appliances that can adjust the temperature and humidity will be activated. In addition, a soil sensor is installed in the pot to detect the temperature in the soil. When the humidity is too low, the sprinkler system will be activated. In this paper, a flame sensor and abnormal gas concentration sensing will be putted in the hazardous area. When the sensor is activated, an alarm will be issued to inform the residents of the danger of disaster. The fish tank is equipped with a water level sensor. When the water level is lower than the standard, it will send a prompt message to inform the user that water needs to be added. The method in this article mainly uses the parity sensor, which can mainly popularize the system in various households. The development process of this article is shown in Fig. 1.

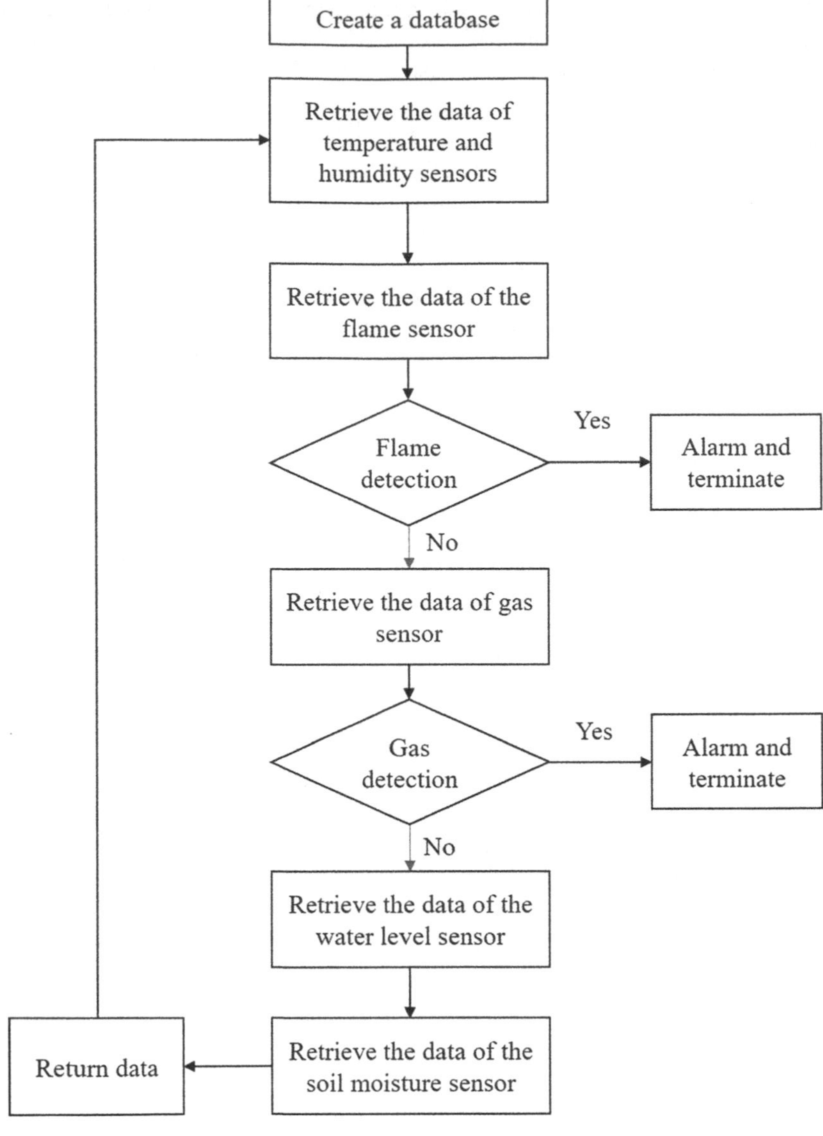

Fig. 1. Development flow chart

4 Experiment Result

There are the proposed system development and testing, as shown in Figs. 2–5. Figure 2 is the environmental numerical values. Figure 3 is the Bluetooth relay. Figure 4 is the remote control. Figure 5 is the physical test.

	ID	Humity	Temperature	Liquid	Mositure	
	Filter	Filter	Filter	Filter	Filter	Fi
1	1	30	24	736	31	0
2	2	30	24	741	383	0
3	3	30	24	664	656	0
4	4	31	24	703	369	0
5	5	30	24	696	401	0
6	6	30	24	699	366	0
7	7	30	24	734	109	0
8	8	59	30	876	105	0
9	9	60	30	635	94	0
10	10	60	28	872	118	0
11	11	60	28	810	126	0
12	12	60	28	892	100	0
13	13	60	28	637	122	0

< 1 - 14 of 460 > Go to: 1

Fig. 2. The environment data

Fig. 3. Bluetooth relay

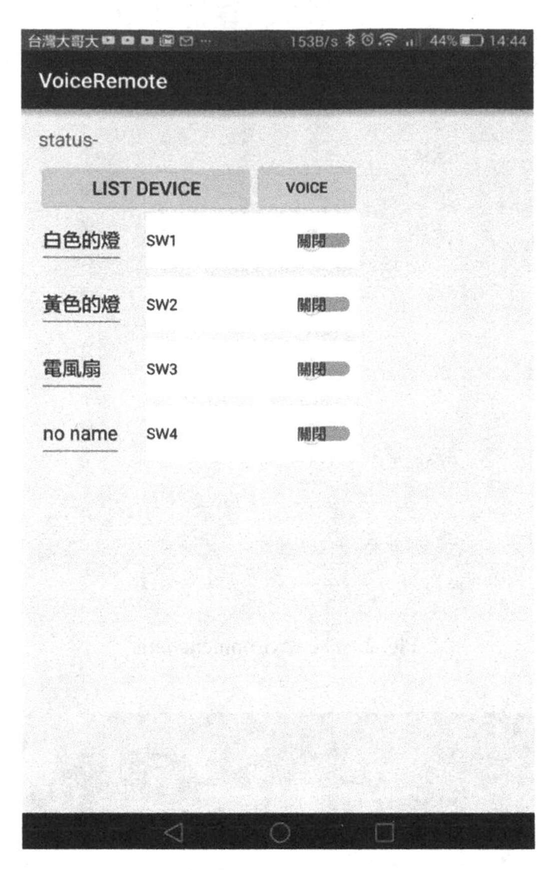

Fig. 4. Remote control

Fig. 5. Physical diagram

5 Conclusion

The system proposed in this article already has temperature and humidity, flame, gas and other sensors. Environmental monitoring of home and integration of mobile devices makes the system monitoring more intelligent. In addition, we also combine Chinese voice synthesis system, let users embrace more functional choices, to provide the safe monitoring for the home environment. The system proposed in this article is mainly based on parity. In the future, the system will be productized and can be introduced into various households.

References

1. Chamoli, S., Singh, S., Guo, C.: Metal-Dielectric-Metal metamaterial-based ultrafast hydrogen sensors in the water transmission window. IEEE Sens. Lett. **1**, 1 (2020)
2. Dargie, W., Wen, J.: A simple clustering strategy for wireless sensor networks. IEEE Sens. Lett. **1**, 1 (2020)
3. England, N.: The graphics system for the 80's. IEEE Comput. Graph. Appl. **40**(3), 112–119 (2020)
4. Sawant, J., Chaskar, U., Ginoya, D.: Robust control of cooperative adaptive cruise control in the absence of information about preceding vehicle acceleration. IEEE Trans. Intell. Transp. Syst. **1**, 1 (2020)
5. Ballo, A., Grasso, A.D., Palumbo, G.: Charge pump improvement for energy harvesting applications by node pre-charging. IEEE Trans. Circuits Syst. II Express Briefs **1**, 1 (2020)
6. Xie, Y., Chen, F.-C., Chu, Q.-X., Xue, Q.: Dual-band coaxial filter and diplexer using stub-loaded resonators. IEEE Trans. Microw. Theory Tech. **1**, 1 (2020)
7. Chen, C., Yang, S., Wang, Y., Guo, B., Zhang, D.: CrowdExpress: a probabilistic framework for on-time crowdsourced package deliveries. IEEE Trans. Big Data **1**, 1 (2020)

Using an Automatic Identification System (AIS) to Develop Wearable Safety Devices

Hsin-Te Wu[1(✉)] and Fan-Hsun Tseng[2]

[1] Department of Computer Science and Information Engineering, National Ilan University, Yilan 26047, Taiwan

[2] Department of Technology Application and Human Resource Development, National Taiwan Normal University, Taipei 10610, Taiwan

p11o0304@mail2000.com.tw

Abstract. The numbers of people who participate in outdoor activities are increasing; however, it has a certain risk of doing outdoor activities, such as wilderness rescue and losing the force while sailing. Therefore, people performing outdoor activities usually wear relevant safety devices. Currently, most users send emergency alarms through mobile phones; yet, the penetration rate of the fourth generation of broadband cellular network technology (4G Network) is not fully equipped in some rural areas, which causes victims failed to send emergency alarms. This study uses an Automatic Identification System (AIS) to develop a wearable safety device because the transmission range of AIS is wider than other types of systems, and it is easier to prepare essential functions in AIS wearable devices. When a victim presses the panic button, the system will send the alarm with a Global Positioning System (GPS) signal; the system will only detect the GPS location once when pressing the button to reduce power consumption, which avoids additional power consumption while repeating the detection. The research conducted practical experiments and discovered that the method is feasible in real-world applications.

Keywords: Automatic Identification System · IoT · Wearable devices

1 Introduction

Many countries promote outdoor activities and aquatic sports despite the activities have some risks. For example, the fatality rate of mountain climbing is 16%, which accounts for the highest among outdoor activities [1]. In Taiwan, there are more than 200 mountains; according to the statistics from the National Fire Agency, the wilderness rescue incidents per year are 185 cases on average [2]. Among the rescue cases, the highest ratio is missing and contact lost, which is 42%. On the other hand, there are diverse types of aquatic sports; it is critical to have relevant safety devices when there are no companies or having to wait for rescue by floating. A safety device cannot be too large to become a burden; hence, it is necessary to develop a wearable device with waterproof and user-friendly functions for carrying around. Additionally, it sometimes takes a long

X. Wang et al. (Eds.): 6GN 2020, LNICST 337, pp. 452–457, 2020.
https://doi.org/10.1007/978-3-030-63941-9_36

time to conduct an outdoor activity, which also needs to consider the power supply of the wearable device to lower the power consumption and avoid the situation of losing power and failing to send a signal.

Currently, many people use the Global System for Mobile Communications (GSM) systems to send emergency alarms while conducting outdoor activities [3]. Nevertheless, the equipment of GSM usually installed at the areas with a higher density of population; many mountains and sea areas do not have relevant GSM systems, which is not capable of sending a signal when encountering emergencies. Moreover, although many places have set large quantities of detectors and relay stations for people to use when doing outdoor activities [4], the transmission ranges are not able to cover fully, and it costs highly to install all of the equipment. Today, rescue teams usually utilize robots and drones to handle searching tasks [5]. With the vast zones in mountains and sea areas, victims should send a signal for the teams to locate the position; nonetheless, many wearable devices send signals via 2.4 GHz band because it is a wireless range opened for the public to use [6]. Unfortunately, 2.4 GHz band is for short-range communications, which is unfavorable for rescue teams; thus, using a wider range for communications is necessary. On the other hand, some emergency systems send signals through satellites [7]; yet, it requires many satellites for developing a complete rescue system, and there are power consumption and installation costs for considering in the meantime.

This study utilizes an Automatic Identification System (AIS) as the foundation for developing a wearable safety device. The typical range of AIS is around 20 to 30 nautical miles, which will be easier for rescue teams to search and locate victims' positions rapidly. The design of the wearable safety device in this research aims to be user-friendly with one panic button, which can reduce the information maladjustment situation for elderly people. When pressing the panic button, the system will detect the location of the victim and send the Global Positioning System (GPS) signal via the AIS communications. The experimental results of this study have proved that the method is feasible.

2 The Proposed Scheme

2.1 System Model

The research uses an AIS system to transmit signals; AIS communications are usually for avoiding collision between merchant ships. The range of AIS is around 20 to 30 nautical miles, which is suitable for receiving the signal sent from outdoor safety devices. As shown in Fig. 1, the system model demonstrates that the system will send an AIS signal regularly when a victim presses the panic button, and the rescue team can locate the position from the AIS signal. Figure 2 shows the diagram of the wearable device, which contains a straightforward panic button for users to understand and operate the device. Additionally, Fig. 3 represents the function diagram that the network layer for AIS communications and the application layer is for GPS and alarm.

2.2 The Proposed Scheme

This research utilizes AIS communications for the wearable device to initiate the GPS and detect the position when victims press the panic button. Further, the system will

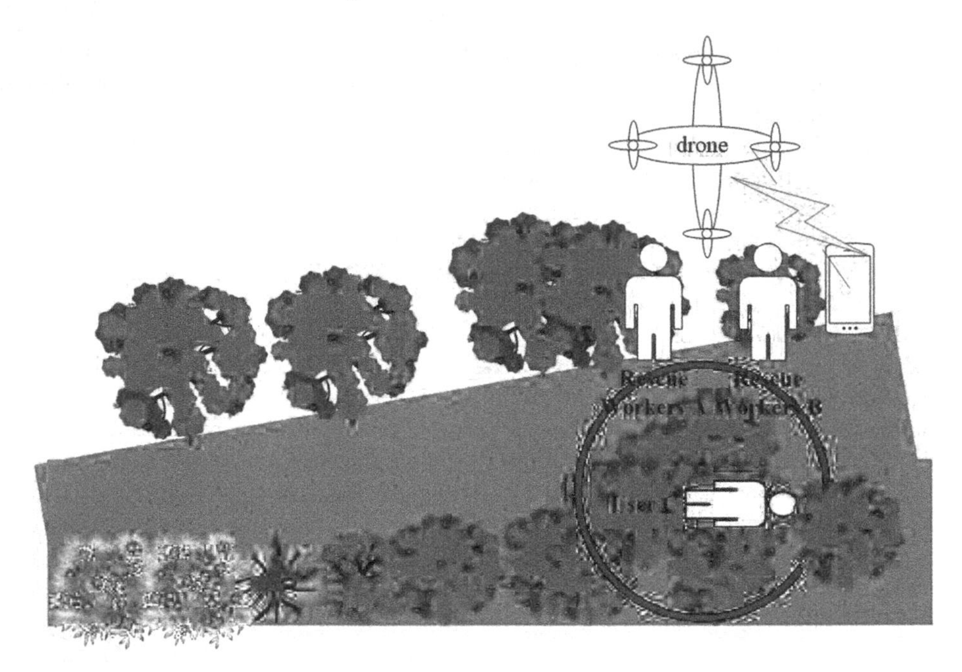

Fig. 1. System model.

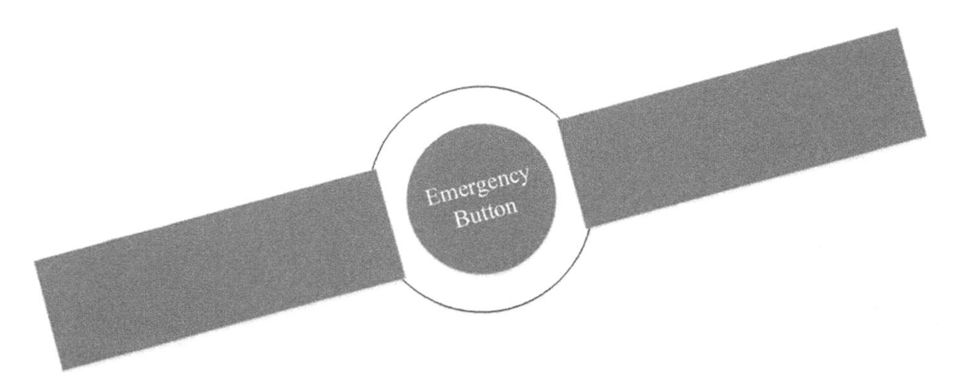

Fig. 2. The designed panic button of the proposed system.

deliver the GPS position with the AIS signal every ten minutes. Considering the overall power consumption, sending out a signal every ten minutes can effectively reduce the energy workload for the wearable device, and the rescue team can receive the signal constantly. The algorithm design is as shown below:

IF pressing the panic button
Sending an emergency alarm every ten minutes
ELSE
Terminate the wearable device
END

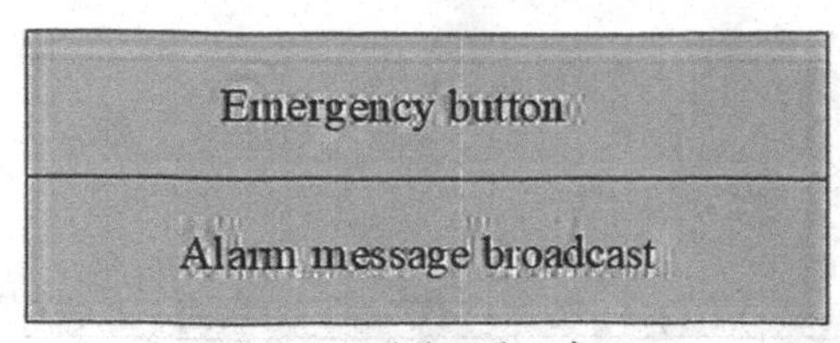

Fig. 3. The function diagram of the proposed system.

In the design, the system will only detect GPS position once when pressing the panic button; afterwards, the system will terminate the detector to avoid large power consumption. Meanwhile, it will also start the buzzer inside the device to make alarm sounds regularly for the rescue team to find the victim rapidly.

3 Performance

The hardware equipment used in the suggested system of this study is as listed in Table 1, which demonstrates the marine AIS processor IC that has the transmission capacity to 20 to 30 nautical miles and uses a GPS detector to check the position. The article applies a portable battery to simulate the experiment, which presents the details in Figs. 4 and 5. Using an Internet of Things (IoT) development board for the software application, confirming through the experiments, the suggested method of this study is feasible.

Table 1. The software and hardware of the proposed system.

Hardware	Software
ASUS X556UR Notebook	Operating System: Windows 10
Drone	Program Language: C Program Development Tools: APP Inventor
Automatic Identification System Chip	

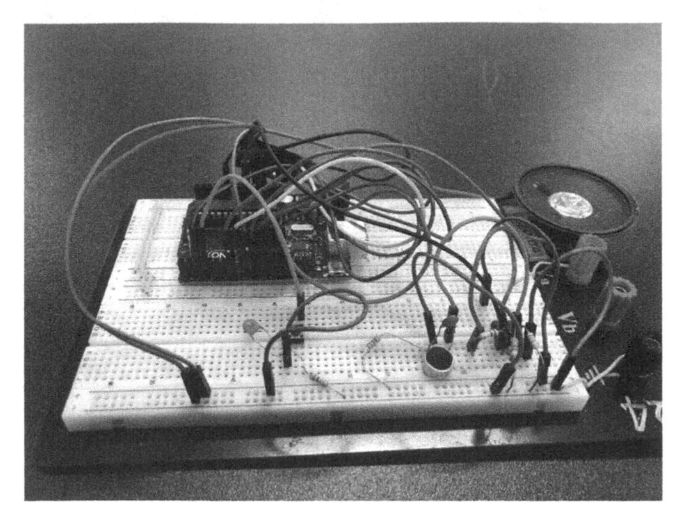

Fig. 4. Experimental trials.

Fig. 5. Portable trials.

4 Conclusions

Today, many countries pay more attention to promote outdoor activities despite there is a certain level of risks for conducting such activities; therefore, wearing suitable safety devices becomes a critical issue. However, some safety devices have smaller transmission ranges that make it fail to send effective signals to rescue teams. This study applies AIS communications that enlarge the transmission ranges and employs a user-friendly operation method for elderly adults. A GPS detector in the system helps the rescue team to confirm the position quickly; moreover, a buzzer makes alarm sounds regularly enables the rescue team to find the victim sooner. The experiment results have proved that the suggested method in this article is capable of enhancing the safety level for people to do outdoor activities.

Acknowledgements. This work was financially supported from the Young Scholar Fellowship Program by Ministry of Science and Technology (MOST) in Taiwan, under Grant MOST109-2636-E-003-001.

References

1. Lin, C., Chen, Y.: A preliminary study on international mountain accident-related policies and systems and Taiwan's search and rescue costs. In: Proceedings of 2017 National Mountain Climbing Conference (2017)
2. Chen, L.: The connection between reformation of mountaineering club organizations and increase in mountain accidents. In: Proceedings of 2015 National Mountain Climbing Conference (2015)
3. Chang, I.-H., Keh, H.-C., Dande, B., Roy, D.S.: Smart hat: design and implementation of a wearable learning device for kids using AI and IoTs techniques. J. Internet Technol. **21**(2), 593–604 (2020)
4. Chou, L.-D., Li, D.C., Chen, W.-S., Yang, Y.-J.: Design and implementation of a novel location-aware wearable mobile advertising system. J. Internet Technol. **18**(5), 963–974 (2017)
5. Ma, Y.-J., Zhang, Y., Dung, O.M., Li, R., Zhang, D.Q.: Health internet of things: recent applications and outlook. J. Internet Technol. **16**(2), 351–362 (2015)
6. Nourbakhsh, I.R., Sycara, K., Koes, M., Yong, M., Lewis, M., Burion, S.: Human-robot teaming for search and rescue. IEEE Pervasive Comput. **4**(1), 72–79 (2005)
7. Lin, C.-C., Chiu, M.-J., Hsiao, C.-C., Lee, R.-G., Tsai, Y.-S.: Wireless health care service system for elderly with dementia. IEEE Trans. Inf. Technol. Biomed. **10**(4), 696–704 (2006)

Using Water Monitoring to Analyze the Livability of White Shrimp

Wu-Chih Hu[1], Hsin-Te Wu[2(✉)], Jun-We Zhan[1], Jing-Mi Zhang[1], and Fan-Hsun Tseng[3]

[1] Department of Computer Science and Information Engineering, National Penghu University of Science and Technology, Penghu, Magong 880011, Taiwan
[2] Department of Computer Science and Information Engineering, National Ilan University, Yilan, Yilan 26047, Taiwan
p11o0304@mail2000.com.tw
[3] Department of Technology Application and Human Resource Development, National Taiwan Normal University, Taipei, Taipei 10610, Taiwan

Abstract. The research develops an intelligent aquaculture system to detect the water quality of a culture pond. Additionally, using Fuzzy Logic to evaluate water quality that influences the aquaculture livability. Each species requires a different environment of water quality; therefore, the study utilizes an intelligent aquaculture system to detect the water quality of white shrimp ponds. After using Fuzzy Logic to analyze water quality, the result is delivered as equally divided into five levels of signals sections. The purpose of the research is to understand whether the aquaculture environment is suitable for white shrimps by detecting the water quality; consequently, through studying the livability to understand the importance of water quality. From the experimental results, the water quality of targeted aquaculture ponds are all within the livability range of white shrimp; the result has shown a livability rate of 33%, which is considered high livability in marine white shrimp farming. Hence, it is concluded that water quality has a high correlation with livability. Moreover, the study demonstrates that water monitoring and water quality analysis are beneficial to monitor the aquaculture environment, which can further increase the livability of white shrimp and boost income.

Keywords: Fuzzy · Water monitoring · Agricultural technology

1 Introduction

Today, the level of reliance on fishery products has been increasing in many countries; however, with the greater demand from human beings, overfishing becomes a major issue that causes the scarcity of marine species. Among which, marine fishing accounts for 1.4% [1] Gross Domestic Product in India. The global demand for prawns becomes extremely large despite the overfishing issue that causes decreasing catches. Aquaculture can fulfill the scarcity of fish and prawns; yet, outdoor aquaculture has to overcome the environmental factors to increase the yield. Therefore, intelligent aquaculture systems

X. Wang et al. (Eds.): 6GN 2020, LNICST 337, pp. 458–463, 2020.
https://doi.org/10.1007/978-3-030-63941-9_37

can monitor diverse environmental factors in culture ponds and send alarms timely to help operators control the pond conditions.

Literature [1] and [7] present an intelligent system to record various environmental factors in the culture pond when inputting feed; further, analyzing the aquaculture skills after collecting the data. Literature [2] mainly uses Fuzzy Logic to judge the water quality of a pond, literature [3] utilizes augmented reality to manage culture ponds, and literature [4] uses an Autonomous Underwater Vehicle for automatic navigation underneath the water to detect water quality and take photos, which helps to understand the conditions of the pond. Additionally, literature [5] applies NB-IoT to detect the water quality of a pond because NB-IoT can reduce the energy consumption of the system. In literature [6], it saves the IoT data on the cloud, which enables the future Big Data analysis. Finally, literature [8–10] construct an intelligent culture system for applying in stock farming and aquaculture; it increases the overall yield of the industry.

This research uses water monitoring to analyze the growth rate of white shrimp; it detects the water quality and uses Fuzzy Logic to analyze the data. The result of Fuzzy Logic is divided into five signals, the optimal quality of water is indicated as 5 and the worst quality of water is 1, when the system detects the signal 1, it would notify the operator to conduct relevant procedures to increase the water quality. At the end of this study, we added all of the signal numbers and averaged the result to compare the overall growth rate; the rate enables us to conclude whether the water quality is proportional to the growth rate. The experimental pond in this study is outdoor; the aquaculture species is white shrimp. From the experimental result, the data shows that the overall water quality is upper-intermediate and the growth rate is over 30%. After consulting with aquaculture experts, the growth rate is considered high livability, which concludes that water quality is proportional to the growth rate.

2 The Proposed Scheme

2.1 System Model

The study uses water monitoring to analyze the growth rate of white shrimp. Firstly, we constructed an outdoor experimental pond and built an intelligent system to monitor the environment in the white shrimp pond. The detector of the monitor controls sea salt, oxygen content, power of hydrogen, and oxidation-reduction potential because these are the main factors that influence the livability and yield of white shrimp. By transferring data to the server through the 4G Internet, the research can analyze whether the environmental data is suitable for white shrimps to survive. The schematic diagram is shown in Fig. 1, the intelligent monitor system would deliver the aquaculture data to the server, which would further define the data into five different signals according to the environmental data of white shrimp and calculate the overall condition. When the signal is too low, the system would notify the operator to adjust the environment, which confirms the method suggested in this study can increase the overall growth rate effectively.

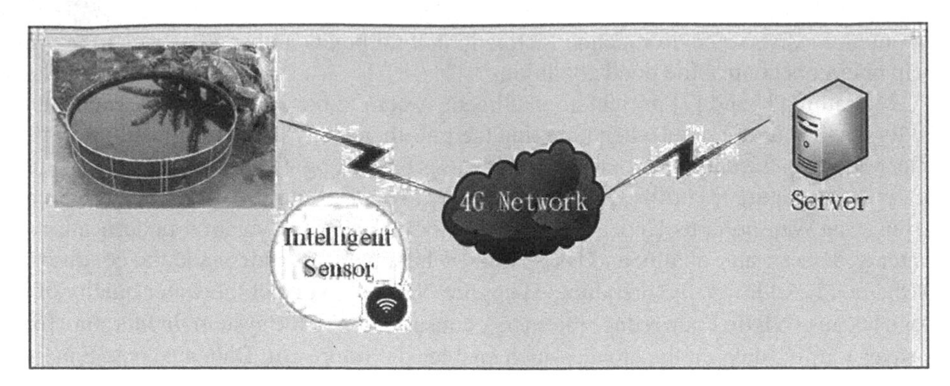

Fig. 1. The schematic diagram of the system.

2.2 Water Monitoring Analysis

The study uses Fuzzy Logic to analyze the quality of water, which is divided into five levels of signals. The sensor of the water quality monitor detects sea salt, oxygen content, power of hydrogen, and oxidation-reduction potential. Set each range of water quality that is suitable for white shrimp as WQi and further separate the range into five sectors; the optimal quality of water means the best living environment for white shrimp while the worst quality of water means the poorest environment for white shrimp. The definitions of the signals are P as 5 points, G as 4 points, N as 3 points, B as 2 points, and NB as 1 point; the formula is defined as below:

For i=1~n

If $WQ_{i,P} <= \lceil WQ_i \rceil > WQ_{i,G}$

$OP_i = 5$

Else IF $WQ_{i,G} <= \lceil WQ_i \rceil > WQ_{i,N}$

$OP_i = 4$

Else IF $WQ_{i,N} <= \lceil WQ_i \rceil > WQ_{i,B}$

$OP_i = 3$

Else IF $WQ_{i,B} <= \lceil WQ_i \rceil > WQ_{i,NB}$

$OP_i = 2$

Else

$OP_i = 1$

End

End

The abovementioned formula mainly calculates and finds out which signal sector the water quality should fit in; when OP_i becomes 1, the system would send out alarms for the operator to notice the changes of water quality. Afterward, to calculate the average numbers of water quality by using the formula, $avg\left(\sum_i OP_i\right)$, which enables us to observe which kind of water quality is the most suitable one. From the formula and the final growth rate calculated by this study, it is concluded that water quality is proportional to the growth rate.

3 Experimental Results

The research offers an intelligent aquaculture system that monitors sea salt, oxygen content, power of hydrogen, and oxidation-reduction potential. In the water quality of aquaculture, the oxidation-reduction potential plays a vital role in the system; therefore, the research aims to normalize the values of oxidation-reduction potential and power of hydrogen, which is shown in Fig. 2 and Fig. 3. Later, to analyze the data by using the formula presented in the study, the result is listed in Table 1. From the values showed in Table 1, it demonstrates that all of the values of the oxidation-reduction potential in the targeted white shrimp ponds are near 5 points, which means the water quality of the experimental white shrimp ponds fulfills the living conditions of white shrimp aquaculture. Finally, the growth rate of white shrimp is more than 30%, after consulting with aquaculture experts; the value is considered high livability. Thus, the study can conclude that water quality is proportional to the growth rate of white shrimp.

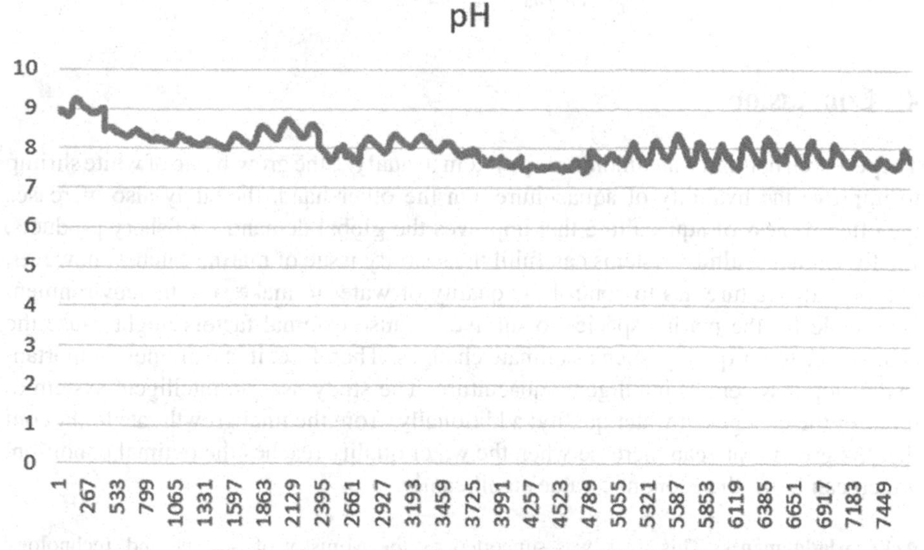

Fig. 2. The normalization of the power of hydrogen.

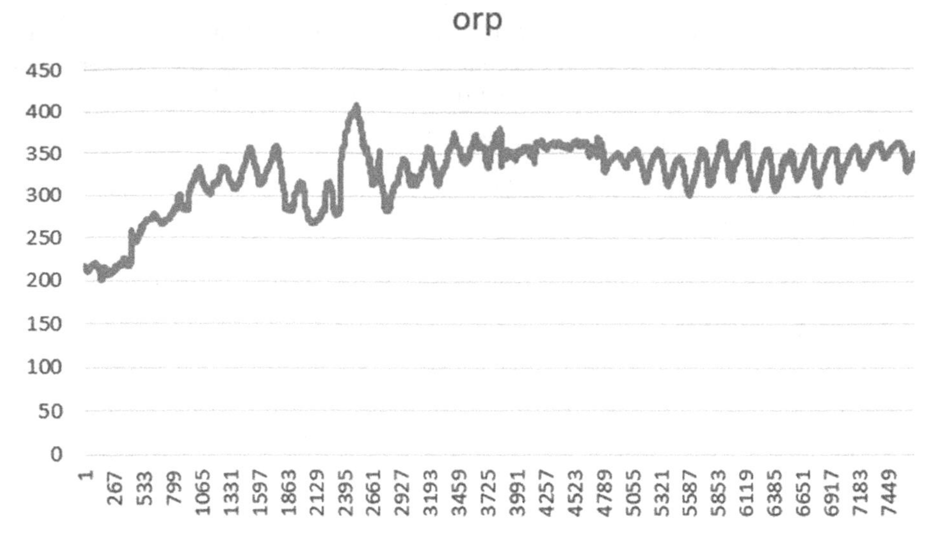

Fig. 3. The normalization of oxidation-reduction potential.

Table 1. Water quality analysis.

Month	Water quality		
	PH	ORP	DO
March	4	5	4
February	4	4	5

4 Conclusions

The research builds a water monitoring system to analyze the growth rate of white shrimp to improve the livability of aquaculture. On the other hand, the study also increases the effectiveness of aquaculture that improves the global demand for fishery products. Intelligent aquaculture systems can fulfil the scarcity issue of marine catches; however, outdoor aquaculture has to control the quality of water to make sure the environment is suitable for the marine species to survive because external factors might cause the changes of water quality, such as climate changes. Therefore, it is extremely important to develop a system of intelligent aquaculture. The study uses an intelligent system to monitor the changes of water quality; additionally, from the final growth rate to find out that the growth rate can increase when the water quality reaches the optimal condition, which can boost the economic value significantly.

Acknowledgements. This work was supported by the Ministry of Science and Technology (MOST), Taiwan, under Grants MOST107-2221-E-346-007-MY2 and MOST109-2636-E-003-001.

References

1. Jing, P., Jing, H., Zhan, F., Chen, Y., Shi, Y.: Agent-based simulation of autonomous vehicles: A systematic literature review. IEEE Access, vol. 1, pp. 1 (2020)
2. Abbassy, M.M., Ead, W.M.: Intelligent greenhouse management system. In: 2020 6th International Conference on Advanced Computing and Communication Systems (2020)
3. Gomathi, S., Prabha, S.U., Sureshkumar, G.: Photovoltaic powered induction motor drive using a high efficient soft switched sazz converter. In: 2020 6th International Conference on Advanced Computing and Communication Systems (2020)
4. Prince Samuel, S., Malarvizhi, K., Karthik, S., Mangala Gowri, S.G.: Machine learning and internet of things based smart agriculture. In: 2020 6th International Conference on Advanced Computing and Communication Systems (2020)
5. Madiwalar, S., Patil, S., Meti, S., Domanal, N., Ugare, K.: A survey on solar powered autonomous multipurpose agricultural robot. In: 2020 2nd International Conference on Innovative Mechanisms for Industry Applications (2020)
6. Vijay, Saini, A.K., Banerjee, S., Nigam, H.: An IoT instrumented smart agricultural monitoring and irrigation system. In: 2020 International Conference on Artificial Intelligence and Signal Processing (2020)
7. Huang, K., et al.: Photovoltaic agricultural internet of things towards realizing the next generation of smart farming. IEEE Access, vol. 1, pp. 1 (2020)
8. Zhu, L., Walker, J.P., Rdiger, C., Xiao, P.: Identification of agricultural row features using optical data for scattering and reflectance modelling over periodic surfaces. IEEE J. Sel. Topics Appl. Earth Obs. Remote Sens., vol. 1, pp. 1 (2020)
9. Jerosheja, B.R., Mythili, C.: Solar powered automated multi-tasking agricultural robot. In: 2020 International Conference on Innovative Trends in Information Technology (2020)
10. Iglesias, N.C., Bulacio, P., Tapia, E.: Internet of agricultural machinery: Integration of heterogeneous networks. In: 2020 IEEE International Conference on Industrial Technology (2020)

Virtual Reality, Augmented Reality and Mixed Reality on the Marketing of Film and Television Creation Industry

Rui Ge[1], Hsuan-Yu Chen[2], Tsung-Chih Hsiao[1,3], Yu-Tzu Chang[4], and Zhen-Yu Wu[5(✉)]

[1] College of Computer Science and Technology, Huaqiao University, Quanzhou, Fujian, China
[2] Department of Radiology, Tri-Service General Hospital, National Defense Medical Center, Taipei, Taiwan
[3] Fujian Key Laboratory of Big Data Intelligence and Security, Huaqiao University, Quanzhou, China
[4] Department of Marketing and Logistics Management, Chaoyang University of Technology, Taichung, Taiwan
[5] Department of Information Management, National Penghu University of Science and Technology, Penghu, Taiwan
zywu@gms.npu.edu.tw

Abstract. Virtual reality and augmented reality technology has been on the road to scientific research for many years since the end of the twentieth century, and although the scope of mixed reality application is relatively narrow compared to the first two applications, we have seen the results. Realistic contents also have exploring progress in the field of film and television creation in recent years, such as virtual reality technology can be utilized to make film clips. However, in fact, these technologies can be applied in film and television creation far more than that. I then illustrate some other aspects which these realistic contents can also be applied to in film and television creation in the future and the influence on the industry.

Keywords: Virtual reality · Augmented reality · Mixed reality · Film and television creation · Film clips

1 Introduction

What is realistic contents? Virtual reality blocks the real world and creates a fully digital, immersive experience for users. Augmented reality, which overlays digital creation content in the user's real world. Mixed reality, including augmented reality and augmented virtuality, is a new visual environment that combines reality and virtual worlds [1] (see Fig. 1).

Realistic contents are now in steady development and have achieved good results in all sectors, but in the film and television industry, their development direction is very

X. Wang et al. (Eds.): 6GN 2020, LNICST 337, pp. 464–468, 2020.
https://doi.org/10.1007/978-3-030-63941-9_38

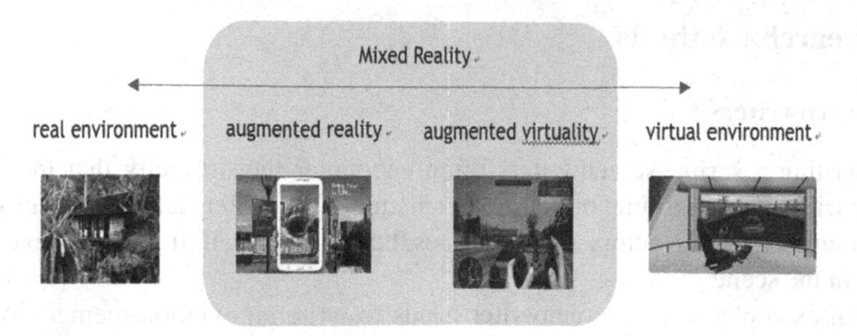

Fig. 1. Reality-virtuality continuum

narrow, with most applications focusing on improving sensory experiences such as VR movies [2], unlike the Comfort that Based on Parallax and Motion 3D movies bring to us [3], VR movies allow viewers to walk into a movie scene to see their surroundings at 360 degrees. What is more, AR immersive panoramic studios and so on.

However, the application of realistic contents in the film and television industry is far more than those, I would like to solve these problems by putting forward existing problems and providing some direction to the film and television industry practitioners to better use this technology.

Here are some examples about what fields these realistic contents have been applied.

(1) The first example is the VR glove, an open source Etextile data glove which is designed to facilitate detailed manipulation of 3d objects in VR, enabling scientists to accomplish a range of spatially complicated molecular manipulation tasks [4].
(2) The second example is also about VR. VR can be used by athletes to prepare for the Olympics. Skiers, for example, can wear VR helmets to simulate a run on the mountain, while ski-shaped balance boards provide some tactile feedback. What is more, it can also applied to the court where witnesses use it to restore the scene, education [5–7] and so on.
(3) The third example is AR mask. Osterhout Design Group, a well-known AR smart glasses manufacturer, has developed a product that uses AR-technology masks that will allow pilots to clearly see other scenes in a smoke-filled cockpit. Designed to help pilots land safely in an emergency, such as in smoke. There are other AR applications, for instance, AR also can enhance awareness on green consumption of electronic devices [8], be applied to sports entertainment [9] and advances designer's understanding of the fabrication equipment as a platform [10].
(4) The fourth example is the application of MR in aviation. Microsoft partnered with Western Michigan University to integrate MR technologies such as Microsoft HoloLens into aviation education. Currently, there are two ways to use it. One is a new simulation that can help pilots prepare for changes in the weather. Another application is an interactive MR application that allows students to explore the various components of an airplane.

2 Research Methods

2.1 Screenwriters

When writing a script, screenwriters often encounter the difficulty that they are not familiar with the background of the subject matter to be written, and cannot fully realize the interaction between actors and set scenes, because after all, it is impossible to write a script in the scene.

As an example, now a screenwriter needs to write an aviation-themed TV series, but he certainly did not have a pilot background, do not know the pilot's work process, and also do not have aviation-related knowledge, then to solve this problem, he can use realistic contents to experience the scene, help to complete the work. What is more, in Harry Potter and the Chamber of Secrets, the scene of Dobby appeared in Harry's home, needs three parts of the interaction, Harry and Dobby, Dobby and items and Harry and the objects, then in this case, if the screenwriter cannot see the layout of the home will easily increase the pressure of the director and props group, so in order to improve efficiency, more natural interaction, the participation of realistic contents is needed.

Another example is when a story completely occurred in a villa, then the screenwriter can use the realistic contents to be in that villa, he can just write the script at home. This method saves time and effort, greatly improve efficiency, and is suitable for achieve the effect of one shot.

2.2 Directors, Anyone Who Wants to Make a Movie (Especially Movie that Requires Fertile Imagination)

Many young people who love movies will have a dream to become a big director like Christopher Nolan and James Cameron, but even if they have great ideas but cannot find a way to achieve, because even though they have learned how to tell stories, even formed their own genres [11], without the help of the ambitious production team, it simply cannot be achieved. For example, the Lord of the Rings, it is impossible to be completed by one person. However, now times are different, with the advent of realistic contents, they can create their own movie world, where they can have unlimited imagination, make their own scenes, actors and so on. This does provide a way for people who don't have a behind-the-scenes team to solve the problems.

Traditional filmmaking requires actors to wear motion-capture suits, and to use green screens and a lot of post-production technology to get the final work by shaping visual effects (VFX) and computer-generated imagery (CGI) characters. The entire production process takes months or even years to create a rendering and output of the movie. Realistic contents can now be used to upend traditions, such as the AR short film NEST mentioned above. It completely subverts tradition. Now, you can put the made CGI characters, import into the camera, mobile phone or other shooting equipment, and then shoot at the real environment. As a director, you can guide its actions in real-time to achieve the combination of virtual characters and real-world environment. Avoiding the limitations of monster characters that can't be shot in real-world scenes and it's also better to realize the simultaneous execution of these thing. What is more, it won't cause the problem which now exists of virtual and real things not to be combined smoothly in the post-production process.

It will achieve the real popularity of film production, no more restrictions to the site, characters and actors, being able to fulfill what users really want to show considerably.

2.3 Film and Television Theme Park

As we all know, Film and television are inspiring, it would want you to try, to do something. For example, watching documentary film may let you be concerned about climate change [12], then watching movies with beautiful scenery, you will want to go to the location to see. Film and television are an effective way of tourism promotion, this kind of marketing method can attract people to the place they have seen on the screen to travel, and the factor is not only because of the attraction of the natural landscape, in fact, the most important factor is its symbolism, storyline, character relationship, star, thrilling scenes and exciting endings, and even some cartoons without exterior view, such as "Beauty and the Beast," "The Lion King" and "Notre Dame", also attract many visitors to the scene of the cartoon.

This has also led to a large number of scholars to research, to find out the different forms and motivations of film and television tourism, and in more than a decade of research, found that most of the film and television cities and theme parks have these problems. First, the sensory nature of environmental experience is too single. Many buildings are only on the level of being taken pictures by tourists, lacking a sense of participation and integration with the real environment. And also the device experience is not interactive.

But if you can show all the film clips taken there at the corresponding shooting venue, then allow visitors to pick and choose their favorite clips, wear glasses, so that fans can actually enter the shooting scene to feel. Such transmission has more effect on people's emotional transmission [13, 14]. This form will greatly enhance their visiting experience and participation, and also let them pay attention to all aspects of the scene, they can choose to pay attention to the main character, or his favorite supporting role, in order to avoid the film shooting's disadvantages of only focusing on the main characters. Moreover, some directors may be very attentive to details but cannot be totally showed in the film, such as some costume drama, in fact, each of the main character's clothing are perfectly restored to the scene at that time, but the film may not reflect this, then this form of visiting will greatly enhance the audience's favorability, so that the audience feel their intentions.

In fact, the film and television theme park does not have to open in the shooting place, can also be completely virtual, like a game, let fans go into the virtual world playing a role [15].

3 Conclusions

In the film and television industry, such as the division of Time and drama of the show, the conduct of marketing means can actually use realistic contents technology through the algorithm to obtain the fastest, the least steps of the solution. In addition to the above proposed 3 points of application, the common benefits are bound to greatly improve work efficiency, complete industrial upgrading.

References

1. Milgram, P., Kishino, F.: A taxonomy of mixed realityvisual display. IEICE Trans. Inf. Syst. **77**(12), 1321–1329 (1994)
2. Dempsey, P.: VR in movies: Hollywood has fully embraced the dawn of the VR age. Eng. Technol. **11**(3), pp. 31 (2016)
3. Tian, F., Xu, H., Feng, X., Sánchez, J.A., Wang, P., Tilling, S.: Comfort evaluation of 3D movies based on parallax and motion. J. Disp. Technol. **12**(12), 1695–1705 (2016)
4. Glowacki, B.R., Freire, R., Thomas, L.M., O'Connor, M., Jamieson-Binnie, A., Glowacki, D.R.: An open source Etextile VR glove for real-time manipulation of molecular simulations (2019). https://www.researchgate.net/publication/330357655_An_open_source_Etextile_VR_glove_for_real-time_manipulation_of_molecular_simulations
5. Merchant, Z., Goetz, E.T., Cifuentes, L., Keeney-Kennicutt, W., Davis, T.J.: Effectiveness of virtual reality-based instruction on students' learning outcomes in K12 and higher education: Ameta-analysis. Comput. Educ. **70**, 29–40 (2014)
6. Lee, J.H., Shvetsova, O.A.: The impact of VR application on student's competency development: A comparative study of regular and VR engineering classes with similar competency scopes. Sustainability, **11**, pp. 2221 (2019)
7. Manuel Sanchez-Garcia, J., Toledo-Morales, P.: Converging technologies for teaching: Augmented reality, BYOD, flipped classroom. Red-Revista De Educacion a Distancia. (55) (2017)
8. Bekaroo, G., Sungkur, R., Ramsamy, P., Okoloa, A., Moedeena, W.: Enhancing awareness on green consumption of electronic devices: The application of augmented reality. Sustain. Energy Technol. Assess. **30**, 279–291 (2018)
9. Mahmood, Z., Ali, T., Muhammad, N., Bibi, N., Shahzad, I., Azmat, S.: EAR: Enhanced augmented reality system for sports entertainment applications. KSII Trans. Internet Inf. Syst. **11**(12), 6069–6091 (2017)
10. Poustinchi, E.: Subtractive digital fabrication with actual robot and virtual material using a MARI platform. Int. J. Archit. Comput. SI. **16**(4), 281–294 (2018)
11. Tseng, C.-I.: Beyond the media boundaries: Analysing how dominant genre devices shape our narrative knowledge. Discourse Context Media **20**, 227–238 (2017)
12. Bieniek-Tobasco, A., McCormick, S., Rimal, R.N., Harrington, C.B., Shafer, M., Shaikh, H.: Communicating climate change through documentary film: Imagery, emotion, and efficacy. Clim. Change. **154**(1–2), 1–18 (2019)
13. Fikkers, K.M., Piotrowski, J.T.: Content and person effects in media research: Studying differences in cognitive, emotional, and arousal responses to media content. Media Psychology. Available online: https://doi.org/10.1080/15213269.2019.1608257. Accessed on 30 Apr 2019
14. Castle, J.J., Stepp, K.: Silver screen sorting: Social identity and selective exposure in popular film viewing. Soc. Sci. J. **55**(4), 487–499 (2018)
15. Timplalexi, E.: Shakespeare in digital games and virtual worlds. Multicult. Shakespear. Transl. Approp. Perform. **18**(1), 129–144 (2018)

VR and AR Technology on Cultural Communication

Minqi Guo[1], Hsuan-Yu Chen[2], Tsung-Chih Hsiao[3], Zhen-Yu Wu[4](✉), Huan-Chieh Tseng[5], Yu-Tzu Chang[6], and Tzer-Shyong Chen[5]

[1] School of Humanities and Social Sciences, Harbin Institute of Technology, Shenzhen, China
[2] Department of Radiology, Tri-Service General Hospital, National Defense Medical Center, Taipei, Taiwan
[3] College of Computer Science and Technology, Huaqiao University, Quanzhou, Fujian, China
[4] Department of Information Management, National Penghu University of Science and Technology, Penghu, Taiwan
zywu@gms.npu.edu.tw
[5] Department of Information Management, Tunghai University, Taichung, Taiwan
[6] Department of Marketing and Logistics Management, Chaoyang University of Technology, Taichung, Taiwan

Abstract. The theme of this platform is built around "the road of Hess", which was initiated in the modern maritime Silk Road capital-Quanzhou, as the core of design and programming. The history of Silk Road will be presented with panorama technology, which is one of the network multimedia technology. In the form of Web APP, the real records and spread of the humanities and scenic sites of Quanzhou will be showed in 720° perspective. An immersive experience of 3D live virtual can be expected. The aim of this study is to simulate and realize the all-around real scene. In the paper, the tasks of the VR panorama fused construction include image acquisition, image registration, image smoothing, image modification by the algorithm and mathematical theory behind splicing fusion, the panoramic camera category and projection model, panoramic transformation and collection principle, the image registration of feature point and transform frequency domain. The product positioning of "maritime silk route culture" will be presented by 18 selected spots in Quanzhou. The study, with an overall design and an organization of connecting different scenes, is for the virtual scene roaming system, which is based on image.

Keywords: Maritime Silk Route · 3D virtual scene · Product design · VR panorama

1 Introduction

With the new-developing technology, virtual reality technology (VR) is applied in user scenarios more frequently and is becoming increasingly important in the computer application field. Through image and video we can realize the 720-degree immersive

X. Wang et al. (Eds.): 6GN 2020, LNICST 337, pp. 469–474, 2020.
https://doi.org/10.1007/978-3-030-63941-9_39

panorama visual experience has been fulfilled by using the Virtual Tour System with 3D Real Scene, that need actual scene to generate [1]. Different from normal viewing angles, VR panoramic warping technology acquire better life-like viewing experience because it demands several steps including collecting all 360° photos, mending them all together and uploading the result on the mobile Internet. Users may feel like they are experiencing the situation themselves and learn new lessons while playing [2–4].

Three methods which distribute 3D life-like scene are described in this research, including the linear fusion algorithm, the future point based image saving algorithm and the gradually fading out algorithm. What makes this paper different from others is applying both the VR and AR on "Maritime Silk Route Culture." [5–7].

2 Method

3D Panoramic Roaming Technology is an advanced virtual reality technology. It only needs some daily-accessible electric devices such as mobile phones or laptops to connect the virtual scene and network whereas VR demands more complex equipment to achieve its goal. Immersion experience can be achieved. It is based on settled spots that include panorama images generation and bring 720-degree full perspective effect [8–10].

The main concept of 3D Panoramic Roaming Technology [11] is how to produce panoramic image fusion. To ensure the quality of image fusion, it is important to pay attention to the panoramic images received. There are three method to achieve the goal, which combine panoramic head and ordinary camera, an panorama camera and applying graphics computing [12]. In this paper, the method is chosen to generate panoramic images with cost-effective equipment.

Therefore, several major contents will be successively analyzed in this paper – the panoramic image acquisition, feature point recognition on image registration task, and use smoothing algorithm on the image fusion process. The framework of image processing is shown in Fig. 1.

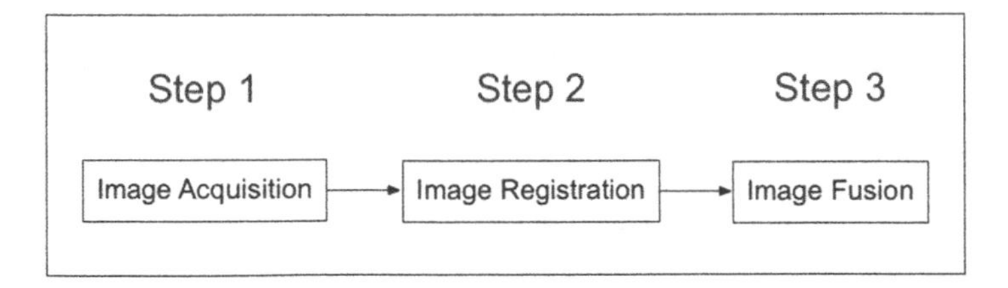

Fig. 1. The framework of image processing

2.1 Acquire Panorama Images

The rule of panoramic image acquisition is to collect more than 30% of the contents with overlapping areas, so that they can identify similar feature points and the later image fusion to generate three-dimensional panorama. Multiple lenses, integrating light and shadow, head and tripod combined with the 360-degree camera are the key tools to fulfill the experiment.

The functions of ordinary camera transformation including translation, rotation, scaling in horizontal and vertical directions, are as shown in Fig. 2 and 3.

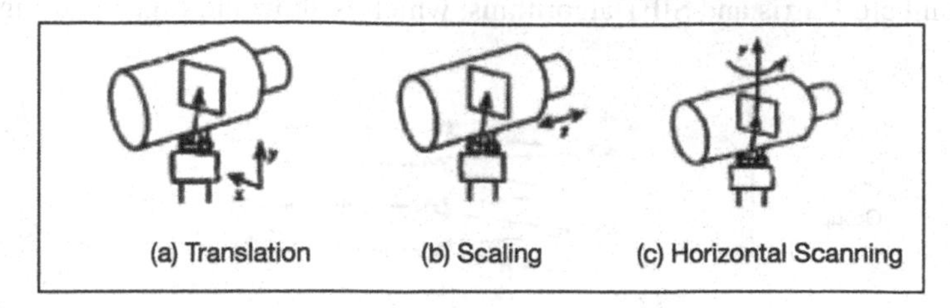

Fig. 2. Camera transformation of (a) (b) (c)

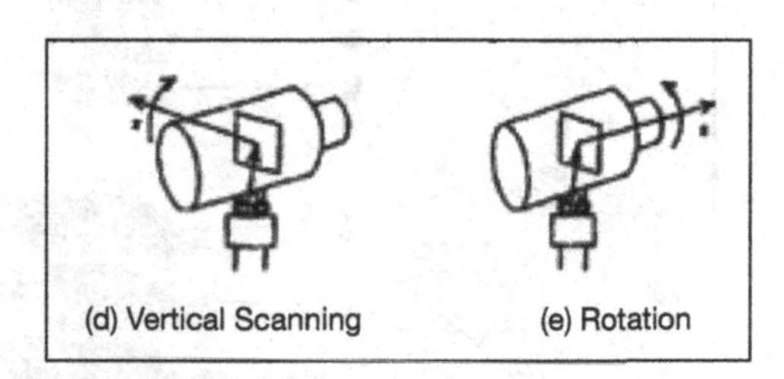

Fig. 3. Camera transformation of (d) (e)

However, in order to simulate the reduction of human fixed visual direction, the panoramic image transformation generated requires the camera in a fixed position and vertical direction without panning and zooming. The idea is to take the center of the camera as the eyes of human. The movement like the translation or rotation of the camera simulates the action of human head. For example, looking up and down, nodding the head and moving the head, these create all-round perspective experience of 720° for users. Figure 4 respectively simulates the rotation principle of the central axis point and plane.

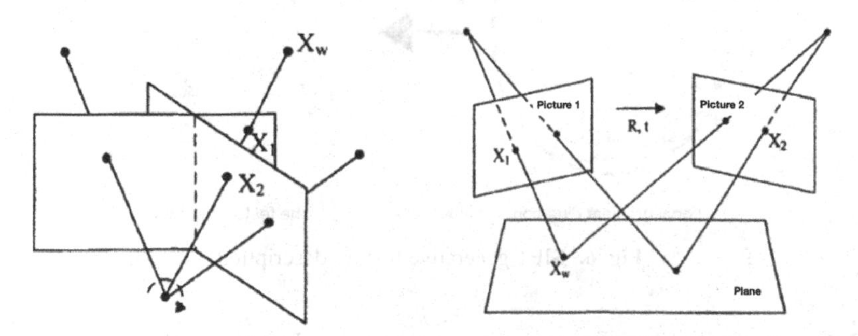

Fig. 4. Arbitrarily rotation with the core point and in the plane

2.2 Feature Point Recognition Methods on Image Registration Task

Firstly, among all methods of image processing, matching feature point is the most famous method. The principle is to label the feature points of the real panorama images with selected important attributes such as edges or lines. The main algorithms to match feature point are Harris and SIFT algorithms, which is shown in Fig. 5 and Fig. 6.

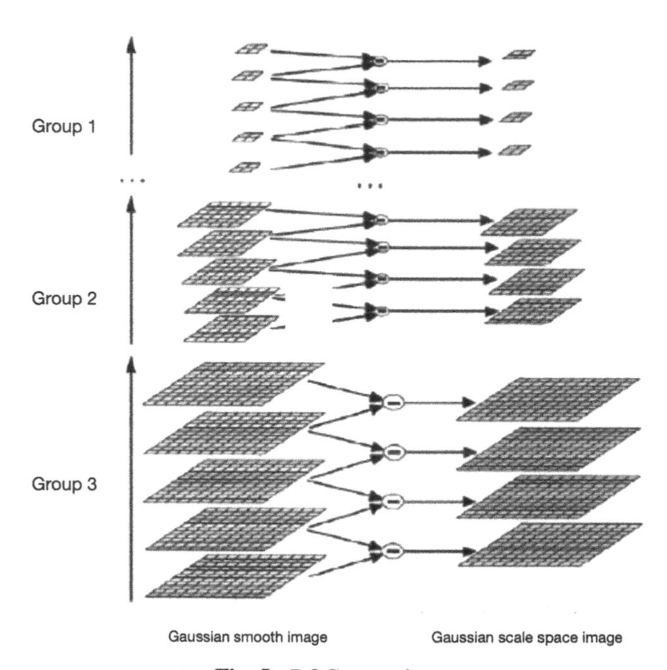

Fig. 5. DOG space image

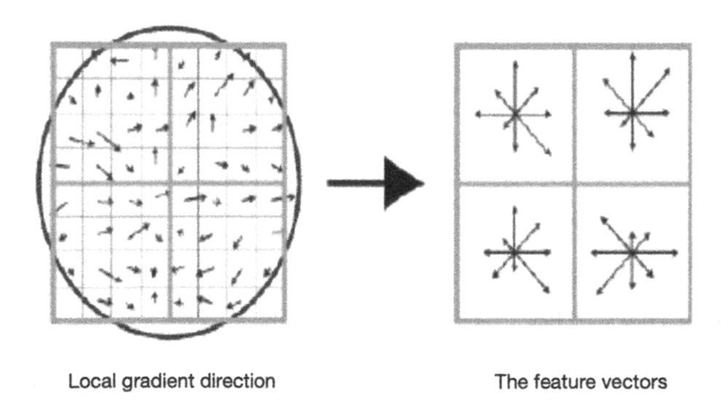

Fig. 6. SIFT generative feature description

Before the image registration being operated, the feature points must be read and calculate first. Feature point matching approach allows the feature point to expand in an easier and a more active way. Moreover, with lower sample size requirement and high flexibility, rich image information and gray coefficient can be reserved to achieve accurate reading. However, the wrong image registration still forbids feature point matching approach its goal. As a result, to avoid image fusion failure, the RANSAC algorithm is needed to check and purify the lapses.

2.3 Use Smoothing Algorithm on Images Fusion Task

Before adjusting image color and linking all images, the images must be enrolled and adjusted. The image may not be able to save accurately if the coordinate pixels is shifted accidently. Furthermore, more missteps including color fading, shapes and lines mangling, and objects dislocating may also show up while restoring the image.

Thus, how to apply the most suitable image synthesis algorithm to not only make the image more vivid but elevate the restoration degree and slash the vague objects. Linear fusion algorithm and fade-in and fade-out algorithm are on the top list of image synthesis algorithms we have known so far.

Linear fusion algorithm demands the user to analyze the pixel value of the images and check if there is any mistake in the saved images and lastly to confirm which side does the pixel overlap appears. If the overlap is on the right side, the image registration task of the left side is assumed to be failed, the right side will obtains full measurement weight. The superposition However, if the overlap is neither on right side or left side, by applying the following 2-1 formula, the measurement index can be found out.

$$P(x) = 0.5 \times P_1(x) + 0.5 \times P_2(x) \quad (2\text{-}1)$$

When images are very distinct from others, the linear fusion algorithm can be applied though edges might be the same. The algorithm allows images to transmit clearly and accurately and decline the lapse showing up from the image.

3 Conclusion

As China has grown into a more significant role on the globe, "One Belt and One Road" has become a popular topic. However, none of the studies has combined this issue with new digital technology. This research is the first to combine the most recognized Web App version of virtual reality panorama with the 3D panoramic roaming system of Marine Silk Road Culture which is at Quanshou. The unique culture is waiting to be explored and documented by applying the 720-degree lifelike panoramic visual experience. Advocating the Marine Silk Road Culture, which may bring more investment to the "One Belt and One Road" tourist cities, to people's acknowledgement is the main focus of this research.

By acquiring the panoramic image, registering image with future-based recognition, mixing the image with suitable algorithm has made this research to be the first one to present the panoramic images successfully. The following step is importing the immersive experience providing technology, 3D panoramic roaming system, on the panoramic images

One limitation of this study is that the sharpness of panoramic technology still has room to be improved. Therefore, as future research, it should pay attention to steps of obtaining panorama such as shooting, registering and fusion.

References

1. Dempsey, P.: VR in movies: hollywood has fully embraced the dawn of the VR age. Eng. Technol. **11**(3), 31 (2016)
2. Glowacki, B.R., Freire R., Thomas, L.M., O'Connor, M., Jamieson-Binnie, A., Glowacki, D.R. : An open source Etextile VR glove for real-time manipulation of molecular simulations (2019)
3. Lee, J.H., Shvetsova, O.A.: The impact of VR application on student's competency development: a comparative study of regular and VR engineering classes with similar competency scopes. Sustainability **11**, 2221 (2019)
4. Manuel Sanchez-Garcia, J., Toledo-Morales, P.: Converging technologies for teaching: augmented reality, BYOD, flipped classroom. Red-Revista De Educacion a Distancia **55**, 22–34 (2017)
5. Mahmood, Z., Ali, T., Muhammad, N., Bibi, N., Shahzad, I., Azmat, S.: EAR: enhanced augmented reality system for sports entertainment applications. KSII Trans. Internet Inf. Syst. **11**(12), 6069–6091 (2017)
6. Ramirez, E.J., LaBarge, S.: Real moral problems in the use of virtual reality. Ethics Inf. Technol. **20**(4), 249–263 (2018). https://doi.org/10.1007/s10676-018-9473-5
7. Bieniek-Tobasco, A., McCormick, S., Rimal, R.N., Harrington, C.B., Shafer, M., Shaikh, H.: Communicating climate change through documentary film: imagery, emotion, and efficacy. Clim. Change **154**(12), 1–18 (2019)
8. Castle, J.J., Stepp, K.: Silver screen sorting: social identity and selective exposure in popular film viewing. Soc. Sci. J. **55**(4), 487–499 (2018)
9. Timplalexi, E.: Shakespeare in digital games and virtual worlds. Multicultural Shakespeare-Transl. Appropriation Perform. **18**(1), 129–144 (2018)
10. Gao, H., et al.: Super-fast refresh holographic display based on liquid crystal films doped with silver nanoparticles. IEEE Photonics J. **11**(3), 1–7 (2019)
11. Van der Ham, I.J.M.: Elapsed time estimates in virtual reality and the physical World: the role of arousal and emotional valence (2019). http://apps.webofknowledge.com/full_record.do?product=UA&search_mode=GeneralSearch&qid=2&SID=8A4hHbvUdc9IuSCQL4b&page=1&doc=1. Accessed 26 Oct 2019
12. Kim, P.W., Lee, S.: Audience real-time bio-signal-processing-based computational intelligence model for narrative scene editing. Multimedia Tools Appl. **76**(23), 24833–24845 (2016). https://doi.org/10.1007/s11042-016-4144-1

Workshop on Enabling Technologies for Private 5G/6G (ETP5/6G)

Research on Time-Sensitive BBU Shaper for Supporting the Transmission of Smart Grid

Jiye Wang[1], Chenyu Zhang[2](✉), Yang Li[1], Baozhong Hao[1], and Zhaoming Lu[2]

[1] Big Data Center of State Grid Corporation of China, Beijing, China
{jiye-wang,yang-li6,baozhong-hao}@sgcc.com.cn
[2] Beijing Laboratory of Advanced Information Networks, Beijing University of Posts and Telecommunications, Beijing, China
{octopus,lzy0372}@bupt.edu.cn

Abstract. Large-scale of data terminals are deployed in the smart grid, especially the power distribution network. The effectiveness of data collection or scheduling for Source-Grid-Load-Storage is affected to some extent by the accuracy of the time synchronization of the terminal equipments. Here we propose a design of BBU Shaper to deal with time synchronization error on the above issues. This article analyzes the impact of synchronization errors between terminal devices within the coverage of a single base station which provides deterministic transmission for the power IoT. For the time-sensitive service flows, when the terminal has different priorities according to the service type, the time synchronization error of the terminal will affect the stability and effectiveness of the time-sensitive service flow transmission on the network. It is mainly reflected in timestamp deviation of the data collection, disrupting the determinism of other service flows and reducing the overall network resource utilization. Based on the analysis of the above problems, we proposes an edge data shaper which can reduces the impact of air interface time synchronization errors on big data services through mechanisms such as "ahead awaits; overtime elevates". Evaluation shows that the Shaper can reduce the impact of time synchronization errors on the average delay of services to a certain extent.

Keywords: Edge shaper · Power big data · Smart grid · Time synchronization

1 Introduction

With the development of 5G-based power IoT, many kinds of power services have benefited from the low latency and high reliability of 5G networks. The research on Ultra-reliable low-latency communication (uRLLC) is continuing to follow up in the field of Smart Grids [1]. Considering the demands of stable transmission in plenty of power IoT scenarios, the Time-Sensitive Networking

X. Wang et al. (Eds.): 6GN 2020, LNICST 337, pp. 477–485, 2020.
https://doi.org/10.1007/978-3-030-63941-9_40

(TSN), which is developed from IEEE AVB, has been discussed to provide the deterministic transmission ensurance for Smart Grid. TSN is a set of standards under development by the IEEE 802.1 working group. It defines mechanisms for the time-sensitive transmission of data over deterministic Ethernet networks [2].

Although the TSN technologies are relatively mature, there are still many problems to be solved before TSN can be perfectly converged with 5G system [3]. TSN can realize the network's deterministic delay based on time synchronization and traffic scheduling technology, ensure the reliability of the network through reliable transmission technology and achieve interoperability between different networks through resource management technology. However, Time Awareness Shaper will cause packet loss due to incorrect gate control status. It highly depends on accurate time synchronization between network nodes [4]. The introduction of the frame preemption mechanism increases the waiting time of low-priority data, it is necessary to balance the average delay and the maximum delay of the network. Due to the uncertainty of wireless channel, i.e. the air interface, determinism at the mobile edge is hard to reach, especially the reliability and stability on uplink. In addition, the accuracy of absolute time synchronization is also restricted by the asymmetry of wireless channel. It brings problems such as timestamp deviation of the data collection, disrupting the determinism of other service flows and reducing the overall network resource utilization. Since the SmartGrid services that use the most wireless connections are the power big data services, which includes various types of power data collection and real-time monitoring, it is most necessary to consider the impact of time synchronization errors on these services. What's more, in the context of the continuous improvement of the electricity market, system power balance can be achieved through demand-side management. To achieve high effect Source-Grid-Load-Storage, coordinated and optimized scheduling is essential, which highly depends on accurate time synchronization.

The remainder of this paper is organized as follows: Sect. 2 shows the related work and the description of our solutions. Section 3 presents the evaluation and future work directions. Finally, we conclude the paper in Sect. 4.

2 Design of the Time-Sensitive BBU Shaper

2.1 Related Work

In our opinion, SmartGrid will adopt 5G & TSN converged network as a wireless business solution for power IoT in the future [5]. Based on the integration of 5G and TSN in the aspect of core network, bearer network and network time synchronization, the power IoT can provide the following features for the big data terminals: service admission, charging, grant of priority and preemption authorizations; real-time deterministic resource allocation based on network status awareness; absolute time synchronization based on gPTP; TSN data adapter and TSN configuration adapter for other parts of the SmartGrid.

At present, the 5G core network, based on network function virtualization and microservices, already has the conditions to implement dynamic network slicing and open network configuration interfaces. In the core network of 5G-TSN converged network, network-aware TSN microservice network elements is able to perform power IoT service management and control based on dynamic network slicing technology to ensure isolation from other services; Grand Master Clock nodes with precise network timing capabilities are introduced to realize that each node can only rely on the network to achieve sub-microsecond level absolute time synchronization [6].

For the absolute time synchronization, traditional TSN network frequently exchange the time synchronization information between the network entities on a regular basis. It will increase the load on the control plane and eventually affect SmartGrid services [7]. The centralized time synchronization system only exchanges messages between the central synchronization controller and each network entity, which can reduce the overhead of the control plane. However, it also may cause a single point failure when the GMC or its service link fails. This will cause the entire network to lose synchronization. Therefore, the 5G-TSN converged network uses IEEE gPTP protocol and a distributed time maintenance architecture which combines the best clock algorithm (BMCA). When a GMC fault occurs, the nodes of the network can still maintain accurate time synchronization for a period of time.

2.2 Problem Description

Most applications/devices of power big data are of large amount, their transmission data is periodic and sensitive to transmission delay, and the size of data transmitted each time is relatively small. Thus they require the following transmission features:

- stable transmission delay;
- accurate timestamp for the data;
- take precedence over normal service transmission;

The terminals have different priorities according to their type. The priority determines the ability of preemption. Normally, each terminal will be assigned to a fixed resource and the BBU does not transmit any other packages even the UE sends no data. Thus the terminals must transmit their data at the right time. Methods of time synchronization used for wireless devices are restricted by the air interface. Although a high synchronization accuracy can be achieved by various of methods, a larger error may still occur [8,9]. The time synchronization error of the terminal, together with the channel delay instability, will cause the data to arrive at the BBU at the wrong time (beyond the resource which they are assigned). This will affect the stability and effectiveness of the time-sensitive service flow transmission on the network. It is reflected in the following aspects:

- time stamp deviation of the data collection;
- excessive transmission delay;
- disrupting the certainty of other service flows;

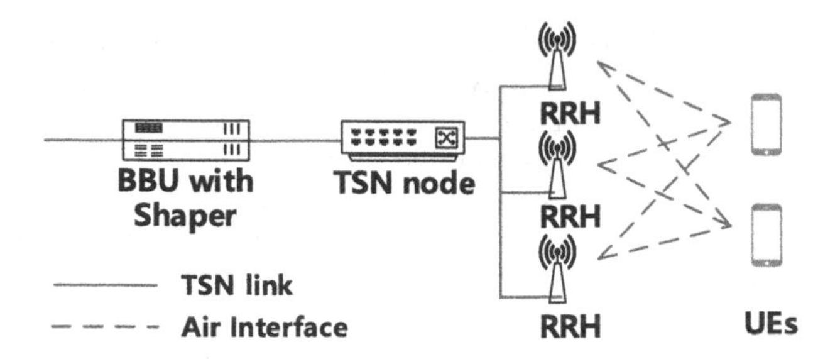

Fig. 1. The application scenario of BBU shaper.

- reducing the overall network resource utilization due to the overhead of preemption.

State Grid Corporation has built a distributed photovoltaic cloud network which realized 1.12 million users' access and full-process one-stop service. However, the application of photovoltaic operation and maintenance cloud platform technology still stays in the passive operation and maintenance of fault detection and resolution, lacking mechanism to prevent malfunction caused by time synchronization.

In summary, it is necessary to discuss the impact of time synchronization errors on the service and the corresponding solutions.

2.3 BBU Shaper

We propose a solution to use time-sensitive Shaper at the edge of the network in response to the requirements of the power big data service for the uplink transmission delay. We choose to shape the upstream at the BBU as it is the entrance of the terminal to access the network. The shaper aims to achieve the abilities of stabilizing the transmission delay of the stream and reduce the delay of the timed-out stream, meanwhile control the reasonable utilization of the network. Figure 1 shows the application scenario of BBU shaper.

The Mechanism of Stabilizing Transmission Delay. First, we set up a threshold for each uplink stream according to their ideal arrival time to make the upstream traffic relatively stable. As shown in Fig. 2, for packages arrives within the threshold, the shaper set up a queue and waits for the threshold. When forwards the data, the shaper checks whether there are multiple queues that need to preempt resources at the current moment.

Reducing Worst Delay. For packages arrives overtime, the shaper temporarily elevates the priority of this stream. Then the shaper forward the streams according to their priorities. By increasing the priority of the timeout stream, we can reduce its waiting delay in other parts of the network. At the same time,

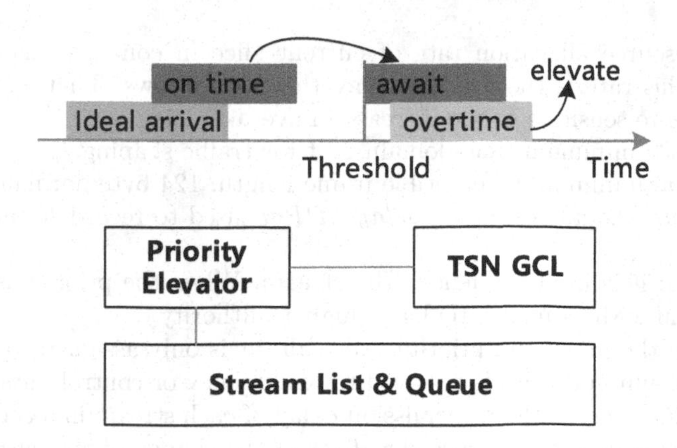

Fig. 2. Stabilizing transmission delay mechanism of BBU shaper.

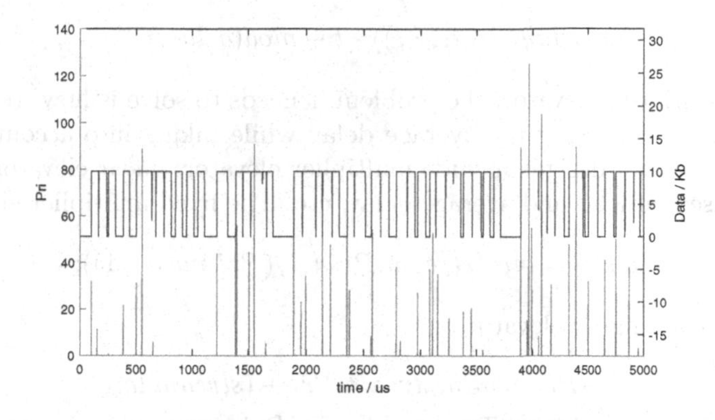

Fig. 3. Schematic diagram of elevator and data forwarding.

although the stream has missed its reserved TSN resources, there is still hope to avoid resource occupancy for subsequent services by occupying resources outside the TSN slice. For the flow with lower priority which has been interrupted, the shaper needs to recalculate the priority according to how much the flow is delayed. The priority elevator need to be carefully designed as unreasonable escalation can lead to large scale of preemption which requires overhead and raises overall delay.

Design of BBU Shaper. The priority elevator is the core function of the BBU Shaper as it directly affects the gate control list (GCL). The shaper also need to form the stream list from the TSN configuration and establish the corresponding queue. Specifically, the following parameters and corresponding functions are essential:

- α_{slice}: resource allocation rate of current slice in converged network. The smaller this rate is the average delay the faster grows. That is, the shaper will be more sensitive to the increase in average delay.
- *slot_length*: minimum time-domain unit for traffic shaping.
- *pmac_min*: minimum preemptible frame length. 124 byte normally.
- *stream.threshold*, *.resource*, *.delay*, *.DPri*: used to record stream information.
- *stream.id*: it is used to identify the stream. When the priority is the same, the stream with a smaller ID has a higher authority.
- *Pri_max*: the maximum priority. Normally it is only assigned to Pri_max-1, the maximum authority is reserved for emergency or control data flow.
- *total_delay*[]: record the transmission delay of each stream in a certain period of time. It is used to measure the effect of the shaper. If the current stream is unable to preempt normal service resource, its final delay will be:

$$fix(delay/(t * \alpha_{slice})) * t + mod(delay, t * \alpha_{slice}) \tag{1}$$

And for the priority elevator, the problem it needs to solve is how to reduce the worst delay by sacrificing the average delay while taking into account network utilization. Typical designs include multiplier elevator, linear elevator, dynamic elevator based on load and average delay, etc. The fixed multiplier elevator:

$$DPri = min(stream.Pri * f, (Pri\_max - 1)); \tag{2}$$

The delay-based linear elevator:

$$\begin{aligned} DPri = min(stream.Pri + (stream.late \\ -stream.Threshold) * k, (Pri\_max - 1)); \end{aligned} \tag{3}$$

The load-based linear elevator:

$$DPri = min(stream.Pri/load * k_l, (Pri\_max - 1)) \tag{4}$$

where k, f and k_l are the scale factors.

3 Evaluation and Future Directions

In this section, we evaluate the design of the BBU shaper with different kind of priority elevators. Also, in order to further evaluate the BBU shaper, we propose a revenue model to evaluate the effect of current traffic shaping.

Assume a base station is connected to *N_devices* terminals. Each terminal is allocated exclusive resources within a slice of one hundred microseconds. All the terminals have a random synchronization error for some reason. Therefore, they will transmit data at the wrong time, and coupled with the unstable delay of the channel, the stream will exceed the expected situation. The uRLLC slice takes $\alpha_{slice} = 0.1$ ratio of time domain with full bandwidth $BW = 100$ Gbps, the terminals has synchronization error within 4 times of their threshold with

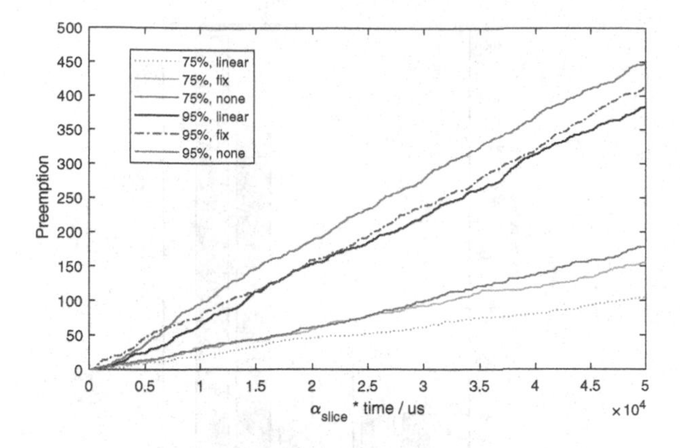

Fig. 4. The amount of preemptions that occur in different situations. $f = 1.5, k = 0.635$

random initial priority from 0 to $Pri_{max} = 127$. Their thresholds are set to $T = 50\,\mu s$ and their data rates are all within 100 Mbps.

Figure 3 shows how the linear elevator works and the corresponding BBU data forwarding output. This kind of elevator increases the priority of those delayed stream with dynamic maximum priority limit.

As shown in Fig. 4, Fig. 5 and Fig. 6, we used delay-based linear elevator and fixed multiplier elevator in the shaper when the uRLLC load is 75% and 95% respectively.

The time synchronization error results in preemption per 4.31 Mb data transmitted when the load is 75%, and the worst delay reaches to about 1 ms which is 20 times longer than the expected transmission delay. We can see that the shaper can reduce the total preemption rate under the same load compared to the

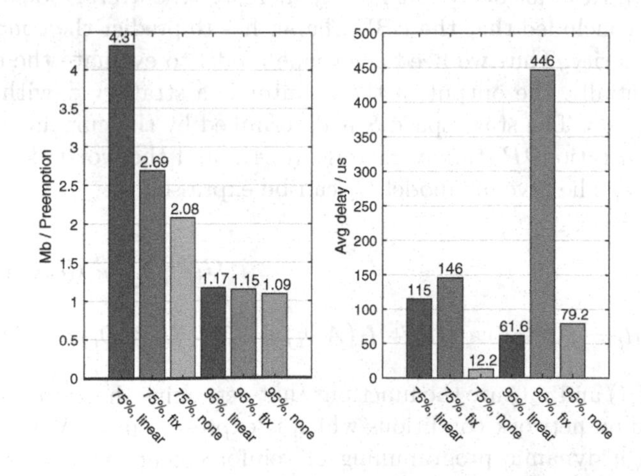

Fig. 5. Network overhead and average delay in different situations.

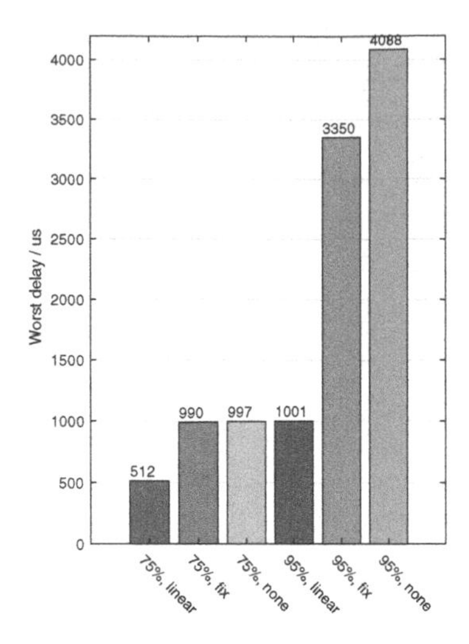

Fig. 6. Worst delay in different situations.

case without the elevator. Under 75% load, the shaper can effectively offset the preemptive signaling overhead required to transmit each MB of business data. But as the load increases to 95%, this effect almost ceases to exist. For average and worst delay, the shaper sacrifices average latency to reduce worst latency. But we also find that unreasonable elevator design will lead to disastrous consequences: the delayed streams preempt the later streams, causing the later streams to get on the priority elevator and to preempt even later streams. The fixed elevator rudely promotes the priority and leads to snowslide of preemption. While linear elevator reduces the worst delay by about 3 ms and only increases the average delay 0.1 ms (even reduce average delay at 95% load), the fixed elevator slightly decreases worst delay but greatly increases the average delay.

It can be concluded that the BBU Shaper has to predict the consequences of promoting priority. Thus we need a revenue model to evaluate the effect of the shaper. Essentially, the output of the elevator is a strategy π with continuous action space $A =$. The state space S is determined by the current network load to preemption ratio DP, the average delay AD and the worst delay WD in a period of time. The revenue model G_t can be expressed as:

$$\begin{gathered} G_t = \sum_{k=0}^{\infty} \gamma^k R_{t+k+1} \\ R_t = f_0(load) * DP_t + f_1(AD_t) * AD + f_2(WD_t) * WD \end{gathered} \tag{5}$$

where $f_0()$, $f_1()$ and $f_2()$ are the functions that determine the corresponding coefficients based on network conditions which is expressed in A. We will solve this model through dynamic programming or reinforcement learning in our future work, so as to obtain a specific and better elevator setting strategy.

4 Conclusion

The devices used for power big data, Source-Grid-Load-Storage, etc. in smart grid have characteristics like large amount, the transmission data is periodic and sensitive to transmission delay, and the size of data transmitted each time is relatively small. The absolute time synchronization error of these devices can lead to increases in preemption, average transmission delay and worst transmission delay. We propose a design of the BBU Shaper to deal with the time synchronization error and uncertainty of channel delay. We propose a mechanism named AAOE to stabilize the transmission delay and reduce worst delay. Evaluation shows the shaper is able to alleviate the negative impact caused by the synchronization error. But there are still some problems remain to be solved such as efficient design of priority elevate. Based on our discussion, we believe that the BBU Shaper will have a role in supporting the transmission of Smart Grid in the future.

Acknowledgment. This work is partly supported by the Big Data Center of State Grid Corporation of China, and partly supported by the Beijing Municipal Education Commission Coconstruction Project under grant 2018BJLAB.

References

1. Zhu, L., Feng, L., Yang, Z., Li, W., Ou, Q.: Priority-based uRLLC uplink resource scheduling for smart grid neighborhood area network. In: Proceedings of the IEEE International Conference on Energy Internet (ICEI 2019), Nanjing, China, pp. 510–515 (2019)
2. IEEE: IEEE 802.1 time-sensitive networking task group. www.ieee802.org/
3. Nasrallah, A., et al.: Ultra-low latency (ULL) networks: the IEEE TSN and IETF DetNet standards and related 5G ULL research. IEEE Commun. Surv. Tutorials **21**(1), 88–145 (2019). Firstquarter
4. Shrestha, D., Pang, Z., Dzung, D.: Precise clock synchronization in high performance wireless communication for time sensitive networking. IEEE Access **6**, 8944–8953 (2018)
5. Li, C.-P., Jiang, J., Chen, W., Ji, T., Smee, J.: 5G ultra-reliable and low-latency systems design. In: Proceedings of the European Conference on Networks and Communications (EuCNC 2017), Oulu, pp. 1–5 (2017)
6. Serrano, J., et al.: The white rabbit project. In: Proceedings of ICALEPCS TUC004, Kobe (2009)
7. Li, Y., Li, C., Wu, G., Zhang, C.: Research on high-precision time distribution mechanism of multi-source power grid based on MEC. In: Proceedings of the IEEE International Conference on Communications, Control, and Computing Technologies for Smart Grids (SmartGridComm 2019), Beijing, China, pp. 1–5 (2019)
8. Zhang, C., Zheng, W., Wen, X., Lu, Z., Wang, L., Wang, Z.: TAP: a high-precision network timing method over air interface based on physical-layer signals. IEEE Access **7**, 175959–175969 (2019)
9. Depari, A., Ferrari, P., Flammini, A., Marioli, D., Taroni, A.: Evaluation of timing characteristics of industrial ethernet networks synchronized by means of IEEE 1588. In: Proceedings of the IEEE Instrumentation and Measurement Technology Conference IMTC 2007, pp. 1–5 (2007)

The 1st International Workshop on 5G/B5G/6G for Underground Coal Mine Safety and Communication (UCMSC)

3D Online Mine Ventilation Simulation System Based on GIS

Hui Liu(✉), Shanjun Mao, Mei Li, and Pingyang Lyu

Institute of Remote Sensing and Geographic Information System, Peking University, Beijing, China
huil@pku.edu.cn

Abstract. An efficient coal mine ventilation system is the guarantee of the safety production in the underground coal mine system. In the procedure of the "intelligent mine" development, the intelligent construction of ventilation system is of primary concern. With the rapid development of the state-of-the-art technology, such as the Internet of Things. This paper aims to implement the online and fast ventilation simulation to provide real-time result for the afterwards decision support. Firstly, the ventilation simulation model based on circuit wind flux method and Scott-Hinsley algorithm is described. Secondly, the architecture of the web system is designed as well as the ventilation network graph model and simulation model is constructed. Thirdly, a prototype web system is developed based on GIS technology with ventilation model integrated at the back-end, which is called 3D VentCloud. The result demonstrated that the system is efficient in providing real-time and online ventilation simulation result, which is potential to guide the fast decision support for coal mine safety production.

Keywords: Coal mine ventilation system · Online ventilation simulation · Safety production · Web system

1 Introduction

Mine ventilation system is one of the crucial systems to ensure the safety production of the underground coal mine, which directly affects the health condition and working efficiency of miners, as well as the economic benefits and sustainable development of the safety production system. With the rapid development of the "intelligent mine" based on advanced technology, it is highly required to realize the intelligent construction of coal mine ventilation system in the complex environment, which aims to provide reliable and accurate decision support for the safety production of the underground coal mine system [1].

The traditional data processing of mine ventilation system is difficult to satisfy the demand of the intelligent construction of mine ventilation system. Many researches and institutes have developed mine ventilation system to realize the informatization of the mine ventilation. For instance, VENTGRAPH and Mine Fire Simulation software, developed by Polish Academy of Sciences, are two widely used systems around the world

X. Wang et al. (Eds.): 6GN 2020, LNICST 337, pp. 489–496, 2020.
https://doi.org/10.1007/978-3-030-63941-9_41

[2, 3]. Ventsim is a software with wider attention, which is developed by Chasm company in Australia [4]. And there are also some software systems such as Ventilation Design and VentPC 2003, VR-MNE and Datamine developed by the US and UK respectively [5]. Besides, some domestic software systems are also developed with complex functions, such as MVSS and MFire developed by the China University of Mining and Technology, VentAnaly developed by Coal science and technology research institute co. LTD. [6].

As is well-known that the mine ventilation system is a typical geographical environment, and the laneway network model has geospatial, attribute data and topological relation. Based on these characteristics of the mine ventilation, its intelligentization is now at the early stage [7]. Correspondingly, the geographic information system (GIS) technology is one of the most effective technology methods to manage the geospatial data of the mine ventilation system. Thus, with the development of GIS technology, more and more ventilation software systems have been developed based on GIS technology. For example, LongruanGIS platform is developed by Beijing LongRuan Technologies Inc. with ventilation function and has its own intellectual property right [8]. LKGIS and VRMine GIS platform are all successful domestic systems developed with ventilation function. Besides, there are also some secondary developed ventilation management systems based on AutoCAD or ArcGIS [9–13].

However, the existing research suggests that the ventilation system based on AutoCAD is difficult to deal with spatial topological relationship and the spatial attribute database. As a consequence, it is difficult to provide decision support with location based information, while the application of GIS combined with ventilation system effectively improved the deficiency of AutoCAD. In addition, in order to satisfy the requirement of the mine ventilation system intelligentization, it is necessary to implement the online ventilation simulation to provide fast or real-time decision support.

Therefore, this study aims to implement the online simulation of ventilation system to provide fast or real-time simulation result for decision support. Firstly, we described the numerical solution based on circuit wind flux method, and the Scott-Hinsley algorithm. Secondly, we designed the data structure of the laneway network model and the ventilation data model. Thirdly, a prototype simulation system of 3D mine ventilation at web end is designed and developed based on GIS technology. Finally, the experiment is conducted, which demonstrated that the system is efficient to provide real-time result for decision support.

2 Ventilation Network Model

Mine ventilation is a daily work in coal mine production, which has to be accurately simulated to master the actual situation of ventilation. The simulation result can be an important guideline to adjust and control the wind, and optimize the ventilation system. This section describes the basic steady state solution of circuit wind flux method with Scott-hinsley algorithm.

The ventilation air flow is assumed to follow the three basic laws, which are the law of ventilation resistance, the law of wind balance at nodes and the law of pressure balance in loops. These three laws reflect the mutual restriction and equilibrium relationship among the three basic ventilation parameters (branch air volume, wind resistance and node pressure) in the ventilation network model [14, 15]. For a simple schematic diagram of ventilation network model shown in Fig. 1. These three airflow laws can be described as:

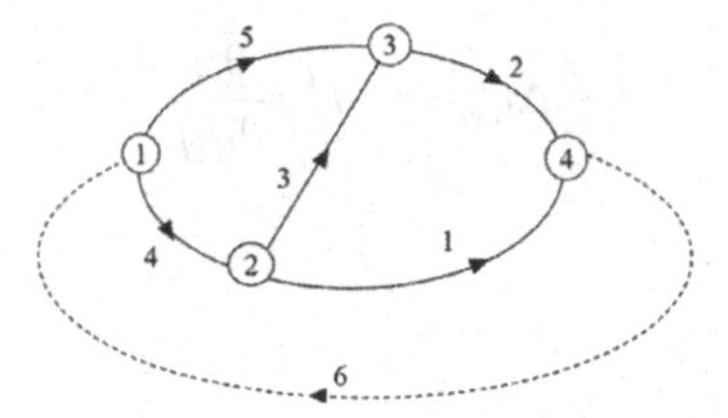

Fig. 1. Schematic diagram of ventilation network.

$$h_i = R_i \cdot |Q_i| \cdot Q_i \ (i = 1, 2, \ldots, 6) \tag{1}$$

$$\begin{cases} Q_6 = Q_1 + Q_2 \\ Q_1 = Q_3 + Q_4 \\ Q_5 = Q_3 + Q_2 \\ Q_6 = Q_4 + Q_5 \end{cases} \tag{2}$$

$$h_2 - H_2 = h_1 - H_1 + h_2 - H_2 \tag{3}$$

Where h_i is the wind pressure or resistance of a wind path or laneway in the ventilation network, with unit Pa; R_i is the wind drag of a wind path, with unit (N • s^2)/m^8; Q_i is the air volume of a wind path, with unit m^3/s. H_1 or H_2 represents the mechanically powered wind pressure provided by the fan or the natural wind pressure [16].

Circuit wind flux method is one of the most widely used numerical calculation methods for complex wind network. The specific calculation idea is as follows:

(1) Take a group of air volume of basic circuit in the ventilation network graph as the unknown variable, establish the basic control equations according to the three laws of air flow.
(2) Solve the circuit air volume.
(3) Calculate the branch air volume, branch ventilation resistance and other unknown variables from the circuit air volume.

In general, for a ventilation network graph with n branches and m nodes, the system equation is:

$$f_i(q_{y1}, q_{y2}, \ldots, q_{yb}) = \sum_{j=1}^{n} C_{ij}R_j \cdot \left|\sum_{s=1}^{b} C_{sj}q_{ys}\right| \cdot \left(\sum_{s=1}^{b} C_{sj}q_{ys}\right) - \sum_{j=1}^{n} C_{ij}(h_{fj} + h_{Nj}) = 0 \tag{4}$$

Where b = n − m + 1 represents the number of independent circuits of the ventilation network; h_{fj} is the fan wind pressure and h_{Nj} is the natural wind pressure; q_{ij} is the air volume for the branch j and circuit i, where i = 1, 2, …, b, j = 1, 2, …, n; $q_{y1}, q_{y2}, \ldots q_{yb}$ cosine tree volume respectively; $(C_{ij})_{bn}$ is the independent circuit matrix.

Based on taylor series expansion and equation simplification, the iteration equation is obtained as (5), which is the Scott-Hinsley algorithm.

$$\Delta q_{yi}^{k} = -f_i^{k} / \frac{\partial f_i}{\partial q_{yi}} \tag{5}$$

3 System Design

The prototype system developed in this study is named as 3D VentCloud. The front-end of the web system is developed by Html, JavaScript and CSS. The three-dimensional visualization function is developed based on three.js. The back-end model and algorithm of the ventilation network is developed by Python and C++. The Web page frame is constructed by Tornado, which is used as the server to transmit data to the front-end of the web system.

As can be seen from Fig. 2, the prototype system consists of three layers from bottom to top: technology layer, service layer and application layer.

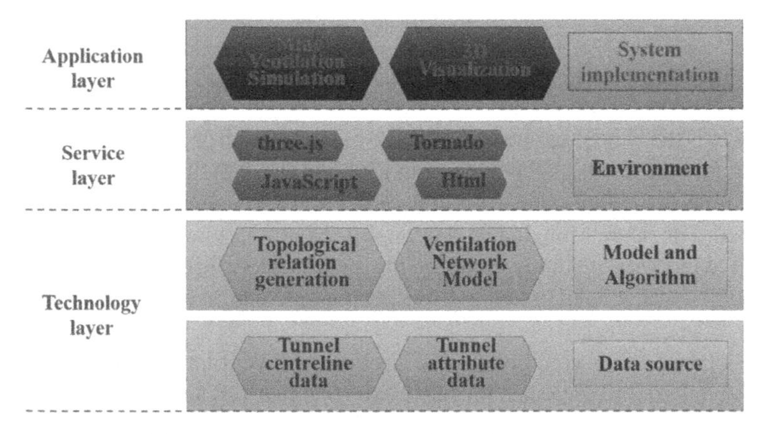

Fig. 2. The architecture of 3D VentCloud system.

The different layers are described as follows:

(1) Technology layer

The technical layer covers all the key technologies in the process of system construction, which includes the data source and mine ventilation model and algorithm. The data source layer includes the centerline data of the whole laneway network, and the geological and attribute data of the laneway and nodes, which are stored in GIS database or stored as GeoJson format. The mine ventilation model includes the topological relation generation model and the ventilation network simulation model. These models jointly build the mine ventilation network solution model library and constitute the technical core of the system.

(2) Service layer

This web system uses Tornado as the service architecture. The WebSocket interface is applied to connect the JavaScript front-end with the back-end Python and C++ program to implement the data transmission.

(3) Application layer

The application layer displays the 3D visualization results in the form of graphical interface at the webpage end. The system can read the geospatial centerline data and attribute data of any mine laneway network, and display the 3D laneway model and generate the stl file of the laneway model. The 3D laneway network model can also be rendered with different colors by the air volume of different laneways, and the geospatial attributes of the laneway can be queried and displayed at the web end.

4 Simulation Result

This study collected the laneway and ventilation network data from Xinqiao coal mine, which is located in Henan province, China. This coal mine was chosen primarily due to the accessibility of continuous monitored data. The Xinqiao coal mine laneway network has 290 laneway branches, and 222 nodes. The geospatial and attribute data of the laneway network includes laneway ID, laneway name, the coordinates and ID of the start node and end node, the area and perimeter of the laneway section, the laneway length, wind drag, ventilation resistance, coefficient of friction resistance and laneway type, etc.

Some of the laneway data and basic ventilation parameters are shown in Table 1. The topological relation for the laneways and nodes are stored as point-line indexed structure commonly used in GIS data organization.

Table 1. Some of the ventilation attribute data for Xinqiao coal mine laneway.

Laneway ID	Start node ID	End node ID	Wind drag (N • s^2/m^8)	Section area (m^2)	Section perimeter (m)	Laneway length (m)
1	1	2	122.336	18.14	15.78	32
2	3	4	114.535	18.67	15.46	40.10
3	5	6	145.384	18.28	15.23	3.30
4	7	8	0.00016	24.00	20.37	612
5	9	10	0.00043	14.06	15.89	584

The 3D laneway network model of Xinqiao coal mine is read and displayed by 3D VentCloud system, as can be seen in Fig. 3.

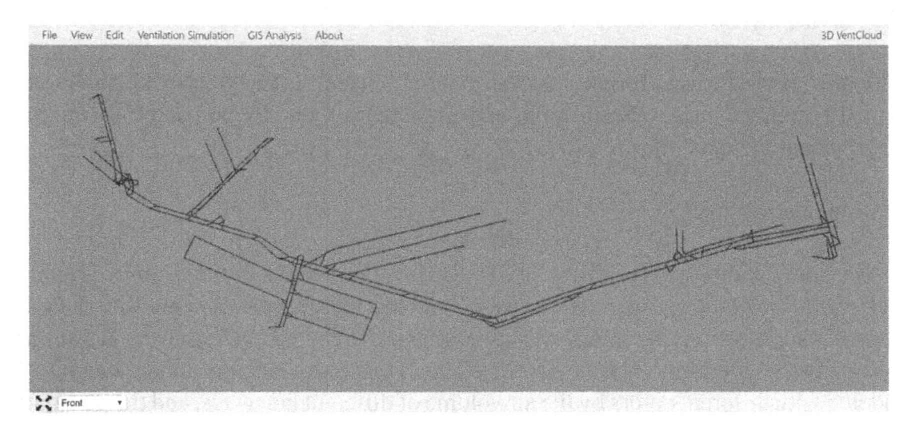

Fig. 3. The 3D laneway network model of Xinqiao coal mine.

By establishing the topological structure of the ventilation network graph of Xinqiao coal mine, the shortest path algorithm is adopted to sort the laneway wind resistance. Thus, the minimum spanning tree and 70 cotree branches of the ventilation network graph are generated. By adding the cotree branches as additional tree branch to the minimum spanning tree, 70 circuits are formed respectively, which is called independent circuit of the ventilation network graph.

Then, based on the ventilation simulation model, the air volume of each laneway is obtained within a few seconds, which can provide fast or real-time result at web end for the decision support afterwards. The 3D laneway network model of Xinqiao coal mine is rendered with simulation result, which is displayed on 3D VentCloud, as shown in Fig. 4. By clicking on different laneway branches, the geospatial query can be conducted and displayed on web page.

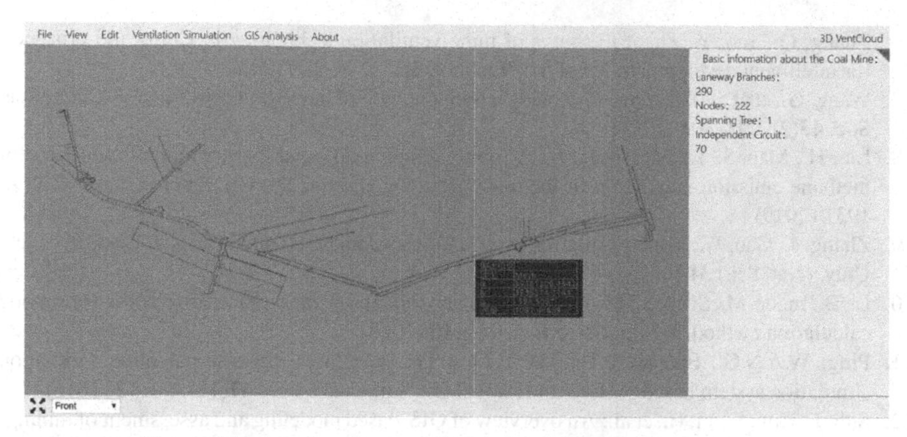

Fig. 4. The 3D visualization of mine ventilation simulation result.

5 Conclusion

This study investigated the mine ventilation simulation model and developed a prototype web system to implement the online ventilation simulation. The system is design and developed based on web technology, and the numerical ventilation solution method is developed and integrated with the web system.

The experimental result demonstrated that this web end system is effective in providing fast or real-time ventilation simulation result, and is expected to guide the real-time decision support for coal mine safety production.

Funding Statement. This work is financially supported by the National Natural Science Foundation of China: [Grant Number 51774281].

Conflicts of Interest. The authors declare that there is no conflict of interest regarding the publication of this paper.

References

1. Ren, C., Cao, S.: Implementation and visualization of artificial intelligent ventilation control system using fast prediction models and limited monitoring data. Sustain. Cities Soc. **52**, 101–860 (2020)
2. Dziurzyński, W., Krach, A., Pałka, T.: A reliable method of completing and compensating the results of measurements of flow parameters in a network of headings. Arch. Min. Sci. **60**(1), 3–24 (2015)
3. Dziurzyński, W., Krause, E.: Influence of the field of aerodynamic potentials and surroundings of goaf on methane hazard in Longwall N-12 in seam 329/1, 329/1-2 in "Krupiński" coal mine. Arch. Min. Sci. **57**(4), 819–920 (2012)
4. Maleki, S.: Application of VENTSIM 3D and mathematical programming to optimize underground mine ventilation network: a case study. J. Min. Environ. **9**, 741–752 (2018)
5. Zielinski, D., Macdonald, B., Kopper, R.: Comparative study of input devices for a VR mine simulation (2014)

6. Zhang, Q., Yao, Y., Zhao, J.: Status of mine ventilation technology in China and prospects for intelligent development. Coal Sci. Technol. **48**(2), 97–103 (2020)
7. Wang, G., et al.: 2025 scenarios and development path of intelligent coal mine. J. China Coal Soc. **43**(2), 295–305 (2018)
8. Liu, H., Mao, S., Li, M., Wang, S.: A tightly coupled GIS and spatiotemporal modeling for methane emission simulation in the underground coal mine system. Appl. Sci. Basel **9**(9), 1931 (2019)
9. Zhang, J., Gao, W.: Mine ventilation information system based on 3D GIS. J. Liaoning Tech. Univ. (Nat. Sci.) **31**(5), 634–637 (2012)
10. Li, B., Inoue, M., Shen, S.: Mine ventilation network optimization based on airflow asymptotic calculation method. J. Min. Sci. **54**(1), 99–110 (2018)
11. Ping, W.A.N.G., Bentao, L.U., Xu, L.I.U.: Two and three dimensional mine ventilation simulation system based on GIS and AutoCAD. Saf. Coal Mines **47**(12), 90–92 (2016)
12. Suh, J., Kim, S., Yi, H., et al.: An overview of GIS-based modeling and assessment of mining-induced hazards: soil, water, and forest. Int. J. Environ. Res. Public Health **14**(12), 1463 (2017)
13. Salap, S., Karslıoğlu, M.O., Demirel, N.: Development of a GIS-based monitoring and management system for underground coal mining safety. Int. J. Coal Geol. **80**(2), 105–112 (2009)
14. Liang, Y., Zhang, J., Ren, T., et al.: Application of ventilation simulation to spontaneous combustion control in underground coal mine: a case study from Bulianta colliery. Int. J. Min. Sci. Technol. **28**(2), 231–242 (2018)
15. Nyaaba, W., Frimpong, S., El-Nagdy, K.A.: Optimisation of mine ventilation networks using the Lagrangian algorithm for equality constraints. Int. J. Min. Reclam. Environ. **29**(3), 201–212 (2015)
16. Wang, X., Liu, X., Sun, Y., et al.: Construction schedule simulation of a diversion tunnel based on the optimized ventilation time. J. Hazard. Mater. **165**(1–3), 933–943 (2009)

A Virtual Reality Simulation System for Coal Safety Based on Cloud Rendering and AI Technology

Mei Li[1(✉)], Zhenming Sun[2], Zhan Jian[1], Zheng Tan[3], and Jinchuan Chen[3]

[1] Institute of Remote Sensing and Geographical Information System, Peking University, Beijing 100871, China
mli@pku.edu.cn

[2] College of Resources and Safety Engineering, China University of Mining and Technology, Beijing 100083, China

[3] Beijing Longruan Technologies, Beijing 100190, China

Abstract. Coal mining, regarded as a high-risk industry, has strongly demand on Virtual Reality (VR) training environment for safety mining and emergency rescue. In order to solve the problems of the current VR platform of lack of immersive experience and interest, a Browser/Client VR simulation system are designed and realized. Three key techniques are studies, including cloud rendering, AI behavior tree and mining disaster animation. Unlike WebGL and HTML5, cloud rendering provides photo-realistic quality. 2 AI characters are designed to guide users to have an overall understanding of the mine and experience the effect mine disasters. The system has been successfully applied in the Virtual Reality Teaching and Experiment Laboratory for undergraduate in China University of Mining & Technology-Beijing. The research, as a new tool for miner training and disaster drilling, has a signification meaning of the work safety IT construction.

Keywords: Coal mine safety · Virtual reality · Game artificial intelligence · Behavior tree · Cloud rendering

1 Introduction

Coal mine, with dangerous working environment and complex production system, has a strong demand for the virtual reality (VR) simulation and drills. Virtual Reality (VR) has its unique advantage for high-risk mining industry, such as miner safety training, emergency rescue drilling, longwall mining and drifting production processes simulation, disaster scenario simulation, mechanical operation training and so on. 5G communication, cloud computing and big data will become IT fundamental infrastructure in the Intelligent Mine construction in the future 3–5 years. A shift from desktop VR to Cloud VR is inevitable, as it becomes the best choice for VR. Research and application on the cloud-based VR simulation system for mine safety are very important in recently year.

X. Wang et al. (Eds.): 6GN 2020, LNICST 337, pp. 497–508, 2020.
https://doi.org/10.1007/978-3-030-63941-9_42

In this article, a Browser/Client VR simulation system for coal mine safety is designed and developed base on the cloud rending technology and game AI technology to solve the problems of the current VR platform, such as lack of immersive experience and interest, high requirement of the client devices performance. Section 2 introduces the related research on virtual reality training of coal mine. Section 3 introduces the system architecture design with cloud rending technology. Section 4 focuses on the Game AI design for general learning and disaster experience. Section 5 introduces the techniques of disaster animation producing. Section 6 shows some system outcome. Finally, Sect. 7 gives some conclusions.

2 Related Works

Developed mining countries such as United States, Australia, and UK have implemented VR as training environment for mining simulation, accident reconstruction and investigation, education and safety training in two decades [1–7].

Commonwealth Scientific and Industrial Research Organization (CSRIO) developed a drilling system for coal mine equipment operation. Artificial Intelligence, Computer Graphics and Virtual Reality (AIMS) in University of Nottingham, UK has a long history of developing and applying VR technology for coal mine safety training. The VR products SafeVR and VRoom are very famous for open pit truck operation training. Researchers at The National Institute for Occupational Safety and Health (NIOSH) are exploring how the mining industry can effectively use "serious games" for mine training. They have used a game engine to create a portion of an underground coal mine. In this virtual mine, trainees have a first-person point as they walk or ride through the mine. The first training package in this virtual mine instructs new miners in how to read mine maps. Marshall University used Unity3D to develop an Interactive Virtual Underground Mine Platform (IVUMP), which can record the messages, information and actions of mine rescue team members. The platform used VFIRE to enhance the ventilation interactive simulation to provide a safety training exercise for emergency response.

In China, some large mining groups and State Administration of Work Safety have been applying the VR hardware and software to safety training and emergency rescue training. For example, two national VR training centers had been established in Shendong Mining Group and Ningxia Mining Group, each of which cost more than 200 million US dollar. Those systems provided the high fidelity for users with the functions of single user, stand-alone version and "Q and A" work safety training. The VR hardware were composed of projection-based panoramic display system, infrared tracking stereoscopic glasses, VR headset display, Pad, touch screens and other devices. Moreover, desktop 3D visualization system are popular in coal mine with the functions of 3D geological model, ventilation simulation, real-time data monitoring [8–11].

There VR applications still have some shortcomings need to improve.

Lack of the Immersive Experience and the Learning Interest. Those coal mine training system have teaching and learning functions with single-player mode and ask-answer pattern, lack of the immersive experience and the learning interest. Most of them did not design AI man-machine interaction functions so that the advantages of immersion, interactivity and imagination of virtual reality could not been fully used.

High Requirement of the Client Devices Performance. Good experiment relies on high-performance local devices. However, traditional VR training applications confront new challenges. VR applications need to support the various light terminals such as PC, mobile phone, PAD, and headset and so on, to have the abilities of mass multiplayer online (MMO), to upgrade the question-answer learning pattern into a serious multiply game experience patter. At the same time, network fluency, interaction ability, user experience and VR quality of VR also need to consider.

Lack of the Gamifying, Intelligent and Diversity. VR learning tend to be gamifying, intelligent and diversity. Non-Player Character (NPC) characters can enrich game content.AI functionalities include Sense/Think/Act cycle. With Game AI design, the VR training of answer-questions turns into interactive with NPC in 3D scenes. This will enhance the user's real experience of learning and training, strengthen their memory and safety awareness.

3 Architecture of the Virtual Reality Simulation System Based on Cloud Rendering Technology

The system is based on Unreal Engine (UE), which is one of the wildly-used engines which has a complete suite of game development tools. The system takes virtual reality visualization as the basic function, uses the NPC game AI to link the knowledgeable sites together to guide the user to experience coal mine. The system architecture is shown as Fig. 1.

A classic Browser/Client architecture of cloud rending is established. The B/S structure means each of individual player connect to the server and get 3D streaming from the server on the cloud. Cloud server handles all the events and the state of the virtual world and transfer the VR render streaming to the client.

Cloud rendering technology refers to run VR rendering program on the virtual server, taking full advantage of virtually infinite computing power to create, capture, compress high-resolution renderings, and sent to the low performance and computing power client. When bandwidth is good, users can interact with VR programs on the client through the network. In this kind of B/S architecture, the client is lightweight, without the need for expensive hardware, responsible only for decoding and displaying 3D scenes, and can both support resource-constrained platforms such as mobile devices. Users can interact with 3D scenes anywhere and anytime [12, 13].

Compared with the other popular Web3D technologies, mainly HTML5 and WebGL, cloud rendering technology has incomparable advantages, mainly because it has no special expensive requirements on the client, and no matter what kind of device can ensure 3D reality and quality. Moreover, since 5G will provide sufficient network bandwidth, online cloud rendering architecture will provide enough computing and rendering power and be a trend in the future.

In this architecture, the whole system is divided into three parts: portal server, game server and client.

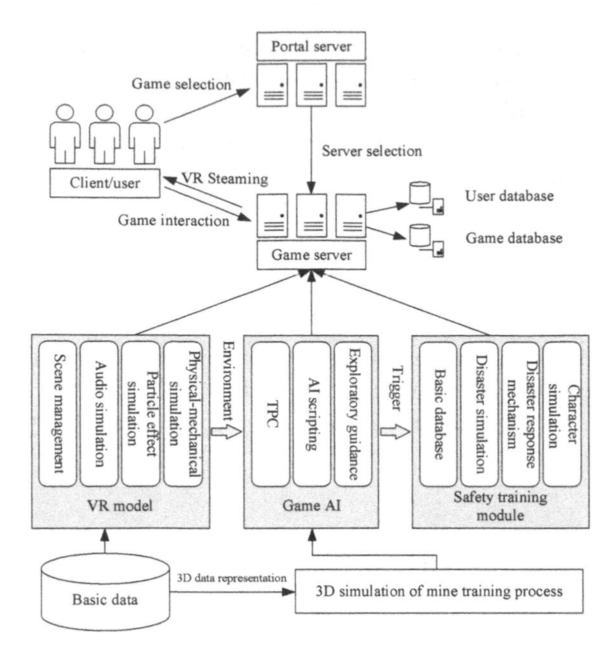

Fig. 1. The architecture of the virtual reality simulation system based on cloud rendering

Portal Server. It is responsible for user login, and according to the user's choice to find the appropriate server, control the server to start the virtual machine (game server). The client and the game server establish a connection when portal server provides the address of the virtual machine to the client, and user can run and interact the 3D program in the remote server through the network.

Game Server. It is responsible for processing the mouse and keyboard input events from the client and translating them and sending them into the interaction module of the 3D program for further parsing. The 3D scenes images are captured, encoded and sent to the client as video streaming. 3D server program includes three parts: virtual reality modeling and visualization module, AI game guidance module and safety training module.

Client. It is responsible for displaying the received 3D scene steaming and sending the user's input such as keyboard and mouse events to the server.

4 Design of Game AI

3D engine, like the automobile engine of the mechanical industry, determines the rendering speed, the sense of reality and immersion, and also affects the convenience and efficiency of the entire 3d application. It combines all the elements in the virtual environment together and coordinate them to work orderly in the background. A typical game engine as Unreal or Unity3D includes a rendering engine ("renderer") for 2D or 3D graphics, a physics engine or collision detection (and collision response), sound, scripting language, animation, artificial intelligence (AI), networking, terrain, streaming, memory management, threading, localization support, and a scene graph.

Artificial intelligence in games (or 'Game AI' for short) is a module of 3D engine. There are similarities and differences between Game AI and traditional AI. Traditional AI typically demonstrate at least some of the following behaviors associated with human intelligence: planning, learning, reasoning, problem solving, knowledge representation, perception, motion, and manipulation. Game AI is emerging and become the narrow branch from traditional artificial intelligence. Game AI refers to Non-Play Characters (NPC) controlled by the computer simulate the intelligent behavior of human being or other creatures, showing a certain amount of intelligent behavior, providing users with a believable challenge to overcome [14, 15]. In our serious game, two NPC are designed to simulate the behavior of miners who guide users to understand the whole underground mine production, to experience the virtual scene and improve the relationship with the user during the process of training, conversation or other tasks.

4.1 The Principle of Game AI

By designing AI behavior, NPCs can automatically analyze and generate the game-oriented effect. When users enter the designed game scenario, they slowly become addicted to NPC's guide-feedback behavior. Compared with traditional learning, the progress of the game itself is also learning, but it is easy to grasp, which will have a positive guiding effect on the user's behavior.

They are two ways of Game AI, the finite state machine (FSM) algorithm and behavior tree algorithm. FSM is to build "state". Transitions between states are triggered by events. A "state" can represent physical conditions that the entity is in, or it can represent emotional states that the entity can exhibit. FSM is a kind of "Event triggered type" to simulate the promptly react to the human player's action with its pre-programmed behavior. An obvious drawback of FSM design is its predictability. All NPCs' behaviors are pre-programmed, the modular design and reusability is not stable when the when games get larger and more complex. A more advanced method used to enhance the personalized experience is the Behavior Tree.

A simplified flow chart of the way behavior tree can be used in such a game is shown in the Fig. 2. Figure 2 shows a sequence of actions of open/close door and open/close window. Behavior tree is mainly composed of leaf node and composite node, which is essentially a tree, that is, acyclic graph.

Leaf Tasks. They are the terminal nodes of the tree and define low level actions which describe the overall behavior. They are typically implemented by user code, maybe in the form of a script, and can be something as simple as looking up the value of a variable in the game state, executing an animation, or playing a sound effect.

Composite Tasks. They provide a standard way to describe relationships between child tasks, such as how and when they should be executed. Contrary to leaf tasks, which are defined by the user, composite nodes are predefined and provided by the behavior tree formalism. They build branches of the tree in order to organize their sub-tasks (the children). Basically, branches keep track of a collection of child tasks (conditions,

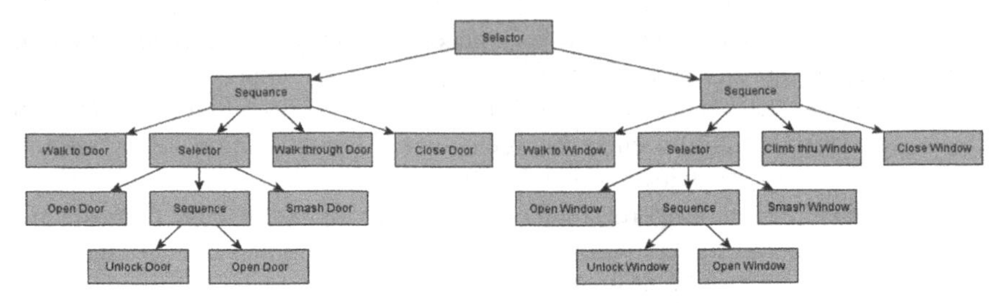

Fig. 2. The type of AI behavior tree

actions, or other composites), and their behavior is based on the behavior of their children. Composite tasks consist of selectors, sequences, parallels and decorators.

A Selector. It is a branch task that runs each of its child behaviors in turn. It will return immediately with a success status code when one of its children runs successfully.

A Sequence. It also executes the node task in turn instead. if one fails the sequence will returns, and the rest task are not executed.

A Decorator. It is a task that has one single child task and modifies its behavior in some way. It will return when a child task output is modified. It can be used to make decisions on whether a branch (or even a single node) in the tree can be executed.

A Parallel. It is composite task to handles "concurrent" behaviors. It's a special branch task that runs all children when stepped. The behavior tree allows parallel node tasks, which may be state machine. Multiple state machines can be executed in parallel.

4.2 Design of AI Behavior

In Unreal engine, Blueprint is used by adding and connecting a series of nodes which have some functionality attached to them to a Behavior Tree Graph. Figure 3 designs a NPC's plot of learning general knowledge and experience the mine disaster. Also, two AI characters are designed to find local navigation grid path automatically.

The First NPC is task-oriented character who explain the common knowledge about the coalmine and guide the user to visit the key underground work site such as longwall working face. Once the user completes the corresponding task and gives feedback can NPC release the next task. The tasks are preparing before entering the underground mine, looking for the entrance, entering the cage, visiting the underground working face. The behavior tree of task-oriented NPC is shown as Fig. 4.

The second NPC is disaster-oriented character who guide user to experience the typical coal mine disaster. The NPC's behavior tree is shown as Fig. 5. Selector, sequence, parallel are designed to trigger the accidents such as gas leakage, roof fall, fire and seepage. The selector firstly judges whether it is dangerous or not, then run each disaster sequence. Sequence executes the disaster animation in turn and test the user's reaction if it is correct or not. NPC also give out the run and waring to the user in parallel node.

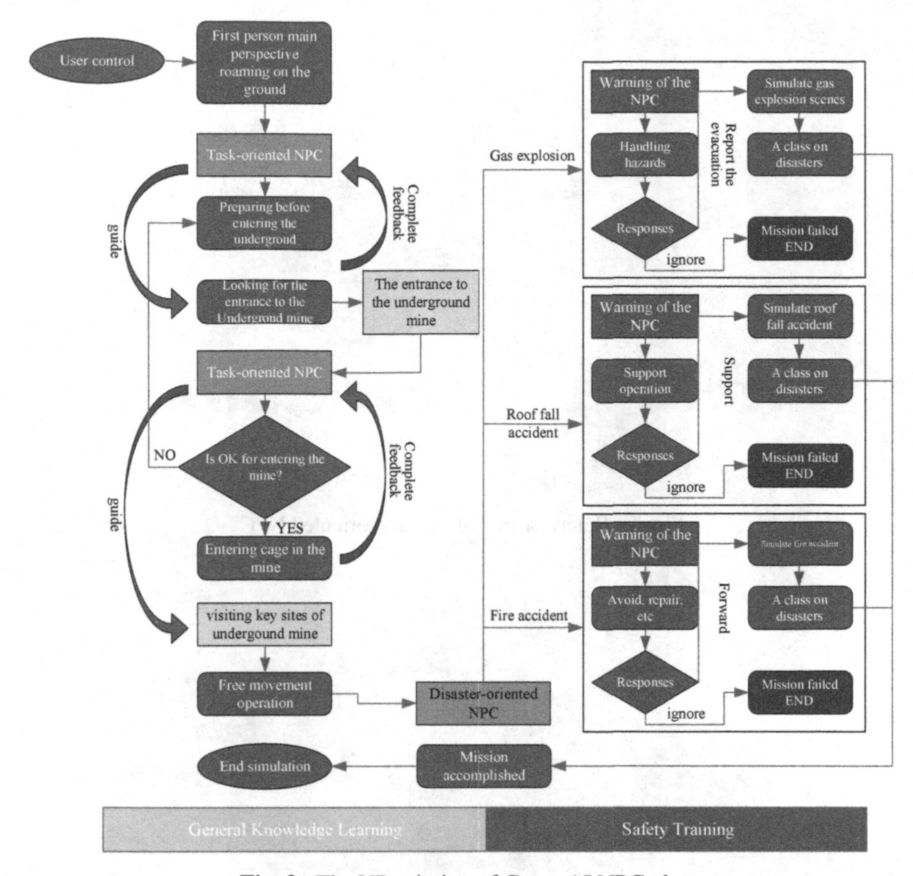

Fig. 3. The VR mission of Game AI NPC plot

5 Key Technique Typical Disaster Simulation Animation

Mining is a high-risk industry, which exist many dangerous accidents such as gas leakage, roof fall, fire and seepage. Once the corresponding animation are triggered by NPC's AI behavior, the system will randomly enter the disaster simulation and it will make the corresponding judgment according to the type of disaster and the handling method of the operator, shown as Table 1.

The disaster simulation module integrates various 3D animation, audio, particle effect, and physical-mechanical models of the 3D engine. 3D model is to establish model of mines, terrain, tunnels, characters, etc. Audio focuses on creating real sound as realistic as possible from the dimension of the source, such as footsteps when walking on the grass and in the tunnel, wheezing when running, mechanical sound when the cage is opened, etc.

The particle effect is to display the dust in the mining face or the dust generated when the roof fall rises. The physical-mechanical model can enhance the virtual character's realistic sense of force, not only the downward acceleration of gravity, but also the sense of gravity when the rock wall is peeling off. Blueprint are used to create those VR effects.

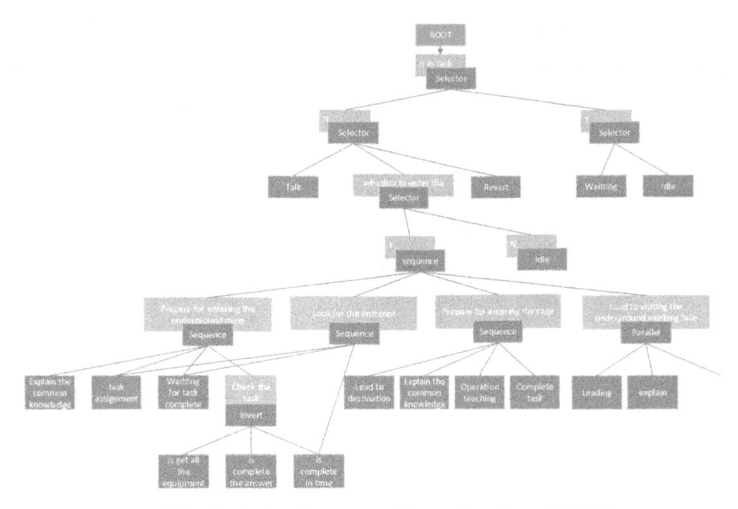

Fig. 4. Behavior tree of the task-oriented NPC

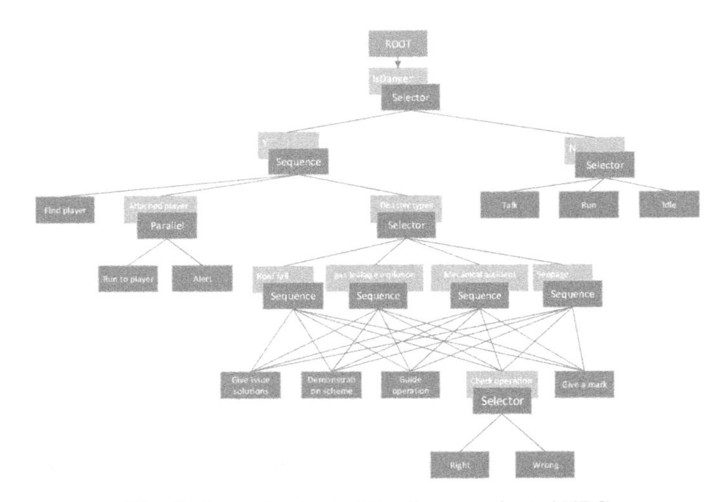

Fig. 5. Behavior tree of the disaster-oriented NPC

5.1 Gas and Coal Dust Explosions

Through this VR system, user will understand boundary conditions of gas and coal dust explosions. By animation technology, a process of coal and gas outburst are established so that user could perceive the disaster from the vision, auditory, and other aspects, and understand the movement and destruction process of the mixed wind and dust. Also, some emergency preparedness such as collaborative gas extraction can be fully displayed.

Table 1. Typical mine disaster simulation

Types of accidents	Characteristics of accidents	Treatments
Gas and dust exploration	Gas leak detector exceeds the upper limit and self-ignite	Report, escape
Roof fall and wall collapse	Rubble drop	Roadway support
Fire accident	Spontaneous combustion	Report escape etc.
Seepage	Water seeps out through a gap or coal seam	Seal

5.2 Roof Fall and Wall Collapse

Through this VR system, user will understand the main factors of roof fall and wall collapse. They can simulate the conditions of various rock burst disasters, simulate the process of ground pressure disasters so that user could understand the form of disaster, causes of these disasters, factors, and the methods of disaster prevention.

5.3 Fire Accident

Through this VR system, user will understand the main form of mine fire such as external fires mainly caused by open flames, blasting, current short circuit, etc., as well as internal fires caused by combustion of coal or other flammable substances due to their own oxidation and heat. VR simulate the conditions of various fires, understand the severity of disasters and master the methods for preventing fires. The system also shows the fire accident treatment, including the closed grouting method and nitrogen injection method.

5.4 Water Seepage

Through the system, user will understand the main factors of mine seepage, such as fissure water, etc. The system can simulate the conditions of various seepage, simulate the process of water inrush, the causes of water inrush and various influencing factors of water inrush so that employees could understand the severity of the disaster accidents and master the methods of seepage prevention and treatment.

6 System Implement

The system has been successfully applied in the Virtual Reality Teaching and Experiment Laboratory for undergraduate in China University of Mining & Technology-Beijing. A cloud rendering cluster is composed by two virtual rendering servers with configuration of M60 8G/ CPU 4Core/32 GB RAM/bandwidth BGP 40 Mbps. A client is required as CPU Intel i7, 16 GB RAM. This hardware environment can only support 6 users at most.

According to former acquired 3D data, such as geographic data, geophysical data, and mining data, the 3D entity model are generated by different modeling method by using 3DMax or Maya. The NPC's characters are made according to the ratio of real people.

NPC Virtual Roaming: NPC will walk around buildings with the user, and enter the underground mine by auxiliary cage, visit longwall mining face or tunneling face, and introduce ventilation system, transportation system, water supply and drainage system and power supply system, etc. This function enables the new miners to have an overall understanding of the mine.

NPC Accident Guidance: Through the dialogue with the AI guider, when the user is working normally, the NPC can have a dialogue with him to judge his behavior, with the function of information prompt, illegal operation prompt. The system triggers disaster simulation according to operating conditions.

In this system, mining safety accidents is concentrated in gas explosion, roof fall, fire and seepage. By analyzing disaster reasons, accident characteristics, occurrence status, and formation impact, a specific disaster simulation knowledge library is constructed, which contains a brief description of various disasters, namely, preventive measures, suggestive prevention on pre-disaster characteristics and the popularization of post-disaster knowledge, and it can combine with system guidance to popularize the consequences of disasters (Figs. 6, 7, 8, 9, and 10).

Fig. 6. Finding entrance of underground mine with task-oriented NPC

Fig. 7. Visiting working face with task-oriented NPC

Fig. 8. Gas explosion animation

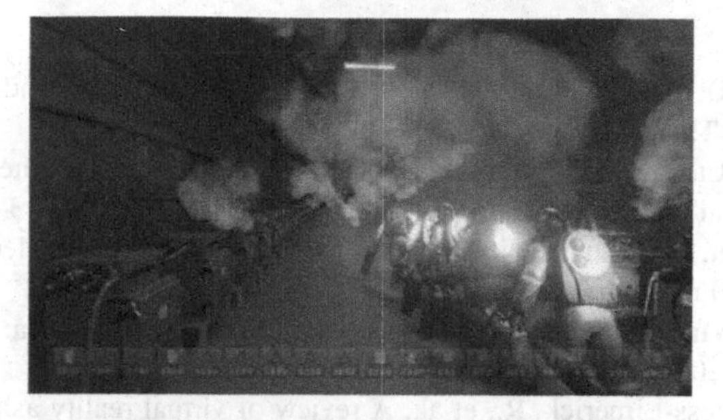

Fig. 9. Fire-extinguishing animation

Fig. 10. Roof accident animation

7 Conclusion

With the development of intelligent mine construction, the development of virtual reality in the mining industry is facing unprecedented opportunities and challenges. Cloud rendering technology will be the mainstream of 3D technology development in the future. The gamification of AI can also attract users to take actions and strengthen their cognition and learning abilities. In this paper, a serious-game virtual reality platform for mine safety training with Unreal engine is designed and developed based on the latest technology. NPC AI guidance are designed to simulate the process of underground roaming and accident experience including gas explosion, roof fall and wall collapse, fire and seepage. The results of the study will provide a new tool for coal mine work safety and training, avoiding high-risk and extreme environment underground. It provides user with a reliable, safe and cheap software platform, which not only meets the urgent needs of the government and enterprises for mine safety training, but also promotes the development of mine safety IT construction.

Acknowledgement. This work is financially supported by the National Key Research and Development Program of China: [Grant Number 2016YFC0803108]. And it is also supported by the Key Research and Development Program of Inner Mongolia titled "technology and application on digital mine resource management and ecological environment monitoring, 2015–2019".

References

1. Penichet, V.M.R., Marin, I., Gallud, J.A., et al.: A Classification Method for CSCW Systems. Electron. Notes Theor. Comput. Sci. **168**, 237–247 (2007)
2. Bednarz, T.P., Caris, C., Thompson, J., et al.: Human-computer interaction experiments immersive virtual reality applications for the mining industry, pp. 1323–1327 (2010)
3. Mallet, L., Unger, R.: Virtual reality in mine training. In: SME Annual Meeting and Exhibition, pp. 1–4 (2007). Preprint 07-031
4. Squelch, A.P.: Virtual reality for mine safety training in South Africa. J. S. Afr. Inst. Min. Metall. **101**(4), 209–216 (2001)
5. Tichon, J., Burgess-Limerick, R., et al.: A review of virtual reality as a medium for safety related training in the minerals industry. J. Health Saf. Res. Pract. **1**(3), 33–40 (2011)
6. Pedram, S., Perez, P., Palmisano, S., et al.: The application of simulation (virtual reality) for safety training in the context of mining industry. In: 22nd International Congress on Modelling and Simulation, Hobart, Tasmania, Australia (2017)
7. Grabowski, A., Jankowski, J.: Virtual reality-based pilot training for underground coal miner. Saf. Sci. **72**(72), 310–314 (2015)
8. Li, M., Sun, Z., Chen, J., Lv, P., Mao, S.: A multiplayer virtual reality training system for fully mechanized coal face. Coal Sci. Technol. **46**(1), 156–161+223 (2018)
9. Zhang, H.: Coal mine safety training system based on virtual reality technology. Ind. Mine Autom. **40**(2), 88–92 (2014)
10. Zhu, H., Qin, X., Yang, C., He, C.: Virtual simulation system of coal and gas outburst accident. Saf. Coal Mines **45**(6), 89–91 (2014)
11. Zhang, X., An, W., Li, J.: Design and application of virtual reality system in fully mechanized mining face. Proc. Eng. **26**(4), 2165–2172 (2011)
12. Li, Q., Wu, W., Cao, Y., Wang, L.: A Task Scheduling strategy with energy optimization for cloud rendering systems. J. Xi'an Jiaotong Univ. **50**(2), 1–6 (2016)
13. Liu, B.: Research on cloud rendering based 3D BIM model visualization techniques. J. Beijing Jiaotong Univ. **41**(6), 107–113 (2017)
14. Yuan, J.: Design and Optimization on Artificial Intelligence of Multiplayer Online Role-Playing Game. Huazhong University of Science and Technology (2018)
15. He, W.: Design and implementation of game AI based on behavior tree. Chengdu University of Technology (2018)

Research on Coal Mine Gas Safety Evaluation Based on D-S Evidence Theory Data Fusion

Zhenming Sun(✉), Dong Li, and Yunbing Hou

School of Energy and Mining Engineering, China University of Mining and Technology (Beijing), Beijing 100083, China
sun@cumtb.edu.cn

Abstract. In order to improve the accuracy of coal mine gas safety evaluation results, a gas safety evaluation model based on D-S evidence theory data fusion is proposed, and multi-sensor fusion of gas safety evaluation is realized. First, the prediction results of the weighted least squares support vector machine are used as the input of D-S evidence theory, and the basic probability assignment function of each sensor is calculated by using the posterior probability modeling method, and the similarity measure is introduced for optimization. Secondly, aiming at the problem of fusion failure in D-S evidence theory when fusing high-conflict evidence, the idea of assigning weights is used to allocate the importance of each evidence to weaken the impact of conflicting evidence on the evaluation results. In order to prevent the loss of the effective information of the original evidence after modifying the evidence source, a conflict allocation coefficient is introduced on the basis of fusion rules. Finally, a gas safety evaluation example analysis is carried out on the evaluation model established in this paper. The results show that the introduction of similarity measures can effectively eliminate high-conflict evidence sources; the accuracy of D-S evidence theory based on improved fusion rules is improved by 2.8% and 15.7% respectively compared to D-S evidence theory based on modified evidence sources and D-S evidence theory; as more sensors are fused, the accuracy of the evaluation results is higher; the multi-sensor data evaluation results are improved by 63.5% compared with the single sensor evaluation results.

Keywords: Data fusion · D-S evidence theory · Gas · Safety evaluation

1 Introduction

Coal mine gas safety evaluation has always been an important means of coal mine safety management. Through the monitoring of environmental data in the coal mine and the correct identification of the gas safety, gas accumulation, outburst, and explosion can be effectively avoided, which has important theoretical significance and practical value for suppressing the occurrence of gas disasters and promoting the safe and sustainable development of the coal industry [1].

X. Wang et al. (Eds.): 6GN 2020, LNICST 337, pp. 509–522, 2020.
https://doi.org/10.1007/978-3-030-63941-9_43

At present, the commonly used safety evaluation methods are probabilistic risk evaluation [2–4], and the use of computer technology and databases to establish disaster databases for casualties. However, in actual evaluation, the structure, indexes and parameters of each evaluation model are very different. The commonly used evaluation methods in China mainly focus on fuzzy comprehensive evaluation [5], gray clustering [6–8], neural network [9–11] and game analysis evaluation based on data mining [12]. Although the qualitative evaluation process is simple, the differences in the professional background and operational capabilities of different participants may lead to differences in accident risk evaluation. The existing gas safety evaluation system only stores information in the database, and does not realize the correlation between the monitoring data of multiple sensors, therefore, a complete and coordinated operating system has not been formed in practice. At the same time, in the analysis of the coal mine gas safety influencing factor system, more studies have magnified the role of people and machines, while neglecting the occurrence of gas accidents mostly is the unfavorable monitoring of environmental factors, the lack of evaluation systems and the insufficient accuracy.

The gas safety evaluation model used in this paper divides the gas safety status into different safety levels. Then, using the various sensor monitoring data collected by the working face monitoring station, the predicted data is obtained based on weighted least squares support vector machine. Finally, multi-sensor data fusion is carried out to realize the evaluation of the gas safety state of the working face at the next moment, so as to realize the early warning of the gas safety state.

2 Weighted Least Squares Support Vector Machine

Suykens [13] proposed a weighted least squares support vector machine (WLSSVM) based on the least squares support vector machine (LSSVM). The Lagrange function of its optimization problem can be described as:

$$L(w, b, \xi, \alpha) = \frac{1}{2}w^T w + \frac{1}{2}\mathrm{C}\sum_{i=1}^{N} v_i \xi_i^2 - \sum\nolimits_{i=1}^{N} \alpha_i \left[w^T \varphi(x_i) + b + \xi_i - y_i \right] \quad (1)$$

In Eq. (1), w is the weight coefficient vector; $\varphi(x_i)$ is the mapping input to the high-dimensional space; C is the regularization parameter; b is the threshold; x_i represents the Lagrange multiplier. According to the KKT (Karush-Khun-Tucker) condition, the function eliminate w, ξ_i, and get Eq. (2):

$$\begin{bmatrix} 0 & l_{1\times N} \\ l_{N\times 1} & R + \frac{1}{\mathrm{C}}V \end{bmatrix} \begin{bmatrix} b \\ \alpha \end{bmatrix} = \begin{bmatrix} 0 \\ y \end{bmatrix} \quad (2)$$

In Eq. (2), $V = \mathrm{diag}\left(v_1^{-1}, v_2^{-1}, \ldots, v_N^{-1}\right)$ is the diagonal matrix, $l_{1\times N}$ is the unit column vector, $R = \left\{K\left(x_i, x_j\right) | i = 1, 2, \ldots, \mathrm{N}\right\}$ is the radial basis kernel function matrix, $\mathrm{y} = \left[y_1, y_2, \ldots, y_N\right]^{\mathrm{T}}$. Equation (2) can be obtained b and α, inputing test samples to get WLSSVM model as follows:

$$y = \sum\nolimits_{i=0}^{n} \alpha K(x_i, x) + b \quad (3)$$

The weight calculation formula is as follows:

$$v_i = \begin{cases} 1, & \left|\frac{\xi_i}{\hat{s}}\right| \le s_1 \\ \frac{s_2 - \left|\frac{\xi_i}{\hat{s}}\right|}{s_2 - s_1}, & s_1 < \left|\frac{\xi_i}{\hat{s}}\right| \le s_2 \\ 10^{-4}, & otherwise \end{cases} \tag{4}$$

In Eq. (4), the values of s_1 and s_2 are 2.5 and 3.0 respectively; $\hat{s}$ is the standard estimated deviation of the error sequence, and its calculation function is as follows:

$$\hat{s} = \frac{\mathrm{IQR}}{2 \times 0.6745} \tag{5}$$

In Eq. (5), IQR is the difference between the third quartile and the first quartile in the sequence of errors ξ_i from small to large.

3 D-S Evidence Theory

3.1 Basic Principles of D-S Evidence Theory

For the reasoning of uncertain problems, Dempster-Shafer (D-S) evidence theory has strong adaptability, and the reasoning process is simpler. Among them, the distribution of belief functions and the fusion of evidence are the basic knowledge of D-S evidence theory. The uncertainty of events can be expressed through the recognition framework and basic belief distribution functions.

Recognition Framework

The recognition framework represents a set X of possible situations of the event, and the elements it contains represent the degree of evaluation of the event status. In the gas safety evaluation system, every possible state is called a hypothesis, and all possible categories constitute a recognition framework. Therefore, the recognition framework contains all possible results of a particular problem. The recognition framework can be expressed in Eq. (6):

$$\mathrm{X} = \{\mathrm{X}_1, \mathrm{X}_2, \mathrm{X}_3, \ldots, \Theta\} \tag{6}$$

In Eq. (6), X_i is called a possible result of the event, and the uncertainty represented by Θ.

Basic Probability Assignment Function (BPA)

Suppose X is a recognition framework, 2^{X} is a power set on X, if m: $2^{\mathrm{X}} \to [0, 1]$, and satisfy Eq. (7).

$$\sum\nolimits_{A \in 2^{\mathrm{X}}} m(A) = 1, m(\Theta) = 0 \tag{7}$$

In Eq. (7), m is called the BPA of the recognition frame X, it also known as the mass function, A is the element in the recognition frame. For $\forall A \subseteq X$, then $m(A)$ is the basic belief, which indicates the degree of trust in proposition A.

Belief Function

If there are A $\in P(X)$ and $B \in$ A, then define the function *Bel* as follows:

$$Bel(A) = \sum_{B \in A} m(B) \tag{8}$$

In Eq. (8), *Bel* represents the belief function, and the Eq. (8) represents the sum of the possibilities of all the subsets of *A*, which represents the overall degree of trust in A, so that it can be inferred that $Bel(\Theta) = 0$ and $Bel(\mathrm{X}) = 1$. The belief function represents the degree of trust of a certain thing. It is incomplete and untrustworthy to only use the belief function to describe the possibility of an event.

Likelihood Function

In D-S evidence theory, the likelihood function is a measure used to express the degree of distrust of an event. Definition: Assuming that X is a recognition framework, m: $2^{\mathrm{X}} \to [0, 1]$ is represented as the basic probability assignment on X. If there are $A \in P(X)$, $B \in A$, then define the function *Pl*: $2^{\mathrm{X}} \to [0, 1]$ as follows:

$$Pl(A) = 1 - Bel\left(\overline{A}\right) = \sum_{B \cap A \neq \Theta} m(B) \tag{9}$$

In Eq. (9), $Pl(A)$ represents that event A is true uncertainty, and $Bel\left(\overline{A}\right)$ represents the trust degree of event $\overline{A}$. The degree of mistrust $Pl(A)$ of A can be calculated by the Eq. (9).

The minimum degree of trust of evidence theory for event *A* is $Bel(A)$, the potential degree of trust in event A is expressed as $Pl(A)$, the support interval of event A can be expressed as $[0, Bel(A)]$, the likelihood interval of event A can be expressed as $[0, Pl(A)]$. When the evidence neither confirms nor denies the occurrence of event A, for this uncertain phenomenon, a trust interval can be used to represent the probability of event A.

3.2 Improved D-S Evidence Theory

D-S evidence theory has strong applicability in data fusion, but in the actual fusion process, there are still some deficiencies in dealing with uncertain problems. It is mainly manifested in the explosive problem, the limited problem of recognition framework, the independent problem between the evidences and the problem of conflicting evidence fusion. In this paper, the improvement of D-S evidence theory is mainly used to solve the problem of conflicting evidence sources.

Evidence-Based Improvements

Modifying the evidence source can reduce the influence of interference factors on the fusion evaluation results and improve the accuracy of the evaluation results. In this paper, the idea of assigning weights is used to allocate the importance of each evidence, which can increase the reliability of the evidence on the decision result and weaken the impact of conflicting evidence. For the method of evidence-based improvement, this paper is called D-S-1 evidence theory.

For an uncertain event, there are n evidences, the corresponding recognition frame X contains N focal elements, and m_i represents the evidence set composed of the basic probability assignment function corresponding to the evidence under each focal element.

$$m_i = [m_i(A_1), m_i(A_2), \ldots, m_i(A_n)]^T, i = 1, 2, \ldots, n \tag{10}$$

Equation (11) is used to calculate the distance between m_i and m_j, d_{ij} represents the distance of m_i and m_j. This distance function has a better reflection in describing the focal element and the reliability between evidences, and can better characterize the conflict between evidences.

$$d_{ij} = d(m_i, m_j) = \sqrt{\frac{1}{2}\left[\|m_i\|^2 + \|m_j\|^2 - 2(m_i, m_j)\right]} \tag{11}$$

The similarity function is further derived from the Eq. (11). The similarity between m_i and m_j can be expressed as S_{ij}. The expression of S_{ij} is as follows:

$$S_{ij} = 1 - d_{ij} \tag{12}$$

The smaller the distance between the evidences, the greater the mutual support. The degree of support for evidence can be expressed by the sum of other evidences, then the degree of support for evidence m_i can be expressed as:

$$T(m_i) = \sum\nolimits_{j=1, j \neq i}^{n} S_{ij}, i = 1, 2, .., n \tag{13}$$

In this paper, the distance similarity matrix between evidences is used to give different weights to each sensor, so as to achieve the purpose of modifying the evidence source. In order to prevent the revised evidence source from being too conservative and losing the advantages of the original evidence, this paper adopts to retain the original set of more correct evidence to ensure the effect of data fusion. Based on the above ideas, according to the ratio of the degree of support of the evidence, under the condition of retaining a good set of evidence sources, the weight β of the evidence is calculated according to the degree of support. The specific formula is as follows:

$$\beta(m_i) = \frac{T(m_i)}{max(T(m_i))} \tag{14}$$

After assigning weights, the modified basic probability assignment function corresponding to the evidence can be expressed as follows:

$$\begin{aligned} m_i'(i) &= \beta(m_i) \cdot m_i \\ m_i'(\Theta) &= \beta(m_i) \cdot m_i + (1 - \beta(m_i)) \end{aligned} \tag{15}$$

Improvements Based on Fusion Rules

In this paper, the time series prediction value of the monitoring data of each sensor is used to calculate the basic probability assignment value. After the value of each sensor is fused, the mine gas safety status is judged. The fusion rules of D-S evidence theory are as follows:

Based on two independent evidences M_1, M_2, the focal elements of the two evidences are B_i and $C_j(i = 1, 2, 3, \ldots, n, j = 1, 2, 3, \ldots, m)$, the basic probability assignment function value after their fusion is $m(A)$:

$$\begin{cases} m(A) = M_1 \oplus M_2 = \frac{1}{1-K} \sum_{B_i \cap C_j = A} m_1(B_i) m_2(C_j) \\ K(M_1, M_2) = \sum_{B_i \cap C_j = \Theta} m_1(B_i) m_2(C_j) \end{cases} \tag{16}$$

In Eq. (16), $K(M_1, M_2)$ is called the conflict coefficient, which represents the degree of conflict between the two evidences M_1, M_2. When the conflict coefficient is 0, there is no conflict between the two evidences; when it is close to 1, the greater the conflict between the two evidences, there is a complete conflict.

Many scholars believe that the fusion rules of evidence theory are imperfect in the processing of evidence, so the reasonable modification of fusion rules can also improve the accuracy of fusion. After modifying the evidence source, simply modifying the evidence source data to prevent high conflicts between the evidences may cause the revised evidence to lose the effective information of the original evidence. The conflict allocation coefficient is introduced on the basis of the fusion rules to improve the accuracy of the decision stage. For the method of modifying the fusion rule, this paper is called D-S-2 evidence theory.

The conflict allocation coefficient $\omega(A_i)$ can be defined as follows:

$$\omega(A_i) = \frac{\sum_{i=1}^{n} m_i'(A_{ij})}{\sum_{i=1}^{n} \sum_{j=1}^{p} m_i'(A_{ij})} \tag{17}$$

The improved formula of D-S evidence theory fusion rule is defined as follows:

$$m(A) = \sum_{B_i \cap C_j = A} m_1(B_i) m_2(C_j) + \mathrm{K} \cdot \omega(A_i) \tag{18}$$

In Eq. (17): set A represents the intersection of focal element B_i and focal element C_j.

4 Construction of Gas Safety Evaluation Model

4.1 Construction of Recognition Framework

From the perspective of D-S evidence theory, the "gas safety state" can be regarded as a judgmental problem, and the summary of hypothetical results can be described as a recognition framework. According to the coal mine safety regulations and the value range of characteristic parameters under specific conditions, the gas safety state is divided into five states: no danger, mild danger, moderate danger, serious danger, and uncertain [14]. No danger indicates that the working face of the coal mine is in a good environment; mild danger indicates that the working face has a certain risk, and this danger value is within the acceptable range; moderate danger indicates that the working face is unsafe, and the indicates value has exceeded the accepted, this danger requires staff to conduct on-site inspection; serious danger indicates that the working face is very bad, and the staff should be evacuated. So the recognition framework of D-S evidence theory can be described as $X = \{X_1$ (no danger), X_2 (mild danger), X_3 (moderate danger), X_4 (serious danger)$\}$.

4.2 Construction of Basic Probability Assignment Function

This paper uses the posterior probability modeling method to find the basic probability assignment function, and introduces the similarity degree to modify the evidence source. The basic probability assignment function characterizes the support degree of each sensor to the safety status of mine gas. In this paper, a time series prediction model is constructed through the WLSSVM, and the prediction model is established with each influence factor as an input to obtain the prediction value of each sensor. The posterior probability modeling method calculates the basic probability assignment function of each sensor.

Taking a single sensor as an example, the basic probability assignment function value obtained by the posterior probability modeling method is y, the recognition framework is $X = \{X_1, X_2, X_3, X_4\}$. The distance between X and y can be expressed as follows:

$$d_i(X_i, y) = |X_i - y| \tag{19}$$

The correlation coefficient between the evidence and X_i can be expressed as follows:

$$c_i = \frac{1/d_i}{\sum_{i=1}^{4}(1/d_i)} \tag{20}$$

Introducing Eq. (20), the basic probability assignment function $m(i)$ and the uncertainty $m(\Theta)$ of the corresponding evidence can be expressed as follows:

$$m(i) = \frac{c_i}{\sum c_i + E}$$

$$m(\Theta) = \frac{E}{\sum c_i + E}$$

$$E = \frac{1}{2}|y - x|^2 \tag{21}$$

In Eq. (21), y is the predicted value of the time series prediction model, and x is the expected output value of the prediction model.

4.3 Construction of Gas Safety Evaluation Model

Five classification indicators of gas safety status can be obtained through Sect. 4.1. The construction process of data fusion model based on D-S evidence theory mainly includes three parts: time series prediction of each sensor, construction of basic probability assignment function, fusion between evidences and decision-making. First, each sensor obtains the predicted value through the time series prediction model. The basic probability assignment function is obtained through the posterior probability modeling method. The similarity degree is introduced to modify the evidence source to obtain the basic probability assignment function. In order to improve the accuracy of decision-making, multi-sensor data fusion was carried out according to the fusion rules. The coal mine gas safety evaluation model based on D-S evidence theory data fusion is shown in Fig. 1.

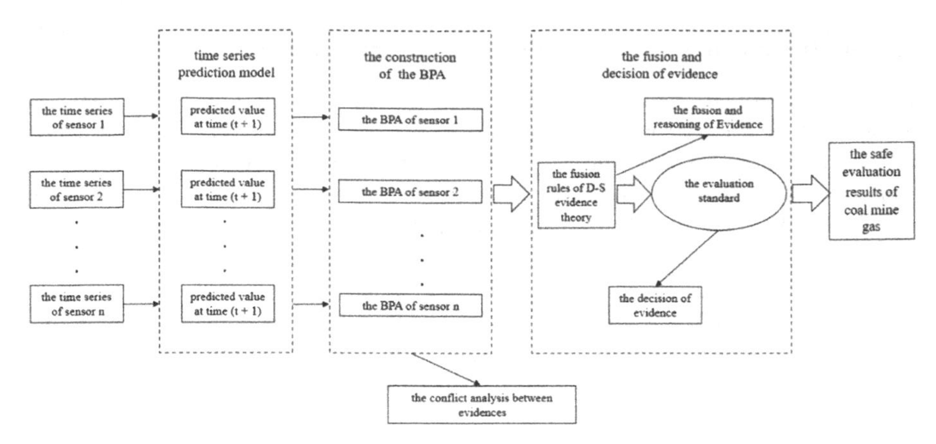

Fig. 1. Gas safety evaluation model

5 Case Analysis

5.1 Data Sources

The data in this paper comes from the gas concentration at the upper corner (No. A02), the gas concentration at the working face 10 m (No. A01), the wind speed (No. A09), the dust (No. A11), the return air 15 m temperature (No. A07), the return air 15 m gas concentration (No. A08) of the a coal mine. The original data sampling interval is 1 min, and the data distribution has obvious jagged characteristics. Therefore, this paper uses 5 min as the sampling interval to obtain 1500 groups of samples, select the first 1400 samples for model training, and the remaining samples for model testing. Some data is shown in Table 1.

Table 1. Sample set of monitoring data.

No.	A02/(%)	A01/(%)	A09/(m/s)	A11/(mg/m^3)	A07/(°C)	A08/(%)
1	0.224	0.262	1.952	0.02	21.332	0.35
2	0.226	0.26	1.992	0.014	21.3	0.342
3	0.218	0.26	1.97	0.08	21.306	0.342
4	0.218	0.27	1.98	0.082	21.3	0.342
5	0.212	0.276	2.016	0.068	21.304	0.34
.	.	.	.	.	.	.
.	.	.	.	.	.	.
1497	0.368	0.408	1.926	0.086	22.026	0.502
1498	0.37	0.406	1.916	0.084	22	0.518
1499	0.362	0.4	1.944	0.076	22	0.496
1500	0.352	0.396	1.944	0.074	22	0.482

5.2 The Predicted Results of the Time Series Prediction Model

This paper uses the multivariable WLSSVM time series prediction model introduced in Sect. 2 to predict the monitoring value of each sensor at the next moment. This paper uses the target sensor as the output and other sensors as the input for model training. SPSS software was used to analyze the Pearson correlation of A02, A01, A09, A11, A07 and A08 monitoring points. The analysis results are shown in Table 2.

Table 2. Correlation analysis results of various influencing factors

	A02	A01	A09	A11	A07	A08
A08	0.572	0.910	0.668	0.324	0.788	1

It can be seen from Table 2 that the correlation coefficients are all greater than 0.3, and it is reasonable for each other sensor to be the input of the target sensor. The prediction results are shown in Table 3.

Table 3. Predicted results of various sensors

	A02	A01	A09	A11	A07	A08
Predicted results	0.380	0.422	1.912	0.094	22.086	0.504

5.3 Experimental Results and Analysis

Contrast Analysis of Conflict Degree

This paper uses the posterior probability modeling method introduced in Sect. 3.2 to calculate the basic probability assignment function of each sensor. The BPA of each sensor is shown in Table 4.

Table 4. Basic probability assignment functions

	A09	A07	A11	A02	A01	A08
X_1	0.0646	0.2057	0.4939	0.5551	0.5664	0.5954
X_2	0.8079	0.2160	0.2358	0.2150	0.2106	0.1979
X_3	0.0557	0.2273	0.1549	0.1333	0.1294	0.1187
X_4	0.0288	0.2399	0.1153	0.0966	0.0934	0.0848
Θ	0.0431	0.1111	0.0000	0.0000	0.0002	0.0032

It can be seen from Table 2 that the results of single sensor recognition are A09 $m(X_2) = 0.8079$, A07 $m(X_4) = 0.2399$, A11 $m(X_1) = 0.4939$, A02 $m(X_1) = 0.5551$, A01 $m(X_1) = 0.5664$ and A08 $m(X_1) = 0.5954$. Obviously, A09 and A07 have a great conflict with other sensors. Using a single sensor evaluation result can not accurately evaluate the safety status of coal mine gas. Therefore, it is necessary to modify the evidence source before fusion.

This paper adopts the improved method of evidence source introduced in Sect. 4.2, redistributes the weights for each sensor according to the BPA in Table 4, the revised BPA is shown in Table 5.

Table 5. Basic probability assignment function after modifying the evidence source

	A09	A07	A11	A02	A01	A08
X_1	0.0369	0.1712	0.4914	0.5551	0.5643	0.5801
X_2	0.4622	0.1797	0.2357	0.2150	0.2098	0.1929
X_3	0.0318	0.1892	0.1541	0.1333	0.1298	0.1157
X_4	0.0165	0.1997	0.1148	0.0966	0.0930	0.0826
Θ	0.4526	0.2602	0.0051	0.0000	0.0040	0.0287

It can be seen from Table 5 that A09 is revised from $m(X_2) = 0.8079$ to $m(X_2) = 0.4622$, and A07 is revised from $m(X_4) = 0.2399$ to $m(X_4) = 0.1997$. The conflict is significantly reduced, indicating that the method of modify the source of evidence is feasible and retains the excellent evidence of A02. At the same time, the Table 5 shows that only using sensors A09 and A07 as evaluation evidence will lead to failure of decision-making, and only using A11, A02, A01 and A08 as evaluation evidence has low recognition accuracy and makes decision reliability low. Therefore, it is not reliable to use only a single sensor to evaluate the safety status of coal mine gas.

Comparative Analysis of Evaluation Results

Through the comparative analysis of the degree of conflict above, we can see that data fusion plays an important role in the decision-making results. Sensors A09, A07, A11, A02, A01, A08 are recorded as evidence e_1, e_2, e_3, e_4, e_5, e_6. The fusion process of multi-sensors is the fusion process of two sensors in sequence. The comparison results of the multi-sensor fusion of the three methods are shown in Table 6, 7, 8, 9 and 10.

Table 6. Comparative analysis of $e_1 e_2$ fusion results

	m(X_1)	m(X_2)	m(X_3)	m(X_4)	m(Θ)	Identify result	Safety status
D-S	0.0822	0.7668	0.0803	0.0573	0.0000	X_2	X_1
D-S-1	0.1346	0.4103	0.1440	0.1412	0.0000	X_2	X_1
D-S-2	0.1023	0.4022	0.1084	0.1026	0.5199	Θ	X_1

Table 7. Comparative analysis of $e_1e_2e_3$ fusion results

	m(X_1)	m(X_2)	m(X_3)	m(X_4)	m(Θ)	Identify result	Safety status
D-S	0.1688	0.7520	0.0517	0.0275	0.0000	X_2	X_1
D-S-1	0.3259	0.4798	0.1118	0.0826	0.0000	X_2	X_1
D-S-2	0.3930	0.3813	0.1262	0.0923	0.0125	X_1	X_1

Table 8. Comparative analysis of $e_1e_2e_3e_4$ fusion results

	m(X_1)	m(X_2)	m(X_3)	m(X_4)	m(Θ)	Identify result	Safety status
D-S	0.3537	0.6103	0.0260	0.0100	0.0000	X_2	X_1
D-S-1	0.5894	0.3361	0.0486	0.0260	0.0000	X_1	X_1
D-S-2	0.6680	0.2511	0.0526	0.0282	0.0000	X_1	X_1

Table 9. Comparative analysis of $e_1e_2e_3e_4e_5$ fusion results

	m(X_1)	m(X_2)	m(X_3)	m(X_4)	m(Θ)	Identify result	Safety status
D-S	0.6012	0.3859	0.0101	0.0028	0.0000	X_1	X_1
D-S-1	0.8056	0.1728	0.0155	0.0061	0.0000	X_1	X_1
D-S-2	0.8578	0.1205	0.0156	0.0061	0.0000	X_1	X_1

Table 10. Comparative analysis of $e_1e_2e_3e_4e_5e_6$ fusion results

	m(X_1)	m(X_2)	m(X_3)	m(X_4)	m(Θ)	Identify result	Safety status
D-S	0.8198	0.1768	0.0028	0.0006	0.0000	X_1	X_1
D-S-1	0.9225	0.0720	0.0042	0.0013	0.0000	X_1	X_1
D-S-2	0.9485	0.0466	0.0038	0.0011	0.0000	X_1	X_1

As shown in Table 6 above, the fusion evidence sources e_1 and e_2 are all highly conflicting evidences, so the decision results of D-S evidence theory and D-S-1 evidence theory are invalidated, and the recognition results of D-S-2 evidence theory are uncertain. After introducing the evidence source e_3 in Table 7, the recognition results of the D-S evidence theory and D-S-1 evidence theory are wrong, and the D-S-2 evidence theory recognition results are accurate, which proves that the improved fusion rule in this paper is effective, and retains the revised evidence source. Effective information in Table 8, $e_1e_2e_3e_4$ fusion, D-S evidence theory recognition result is wrong, D-S-1 evidence theory and D-S-2 evidence theory recognition results are accurate, which proves that the modified method of the evidence source improved in this paper is correct, eliminating the inter-evidence Highly conflicting. Tables 9 and 10 show that the D-S-2 evidence theory method for the modification of evidence source and fusion rules in this paper is reasonable. The recognition accuracy of D-S-2 evidence theory is higher than that of D-S evidence theory and D-S-1 evidence theory. The accuracy rate of mine gas safety status recognition has been improved. At the same time, the fusion rule satisfies the exchange law, and it can be concluded that as the evidence increases during the fusion process, the accuracy of the identification in the decision stage is higher. The problem that the single sensor is difficult to accurately characterize the gas safety state is solved.

Through the above analysis, it can be concluded that the multi-sensor data fusion gas safety status evaluation system proposed in this section has high practical value in field applications, and has important theoretical significance for suppressing the occurrence of gas disasters and promoting the safe and sustainable development of the coal industry. In Table 10, the accuracy of D-S evidence theory based on improved fusion rules is improved by 2.8% and 15.7% respectively compared to D-S evidence theory based on modified evidence sources and D-S evidence theory, as more sensors are fused, the accuracy of the evaluation results is higher; the multi-sensor data evaluation results are improved by 63.5% compared with the single sensor evaluation results.

Model Uncertainty Measure

This paper uses Shannon entropy to measure the uncertainty of the above three D-S evidence theories. Let n signal sources make up the signal $X = \{x_1, x_2, x_3 \ldots, x_n\}$, the probability that each signal source provides corresponding information for an event is $P = \{p(x_1), p(x_2), p(x_3), \ldots, p(x_n)\}$, then the system structure S of the signal can be expressed as:

$$S = \begin{pmatrix} X \\ P \end{pmatrix} = \begin{pmatrix} x_1 & x_2 & \ldots & x_n \\ p(x_1) & p(x_2) & \ldots & p(x_n) \end{pmatrix} \tag{22}$$

Therefore, the Shannon entropy of the signal is expressed as follows:

$$H(x) = -\sum\nolimits_{i=1}^{n} p(x_i) \ln p(x_i) \tag{23}$$

The uncertainty of D-S evidence theory fusion is:

$$\begin{aligned} &- 0.8198 * \ln 0.8198 - 0.1768 * \ln 0.1768 - 0.0028 * \ln 0.0028 \\ &- 0.0006 * \ln 0.0006 = 0.4901 \end{aligned}$$

The uncertainty of D-S-1 evidence theory fusion is:

$$\begin{aligned} &- 0.9225 * \ln 0.9225 - 0.0720 * \ln 0.0720 - 0.0042 * \ln 0.0042 \\ &- 0.0013 * \ln 0.0013 = 0.2955 \end{aligned}$$

The uncertainty of D-S-2 evidence theory fusion is:

$$\begin{aligned} &- 0.9485 * \ln 0.9485 - 0.0466 * \ln 0.0466 - 0.0038 * \ln 0.0038 \\ &- 0.0011 * \ln 0.0011 = 0.2217 \end{aligned}$$

From the comparison of the above results (Fig. 2), we can see that the improved D-S-2 evidence theory has lower uncertainty than D-S evidence theory and D-S-1 evidence theory, and can better evaluate the safety of coal mine gas.

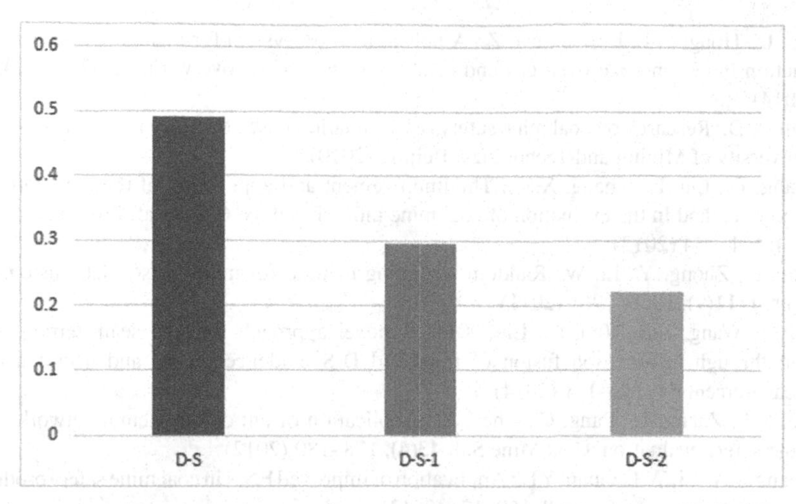

Fig. 2. Uncertainty for three types of D-S evidence theory

6 Conclusion

(1) According to the characteristics of coal mine monitoring data, an index system is constructed. By acquiring the predicted values of each sensor, the basic probability assignment function of each sensor is calculated using the posterior probability modeling method.

(2) A safe evaluation model of coal mine gas status is constructed, and multi-sensor data fusion is realized. As more sensors are fused, the evaluation results are more accurate. The model in this paper effectively solves the problem that it is difficult for a single sensor to accurately characterize the gas safety state.

(3) Aiming at the problem of evidence fusion failure caused by high conflict data, this paper introduces the similarity to modify the evidence source of conflict data, which effectively reduces the conflict between the evidence. At the same time, in order to prevent distortion of evidence sources, the conflict allocation coefficients are introduced to improve the fusion rules, and the accuracy of evaluation results is improved. It proves that the improved D-S evidence theory has higher accuracy and better generalization ability for coal mine gas safety evaluation, which can provide theoretical basis for gas disaster accident prevention.

References

1. Sun, Q.G.: Current situation of coal mine gas disasters in China and countermeasures. China Coal **40**(3), 116–119 (2014)
2. Pejic, L.M., Torrent, J.G., Querol, E., Lebecki, K.: A new simple methodology for evaluation of explosion risk in underground coal mines. J. Loss Prev. Process Ind. **26**, 1524–1529 (2013)
3. Ghasemi, E., Ataei, M., Shahriar, K., Sereshki, F., Jalali, S.E., Ramazanzadeeh, A.: Assessment of roof fall risk during retreat mining in room and pillar coal mines. Int. J. Rock Mech. Min. Sci. **54**, 80–89 (2012)

4. Hu, L., Hong, G.J., Lin, G., Na, Z.: A polygeneration system for methanol and power production based on coke oven gas and coal gas with CO^2 recovery. Energy **74**(2), 143–149 (2014)
5. Sun, X.D.: Research on coal mine safety risk evaluation based on fuzzy information, 4. China University of Mining and Technology, Beijing (2010)
6. Wang, D., Liu, L., Zhang, X.M.: The improvement and application of the grey correlation degree method in the evaluation of coal mine intrinsic safety. China Saf. Prod. Sci. Technol. **9**(1), 151–154 (2013)
7. Gao, S., Zhong, Y., Li, W.: Random weighting method for multi-sensor data fusion. IEEE Sens. J. **11**(9), 1955–1961 (2011)
8. Si, L., Wang, Z.B., Tan, C., Liu, X.H.: A novel approach for coal seam terrain prediction through information fusion of improved D-S evidence theory and neural network. Measurement **54**, 140–151 (2014)
9. He, R.J., Zhang, L., Pang, C., Chen, X.: Application of ant colony neural network in coal mine safety evaluation. Coal Mine Saf. **43**(4), 178–180 (2012)
10. Zhang, J.N., Li, W.J., Guan, Y.L.: Application of improved FNN in coal mine safety production warning system. Coal Eng. **8**, 168–171 (2013)
11. Li, X., Li, N.W., Yang, Z.: Coal mine safety evaluation model based on quantum genetic algorithm. Comput. Syst. Appl. **21**(7), 101–105 (2012)
12. Li, P.L., Duan, J.: Game model of coal mine safety production. J. Xi'an Univ. Sci. Technol. **33**(1), 72–76 (2013)
13. Suykens, J.A.K., Brabante, J.D.E., Lukas, L., Vandewalle, J.: Weighted least squares support vector machines: robustness and sparse approximation. Neurocomputing **48**(1), 85–105 (2002)
14. Bao, Y.: Application of Multi-Sensor Information Fusion in Coal Mine Environmental Monitoring System. China University of Mining and Technology Library, Xuzhou (2007)

Rockburst Prediction of Multi-dimensional Cloud Model Based on Improved Hierarchical and Critic

Xiaoyue Liu and Wei Yang(✉)

North China University of Science and Technology, Tangshan 063200, Hebei, China
1417705803@qq.com

Abstract. In high terrestrial stress regions, rockburst is a major geological disaster influencing underground engineering construction significantly. How to carry out efficient and accurate rock burst prediction remains to be solved. Comprehensively consider the objective information of the index data and the important role of subjective evaluation and decision-making in rockburst prediction, and use the improved analytic hierarchy process and the CRITIC method based on index correlation to obtain the subjective and objective weights of each index, and obtain comprehensive weights based on the principle of minimum discriminant information. The original cloud model and the classification interval of the forecast index were modified to make up for the lack of sensitivity of the original cloud model to the average of the grade interval. A hierarchical comprehensive cloud model of each index was generated through a cloud algorithm. Finally, the reliability and effectiveness of the model were verified through several sets of rockburst examples, and compared with the entropy weight-cloud model, CRITIC-cloud model and set pair analysis-multidimensional cloud model. The results show that the model can describe various uncertainties of interval-valued indicators, quickly and effectively determine rockburst severity.

Keywords: Rockburst · Prediction · Analytic hierarchy process · CRITIC method · Multidimensional cloud model

1 Introduction

With the continuous development of tunnels and underground engineering, rock bursts are sudden, difficult to control and highly destructive, which seriously threatens the lives of workers, delays construction periods and causes huge economic losses. It has become a major problem urgently to be solved in international deep mining engineering and underground space development engineering, and it is urgent to find a more effective method for rockburst prediction.

Rockburst prediction includes long-term prediction before construction and short-term prediction of construction process. Short-term prediction generally uses microseisms [1], infrared radiation [2], acoustic emission [1, 2], and ultrasonic methods to

X. Wang et al. (Eds.): 6GN 2020, LNICST 337, pp. 523–537, 2020.
https://doi.org/10.1007/978-3-030-63941-9_44

make real-time early warning of the exact location and time of rockburst. Among them, microseisms and acoustic emissions are most commonly used in engineering. The macro-prediction of the existence and intensity level of rockburst before construction has guiding significance for the feasibility study stage of the project. The long-term prediction methods are mainly theoretical analysis and prediction. At present, the commonly used processing methods include mathematical comprehensive processing analysis method, model test verification method, and numerical simulation analysis verification method. Among them, the mathematical comprehensive processing analysis method has achieved good prediction results in rockburst prediction and has been successfully applied to practical engineering, such as fuzzy mathematical comprehensive evaluation method [3], generalized artificial neural network [4], particle swarm algorithm [4], Probabilistic Neural Network [5], Support Vector Machines [6, 7], Decision Tree [7], Multilayer Perceptron (MLP) [7, 8], K-Nearest Neighbor (KNN) [7, 8], Rough Set Theory [9], cloud model [9, 12], etc. It should be noted that different criteria and theoretical analysis methods have their own limitations, such as the slow convergence rate of artificial neural networks; the comprehensive evaluation method of fuzzy mathematics cannot reflect the randomness of the system, and the distance discrimination method is highly dependent on samples.

In terms of weight assignment, the expert-based subjective weighting method has obvious shortcomings due to the complex factors affecting the rockburst mechanism and has not yet formed a perfect system; The objective weighting method does not consider the correlation between indicators, and ignores the role of subjective decision-making in practical applications; The analytic hierarchy process is too subjective and may not satisfy the judgment matrix. The single weight assignment cannot accurately measure the influence of various factors, which makes the prediction result deviate from the actual result, and the combination weight lacks the corresponding basis. The cloud model has certain advantages for rockburst prediction due to its ambiguity and randomness. However, with the increase of indicators, the calculation process of the one-dimensional cloud model is complicated, and it cannot reflect the interaction between multiple factors.

This paper adopts a combination weighting method combining improved analytic hierarchy process and CRITIC (Criteria Importance Through Intercriteria Correlation) method based on index correlation, and combines subjective and objective weights to obtain combined weights based on the principle of minimum discriminant information. Make full use of subjective and objective factors to make empowerment more reasonable; The multi-dimensional cloud model is used to predict the rockburst level, which reflects the comprehensive influence of various indicators and simplifies the calculation process of the model. The original cloud model and the classification interval of the predictive indicators were modified to make up for the lack of sensitivity of the original cloud model to the mean of the grade interval. Finally, the established model is used to verify the reliability of the model in the application of rockburst examples in related literature.

2 Combination Empowerment

2.1 Improved Analytic Hierarchy Process

This paper uses the scale construction method to construct the judgment matrix, thereby improving the subjective weight calculation of AHP and avoiding the consistency check. The judgment matrix $R = \left[r_{ij}\right]$ satisfies the following conditions: 1) $r_{ij} > 0$; 2) $r_{ii} = 1$; 3) $r_{ij} = \frac{1}{r_{ij}}$; 4) $r_{ij} = r_{ik}r_{kj}.r_{ij}$ is the scale value of the first indicator relative to the j-th indicator. The meaning of the standard values is shown in Table 1.

Table 1. Meaning of scale values

Scale value	Meaning	Scale value	Meaning
1.0	Equally important	1.6	Obviously important
1.2	Slightly important	1.8	Absolutely important
1.4	Strongly important		

There are n indicators $x_1, x_2, \ldots, x_n$, subjectively rank the indicators according to the principle of undiminished importance, determine the scale value and record the corresponding scale as t_i. Other elements in the judgment matrix are obtained according to the degree of transitivity, and then the final judgment matrix R is:

$$R = \begin{bmatrix} 1 & t_1 & t_1 t_2 & \cdots & \prod_{i=1}^{n} t_i \\ \frac{1}{t_1} & 1 & t_2 & \cdots & \prod_{i=2}^{n-1} t_i \\ \frac{1}{t_1 t_2} & \frac{1}{t_2} & 1 & \cdots & \prod_{i=3}^{n-1} t_i \\ \vdots & \vdots & \vdots & & \vdots \\ \frac{1}{\prod_{i=1}^{n} t_i} & \frac{1}{\prod_{i=2}^{n-1} t_i} & \frac{1}{\prod_{i=3}^{n-1} t_i} & \cdots & 1 \end{bmatrix}$$

α_i is the weight value of the i-th index; $\prod_{j=1}^{n} r_{ij}$ represents the product of all elements in the i-th row of the matrix R. From this, the subjective weight of each indicator in the rockburst prediction can be quantitatively determined as:

$$\alpha_i = \left(\prod\nolimits_{j=1}^{n} r_{ij}\right)^{\frac{1}{n}} / \sum\nolimits_{i=1}^{n} \left(\prod\nolimits_{j=1}^{n} r_{ij}\right)^{\frac{1}{n}} \tag{1}$$

2.2 CRITIC

The CRITIC method is an objective weighting method based on evaluation indicators. It takes into account the comparative strength of the sample and the conflict between the indicators, and the calculation results are more objective and reasonable. Suppose there are m samples and n indicators, and x_{ij} represents the value of the j-th evaluation index of the i-th sample. The evaluation matrix can be expressed as:

$$X = \begin{bmatrix} x_{11} & x_{12} & \cdots & x_{1n} \\ x_{21} & x_{22} & \cdots & x_{2n} \\ \vdots & \vdots & & \vdots \\ x_{n1} & x_{n2} & \cdots & x_{nn} \end{bmatrix}$$

The calculation steps of objective weight are as follows:

1. Normalization of indicators.

The larger and better indicators are:

$$y_{ij} = \frac{x_{ij} - \min(x_{ij})}{max(x_{ij}) - min(x_{ij})} \tag{2}$$

The smaller and better indicators are:

$$y_{ij} = \frac{x_{ij}}{max(x_{ij}) - min(x_{ij})} \tag{3}$$

The normalized matrix Y is calculated.

2. Calculate the mean x and standard deviation s:

$$\bar{x}_j = \frac{1}{m}\sum\nolimits_{i=1}^{m} x_{ij} \tag{4}$$

$$s_j = \sqrt{\frac{1}{m}\sum\nolimits_{i=1}^{m} \left(x_{ij} - \bar{x}_j\right)^2} \tag{5}$$

3. Calculate the coefficient of variation:

$$v_j = \frac{s_j}{\bar{x}_j} \tag{6}$$

4. Calculate the correlation coefficient matrix:

$$\rho_{ij} = cov(y_k, y_l)/\left(s_k s_j\right)(k = 1, 2, \ldots, n; l = 1, 2, \ldots, n) \tag{7}$$

ρ_{ij} is the correlation coefficient between the k-th index and the l-th index, and $cov(y_k, y_l)$ is the covariance between the k-th index and the l-th index.

5. Calculate the amount of information contained in the indicator:

$$\eta_j = v_j \sum\nolimits_{k=1}^{n} (1 - \rho_{ij})(j = 1, 2, \ldots, n) \tag{8}$$

6. Determine the objective weights as:

$$\beta_{ij} = \frac{\eta_j}{\sum_{i=1}^{n} \eta_j}(j = 1, 2, \ldots, n) \tag{9}$$

2.3 Comprehensive Weight

In order to make the comprehensive weight ω_i as close to a and b as possible without biasing any one of them, the comprehensive weight ω_i is obtained according to the principle of minimum discriminant information, and the objective function is [13]:

$$\left\{ \begin{array}{l} \min J(\omega) = \sum\limits_{i=1}^{n} (\omega_i \ln \frac{\omega_i}{\alpha_i} + \omega_i \ln \frac{\omega_i}{\beta_i}) \\ s.t. \sum\limits_{i=1}^{n} \omega_i = 1, \omega_i \geq 0 (i = 1, 2, \ldots, n) \end{array} \right\} \tag{10}$$

Solving this optimization model, the comprehensive weights are:

$$\omega_i = \frac{\sqrt{\alpha_i \beta_i}}{\sum_{j=1}^{n} \sqrt{\alpha_i \beta_i}} \tag{11}$$

3 Multidimensional Cloud Model

3.1 Multidimensional Cloud Model Definition and Digital Features

A multi-dimensional cloud model definition is introduced on the definition of a one-dimensional cloud model as follows [11]: Let C be a qualitative concept on the quantitative field $U\{X_1, X_2, \ldots, X_n\}$. If $x(x_1, x_2, \ldots, x_n)$ is a random realization of the concept, the degree of certainty U of x on $\mu(x(x_1, x_2, \ldots, x_n)) \in [0, 1]$ is subject to the normal The distribution $x(x_1, x_2, \ldots, x_n) \sim N(E_x(E_{x1}, E_{x1}, \ldots, E_{xn}))$ satisfies the normal distribution:

$$Ex_j^i = \frac{C_{min\,j}^i + C_{max\,j}^i}{2} \tag{12}$$

$$En = \frac{a_{ij}}{3} \tag{13}$$

$$He = \beta \tag{14}$$

In the formula, $C_{max\,j}^i$c and $C_{min\,j}^i$ represent the maximum and minimum values of the i-th index interval of the jth index; a_{ij} is the width of the left and right half branches connecting the cloud; E_x represents the basic certainty of the qualitative concept, and is the spatial distribution of cloud drops in the universe of discourse. E_n entropy represents the uncertainty measure of qualitative concepts; H_e superentropy represents the uncertainty of entropy, and reflects the degree to which the random variable corresponding to the qualitative concept deviates from the normal distribution.

3.2 Forecast Indicators and Classification

The mechanism of rockburst occurrence is complex and there are many influencing factors. The selection of indicators is very important for the accuracy of prediction results. Based on the existing research results of rockburst [4, 7, 8, 10–12, 14], Considering internal and external factors, $\frac{\sigma_c}{\sigma_t}$, σ_θ/σ_c, w_{et} and k_s are selected as the main evaluation factors.

Tangential stress refers to the force acting on the bearing surface of the rock mass and parallel to the bearing surface. The greater the stress on the rock mass, the easier it is to destroy the rock mass. The ratio of tangential stress to the uniaxial compressive strength of the rock σ_θ/σ_c reflects the strength conditions of the rock mass and determines the lower limit of the energy required to destroy the rock mass.

Professor Lu Jiayou believes that the occurrence of rock bursts and their intensity are related to the nature of the rock mass. The uniaxial compressive strength of the rock reflects the hardness and lithology characteristics of the rock mass. The tensile strength of the rock is the effect of the uniaxial tension of the rock. The maximum tensile stress that can be withstood during failure is reached, and both are important indicators for judging the stability of rock mass engineering. Therefore, the ratio coefficient of $\frac{\sigma_c}{\sigma_t}$ reflects the lithology of the surrounding rock and can further reflect the integrity of the rock mass.

The rock brittleness index w_{et} is an inherent property of the rock when it is damaged under force. It reflects that the rock has a small strain before macroscopic failure under the action of force, and it is all released in the form of elastic potential energy when it is broken. The degree of difficulty of instantaneous fracture of a rock before failure. The more brittle the rock, the greater the possibility of the rock releasing energy. In a rock mass with the same energy level, the greater the rock brittleness index is the energy released in the same time The bigger it is, the greater the impulse generated during the rock ejection process, and the greater the destructive force caused by the rock ejection, which is an important basis for measuring the intensity of rock bursts.

The elastic deformation index k_s represents the ratio of the elastic strain energy accumulated before the rock reaches the maximum ultimate strength during the compressive deformation process and the loss strain energy after unloading, and reflects the rock's ability to store the elastic deformation potential energy, whether rockburst occurs or not The internal dominant factors of its intensity and its magnitude, the greater the internal energy, the greater the ability to destroy the rock mass, and the greater the probability of rock burst ejection damage.

The one-dimensional cloud model requires that indicators follow a normal distribution within infinite intervals. In fact, the measured values of indicators are usually vague and randomly distributed within a finite interval. This may bc inconsistent with the actual distribution of the indicators, leading to deviations from the actual results. According to the research work of Wang.et al [2] and others and formula (15), the standard interval for predicting rockburst propensity indicators was revised, as shown in Table 2.

$$\begin{aligned} C^n_{\max} &= Ex^{n-1} + (Ex^{n-1} - C^{n-2}_{\min}) \\ C^1_{\max} &= Ex^2 + (Ex^2 - C^3_{\min}) \end{aligned} \tag{15}$$

Table 2. Modified rockburst tendency prediction index

Rockburst level	I no rock burst	II weak rock burst	III medium rock burst	IV strong rockburst
σ_c/σ_t	40.00–52.2	26.7–40.00	14.5–26.7	0–14.5
σ_θ/σ_c	0–0.3	0.3–0.5	0.5–0.7	0.7–0.9
w_{et}	0–2.00	2.00–4.00	4.00–6.00	6.00–8.00
k_S	0–0.555	0.55–0.65	0.65–0.75	0.75–0.85

3.3 Determine Rockburst Level

To determine the level of rockburst:

1. According to the revised prediction index of rockburst propensity (Table 2), substitute the formula (12) to obtain the digital feature *Ex* of the multi-dimensional cloud model;
2. Find a_j^i according to formulas (16) to (17), substitute it into formula (13), and find the digital feature *En*. The fixed value of; *He* is 0.01;

   Growth indicator [1]:

$$a_{jleft}^{i} = Ex_j^i - C_{min\,y}^{i-1},\ a_{jright}^{i} = C_{max\,j}^{i-1} - Ex_j^i \tag{16}$$

   Consumption indicators [1]:

$$a_{jleft}^{i} = Ex_j^i - C_{min\,y}^{i-1},\ a_{jright}^{i} = C_{max\,j}^{i-1} - Ex_j^i \tag{17}$$

3. The obtained *En* is substituted into formula (19) to generate a random number En' that obeys the normal distribution;

$$E_n^{'}\left(E_{n1}^{'}, E_{n2}^{'}, \ldots, E_{nn}^{'}\right) \sim N(E_n(E_{n1}, E_{n1}, \ldots, E_{nn})) \tag{18}$$

4. According to formula (20), obtain k_j^i and rockburst instance data and substitute formula (21) to obtain a certain degree of membership of the sample;

$$k_j^i = \frac{ln(\frac{ln4}{9})}{ln\left|\frac{C_j^i - Ex_j^i}{3En_j^i}\right|} \tag{19}$$

   In the formula, C_j^i represents $C_{max\,j}^i$ or $C_{min\,j}^i$

$$\mu^i\left[x^i\left(x_1^i, x_2^i, \ldots, x_n^i\right)\right] = \exp(-\frac{9}{2}\sum_{j=1}^{m}\left|\frac{x_j^i - Ex_j^i}{3En_j^{\prime i}}\right|^{k_j^i}) \tag{20}$$

5. Repeat the above steps to obtain the membership of each grade of the sample, and determine the rockburst grade according to the principle of maximum membership.

4 Case Analysis

4.1 Rockburst Case

The 31 cases of rockbursts in this paper are all from published articles, 1–12 groups of data come from literature [11], and 13–31 groups of data come from literature [10].

4.2 Determine Weight

Calculating Subjective Weights

According to the improved analytic hierarchy process (AHP) according to Sect. 2.1, combined with the literature [14], the evaluation matrix (in the order of $\frac{\sigma_\theta}{\sigma_c}$, w_{et}, k_s, $\frac{\sigma_c}{\sigma_t}$) of the prediction indicators in Table 2 is:

$$R = \begin{bmatrix} 1 & 1.8 & 1.7 & 1.6 \\ \frac{1}{1.8} & 1 & \frac{1}{1.3} & \frac{1}{1.4} \\ \frac{1}{1.7} & 1.3 & 1 & \frac{1}{1.2} \\ \frac{1}{1.6} & 1.4 & 1.2 & 1 \end{bmatrix}$$

According to formula (1), the final judgment matrix R is:

$$R = \begin{bmatrix} 1 & 1.8 & 1.7 * 1.8 & 1.6 * 1.7 * 1.8 \\ \frac{1}{1.8} & 1 & \frac{1}{1.3} & \frac{1}{1.4} * \frac{1}{1.3} \\ \frac{1}{1.7 * 1.8} & 1.3 & 1 & \frac{1}{1.2} \\ \frac{1}{1.6 * 1.7 * 1.8} & 1.4 * 1.3 & 1.2 & 1 \end{bmatrix}$$

Obtain the subjective weight (in the order of $\frac{\sigma_\theta}{\sigma_c}$, w_{et}, k_s, $\frac{\sigma_c}{\sigma_t}$) according to formula (1) as:

$$\alpha = \begin{bmatrix} 0.4994 & 0.1791 & 0.1525 & 0.1690 \end{bmatrix}$$

Calculate Objective Weights

According to the CRITIC method in Sect. 2.2, normalize the sample data (Table 3) (only $\frac{\sigma_c}{\sigma_t}$ is the larger the better), and bring the sample data into formulas (4) to (6) to obtain the average, variance The coefficient of variation is:

$$\bar{x} = \begin{bmatrix} 24.1388 & 0.4506 & 4.4531 & 0.6750 \end{bmatrix}$$

$$s = \begin{bmatrix} 7.7175 & 0.1755 & 1.3942 & 0.1093 \end{bmatrix}$$

$$v = \begin{bmatrix} 0.3197 & 0.3895 & 0.3131 & 0.1619 \end{bmatrix}$$

Table 3. Engineering example data

Sample	σ_c/σ_t	σ_θ/σ_c	w_{et}	k_s
1	13.20	0.58	6.3	0.79
2	17.50	0.45	5.1	0.68
3	20.90	0.39	4.6	0.65
4	41.00	0.20	1.7	0.50
5	13.20	0.66	6.8	0.82
6	15.00	0.53	6.5	0.70
7	21.70	0.42	4.5	0.67
8	21.70	0.39	5.0	0.73
9	26.90	0.44	5.5	0.78
10	18.50	0.81	3.8	0.68
11	29.40	0.41	7.3	0.64
12	19.70	0.38	5.0	0.69
13	28.40	0.38	5.3	0.58
14	22.30	0.66	3.2	0.88
15	29.73	0.37	3.5	0.68
16	32.77	0.42	3.0	0.71
17	42.73	0.28	2.2	0.49
18	20.13	0.49	3.8	0.91
19	28.77	0.38	3.0	0.70
20	27.52	0.72	4.3	0.73
21	16.55	0.69	5.7	0.90
22	15.50	0.42	3.2	0.62
23	30.12	0.58	4.5	0.64
24	36.42	0.22	1.8	0.46
25	19.35	0.62	4.50	0.88
26	31.20	0.57	3.15	0.58
27	12.36	0.65	5.41	0.91
28	18.75	0.59	4.20	0.84
29	29.70	0.73	3.82	0.70
30	42.30	0.37	2.75	0.36
31	37.35	0.37	3.08	0.66

The correlation coefficient matrix is:

$$R = \begin{bmatrix} 1 & -0.5606 & -0.6533 & -0.5166 \\ -0.5606 & 1 & 0.4034 & 0.5850 \\ -0.6533 & 0.4034 & 1 & 0.3603 \\ -0.5166 & 0.5850 & 0.3603 & 1 \end{bmatrix}$$

The objective weights (in the order of $\frac{\sigma_\theta}{\sigma_c}$, w_{et}, k_s, $\frac{\sigma_c}{\sigma_t}$) according to formulas (8) and (9) are:

$$\beta = \begin{bmatrix} 0.3943 & 0.2612 & 0.2359 & 0.1985 \end{bmatrix}$$

Calculate the Overall Weight

According to the principle of minimum information discrimination, substituting α and β into formula (11) to obtain the integrated weight (in the order of $\frac{\sigma_\theta}{\sigma_c}$, w_{et}, k_s, $\frac{\sigma_c}{\sigma_t}$) is:

$$\omega = \begin{bmatrix} 0.4517 & 0.2032 & 0.2032 & 0.1419 \end{bmatrix}$$

4.3 Forecast Model and Results

According to the step of determining the rockburst grade in Sect. 3.3, substitute the rockburst prediction index and grade interval in Table 1. The values of the numerical characteristics *Ex*, *En* and k_j^i are shown in Table 3. According to the digital characteristics substituted into the formula (20) to generate each index cloud model, see Fig. 1; According to the above results, the actual measured values of rock burst examples are brought into the constructed model for prediction, and compared with the actual rock burst level, entropy weight-cloud model, Critic-multidimensional cloud model and analysis-multidimensional cloud model. The specific results are shown in Table 4 (Table 5).

The results show that the prediction results in this paper are basically consistent with the actual rockburst grade, and are not much different from the prediction results of other models, indicating that the proposed multi-dimensional cloud model based on the improved hierarchy method and the CRITIC method is reasonable and effective. The improved analytic hierarchy process gives the weight based on the subjectivity of the decision, the CRITIC method gives the weight value based on the data of the rock burst instance, and the integrated weight obtained based on the fusion of the principle of minimum information identification is more reasonable and improves the reliability of the prediction. The multi-dimensional cloud model reflects the uncertainty of rockburst grade prediction with ambiguity and randomness, and is simpler than the one-dimensional cloud model calculation process; The left and right parts of the normal cloud respectively give the characteristic values of the cloud, optimize the characteristic interval of the multi-dimensional cloud model, and increase the prediction accuracy of the cloud model, especially for the first-level rockburst and second-level rockburst.

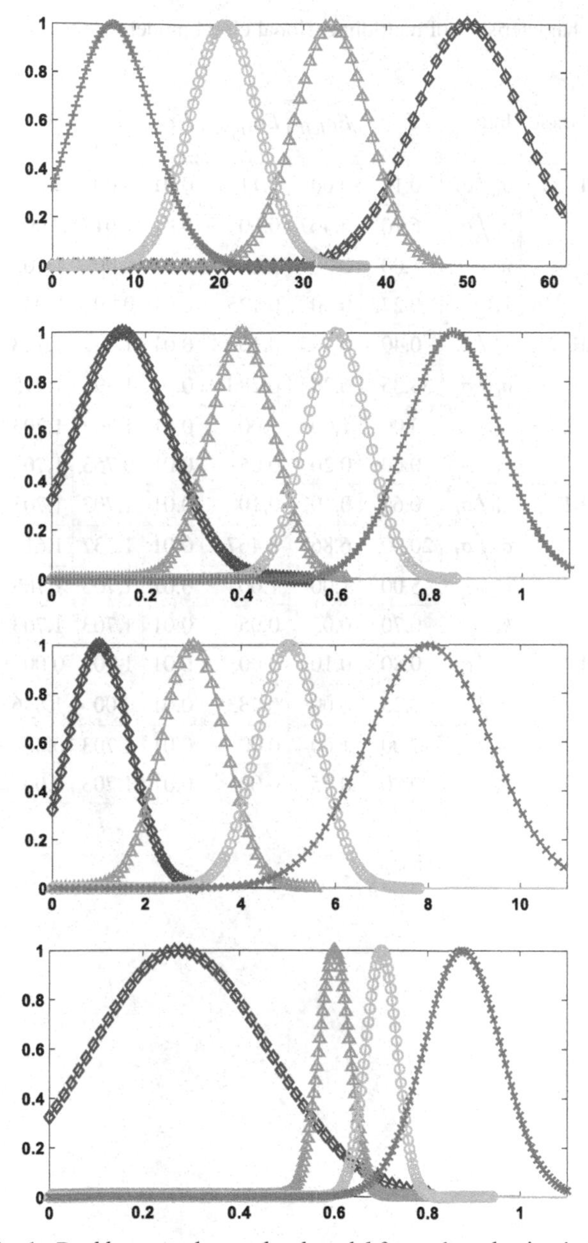

Fig. 1. Rockburst tendency cloud model for each evaluation index

Table 4. Digital characteristics of multidimensional cloud model for rock burst indicators at all levels

Grade	Index	Ex	En_{left}	En_{right}	He	k^{i}_{jleft}	k^{i}_{jright}
I	σ_θ/σ_c	0.15	0.00	0.117	0.01	0.00	2.208
	σ_c/σ_t	46.10	6.467	0.00	0.01	1.617	0.00
	w_{et}	1.00	0.00	1.00	0.01	0.00	1.703
	k_s	0.275	0.00	0.125	0.01	0.00	6.031
II	σ_θ/σ_c	0.40	0.133	0.100	0.01	1.349	1.703
	σ_c/σ_t	33.35	6.283	6.283	0.01	1.795	1.795
	w_{et}	3.00	1.00	1.00	0.01	1.703	1.703
	k_s	0.60	0.20	0.05	0.01	0.753	1.703
III	σ_θ/σ_c	0.60	0.10	0.10	0.01	1.703	1.703
	σ_c/σ_t	20.60	6.867	6.467	0.01	1.537	1.617
	w_{et}	5.00	1.00	1.00	0.01	1.703	1.703
	k_s	0.70	0.05	0.05	0.01	1.703	1.703
IV	σ_θ/σ_c	0.80	0.10	0.00	0.01	1.703	0.00
	σ_c/σ_t	7.25	0.00	6.483	0.01	0.00	1.896
	w_{et}	7.00	1.00	0.00	0.01	1.703	0.00
	k_s	0.80	0.05	0.00	0.01	1.703	0.00

Table 5. Prediction results and comparison of rock burst grade

Sample	Predicted value of each level				Actual level	Prediction level	Entropy weight-cloud model [11]	Critic-multi-dimensional cloud model [12]	Analysis-multi-dimensional cloud model [10]
	I	II	III	IV					
1	0	0.0132	0.3658	0.4326	IV	IV	IV	IV	IV
2	0.0022	0.1090	0.6238	0.0704	III	III	III	III	III
3	0.0107	0.2512	0.4872	0.0210	III	III	III	III	III
4	0.8819	0.2796	0.0037	0	I	I	I	I	I
5	0	0.0065	0.2581	0.6102	IV	IV	III-IV*	IV	IV
6	0	0.0279	0.5320	0.2273	III	III	III	III	III
7	0.0091	0.2723	0.5693	0.0249	III	III	III	III	III
8	0.0024	0.1881	0.5300	0.0351	III	III	III	III	III
9	0	0.2087	0.3702	0.0268	III	III	III	III	III
10	0	0.0577	0.4365	0.1129	III	III	III	III	III
11	0.0054	0.1509	0.1676	0.0086	II	II-III*	II	II	II
12	0.0041	0.1620	0.5489	0.0358	III	III	III	III	III
13	0.0360	0.4433	0.2213	0.0038	II	II	II	II	II
14	0	0.0815	0.2768	0.0456	III	III	III	III	III
15	0.0541	0.7042	0.2156	0.0025	II	II	II	II	II
16	0.0449	0.7453	0.1626	0.0015	II	II	II	II	II
17	0.7141	0.3198	0.0055	0	I	I	I	I	I
18	0	0.0754	0.2620	0.0372	III	III	III	III	III
19	0.0400	0.6599	0.2245	0.0028	II	II	II	II	II
20	0	0.1882	0.4646	0.0394	III	III	III	III	III
21	0	0.0125	0.2786	0.2988	IV	IV	IV	IV	IV
22	0.0089	0.1496	0.2697	0.0181	II	III*	III*	III*	III*
23	0.0177	0.4541	0.4021	0.0092	II	II	II	II	II
24	0.6896	0.3644	0.0070	0	I	I	I	I	I
25	0	0.0472	0.4150	0.1227	III	III	III	III	III
26	0.0605	0.6703	0.1910	0.0020	II	II	II	II	II
27	0	0.0069	0.2110	0.3228	IV	IV	IV	IV	IV
28	0	0.0655	0.4803	0.1188	III	III	III	III	III
29	0.0028	0.2616	0.3697	0.0170	III	III	III	III	III
30	0.4963	0.2371	0.0048	0	I	I	I	I	I
31	0.1782	0.7515	0.0651	0	II	II	II	II	II

5 Conclusion

1. Choose four indexes: the ratio of the maximum principal stress to the rock uniaxial tensile strength σ, the ratio of the maximum tangential stress to the maximum principal stress σ, the rock elasticity index w and the rock integrity coefficient k, and revise the upper limit of the infinite interval of the index, Establish a multi-index forecasting standard for propensity. AHP and CRITIC method are used to obtain subjective weight and objective weight respectively, and the comprehensive weight is obtained according to the principle of minimum identification information.
2. A multi-dimensional cloud model is adopted to establish a graded comprehensive cloud for rockburst propensity prediction. The asymmetric interval in the typical multi-dimensional cloud model is divided into two parts. The data is verified by 31 sets of rockburst engineering examples. The rationality and effectiveness of propensity forecasting, compared with other forecasting methods, shows the applicability of this model.
3. Compared with other methods, the cloud model can reflect the uncertainty of multi-index forecasting and visually display the forecasting process. The establishment process of the one-dimensional cloud model is complicated and the calculation time is long, but the establishment process of the multi-dimensional cloud model is simple, the calculation time is short, and the prediction results are more accurate; the selection of the digital features of the multi-dimensional cloud model is conducive to improving the accuracy of rockburst prediction and the impact. The index division of rockburst grading can further improve the cloud model for rockburst prediction, and the prediction result will be more in line with reality.

References

1. He, S., Song, D., Li, Z., et al.: Precursor of spatio-temporal evolution law of MS and AE activities for rock burst warning in steeply inclined and extremely thick coal seams under caving mining conditions. Rock Mech. Rock Eng. **52**(7), 2415–2435 (2019)
2. Xiao, F., He, J., Liu, Z.: Analysis on warning signs of damage of coal samples with different water contents and relevant damage evolution based on acoustic emission and infrared characterization. Infrared Phys. Technol. **97**, 287–299 (2019)
3. Wang, Y., Li, W., Li, Q., et al.: Method of fuzzy comprehensive evaluations for rockburst prediction. Chin. J. Rock Mech. Eng. **17**(5), 493–501 (1998)
4. Jia, Y., Lu, Q., Shang, Y.: Rockburst prediction using particle swarm optimization algorithm and general regression neural network. Chin. J. Rock Mech. Eng. **32**(2), 343–348 (2013)
5. Wu, S., Zhang, C., Cheng, Z.: Classified prediction method of rockburst intensity based on PCA-PNN principle. Acta Coal Min. **44**(09), 2767–2776 (2019)
6. Pu, Y., Apel, D.B., Wei, C.: Applying machine learning approaches to evaluating rockburst liability: a comparation of generative and discriminative models. Pure. appl. Geophys. **176**(10), 4503–4517 (2019). https://doi.org/10.1007/s00024-019-02197-1
7. Afraei, S., Shahriar, K., Madani, S.H.: Developing intelligent classification models for rock burst prediction after recognizing significant predictor variables, Section 2: Designing classifiers. Tunn. Undergr. Space Technol. **84**, 522–537 (2019)

8. Tang, Z., Xu, Q.: Research on rockburst prediction based on 9 machine learning algorithms. Chin. J. Rock Mech. Eng., 1–9 (2020)
9. Liu, R., Ye, Y., Hu, N., Chen, H., Wang, X.: Classified prediction model of rockburst using rough sets-normal cloud. Neural Comput. Appl. **31**(12), 8185–8193 (2018). https://doi.org/10.1007/s00521-018-3859-5
10. Wang, M., Liu, Q., Wang, X., Shen, F., Jin, J.: Prediction of rockburst based on multidimensional connection cloud model and set pair analysis. Int. J. Geomech. **1**(20), 1943–5622 (2020)
11. Zhou, K., Lin, Y., Hu, J., et al.: Grading prediction of rockburst intensity based on entropy and normal cloud model. Rock Soil Mech. **37**(Supp. 1), 596–602 (2016)
12. Guo, J., Zhang, W., Zhao, Y.: Comprehensive evaluation method of multidimensional cloud model for rock burst prediction. J. Rock Mech. Eng. **37**(5), 1199–1206 (2018)
13. Zhao, S., Tang, S.: Comprehensive evaluation of transmission network planning based on improved analytic hierarchy process CRITIC method and approximately ideal solution ranking method. Electr. Power Autom. Equip. **39**(03), 143–148+162 (2019)
14. Yin, X., Liu, Q., Wang, X., Huang, X.: Application of attribute interval recognition model based on optimal combination weighting to the prediction of rockburst severity classification. J. China Coal Soc., 1–9 (2020)

Sports Pose Estimation Based on LSTM and Attention Mechanism

Chuanlei Zhang, Lixin Liu(✉), Qihuai Xiang, Jianrong Li, and Xuefei Ren

School of Computer Science and Information Engineering, Tianjin University of Science and Technology, Tianjin 300457, China
18232172760@163.com

Abstract. In our life, we often need to estimate the accuracy of sports pose, which usually costs a lot of time and human resources. To solve the problem, we propose a LSTM-Attention model. In spatial dimension, we use the two-branch multi-stage CNN to extract human joints as features, which not only guarantees the real-time performance, but also ensures the accuracy. For the time dimension, the extracted joint features sequence is input into the LSTM-Attention model for training. In order to verify the effectiveness of our proposed method, we collected data for processing and trained with the proposed model. The experimental results show that our method has a high performance.

Keywords: Sports pose estimation · Two-branch multi-stage CNN · LSTM-attention mechanism

1 Instruction

As one of the research hot spots in the field of computer vision, human action recognition based on video technology has high scientific research value and application value. It includes automatic human behavior detection, recognition and understanding of image sequence of in video. At present, there are a lot of researches on human behavior, but few in motion pose estimation. In the army, it is necessary to conduct physical training and assessment on soldiers regularly, including push-ups, pull-ups and sit-ups. If people supervise and assess them artificially, it will cause the waste of human resources. And the assessment result will have certain emotional color. However, if we record the motion through a camera during the assessment, and then identify and score them, it can save a lot of time and human resources. By analyzing the data, it can reach high accuracy and find many problems that cannot be found by human, so we can correct them in time. Therefore, the research of motion pose recognition has important application value.

Before the emergence of deep learning methods, most of the traditional behavior recognition methods are divided into three steps: (1) Behavior feature extraction. Spare spatial temporal interest points are extracted, for example, Harris corner detection [11] is applied to three-dimensional spatial temporal domain. In addition, there are methods to extract local dense visual features from video data. (2) Description of behavior features.

X. Wang et al. (Eds.): 6GN 2020, LNICST 337, pp. 538–550, 2020.
https://doi.org/10.1007/978-3-030-63941-9_45

The extracted behavior features need to be combined into a standard video description, in which the word bag model is a more commonly used feature description model. (3) The feature description is classified by Support Vector Machine (SVM).

Laptev [12] et al. proposed to extend 2D Harris corner detection operator to three-dimensional domain to extract spatial temporal interest points from video. However, the 3D Harris interest point detection operator is too sparse to describe the behavior accurately, and its robustness is poor, which cannot solve the common problems, such as occlusion, illumination and perspective change in video data. For this reason, Dollar [13] et al. put forward corresponding improved methods using Gabor filter and Gaussian filter to detect the spatial temporal position of interest points in the spatial temporal domain to improve the density of interest points. Although the corresponding problems have been improved, they are not solved perfectly. In 2013, Wang [14, 15] et al. proposed a dense trajectory algorithm and an improved dense trajectory algorithm (IDT) based on the original algorithm. The algorithm is based on the shape characteristics of the trajectory, and integrates the characteristics of hog, HoF and MBF. Although the improved dense trajectory algorithm is a very classical algorithm for manual feature extraction and has good recognition effect, there are problems in practical application: the algorithm speed is relatively slow due to the intensive calculation, and it is difficult to deal with large-scale data sets.

In recent years, researchers have been trying to apply Convolutional Neural Network to video behavior recognition. In 2014, Karpathy [16] et al. proposed to use the pretrained 2D Convolutional Neural Network to extract the spatial features of each frame, and in the final stage, the spatial features of continuous frames were fused to get the classification results, and several fusion methods were investigated. Although the method of deep learning is applied, the experimental results are significantly worse than the algorithm based on artificial design features. There are two main reasons for the failure: the lack of diverse data sets and the inability of network models to effectively extract dynamic features. Simmoyan and Zisserman [3], based on the previous experience of Karpathy and other, proposed a Two-Stream Convolutional Neural Network with spatial network and temporal network. The architecture is no longer a single network to extract spatial features, but has two independent networks, which has a profound impact on the follow-up research. The temporal network extracts the dynamic features of the behavior with the input of stacked dense optical flow vectors. The spatial network extracts the behavior static features from the single video frame as the input, and finally obtains the results through SVM classification. Although this kind of Two-Stream Convolutional Neural Network has good performance, the training process of the two networks is separate, not the end-to-end training process. Du Tran [6] et al. proposed a convolution network of C3D. The network structure no longer uses 2D convolution, but extends to 3D convolution which can deal with temporal information, and computes features simultaneously in the spatiotemporal dimension of video data. The C3D Convolutional Network is pre-trained on Sports-1 M, and then the pretrained model is used on other data sets and can achieve better results. And it was found in the experiments that if artificially designed features such as IDT were used, the model would perform better. It is worth noting that the main advantages of C3D are its operating speed and processing efficiency, which makes it a good application prospect. Feichtenhofer [8] et al. further use 3D convolution kernel to fuse spatial and temporal networks on the basis of Two-Stream Convolution Neural Network. Limin Wang [17] et al. of the Chinese University of Hong Kong proposed a TSN network structure in 2016. This network structure can extract K short video fragments with the same time in a long video by sparse sampling method, and then randomly sample the fragments from the K fragments. The rest steps are similar to the Two-Stream Convolutional Neural Network, and have achieved better recognition results. In addition, Recurrent Neural Networks (RNN) are also attracting attention because of its ability to process temporal series data. For example, Ng [18, 19] et al. applied Long Short-Term Memory, (LSTM) to the fusion of temporal domain information in a Two-Stream Neural Network, but the effect is average. Long-term Recurrent Convolutional Network (LRCN) [9] extracts feature from single frame image information through convolutional network, and then outputs the features through LSTM in chronological order. The whole architecture is an end-to-end training process. The author also compares RGB and optical flow as input, and finds that the best recognition effect can be obtained by weighting the prediction based on the two inputs.

Motivated by these facts, we proposed an attention-based LSTM architecture for motion pose assessment in videos, which effectively determines the accuracy of motion posture. We take the sequence of the joint point features as input and input it into the LSTM-Attention model, and then take the output of LSTM-Attention model through Softmax as result. The advantage of this paper is that using the architecture of two-branch multi-stage CNN [1] which can accurately and effectively extract the features of human joint points in the video. Inputting the extracted features into the LSTM network can express the serialized features well. Adding the Attention model on the one hand improves the performance of the last model. On the other hand, using the attention mechanism can facilitate the observation of how the information in the input sequence affects the final output sequence, which helps to better understand the internal working mechanism of the model.

The remainder of this paper is organized as follows. Section 2 expounds the theory and design of LSTM-Attention network model. Section 3 designs experiment to verify the feasibility of the proposed method, and analyzes the experimental results. Finally, the conclusions based on this paper are given in Sect. 4.

2 Model Design

For spatial dimension, we use the two-branch multi-stage CNN to extract the joint points. For temporal dimension, we input the sequence of joint feature to obtain the temporal feature. The LSTM-Attention architecture is shown in Fig. 1. There are about four major modules, and we intend to discuss them in details in the following.

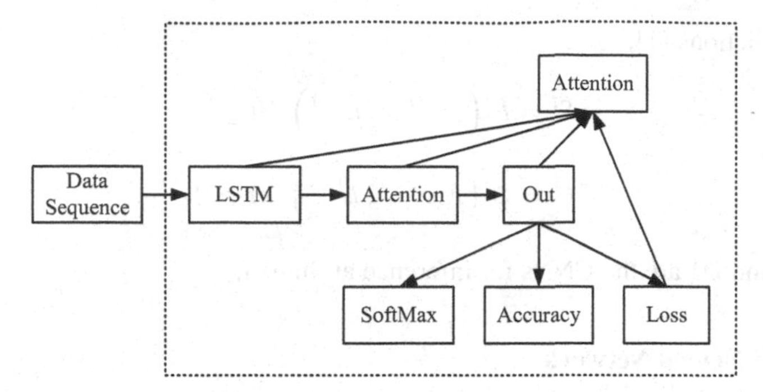

Fig. 1. The LSTM-attention architecture

2.1 The Two-Branch-Multi-stage CNN

The traditional method of pose estimation is top-down, which refers to detecting the human body area first, and then detecting the key points of the human body in the area. Because it is necessary to perform forward key point detection for each detected human body area, the speed is slow. Therefore, the Real-time Multi-Person 2D Pose Estimation [1] presents the first bottom-up representation of association scores via Part Affinity Fields (PAFs), a set of 2D vector fields that encode the location and orientation of limbs over the image domain. Based on the detected joint points and Part Affinity Fields, using the greedy inference algorithm, these joint points can be mapped to different individuals. The network structure [1] is shown in Fig. 2.

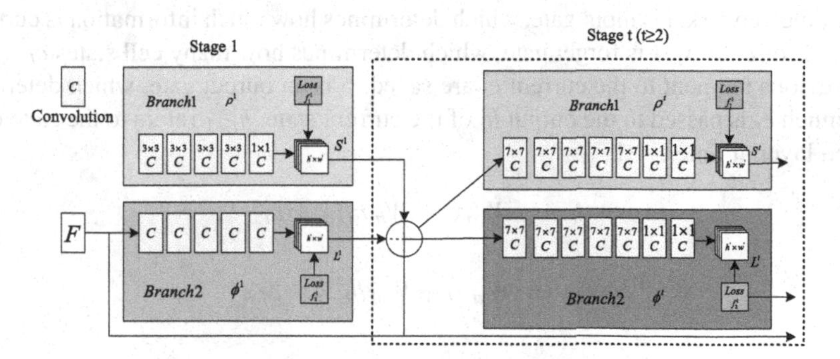

Fig. 2. The architecture of two-branch multi-stage CNN

The network is divided into two branches: the top branch predicts the confidence maps, and the bottom branch predicts the affinity fields. The image is first input to VGG-19, generating a set of feature maps F, which is input to the first stage of each branch. At the first stage, the network produces a set of detection confidence map $S^1 = \rho^1(F)$ and a set of part affinity fields $L^1 = \varnothing^1(F)$, where ρ^1 and $\varnothing^1$ are the CNNs for inference at Stage 1. In each subsequent stage, the predictions from both branches in the previous stage, along with the original image features F, are concatenated and used to produce refined predictions [1],

$$S^t = P^t\left(F, S^{t-1}, L^{t-1}\right), \forall t \geq 2 \tag{1}$$

$$L^t = \varnothing^t\left(F, S^{t-1}, L^{t-1}\right), \forall t \geq 2 \tag{2}$$

Where ρ^1 and $\varnothing^1$ are the CNNs for inference at Stage t.

2.2 LSTM Neural Network

The recurrent neural network (RNN) is the network structure which can express the time sequence well in deep learning, and the best one is LSTM. Because LSTM operates on sequences, multi-layer LSTM stacking can increase the level of abstraction of the input. When time t increases, the stool can be observed in blocks, or the representation problem on different time scales can make the network extract more abstract features. Therefore, this paper uses multi-layer LSTM stacking to extract features in temporal domain. The motion posture evaluation problem we studied is a typical timing problem, that is, the value of a certain moment is affected by the previous moment or several moments, so we choose the LSTM model.

LSTM is a time-series convolutional neural network, which is derived from recurrent neural networks. By introducing structures called gates, it can mine the time series rules of relatively long intervals and delays in time series. The internal structure of LSTM [2] is shown in Fig. 3. Among them, x_t is he t-th input sequence element value. c is the cell state or memory cell, which controls the transmission of information, and is also the core of the network. i is input gate, which determines how much information is currently reserved for c_t by x_t. f is forget gate, which determines how many cell states c_{t-1} from the previous moment to the current c_t are saved. o is an output gate, which determines how much c_t is passed to the output h_t of the current state. h_{t-1} refers to the state of the hidden layer at time t − 1.

$$i_t = \sigma(W_{xi}x_t + W_{hi}h_{t-1} + b_i) \tag{3}$$

$$f_t = \sigma(W_{xf}x_t + W_{hf}h_{t-1} + b_f) \tag{4}$$

$$o_t = \sigma(W_{xo}x_t + W_{ho}h_{t-1} + b_o) \tag{5}$$

$$\overline{c_t} = \tanh(W_{xc}x_t + W_{hc}h_{t-1} + b_c) \tag{6}$$

$$c_t = f_t \bullet c_{t-1} + i_t \bullet \overline{c_t} \tag{7}$$

$$h_t = o_t \bullet \tanh(c_t) \tag{8}$$

Among them, W_{xi}, W_{xf}, W_{xo} and W_{xc} are the weight vectors from the input layer to the input gate, the forget gate, the output gate and the cell state. W_{hi}, W_{hf}, W_{ho} and

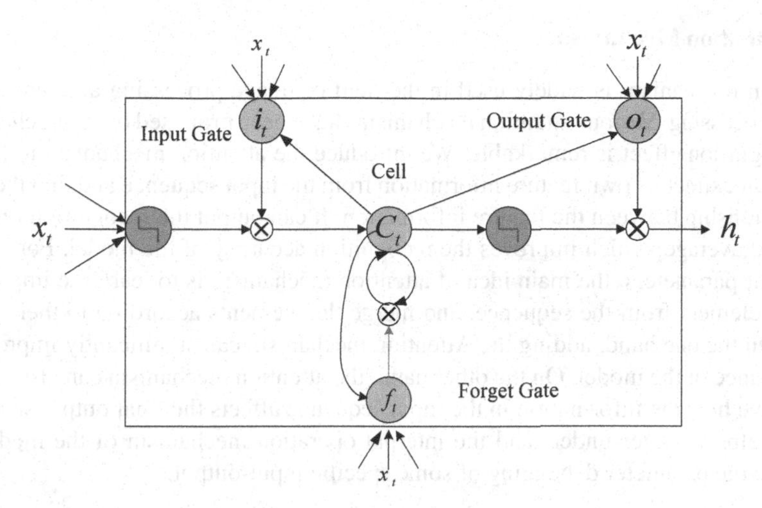

Fig. 3. LSTM internal structure

W_{hc} are the weight vectors from the hidden layer to the input gate, the forget gate, the output gate and the cell gate. b_i, b_f, b_o and b_c are the bias from the input gate, the forget gate, the output gate and the cell gate. $\sigma(\cdot)$ is Sigmoid activation function. tanh() means hyperbolic tangent activation function, which represents vector element multiplication.

Figure 4 shows the LSTM classification model, in which the input layer is $x_0, x_1, x_2, \cdots, x_t$ the corresponding video frame vector, and the upper layer of the input layer is the forward LSTM layer, which is composed of a series of LSTM units. The results of the addition and averaging of the LSTM outputs at all times are then used as the upper-layer representation. Finally, through the softmax layer, the full connection operation is carried out, then the predicted category y is obtained.

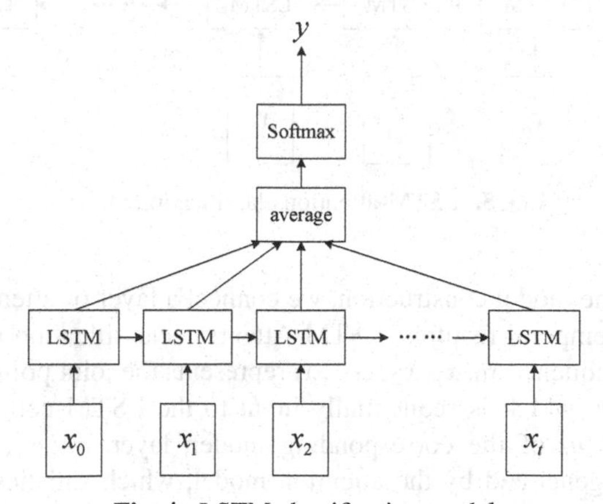

Fig. 4. LSTM classification model

2.3 Attention Mechanism

Attention mechanism is widely used in the field of image processing and natural language processing. Various attention mechanisms have been proposed by researchers, and the recognition effect is remarkable. We introduce the attention mechanism to LSTM, which can extract its own feature information from the input sequence and find the internal relationship between the feature information. It can output the recognition result by weighted average, which improves the recognition accuracy of the model. For a series of weight parameters, the main idea of attention mechanism is to learn the importance of each element from the sequence, and merge the elements according to their importance. On the one hand, adding the Attention mechanism can significantly improve the performance of the model. On the other hand, the attention mechanism can also be used to observe how the information in the input sequence affects the final output sequence, which helps to better understand the internal operation mechanism of the model and facilitate the parameter debugging of some specific input-output.

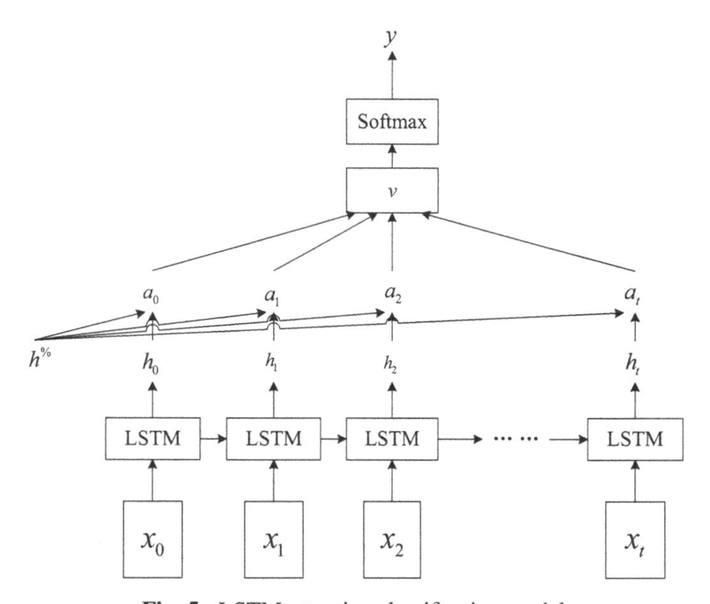

Fig. 5. LSTM-attention classification model

Therefore, in the model construction, we connect a layer of attention network after LSTM to extract temporal features. LSTM-Attention classification model is shown in Fig. 5. The input sequence $x_0, x_1, x_2, \cdots, x_t$ represents the joint point feature sequence of the video frame, which is sequentially input to the LSTM cell to obtain the output $h_0, h_1, h_2, \cdots, h_t$ of the corresponding hidden layer. $\alpha_0, \alpha_1, \alpha_2, \cdots, \alpha_t$ is the weight parameter generated by the attention model, which satisfies the constraint of $\sum_{t=1}^{T} a_t = 1$. h_i is the output state of the hide layer at the i-th time, and h is the feature representation vector one level higher than the video frame. h is initialized randomly as a parameter, which is updated gradually in the training process. The attention parameter can be computed with the following equation

$$\alpha_i = \frac{\exp(\beta_i)}{\sum\limits_{j=1}^{n} \exp(\beta_j)} \tag{9}$$

where β_j represents the score of the i-th hidden layer output h_i in the video frame representation vector $\bar{h}$. The larger β_j is, the greater the attention of the input in the whole at this moment. It can be computed with the following equation

$$\beta_i = V^T \tanh(W\bar{h} + Uh_i + b) \tag{10}$$

The attention vectors can be obtained according to the outputs of the LSTM network and the temporal attention weight values at each running step with this equation.

$$v = \sum_{j=1}^{t} \alpha_j h_j \tag{11}$$

Finally, the prediction category y can be obtained after the softmax classification function, the formula is as follows:

$$y = soft\max(W_v v + b_v) \tag{12}$$

2.4 Loss Function

The loss function used we use is cross-entropy, which comes from information theory. In order to solve the problem of information measurement, we use the concept of "entropy" in physics to describe the average amount of information contained in the received message. In information theory, the larger the entropy of a message, the larger the information it carries. Simply speaking, in deep learning, cross entropy is to measure the similarity between two probability distributions p and q, which is more suitable to measure the distribution difference between two probabilities. The formula is as follows:

$$H_y \cdot (y) = \sum_i y_i' \log y_i \tag{13}$$

where y is the predicted probability distribution vector of the model output, and y′ is the true distribution. y_i is element 0 or 1 in vector y, which needs to be distinguished from the discrete value of sample i category, that is, y. In vector y, only the y-th element y_y is 1, and the rest are all 0 (one-hot coding). That is to say, the cross entropy only relates to the prediction probability of the correct result, as long as its value is large enough, it can also ensure that the classification result is correct.

3 Experiment and Analysis

To evaluate the effectiveness of LSTM-Attention model, we train the architecture and test the well-trained model on our dataset. We next describe the implementation details of our algorithms and discuss the experiment results.

3.1 Data Collection and Processing

The data of training set, test set and verification set used in our paper are collected by our team. The steps to obtain the data are as follows:

(1) Record all the actions into video. The collected video data is divided into 6 categories, which are push-up front, push-up side, sit-up front, sit-up side, pull-up front, pull-up side. The sample pictures of the dataset are shown in Fig. 6;
(2) Extract a frame every 6 frames of the video, that is, extract about 5 frames of images per second;
(3) Cut out the main part of the human behavior in the picture.

Fig. 6. The sample pictures of the dataset

After data collection, we need to process the collected data to facilitate the extraction of feature points. The steps to process the data are as follows:

(1) We need to classified the images. The captured images are manually marked, and the sequence of pictures marked as the same action is placed in the same folder.
(2) We mark each type of action, and divide it into standard action and non-standard action;

(3) We use the model of the two-branch multi-stage CNN to extract 17 joint features of each image into one-dimensional vector and store them in the CSV file. The extracted information includes joint points identification, relative coordinates, and confidence, so that 68 nodes of information stored into one-dimension vector can be extracted from a picture;
(4) we take 10 frames as a sequence to make the dataset of the model, That is, the joint points data of every ten frames is used as the input of the model.

Finally, the amount of each type of data sequence is shown in Table 1. We have cropped 8044 data sequences in total. We randomly divide the verification set and training set according to the ratio of 1:9. The joint information we extract is relative coordinates, that is, the coordinates relative to the length and width of the picture. In order to improve the generalization ability of the model, we cut the sequence pictures randomly during the process of extracting joint points to ensure that the relative positions of joint points change in turn, so as to expand the data set.

Table 1. The amount of data sequence

	Push-up front	Push-up side	Pull-up front	Pull-up side	Sit-up front	Site-up side
Standard	72	606	139	331	290	411
Non-standard	279	1896	450	1073	827	1670

3.2 Implement Details

We perform the experiment with the following implement details. First, we read the data from the Tfrecord file and put it into the memory buffer, then read a training group randomly with the batch size of 128. This training group is put into the LSTM attention model to perform training operations. Then we calculate the loss and use the optimizer back propagation to reduce the loss and adjust the network parameters of each layer. The optimizer we use is the Adam optimizer [20] provided by tensorflow. The initial value of learning rate is set to 0.0001, and we optimize the model by using learning rate exponential decay.

3.3 Experiment Results and Analysis

First, we train the model on the data set according to the above training process. The accuracy changes and loss changes are shown in Fig. 7 and Fig. 8 respectively. Among them, the blue line represents the training operation, and the orange line represents the test operation. As is shown in the Figures, we can find that after about 2000 batches the model begins to coverage. The accuracy gradually increases, and the loss gradually decrease.

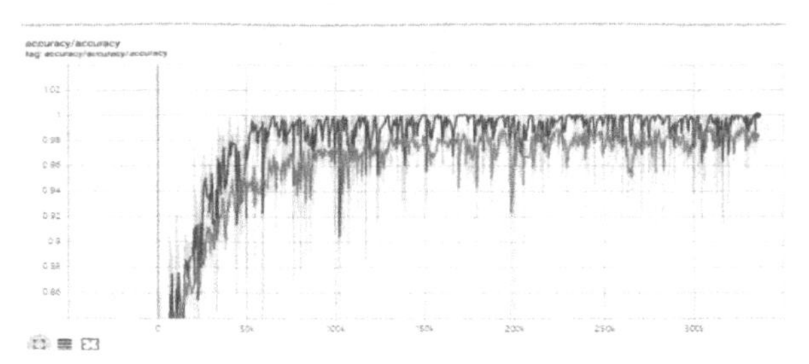

Fig. 7. Accuracy changes

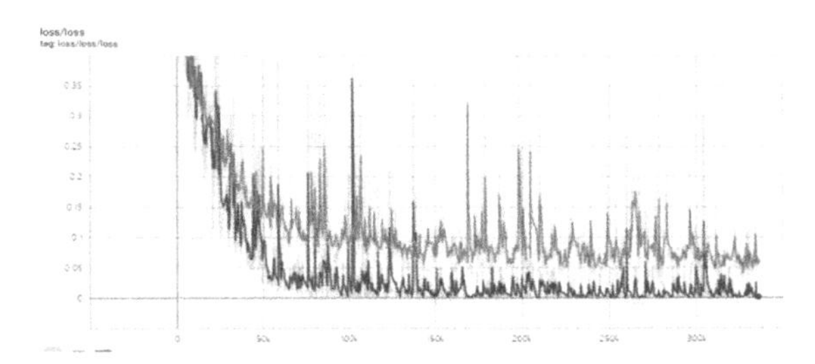

Fig. 8. Loss changes

After 300000 batches, the loss of training set converges to about 0.01, and the accuracy reaches about 0.99; the loss of verification set converges to about 0.09, and the accuracy reaches about 0.97. After training model, we test it in the video. The result is shown in Fig. 9 and Fig. 10.

Fig. 9. Standard sports pose

The upper left corner of the picture shows the probability that this motion pose is standard. If the probability is greater than 0.5, it instructs that the motion pose is standard.

Fig. 10. Non-standard sports pose

4 Conclusion

We propose a sports pose estimation method based on LSTM-Attention network structure. Firstly, we use the two-branch multi-stage CNN to extract human joints as a spatial features. Secondly, the extracted joint features sequence is input to LSTM-Attention model to get the temporal features. The attention Mechanism which can adaptively learn detailed spatial-temporal attention feature to enhance the action recognition at each step of LSTM. Finally, we do some experiment to verify our proposal. The result proves that the recognition accuracy and loss of this method can reach a good state, which proves that the method proposed in this paper has certain significance and value. Later, we will further improve the performance of the method for video data in complex environment. We can expand the training set by collecting data sets in a variety of complex environments, and try to solve the problem of insufficient generalization ability by enhancing the pictures of the training set.

References

1. Cao, Z., Simon, T., Wei, S.-E.: Realtime multi-person 2D pose estimation using part affinity fields. In: CVPR (2017)
2. Dai, C., Liu, X., Lai, J.: Human action recognition using two-stream attention based LSTM. Appl. Soft Comput. J. **86**, 105820 (2019)
3. Simonyan, K., Zisserman, A.: Two-stream convolutional networks for action recognition in videos (2014)
4. Liu, J., Shahroudy, A., Xu, D., et al.: Spatio-temporal LSTM with trust gates for 3D human action recognition (2016)
5. Shi, X., Chen, Z., Hao, W., et al.: Convolutional LSTM Network: a machine learning approach for precipitation nowcasting. In: International Conference on Neural Information Processing Systems (2015)
6. Tran, D., Bourdev, L., Fergus, R., et al.: Learning spatiotemporal features with 3D convolutional networks (2014)
7. Xu, H., Das, A., Saenko, K.: R-C3D: region convolutional 3D network for temporal activity detection (2017)
8. Feichtenhofer, C., Pinz, A., Zisserman, A.: Convolutional two-stream network fusion for video action recognition. In: Proceedings of the IEEE Conference on Computer Vision and Pattern Recognition, pp. 1933–1941 (2016)

9. Donahue, J., Anne Hendricks, L., Guadarrama, S., et al.: Long-term recurrent convolutional networks for visual recognition and description. In: Proceedings of the IEEE Conference on Computer Vision and Pattern Recognition, pp. 2625–2634 (2015)
10. Carreira, J., Zisserman, A.: Quo vadis, action recognition a new model and the kinetics dataset. In: 2017 IEEE Conference on Computer Vision and Pattern Recognition (CVPR), pp. 4724–4733. IEEE (2017)
11. Harris, C.G., Stephens, M.: A combined corner and edge detector. In: Alvey Vision Conference, vol. 15, no. 50, pp. 5234–5244 (1988)
12. Laptev, I., Marszalek, M., Schmid, C., et al.: Learning realistic human actions from movies. In: IEEE Conference on Computer and Pattern Recognition, CVPR 2008, pp. 1–8. IEEE (2008)
13. Dollar, P., Rabaud, V., Conttrell, G., et al.: Behavior recognition via sparse spatio-temporal feature. In: 2nd Joint International Workshop on Surveillance and Performance Evolution of Tracking and Surveillance, 2005, pp. 65–72. IEEE (2005)
14. Wang, H., Schmid, C.: Action recognition with improved trajectories. In: IEEE International Conference on Computer Vision (ICCV), pp. 3551–3558 (2014)
15. Wang, H., Kläser, A., Schmid, C., et al.: Action recognition by dense trajectories. In: IEEE Conference on Computer Vision and Pattern Recognition (CVPR), pp. 3169–3176 (2011)
16. Karpathy, A., Toderici, G., Shetty, S., et al.: Large-scale video classification with convolutional neural networks. In: IEEE Conference on Computer Vision and Pattern Recognition, pp. 1725–1732. IEEE Computer Society (2014)
17. Wang, L., et al.: Temporal segment networks: towards good practices for deep action recognition. In: Leibe, B., Matas, J., Sebe, N., Welling, M. (eds.) ECCV 2016. LNCS, vol. 9912, pp. 20–36. Springer, Cham (2016). https://doi.org/10.1007/978-3-319-46484-8_2
18. Ng, Y.H., Hausknecht, M., Vijayanarasimhan, S., et al.: Beyond short snippets: deep networks for video classification, vol. 16, no. 4, pp. 4694–4702 (2015)
19. Sharma, S., Kiros, R., Salakhutdinov, R.: Action recognition using visual attention. arXiv preprint arXiv:1511.04119 (2015)
20. Abadi, M.: TensorFlow: learning functions at scale. ACM SIGPLAN Not. **51**(9), 1 (2016)
21. Yi, Y., Lin, M.: Human action recognition with graph-based multiple-instance learning. Pattern Recognit. **53**, 143–162 (2016)
22. Pfifister, T., Charles, J., Zisserman, A.: Flowing ConvNets for human pose estimation in videos. In: ICCV (2015)
23. Papandreou, G., et al.: Towards accurate multi-person pose estimation in the wild. arXiv preprint arXiv:1701.01779 (2017)
24. Pishchulin, L., et al.: DeepCut: joint subset partition and labeling for multi person pose estimation. In: CVPR (2016)
25. Tang, P., Wang, H., Kwong, S.: Deep sequential fusion LSTM network for image description. Neurocomputing **312**, 154–164 (2018)
26. Liu, Z., Tian, Y., Wang, Z.: Improving human action recognition by temporal attention, In: 2017 IEEE International Conference on Image Processing, Beijing, China, September 2017, pp. 870–874 (2017)
27. Tompson, J.J., Jain, A., LeCun, Y., Bregler, C.: Joint training of a convolutional network and a graphical model for human pose estimation. In: NIPS (2014)
28. Ramakrishna, V., Munoz, D., Hebert, M., Andrew Bagnell, J., Sheikh, Y.: Pose machines: articulated pose estimation via inference machines. In: Fleet, D., Pajdla, T., Schiele, B., Tuytelaars, T. (eds.) ECCV 2014. LNCS, vol. 8690, pp. 33–47. Springer, Cham (2014). https://doi.org/10.1007/978-3-319-10605-2_3

Author Index

Printed in the United States
By Bookmasters